FULL FIELD INVERSION METHODS IN OCEAN AND SEISMO-ACOUSTICS

MODERN APPROACHES IN GEOPHYSICS
formerly *Seismology and Exploration Geophysics*

VOLUME 12

The titles published in this series are listed at the end of this volume.

FULL FIELD INVERSION METHODS IN OCEAN AND SEISMO-ACOUSTICS

edited by

O. DIACHOK

SACLANT Undersea Research Centre,
La Spezia, Italy

A. CAITI

University of Genova, Italy

P. GERSTOFT

SACLANT Undersea Research Centre,
La Spezia, Italy

and

H. SCHMIDT

Massachusetts Institute of Technology,
Cambridge, U.S.A.

KLUWER ACADEMIC PUBLISHERS

DORDRECHT / BOSTON / LONDON

Library of Congress Cataloging-in-Publication Data

Full field inversion methods in ocean and seismo-acoustics / edited by
O. Diachok ... [et al.].
p. cm. -- (Modern approaches in geophysics ; v. 12)
Includes index.
ISBN 0-7923-3459-0 (acid free paper)
1. Underwater acoustics. 2. Inversion (Geophysics) I. Diachok,
O. II. Series.
QC242.2.F85 1995
620.2'5--dc20 95-11663
CIP

ISBN 0-7923-3459-0

Published by Kluwer Academic Publishers,
P.O. Box 17, 3300 AA Dordrecht, The Netherlands.

Kluwer Academic Publishers incorporates
the publishing programmes of
D. Reidel, Martinus Nijhoff, Dr W. Junk and MTP Press.

Sold and distributed in the U.S.A. and Canada
by Kluwer Academic Publishers,
101 Philip Drive, Norwell, MA 02061, U.S.A.

In all other countries, sold and distributed
by Kluwer Academic Publishers Group,
P.O. Box 322, 3300 AH Dordrecht, The Netherlands.

Printed on acid-free paper

Printed in the Netherlands

CONTENTS

PREFACE

In the past few years major advances in several fields have converged, resulting in a new approach to acoustic sensing of the ocean: full-field inversion methods. These methods exploit amplitude and phase information of the acoustic field, thereby extending traditional inversion schemes based on time-of-flight measurements. Most generally the scheme is based on matching, i.e., correlation of modes or interference patterns measured on large aperture array(s) through iteration of ocean and marine geological parameters. The major components of the inversion concept include:

(a) low-frequency source and large aperture receiving array technology and instrumentation for precisely localizating sources, the array and the elements within the array;

(b) an aperture long enough to measure spatial diversity in the field;

(c) a set of parameters that efficiently and accurately describes the physical properties of the ocean and/or bottom;

(d) a high speed, precise acoustic field prediction algorithm;

(e) an effective algorithm for correlating measured and predicted fields; and

(f) a search algorithm that reliably and efficiently finds the global maximum in the correlation search space of the unknown ocean and/or marine geological parameters.

To review the state of the art of this method for probing the ocean environment, NATO's SACLANT Undersea Research Centre hosted a conference entitled *Full Field Inversion Methods in Ocean and Seismo-Acoustics* which was held in Lerici, Italy in June 1994.

Over sixty papers were presented covering analytical procedures and experimental strategies for applying full field methods for inversion of ocean and marine geological parameters. The Conference brought together a diverse array of specialists from western nations, and for the first time at a NATO-sponsored Conference, scientists from former Eastern bloc nations. All of the papers were peer reviewed.

The papers in this volume are grouped into five sections:

1. Speed and Accuracy of Replica Field Computations
2. Signal Processing and Global Search Strategies
3. Source and Receiving Array Technology
4. Ocean Parameter Estimation
5. Geoacoustic Parameter Estimation

Each section consists of papers loosely grouped by subject matter. For example, papers concerned with neural net approaches to global search strategies are grouped together in Section 2. A large number of papers, however, cover many of the topics listed above. Consequently, the reader should not assume that papers assigned to a particular Section are necessarily restricted to the stated topic of that Section. The index provides a list of keywords referenced to the first pages of the papers.

The papers on ocean parameter estimation were generally concerned with the deep ocean environment, whereas the papers on geoacoustic parameter estimation were primarily concerned with the littoral environment. One of the notable exceptions to these trends was a paper which presented results on inversion of the temporal evolution of mesoscale ocean parameters in a shallow-water environment, possibly an indicator of the future of Full Field Methods.

The large number of papers on inversion of geoacoustic parameters in shallow water, to some extent, reflects NATO's interest in littoral environments. Other potential beneficiaries of the research presented in this volume are research programmes concerned with (1) the ecology of littoral seas, which must be conducted over a broad range of scales, including mesoscale, in physically well-calibrated environments, and (2) global environmental changes.

ACKNOWLEDGEMENTS

We wish to thank the advisory committee, viz. Leif Bjørnø, Michael Buckingham, Ross Chapman, Michael Collins, Jens Hovem and Peter Worcester for their recommendations during the planning phase of this conference; and the session chairmen, viz. Finn Jensen, Ed Sullivan, Art Baggeroer, Jens Hovem, Sergio Jesus, Nick Makris, Hans Schneider, Don Gingras, Michael Buckingham, Jean Pierre Hermand, Aldo Vesnaver, Bill Hodgkiss, Nick Chotiros, Bob Weatherburn, Peter Mignerey and Bill Carey for reviewing the papers.

We are also sincerely grateful to John Foxwell, the former director, and David Bradley, the current director, of SACLANT Undersea Research Centre respectively, for their thoughtful guidance and encouragement; and to all the employees of the SACLANT Undersea Research Centre, who participated in the planning and conduct of the conference and the preparation of these proceedings. In particular we thank Adolf Legner, Jane Bragge, Peter Thorpe, Checco Faggioni and Antonio Di Prisa for preparing pre-conference documents and for their assistance during the editing process, to Edoardo Bovio and Tony Hammond for advertising the conference through the World Wide Web on Internet, to Sabrina Matteucci and Nicky Watkins for their assistance during registration, and to Roberto Albini, John Jenkins, June Waller, Heather Barrett and in particular to Mario Del Vento for admistrative assistance.

Finally, we wish to express special gratitude to Anna Bizzarri, the conference secretary. Due to her tireless efforts, professional skills and easy manner the conference was a highly rewarding scientific and social experience.

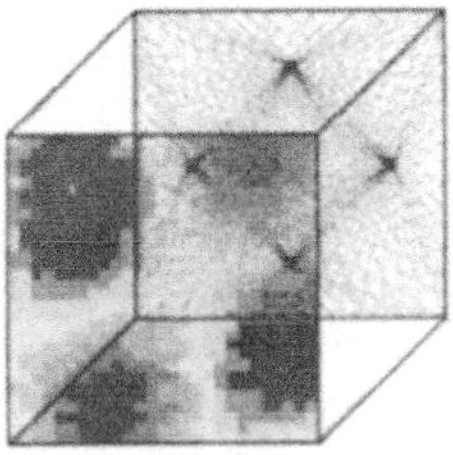

section 1

Speed and Accuracy of Replica Field Computations

MODEL VALIDATION FOR DIRECT AND INVERSE PROBLEMS

M.B. PORTER
Center for Applied Mathematics and Statistics
New Jersey Institute of Technology, Newark, NJ 07102

ABSTRACT Sophisticated acoustic models form the inner loop in most inversion schemes. Trust in these models has come from comparing their results to scale-model and at-sea experiments. The modeling is challenging, partly because ocean acoustic problems span a wide range of frequencies. At the high end we have sources squealing at megahertz frequencies, propagating sound just a few millimeters. This is the scenario for sing-around velocimeters used to measure the speed of sound in the ocean. At the low end sources rumble like the deepest organ notes at tens of Hertz making sound heard around the world. Such problems emerge for a new type of sing-around velocimeter proposed for monitoring global warming. Important decisions are made based on the models. Can they be trusted? We briefly review some of the efforts that have been made to answer this question.

1. Introduction

As research in ocean acoustics has changed, the demands on the models have changed accordingly. The models are applied to new areas and must be validated anew. The earliest benchmarking efforts related to the once dominant ray models. Later, parabolic equation (PE) models emerged and were compared in the first PE workshop[1]. Initially the focus was on range-independent applications. Later the much more challenging range-dependent problems were studied in depth[2, 3, 4]. Careful tests have also been done to compare blind acoustic-model predictions to at-sea experiments[5].

The problem of knowing the environment is often cited as the biggest problem for acoustic modeling. Indeed it is often argued that the models are now so accurate that further improvements are not worthwhile. However, as the papers in this volume show there has been much progress on the important problem of using sound to sense the environment. Furthermore, many inversion schemes work by repeatedly solving the forward problem as the environment is brought into focus: If the acoustic models are inaccurate then they are again the limiting factor.

There have been few efforts to do careful performance comparisons of different algorithms for inverse problems. However, in a way the process is easier. In benchmarking solutions to forward problems the researcher often has the opportunity to do his own intermodel comparisons to find the 'exact' solution. He may then adjust 'knobs' in a program to get the best fit. This has been an issue in both early PE studies (where the reference sound speed was adjusted) and in Gaussian beam models (where the initial beam width and curvature could be tuned).

However, for inversion schemes one may give the data required for an inversion without the solution. The data may also be corrupted by noise to provide a test of the robustness of the inversion scheme. This approach was used very successfully in a recent matched-field

O. Diaschok et al. (eds.), Full Field Inversion Methods in Ocean and Seismo-Acoustics, 3–8.

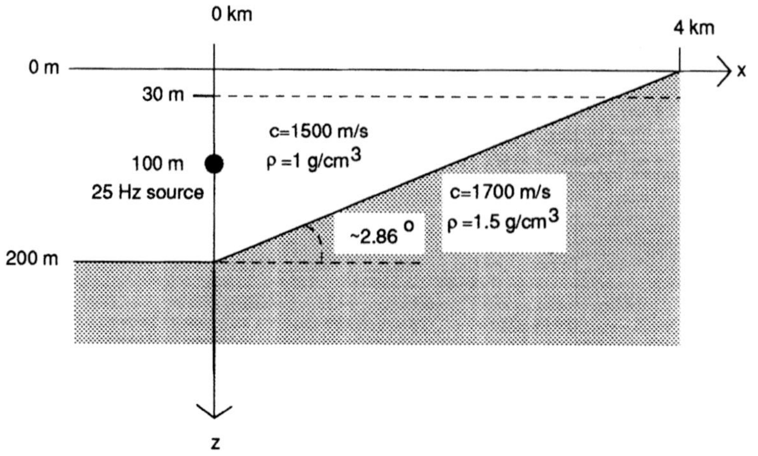

FIGURE 1. Schematic of the wedge problem.

processing workshop[6] to compare different approaches to the inverse problem of finding the source position.

In the following sections we shall review a couple of interesting benchmarking problems. Typically these sorts of reviews focus on the successes. Here instead, we shall examine the notable failures and the lessons learned from them. For details on the models approaches we refer the reader to Ref. [7].

2. The Wedge

The problem of sound propagating in a wedge is one of a few canonical problems in underwater acoustics which have been studied in depth. When the boundaries of the wedge are perfectly reflecting the problem may be solved analytically. In 1980 Jensen and Kuperman[8] used the PE to study propagation in a wedge-shaped ocean with a *penetrable* bottom. This problem shows some very interesting physical phenomena including the cutoff of a mode propagating upslope and the resulting injection of a beam into the ocean bottom.

A version of the penetrable wedge problem was chosen for the ASA benchmark tests in 1990. The particular scenario is shown in Fig. 1. Some very curious results emerged from those comparisons as shown in Figure 2. If we look first at the comparison between one-way coupled-mode and PE results we see excellent agreement— the lines are almost indistinguishable on the plot. One tends to conclude from this that both models are correct. However, the opposite is true! The correct answer is obtained by a full two-way coupled mode calculation, that is, one that allows for back-scatter. This is shown as the solid line in Fig. 2.

Initially this discrepancy was attributed to back-scatter. However, a ray model[9] gave the correct result without backscatter. Furthermore some PE's gave very excellent results.

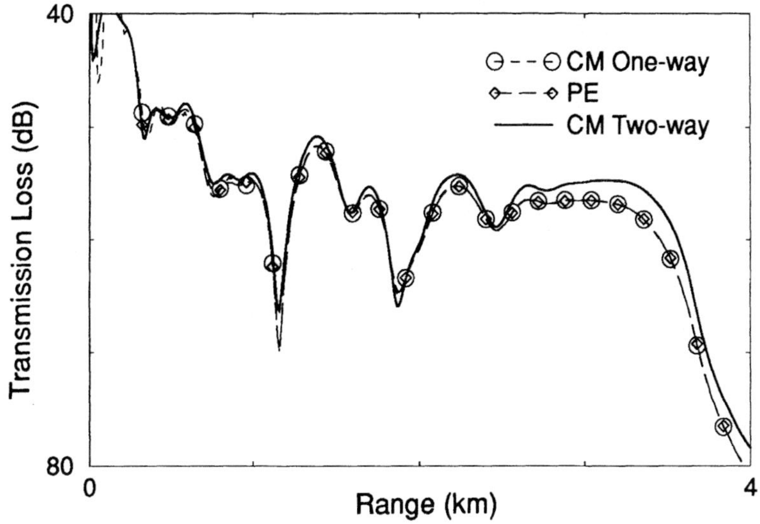

FIGURE 2. Coupled-mode and PE results for the wedge.

Lastly, it was found that when the back-scattered component was extracted from the full two-way coupled-mode solution, it was negligible.

Thus, it seemed it should be possible to construct a one-way model that gave the exact outgoing solution. On the other hand, none of the mode or PE models were doing this well. The models that did poorly all used a stairstep approximation to the bottom and were losing a small amount of energy at each step. Over range,these errors accumulated and became significant. Fortunately, a simple correction was all that was required to resolve these problems in both PE and coupled mode codes[10, 11].

3. The Leaky Surface Duct

Another unexpected failure occurred in the leaky duct problem[12] shown in Fig. 3. This became Test Case 7 in the second PE workshop. Fig. 4 shows TL for a source at 25 m and a receiver at 100 m depth (both in the surface duct). The source frequency is 80 Hz. The exact solution is given by a normal mode code yielding the solid line. Considering the narrow angle of propagation and the gradual variation in transmission loss, one expects that a PE should give a nearly perfect answer. To be still more certain, we initially used a modern wide-angle PE due to Thomson and Chapman.

Curiously, the Thomson-Chapman PE did extremely well out to the first convergence zone (CZ) at 50 km. Beyond that it looks as if perhaps something is triggered in the PE causing it to dump energy out of the surface duct. Then, after the second CZ the levels are close to the true values again. When the problem was rerun using the original PE developed by Tappert in 1973 (and now seldom used) excellent results were obtained, adding further to the confusion.

Through a careful broadband study the cause of this problem emerged. We tend to think

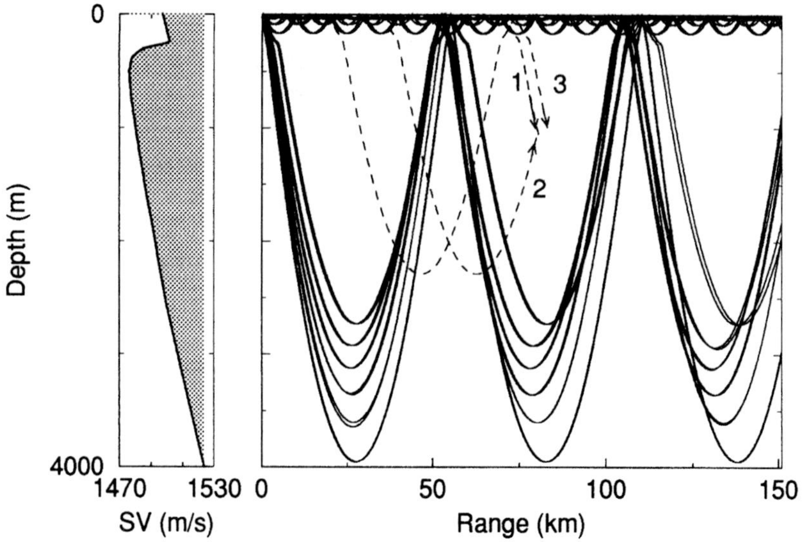

FIGURE 3. Sound speed profile and corresponding ray trace for the surface-duct problem.

of the field in the surface duct as dominated by ray paths that are trapped in the surface duct. Here, it turned out that ray paths associated with leakage out of the duct were strong. (These paths are shown by the dashed lines in Fig. 3.) Beyond the first CZ these leakage paths refocus in the surface duct and interfere constructively or destructively with the ducted paths. An error in the predicted path-length of just half a wavelength is sufficient to cause destructive interference and thus the dropout between the CZ's. All the PE's make such small errors in the phase of the various paths; however, some PE's made consistent errors for the interfering ray paths and therefore preserved the *relative* phase. This explains the fact that some PE's worked well and others failed.

The Thomson-Chapman PE used here has many useful properties. As such, work is ongoing to produce a modified version that is immune to this problem. However, there is a vastly more important aspect to this test problem: it is intrinsically ill-posed in the sense that the TL is extremely sensitive to small changes in the environment. A change of just 0.1 m/s in the mean sound speed in the duct is sufficient to cause a similar dropout. We may think of this environment as just one snapshot in a time-evolving ocean. Then we must conclude that a TL calculation from a single realization is not meaningful.

4. Conclusions

In the last few years the applications have changed significantly. For instance, interest in acoustic transients has led to an increased emphasis on broadband modeling[13]. Similarly, a renewed interest in mine countermeasures has prompted work in the very-high frequency

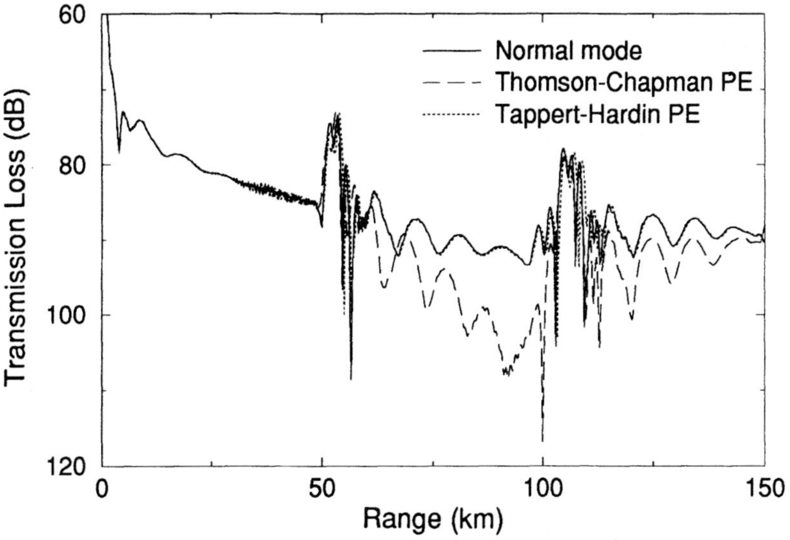

FIGURE 4. Comparison of mode and PE solutions for the surface-duct problem.

regime.

For years, reverberation modeling was dominated by heuristic generalizations of ray-based models. However, lately more formal full-wave approaches have become practical, thus motivating another recent workshop[14]. Lastly, sound is now being contemplated as a tool to probe the ocean on a transoceanic scale to monitor global warming[15, 16]. To date there have been few systematic intermodel comparisons over such distances[17].

In response to these challenges, acoustic models have continued to improve over time. Of course, the models have also benefited greatly from the faster heartbeat of modern computers. Not surprisingly, demands are being placed on them to match their increased capability.

It may be argued that the models are now mature: for fleet applications it is true that details of the environment often affect the results much more than numerical errors in the models. On the other hand, the models themselves will be used to probe the environment, for instance in acoustic tomography. In such applications one envisions acoustic models run across a sweep of frequencies for 3D environments. Furthermore, the models will be run repeatedly as the environmental is brought into focus. These sorts of problems are the 'grand challenges' for acoustic modelers in the next decade.

References

1. Davis, J.A., White D., and Cavanagh R.C. (Eds.) (1981) *NORDA Parabolic Equation Workshop*. NORDA Technical Note 143, Stennis Space Center.

2. McGirr, R.W., King, D.B., Davis, D.B., Campbell, J. (1985) "An evaluation of range-dependent ray theory models", NORDA Technical Report 115, Stennis Space Center.
3. Felsen, L.B. (1990) "Benchmarks: An option for quality assessment," *J. Acoust. Soc. Am.* **87,** 1497–1498.
4. Chin-Bing, S.A., King, D.B., Davis J.A. and Evans, R.B. (Eds.) (1993) *Proceedings of the Second PE Workshop*. Naval Research Laboratory, Stennis Space Center.
5. Thiele, R. (1989) "NATO model study," SACLANT Undersea Research Report, SR-149.
6. Porter, M.B. and Tolstoy, A. (1994) "The Matched-Field Processing Benchmark Problems," *Journal of Computational Acoustics*, (to appear).
7. Jensen, F., Kuperman, W., Porter, M. and Schmidt, H. (1994) *Computational Ocean Acoustics*. American Institute of Physics, New York.
8. Jensen, F.B. and Kuperman, W.A. (1980) "Sound propagation in a wedge-shaped ocean with a penetrable bottom," *J. Acoust. Soc. Am.*, **57,** 1564–1566.
9. Westwood, E.K. (1990) "Ray model solutions to the benchmark wedge problems," *J. Acoust. Soc. Am.*, **87**, 1539–1545.
10. Porter, M.B., Ferla, C.M. and Jensen, F. (1991) "The Problem of Energy Conservation in One-Way Equations", *J. Acoust. Soc. Am.*, **89**(3):1058–1067.
11. Collins, M.D., and Westwood, E.K. (1991) "A higher-order energy-conserving parabolic equation for range-dependent ocean depth, sound speed, and density," *J. Acoust. Soc. Am.*, **89**(3), 1051–1057.
12. Porter, M.B. and Jensen, F.B. (1993) "Anomalous PE results for propagation in leaky surface ducts", *J. Acoust. Soc. Am.*, **94**(3):1510–1516.
13. D. Bradley (Ed.) (1989) *Proceedings of the Transients Workshop*. Naval Research Laboratory publication 152-5100, Washington, D.C.
14. Chin-Bing, S., Evans, R. (Eds.) (1995) *Proceedings of the Reverberation Workshop* . Naval Research Laboratory publication (in prep.).
15. Munk, W., Spindel, R., Baggeroer, A. and Birdsall, T. (1995) "The Heard Island Feasibility Test," in press, *J. Acoust. Soc. Am.*,(in press).
16. Spiesberger, J.L., Metzger, K. and Furgerson, J.A. (1992) "Listening for climatic temperature change in the northeast Pacific: 1983-1989" *J. Acoust. Soc. Am.*, **92,** 384–296.
17. Boden, L., Bowlin, J.B., and Spiesberger, J.L. (1991) "Time domain analysis of normal mode, parabolic and ray solutions of the wave equation," *J. Acoust. Soc. Am.*, **90,** 954–958.

PARALLELISM IN OCEAN ACOUSTIC COMPUTATIONS

PETER C. MIGNEREY
Naval Research Laboratory, Acoustics Division 7120
Washington, DC 20375-5350

ABSTRACT
Many computational problems in ocean acoustics are intrinsically parallel on a large scale. Exploitation of parallelism is discussed within the context of matched-field processing and transient deconvolution. The Connection Machine and Fortran 90 language are used to illustrate the nature of parallel algorithms for obtaining the acoustic field. A factorization of the field is introduced as the basis for an efficient parallel implementation of acoustic signal processors.

1. Introduction

Neither parallel computing nor the new Fortran 90 language has been embraced by the underwater acoustics community. This is unfortunate because the parallel implementation of multiple tasks is the key to solving large computational problems in underwater acoustics.

Indeed some ocean-acoustic computations are so large that the current generation of parallel computers have too few processors to handle the entire task. To overcome this limitation each physical processor must loop over several independent program segments each handled by a virtual processor. This approach even supports the implementation of parallel constructs on standard serial computers. Such scalability across a wide range of computers establishes parallelism as the computational basis of the future.

The parallel implementation of algorithms gives rise to new ideas and problems. New semantic constructs are introduced in parallel versions of the Fortran 90 and C languages to express such ideas. New sources of computational slowdown are found to be the result of communication bottlenecks as the data is moved into place for parallel processing.

A parallel approach to the computation of the acoustic field via normal modes [1, 2] follows from an analysis of the problem independent of whether the implementation is serial or parallel. This is because the problem is naturally parallel. Eigenvalues of the Helmholtz equation are obtained by dedicating individual processors to each mode with the set of processors spanning the entire Nyquist frequency band for numerous sound-speed profiles.

Under a reasonable assumption for long range propagation, the Green's function for the range independent Helmholtz equation has an efficient factorization. This factorization is then used to obtain parallel algorithms for the matched-field and deconvolution processors.

2. Parallel Programming

In this section the notion of parallel vs. serial programming is discussed. New ideas are introduced such as: multitasking, virtual processors, array shape, logical masking, process control, data organization and communication bottlenecks. They provide the framework for

O. Diaschok et al. (eds.), Full Field Inversion Methods in Ocean and Seismo-Acoustics, 9–14.

a new way of thinking about program organization. The programmer must now distinguish those parts of his algorithm that are serial from those that are parallel. Data must be organized in new ways to facilitate parallel processing while avoiding unnecessary movement.

2.1. SERIAL VS. PARALLEL ALGORITHMS

Algorithms fall into several categories from the intrinsically serial to the obviously parallel. Examples of serial processes include the real-time acquisition and processing of a time-series, iteration until convergence, and the evaluation of recursion relations. For such algorithms each step depends on either newly acquired data or the complete history of computations.

Obviously parallel processes involve computations which are completely independent. For example in a broadband matched-field processor vector inner products must be computed for many frequency components. But there is no dependence between frequency components and the algorithm is identical for all components. Even the inner products themselves are parallel because the required multiplications are independent. No special effort is required of the programmer to exploit such parallelism. Languages like Fortran 90 [3, 4, 5] support the expression of parallel computations while the implementation details are left up to the compiler. These parallel languages allow the elimination of almost all DO loops. Only those loops which control intrinsically serial processes remain in a parallel program.

Between the extremes of obviously serial and parallel algorithms lie non-obviously parallel approaches to problems which seem to be hopelessly serial [6, 7] . For example the determinant of a tri-diagonal matrix may be computed by a recursion relation [8] starting at the upper-left corner. But the matrix may be cut in half, the determinant of each half computed and the results pasted together. Furthermore the cutting process can be continued until little submatrices are obtained. Pairs of submatrices can then be pieced together in parallel. The net result is a mixed parallel/serial algorithm composed of a much shorter recursion (order log matrix dimension) of parallel operations [9].

2.2. MULTITASKING

Related to the parallel nature of a task is the issue of multitasking. Trivially parallel processes usually require the application of one algorithm to large quantities of data all having similar properties. Such organization, called Single Instruction Multiple Data (SIMD), is the design basis of the massively parallel Connection Machines (CM-2) and vector pipeline Crays. Fortran 90 allows the SIMD structure of such algorithms to be expressed clearly.

Independent tasks requiring separate algorithms are examples of Multiple Instruction Multiple Data (MIMD) computation which requires nontrivial multitasking. The CM-5 is capable of such processing. MIMD computers and languages are not yet as highly developed as those for SIMD. Thus no standard language has emerged which is portable across most computer systems. However the more general nature of this approach gives it an encouraging future.

Some tasks are a mixture of SIMD and MIMD. Consider computing the eigenvalues of the Helmholtz equation. The basic algorithm is the same for all modes. But the number of modes is frequency dependent and some modes require more iterations than others. Although in a narrow sense this is a MIMD process, it may be handled adequately by a SIMD architecture because only one algorithm is used. This leads to the notion of processor idling. Those processors that have completed their task lie idle while the remaining processors finish. Although a waste of resources, this is not necessarily a waste of time on a parallel computer.

However when the same program is compiled and run on a serial machine, time gets wasted skipping the idle steps. Clearly a MIMD approach would be more efficient here.

2.3. VIRTUAL PROCESSORS

Parallel programs can require very large arrays for two reasons. First, it is sometimes desirable to allocate processors to nonexistent data. For example, if all frequency components are allocated as many processors as needed to accommodate the highest number of modes, too many will have been allocated for the lowest. Second, any intermediate temporary variables must be parallel arrays of the correct shape.

Large arrays are accommodated by virtual processors which subtask each physical processor to handle several subprocessors. Each physical processor implicitly hides sequential loops similar to the vector pipelines used by Cray. Indeed the design philosophy of the CM-5 moves away from the massive parallelism of the CM-2 in a direction that may be viewed as a parallel set of vector pipelines.

Although virtual processors enable large parallel programs, they slow processing below a full parallel implementation in proportion to the ratio of virtual to physical processors. But this approach has an important benefit also. It allows parallel programs to be run on existing serial computers.

2.4. DATA SHAPE

Parallel computer languages require a notion of data shape. Usually parallel operations combine data with the same shape only. In its simplest form shape encompasses rank, dimension and size. Rank is the number of array axes, dimension is the number of elements along each axis and size is the product of the dimensions. Shape is a property needed to support features such as the run-time allocation of arrays and the passing of array arguments without explicitly listing all of the dimensions. Functions may now return arrays for use in assignments or as arguments to other functions in a manner that has historically been possible in Fortran with scalars only.

Differently shaped arrays can be made conformal by reshaping, spreading or gathering. Reshaping maintains the size of an array while changing the rank and dimensions. Spreading is a process in which the rank is increased and the original array is copied everywhere along the new axis. Gathering reduces the rank by combining the elements along one axis according to some rule. Fortran 90 supports several gathering rules such as sum, product, minimum/maximum values and the logical AND/OR operations.

2.5. COMMUNICATION

Spreading and gathering require communication between processors while reshaping does not. During a spread each processor gets data from another processor. Conversely during a gather each processor puts data into a collection processor. Because many processors might put data into the same collection processor, the gathering rule is needed to resolve collisions. Gathering and spreading are examples of communication among processors. In general they are more efficient than arbitrary permutations because the communication is one way along a single array axis and often implemented using adjacent neighbors only.

Arbitrary communications require interconnections between all of the processors. A data network is the only feasible way to link a large number of processors. Such networks can be the site of bottlenecks when many processors simultaneously send messages. Arbitrary communications are also difficult to express in Fortran 90 because permutations are only allowed along one axis. For example, it is easy to permute the rows of a matrix but difficult to transpose. However Fortran 90 does have a transpose function.

For scientific programming the permutation limitation is of little importance. Consider the finite difference approximation of a derivative for which each processor requires neighboring data. To get this information each processor uses an index-array $i(p) = p$, $(p = 1, \ldots, N)$ to access neighbors, e.g. $A(i+1) - A(i-1)$ means $A(i(p)+1) - A(i(p)-1) \forall p$. Problems occur at the boundaries where one neighbor does not exist. Here a new form of conditional processing must be used.

2.6. CONDITIONAL PROCESSING

In parallel programs there are two distinctly different types of logical control. As in serial programs, IF blocks control which program instructions are executed. But in parallel programs the IF blocks *do not* control which array elements are to have their values changed. A new construct, the WHERE mask is used for that.

Where the data meets a logical condition the processors are turned on to compute and assign new array values. Where the condition is not met the processors remain idle and the data are left unchanged. Unfortunately in Fortran 90 the WHERE blocks may not be nested. Instead all of the nested conditions must be encoded in a parallel logical variable which is used to control a single WHERE block.

3. Parallel Acoustic Signal Processing

In this section the Green's function of the Helmholtz equation for a range independent environment is factored for efficient parallel processing. This factorization allows the Green's function to be expressed as the product of three arrays dependent on hydrophone position, source range and source depth. The matched-field and deconvolution processors may then be written as matrix products possessing well known parallel algorithms.

3.1. GREEN'S FUNCTION FACTORIZATION

The normal mode solution for the Green's function [1, 2] connecting a source emitting a frequency component f at location $\mathbf{a}_o \equiv (r_o, z_o)$, with the pth hydrophone located at $\mathbf{a}_p \equiv (r_p, z_p)$, is given by a weighted sum of M normal modes $\phi_m(z, f)$ propagating with horizontal wavenumber $\kappa_m(f)$,

$$G_p(\mathbf{a}_o, \mathbf{a}_p, f) = \frac{e^{i\pi/4}}{2\sqrt{2\pi}} \sqrt{\frac{1}{1 - r_p/r_o}} \sum_{m=1}^{M} \left[\phi_m(z_p, f)\, e^{-i\kappa_m r_p}\right] \left[\frac{\phi_m(z_o, f)}{\rho(z_o)}\right] \left[\frac{e^{i\kappa_m r_o}}{\sqrt{\kappa_m r_o}}\right]. \quad (1)$$

Note that the complex exponential $\exp i(\kappa_m(r_o - r_p))$ has been split into hydrophone and source dependent factors for computational efficiency. For a perfectly straight vertical array $r_p = 0 \rightarrow \sqrt{\frac{1}{1 - r_p/r_o}} = 1$, $\forall p$, which further simplifies the problem. Unfortunately vertical arrays can not be maintained in a perfectly straight configuration, nor is this assumption obviously valid for a horizontal array. However it is usually true that the horizontal scale of the array is only several wavelengths (λ) and yet small compared with the source range,

$$r_p \approx \lambda \ll r_o. \quad (2)$$

Under these conditions the radical can be safely neglected. Although hydrophone range has only a small affect on the field amplitude, it has a large influence on the phase because $r_p \approx \lambda$ can make $\exp(-i\kappa_m r_p)$ large.

The Green's function can then be written as a matrix-vector product,

$$[G_p](\mathbf{a}_o, \mathbf{a}_p, f) = C\,[E_{pm}](f)\,[A_m B_m](r_o, z_o, f), \tag{3}$$

in which

$$\begin{aligned} C &\equiv e^{i\pi/4}/2\sqrt{2\pi}, & (4)\\ [E_{pm}](f) &\equiv \phi_m(z_p, f)\,e^{-i\kappa_m r_p}, & (5)\\ [A_m](r_o, f) &\equiv e^{i\kappa_m r_o}/\sqrt{\kappa_m r_o}, & (6)\\ [B_m](z_o, f) &\equiv \phi_m(z_o, f)/\rho(z_o). & (7) \end{aligned}$$

Here $[A_m B_m](r_o, z_o, f)$ denotes an outer product over all source ranges and depths where each product is itself a vector of modal components. In principle the outer product may be expressed in Fortran 90 and computed in parallel by spreading the data followed by a vector multiply. In practice the problem is often too big to fit in the computer all at once. So loops over source range, depth and frequency must be used in place of virtual processors (which are loops anyway). Good parallel algorithms also exist for the matrix-vector product. Computation of $[E_{pm}](f)$, $[A_m](r_o, f)$ and $[B_m](z_o, f)$ is efficient and takes up minimal storage because the hydrophone positions, source ranges and source depths have been decoupled. They may be computed once when an acoustic processor is first started and then saved for later use.

3.2. PARALLEL MATCHED FIELD PROCESSING

The factorization of the preceding section is the basis for efficient construction of the matched-field [10, 11] and matched-mode processors [12, 13]. A parallel version was fully implemented on a Connection Machine to do transient deconvolution and both coherent and incoherent broadband matched-field processing [14, 15].

With this factorization, the matched-field processor may be written as

$$P_{MFP}(r_o, z_o, f, T) = \left| C^*\,[A_m B_m]^\dagger\,[E_{pm}]^\dagger\,[R_p] \right|^2, \tag{8}$$

where $[R_p](f, T)$ represents array data in the frequency domain obtained at time T, and $(\dagger)$ denotes the conjugate transpose. Likewise the matched-mode processor is

$$P_{MMP}(r_o, z_o, f, T) = \left| C^*\,[A_m B_m]^\dagger\,[F_{pm}]^\dagger\,[R_p] \right|^2, \tag{9}$$

where

$$[F_{pm}]^\dagger \equiv ([E_{pm}]^\dagger\,[E_{pm}])^{-1}\,[E_{pm}]^\dagger \tag{10}$$

is the Moore-Penrose left inverse of $[E_{pm}]$. The Gauss-Markov form of the transient deconvolution processor is slightly less efficient, even in the presence of uncorrelated noise

$$\hat{S}_{GM}(r_o, z_o, f, T) = \frac{[A_m B_m]^\dagger\,[E_{pm}]^\dagger\,[R_p]}{C\,\|[E_{pm}]\,[A_m B_m]\|^2}. \tag{11}$$

For all of these processors the total number of arithmetic operations depends on whether the matrix product is left associated with the mode amplitude vector (*phone space*),

$$P_{phone} = \left| C^* \left\{ [A_m B_m]^\dagger [E_{pm}]^\dagger \right\} [R_p] \right|^2 , \tag{12}$$

or right associated with the data vector (*mode space*),

$$P_{mode} = \left| C^* [A_m B_m]^\dagger \left\{ [E_{pm}]^\dagger [R_p] \right\} \right|^2 . \tag{13}$$

Computation in mode space generally has fewer operations than in phone space because there are usually more mode vectors dependent on source position, $[A_m B_m](r_o, z_o, f)$, than data vectors dependent on time, $[R_p](f, T)$. Furthermore large arrays in shallow water usually have more phones than modes, which contributes to the efficiency of mode space.

4. Conclusion

Acoustic field computations and signal processors are well suited for implementation on massively parallel computers. Indeed parallel algorithms have been successfully used for the deconvolution of acoustic transients in a deep ocean environment [14, 15]. Scalable parallel architectures with Fortran 90 as a supporting language are now available on a wide range of computers from the Connection Machine to desk-top workstations. Parallel computing is an exciting new approach to acoustic analysis. It provides the only realistic way to achieve the large computational speeds needed to solve tomorrow's increasingly complex problems.

References

1. Boyles, C. A., (1984) *Acoustic Waveguides*, Wiley, New York.
2. Porter, M. B. and Reiss, E. L. (1985) A numerical method for bottom interacting ocean acoustic normal modes, *J. Acoust. Soc. Am.*, **76**, 1760-1767.
3. Metcalf, M. and Reid, J. (1990) *Fortran 90 Explained.* Oxford University Press, Oxford.
4. Kerrigan, J. F. (1993) *Migrating to Fortran Ninety.* Loukides, M. ed., O'Reilly.
5. Brainerd, W. S., Goldberg, C. H., Adams, J. C. (1990) *Programmers Guide to Fortran 90.* Mc Graw Hill, New York.
6. Leighton, F. T. (1992) *Introduction to parallel algorithms and architectures*, Morgan Kaufmann, San Mateo
7. Smith, J. R. (1993) *The design and analysis of parallel algorithms*, Oxford University Press, Oxford
8. Golub, G. H. and Van Loan, C. F. (1983) *Matrix Computations.* The Johns Hopkins University Press, Baltimore.
9. Amodio, P., Brugnano, L. and Politi, T. (1993) Parallel factorizations for tridiagonal matrices, *SIAM J. Numer. Anal.*, **30**, 813-823.
10. Bucker, H. P. (1976) Use of calculated sound fields and matched field detection to locate sound sources in shallow water, *J. Acoust. Soc. Am.*, **59**, 368-373.
11. Tolstoy, A. (1993) *Matched Field Processing for Underwater Acoustics*, World Scientific, Singapore
12. Yang, T.C. (1987) A method of range and depth estimation by modal decomposition, *J. Acoust. Soc. Am.*, **82**, 1736-1745.
13. Yang, T.C. (1993) Broadband source localization and signature estimation, *J. Acoust. Soc. Am.*, **93**, 1797-1806.
14. Mignerey, P. C. and Finette, S. (1992) Multichannel deconvolution of an acoustic transient in an oceanic waveguide, *J. Acoust. Soc. Am.*, **92**, 351-364.
15. Finette, S., Mignerey, P. C., Smith, J. F. and Richmond, C. D. (1993) Broadband source signature extraction using a vertical array, *J. Acoust. Soc. Am.*, **94**, 309-318.

PROPAGATION AND INVERSION IN COMPLEX OCEAN ENVIRONMENTS

MICHAEL D. COLLINS
Naval Research Laboratory
Washington, DC 20375, USA

W.A. KUPERMAN
Scripps Institution of Oceanography
La Jolla, California 92093, USA

and

WILLIAM L. SIEGMANN
Rensselaer Polytechnic Institute
Troy, New York 12180, USA

ABSTRACT Topics in propagation and inversion are discussed. The multivalued Bartlett processor is useful for surgically extracting signals from noisy data. The split-step Padé solution and the poro-elastic parabolic equation are forward models that are useful tools for solving inverse problems. The covariance matrix of the gradient of a cost function contains a great deal of useful information about a parameter landscape.

1. Introduction

In this paper, we discuss topics in propagation and inversion. The quality of the output of an inversion algorithm depends on the quality of the input data. In Section 2, we describe an eigen-processing technique for surgically extracting signals from noisy data [1]. The forward model should be efficient because inversion algorithms may require thousands of replica fields (i.e., solutions of the wave equation). In Section 3, we describe an efficient forward model based on the split-step Padé solution [2,3] and the self-starter [4]. The forward model should account for the relevant physics. In Section 4, we describe the poro-elastic parabolic equation (PE) [5], which is a forward model that handles range dependence and poro-elastic [6] ocean bottoms. The parametrization and inversion algorithm should be efficient. In Section 5, we describe an approach for identifying the key underlying parameters and improving the efficiency of search algorithms.

2. Extracting High-Quality Data

In this section, we describe an eigen-processing technique for extracting high-quality data from the covariance matrix K for a vertical array. Eigen-processors [7,8] are based on the representation,

$$K = \sum_{i=1}^{n} \lambda_i \mathbf{v}_i \mathbf{v}_i^*, \tag{1}$$

O. Diaschok et al. (eds.), Full Field Inversion Methods in Ocean and Seismo-Acoustics, 15–20.

where λ_i and $\mathbf{v}_i$ are the ith eigenvalue and eigenvector of K. The acoustic fields due to different discrete processes (e.g., sources at different locations) tend to be spatially uncorrelated over an array of receivers. The energy from different sources therefore tends to be partitioned into different eigenvectors. Energy due to distributed processes such as ambient noise tends to be distributed among the eigenvectors. This partitioning of energy can be exploited to improve the performance of matched-field methods for problems involving interference due to multiple sources and noise. If there are no discrete noise sources present, the eigenvector corresponding to the largest eigenvalue usually provides good input data. For an example in Ref. 1, a signal is surgically extracted from 5 dB below surface generated ambient noise.

Eigen-processing is also applicable when discrete noise sources (e.g., passing ships) are present. The multivalued Bartlett (MVB) processor is an effective eigen-processor for ocean acoustics applications [1]. The MVB processor may perform poorly for fixed sources because the energy due to different sources does not partition favorably for all possible source locations. For an array of sufficient aperture, however, the energy partitioning has a high probability of being favorable for randomly selected source locations. The MVB processor is therefore a reliable approach for determining the tracks of moving sources [1]. For an inversion configuration involving a fixed source, it is necessary to wait for the noise sources to move through the waveguide until the partitioning is favorable. This is determined by looking for an eigenvector that is reproducible.

3. The Split-Step Padé Solution

Solving inverse problems may require thousands of replica fields. It is therefore important to have an efficient forward model. When the ocean environment may be regarded as range independent over a relatively short range interval, matched-field inversion for ocean-bottom properties may be performed effectively using a vertical array of receivers located a relatively short distance from a source. One of the advantages of short-range data is that it contains information about the acoustic properties deep within the sediment. For this inversion problem, it is important to use a propagation model that handles both trapped and leaky modes and energy that propagates at steep angles. In this section, we describe an efficient approach for solving this problem that can exploit parallel processing. We work in cylindrical coordinates with z being the depth below the ocean surface and the range r being the horizontal distance from the source.

By combining the split-step Padé solution [2,3] and the self-starter [4], we obtain

$$p(r_0, z) = (k_0 r_0)^{-1/2} (1+X)^{-1/4} \exp\left[ik_0 r_0 (1+X)^{1/2}\right] \delta(z - z_0), \tag{2}$$

$$k_0^2 X = \rho \frac{\partial}{\partial z} \frac{1}{\rho} \frac{\partial}{\partial z} + k^2 - k_0^2, \tag{3}$$

where p is the complex pressure, k is the complex wave number, ρ is the density, k_0 is a representative wave number, r_0 is the range of the array, and z_0 is the source depth. We apply the Padé approximation,

$$(k_0 r_0)^{-1/2} (1+X)^{-1/4} \exp\left[ik_0 r_0 (1+X)^{1/2}\right] \cong \gamma \prod_{j=1}^{n} \frac{1 + \alpha_{j,n} X}{1 + \beta_{j,n} X}, \tag{4}$$

where $\alpha_{j,n}$, $\beta_{j,n}$, and γ are constants that depend on $k_0 r_0$, to obtain

$$p(r_0, z) = \gamma \prod_{j=1}^{n} \frac{1 + \alpha_{j,n} X}{1 + \beta_{j,n} X} \delta(z - z_0). \tag{5}$$

The numerical implementation of (5) is described in Ref. 4. The coefficients of the rational approximation must be selected with care to provide efficiency and at the same time avoid numerical difficulties associated with the singularity at the source location. The existence of a set of coefficients that provide stability was demonstrated in Ref. 4. The approach described in Ref. 3 can be used to obtain coefficients that provide both efficiency and stability.

On a multi-processor computer, $p(r_0, z)$ is obtained by expanding the rational function into partial fractions and solving n tridiagonal systems in parallel on different processors. This is the most efficient existing technique for obtaining short-range solutions to problems involving arbitrary depth dependence in the acoustic parameters (even without parallel processing). Normal-mode and wave-number integration techniques are based on solving similar tridiagonal systems. For the values of $k_0 r_0$ that are appropriate for inversion with a vertical array, the PE method requires the solution of $O(10)$ tridiagonal systems (computers with $O(10)$ processors have recently become widely available). The separation of variables methods typically require the solution of $O(10^3)$ to $O(10^4)$ tridiagonal systems. The split-step Padé solution is also very efficient for solving range-dependent problems out to long ranges [2,3].

4. The Poro-Elastic PE

In this section, we discuss the poro-elastic PE [5]. Three types of body waves exist in poro-elastic media [6], slow and fast compressional waves and shear waves. The porosity α is the fraction of the sediment that consists of pore spaces. The densities of the solid and fluid components of the sediment are denoted by ρ_s and ρ_f. The displacements of the solid and fluid components are denoted by $\mathbf{u}$ and $\mathbf{u}_f$. We define $\mathbf{U} = \alpha(\mathbf{u}_f - \mathbf{u})$, $\Delta = \nabla\cdot\mathbf{u}$, $\zeta = \nabla\cdot\mathbf{U}$, and $\rho = (1-\alpha)\rho_s + \alpha\rho_f$. The equations of motion derived in Ref. 5 are

$$\mu\frac{\partial^2 u}{\partial r^2} + \mu\frac{\partial^2 u}{\partial z^2} + \rho\omega^2 u + (\lambda+\mu)\frac{\partial\Delta}{\partial r} + \frac{\partial\mu}{\partial z}\frac{\partial u}{\partial z} + \frac{\partial\mu}{\partial z}\frac{\partial w}{\partial r} + C\frac{\partial\zeta}{\partial r} + \rho_f\omega^2 U = 0, \tag{6}$$

$$\mu\frac{\partial^2 w}{\partial r^2} + \mu\frac{\partial^2 w}{\partial z^2} + \rho\omega^2 w + (\lambda+\mu)\frac{\partial\Delta}{\partial z} + 2\frac{\partial\mu}{\partial z}\frac{\partial w}{\partial z} + \frac{\partial\lambda}{\partial z}\Delta + \frac{\partial}{\partial z}(C\zeta) + \rho_f\omega^2 W = 0, \tag{7}$$

$$C\frac{\partial\Delta}{\partial r} + M\frac{\partial\zeta}{\partial r} + \rho_f\omega^2 u + \rho_c\omega^2 U = 0, \tag{8}$$

$$\frac{\partial}{\partial z}(C\Delta) + \frac{\partial}{\partial z}(M\zeta) + \rho_f\omega^2 w + \rho_c\omega^2 W = 0, \tag{9}$$

where λ, μ, C, M, and ρ_c depend on the porosity, wave speeds, densities, and attenuations. For the case $\alpha = 0$, (6) and (7) reduce to the equations of motion for an elastic medium [9].

Differentiating (6) and (8) with respect to r and (7) and (9) with respect to z and summing and combining with (7) and (9), we obtain the system of equations,

$$R\frac{\partial^2\mathbf{q}}{\partial r^2} + S\mathbf{q} = \mathbf{0}, \tag{10}$$

where R and S are depth operators and $\mathbf{q} = (\Delta, w, \zeta)^T$. Factoring (10), we obtain the outgoing poro-elastic wave equation,

$$\frac{\partial \mathbf{q}}{\partial r} = i\left(R^{-1}S\right)^{1/2}\mathbf{q}, \tag{11}$$

which is solved with standard PE techniques.

5. Navigating Parameter Landscapes

In this section, we describe an approach for identifying the important parameters in a parameter space and an efficient set of generators for navigating a parameter space. We randomly select $\mathbf{x}_i$ uniformly from the parameter space Ω and define $\mathbf{g} = \nabla E$ and

$$\mathbf{g}_i = \nabla E \,|_{\mathbf{x}=\mathbf{x}_i} \tag{12}$$

for $1 \leq i \leq N$, where E is the cost function. Dimensions are removed by dividing the jth entry x_j of $\mathbf{x}$ by a representative value so that it is meaningful to compare entries of $\mathbf{g}_i$.

The covariance matrix is defined by

$$K = \langle \mathbf{g}\,\mathbf{g}^* \rangle_\Omega \cong \frac{1}{N}\sum_{i=1}^{N} \mathbf{g}_i \mathbf{g}_i^*. \tag{13}$$

For many problems, a good estimate of K may be obtained by taking N to be on the order of the number of entries in x. Since the eigenvectors of a covariance matrix are an optimal basis for compressing data [10], the $\mathbf{v}_i$ are an optimal basis for ∇E over Ω. Since ∇E is used to guide optimization algorithms, the $\mathbf{v}_i$ are in some sense an optimal set of generators for navigating the parameter landscape. The ideas presented in this section are related to Cramer-Rao bounds [11]. The main difference is that Cramer-Rao bounds provide local information whereas the optimal generators provide global information.

We illustrate the optimal generators for a two-parameter problem that involves the experimental configuration of Ref. 12, with a fixed receiver at $z = 50$ m and a towed 50-Hz source at $z = 50$ m. This convenient configuration involves a limited amount of hardware and is effective for range-dependent problems [12,13]. The data consists of the amplitude of the complex pressure sampled every 10 m in range on a synthetic aperture horizontal array that spans 0.5 km $\leq r \leq$ 5 km. We use the cost function that was used in Ref. 13. The speed of sound is 1500 m/s in the 200-m deep water column. The ocean bottom consists of a 50-m thick sediment layer overlying a homogeneous basement. The sediment sound speed is a linear function of depth defined by the values x_1 =1600 m/s at z =200 m and x_2 =1650 m/s at z =250 m. The sediment density and attenuation are 1.2 g/cm^3 and 0.2 dB/λ. In the basement, the sound speed, density, and attenuation are 1750 m/s, 1.5 g/cm^3, and 0.5 dB/λ.

The cost function has a long valley that is obliquely oriented with respect to the parameter axes. The first eigenvector is approximately normal to the valley. The second eigenvector is approximately parallel to the valley. We solved this problem using simulated annealing [14,15] with an initial guess of x_1 =1700 m/s and x_2 =1560 m/s, which is in the valley but far from the global minimum. In order to focus on navigation via the gradient, we used quenched simulated annealing (i.e., set the temperature to zero) so that only downhill

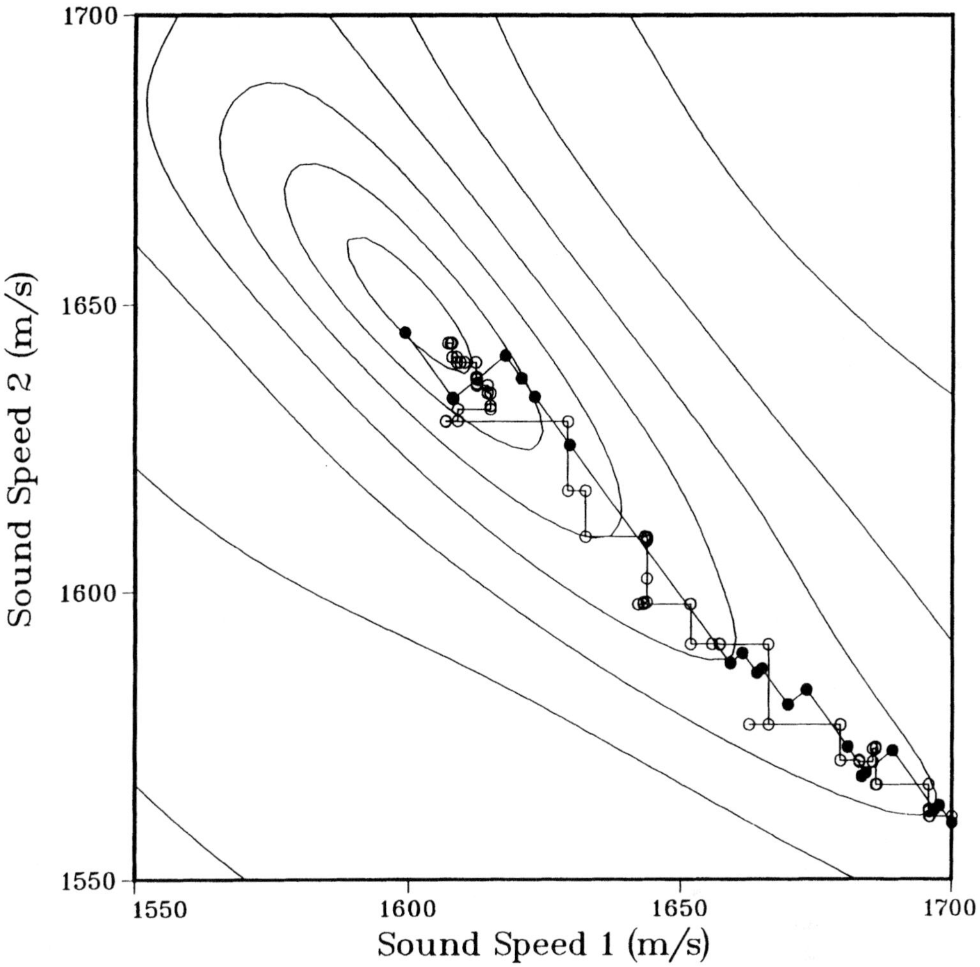

FIGURE 1. Typical parameter searches along a valley that is oriented obliquely to the parameter axes. The contours indicate levels of E in multiples of 0.1. The path marked with solid dots uses the optimal generators, which are oriented approximately normal and parallel to the valley. The path marked with circles uses the coordinate generators, which are parallel to the parameter axes.

perturbations were accepted. The perturbations were selected from a cubic distribution [13], with the largest possible perturbations selected to be 75 m/s.

We performed 1000 searches using both the optimal generators and the coordinate generators. We terminated each search when $\mathbf{x}$ came within a radius of 10 m/s of the correct parameter vector. On the average, quenched simulated annealing converged after 23 iterations for the optimal generators and 183 iterations for the coordinate generators. Typical searches are illustrated in Figure 1 for both of the generator sets. The numbers of points along these paths are less than the numbers of iterations because all uphill perturbations were rejected. With the optimal generators, the search proceeds quickly along the valley.

With the coordinate generators, the search slowly zigzags along the valley.

6. Conclusions

We have discussed four topics in propagation and inversion in complex ocean environments. The MVB processor can be used to surgically extract signals from noisy data. The split-step Padé solution and self-starter can be combined to obtain an efficient forward model. The poro-elastic PE handles realistic ocean bottoms. The covariance matrix of the gradient of the cost function contains a great deal of information about a parameter space.

References

[1] M.D. Collins, L.T. Fialkowski, W.A. Kuperman, and J.S. Perkins, "The multivalued Bartlett processor and source tracking," J. Acoust. Soc. Am. (in press).
[2] M.D. Collins, "A split-step Padé solution for the parabolic equation method," J. Acoust. Soc. Am. 93, 1736–1742 (1993).
[3] M.D. Collins, "Generalization of the split-step Padé solution," J. Acoust. Soc. Am. 96, 382–385 (1994).
[4] M.D. Collins, "A self-starter for the parabolic equation method," J. Acoust. Soc. Am. 92, 1357–1368 (1992).
[5] M.D. Collins, W.A. Kuperman, and W.L. Siegmann, "A parabolic equation for poro-elastic media," J. Acoust. Soc. Am. (in press).
[6] M.A. Biot, "Theory of propagation of elastic waves in a fluid-saturated porous solid," J. Acoust. Soc. Am. 28, 168–191 (1956).
[7] R.O. Schmidt, "A signal subspace approach to multiple emitter location and spectral estimation," Ph.D. dissertation, Stanford Univ., Stanford, California (1981).
[8] D.H. Johnson, "The application of spectral estimation methods to bearing estimation problems," Proc. IEEE 70, 1018–1028 (1982).
[9] M.D. Collins, "Higher-order parabolic approximations for accurate and stable elastic parabolic equations with application to interface wave propagation," J. Acoust. Soc. Am. 89, 1050–1057 (1991).
[10] R.E. Davis, "Predictability of sea surface temperature and sea level pressure anomalies over the North Pacific Ocean," J. Phys. Ocean. 6, 249–266 (1976).
[11] A.B. Baggeroer, W.A. Kuperman, and H. Schmidt, "Matched field processing: Source localization in correlated noise as an optimum parameter estimation problem," J. Acoust. Soc. Am. 83, 571–587 (1988).
[12] J.F. Lynch, S.D. Rajan, and G.V. Frisk, "A comparison of broadband and narrow-band modal inversions for bottom properties at a site near Corpus Christi, Texas," J. Acoust. Soc. Am. 89, 648–665 (1991).
[13] M.D. Collins, W.A. Kuperman, and H. Schmidt, "Nonlinear inversion for ocean-bottom properties," J. Acoust. Soc. Am. 92, 2770–2783 (1992).
[14] N. Metropolis, A.W. Rosenbluth, M.N. Rosenbluth, A.H. Teller, and E. Teller, "Equations of state calculations by fast computing machines," J. Chem. Phys. 21, 1087–1091 (1953).
[15] S. Kirkpatrick, C.D. Gellatt, and M.P. Vecchi, "Optimization by simulated annealing," Science 220, 671–680 (1983).

FAST AND ACCURATE FINITE ELEMENT METHODS FOR THE NUMERICAL PREDICTION OF THE ACOUSTIC FIELD

N. A. KAMPANIS
Institute of Applied and Computational Mathematics, FORTH,
71110 Heraklion, Greece

ABSTRACT
The standard parabolic approximation of the reduced wave equation, used to describe sound propagation in a multi–layered, range–dependent ocean, is solved numerically using finite element methods to discretize in depth and an implicit range–stepping method of Runge–Kutta type. Efficiency tests performed ensure that in terms of computational effectiveness (achieved accuracy vs. computational work), the resulting higher–order accurate, numerical methods compare favorably with classical, second–order accurate, implicit finite difference schemes. They are also faster in obtaining equally accurate results when applied to standard benchmark problems of underwater acoustics.

1. Introduction

Let $0 < z_* < z_b < z_{\max}$ and denote $I_1 = (0, z_*)$, $I_2 = (z_*, z_b)$, $I_3 = (z_b, z_{\max})$, and $I = (0, z_{\max})$. Given $R > 0$ seek a complex–valued function $u = u(z,r)$, for $0 \le z \le z_{\max}$, $0 \le r \le R$, satisfying

$$u_r = i\alpha u_{zz} + i[\beta(z,r) + i\nu(z,r)]u + g(z,r), \qquad z \in I_1 \cup I_2 \cup I_3, \qquad 0 \le r \le R, \quad (1)$$

where $\alpha > 0$ is a given constant, $\beta(z,r)$ is a real–valued function, smooth for $r \ge 0$ on $[0, z_*]$ and on $[z_*, z_{\max}]$ with a possible jump discontinuity at $\{z_*\} \times [0, R]$, and $\nu(z,r)$ is a nonnegative function, smooth in each strip $\bar{I}_i \times [0, R]$, with possible jump discontinuities at z_* and z_b, for $0 \le r \le R$. Across $z = z_*$, u is continuous and satisfies the interface condition $u_z(z_*-, r) = \rho u_z(z_*+, r)$, with $\rho > 0$. We also assume that $u(\cdot, r)$ is C^1 across $z = z_b$, satisfies a zero boundary condition at $z = 0$ and $z_{\max}$, and that $u(z, 0) = u^0(z)$, $z \in \bar{I}$, where u^0 and g are given functions.

(1) is the standard PE, [11], with u representing the acoustic field in an ocean consisting of three horizontal layers: water of density ρ_1, sediment of density ρ_2, and an artificial absorbing layer of density ρ_2, occupying the strips I_1, I_2 and I_3, respectively. The speed of sound $c(z,r)$ is depth– and range–dependent and the field is generated by a point harmonic source of frequency f. In this context, we let $g = 0$, $\alpha = 1/2k_0$, where $k_0 = 2\pi f/c_0$ is the wave number associated with a reference sound speed c_0, $\rho = \rho_1/\rho_2$, $\beta(z,r) = k_0(\eta^2(z,r) - 1)/2$, with $\eta = c_0/c(z,r)$, and $\nu(z,r)$ is a dissipative term, [5].

Since [11] many works have appeared on the numerical solution of the PE and its wide angle extensions, cf. e.g. [10], [7], [6]. In particular, in [6] Collins considers a sequence of very wide angle extensions of the PE, which he solves numerically by piecewise linear finite elements in depth and the ADI-Crank -Nicolson method in range. In the present paper,

O. Diaschok et al. (eds.), Full Field Inversion Methods in Ocean and Seismo-Acoustics, 21–26.

we consider the standard PE and discretize it with a high order accurate finite element technique in depth coupled with implicit Runge–Kutta methods in range.

2. Numerical Schemes

We let $\{z_0, z_1, \ldots, z_M\}$ be a (not necessarily uniform) partition of $\bar{I}$ such that $z_0 = 0$, $z_M = z_{\max}$, having z_* and z_b as nodes, and denote $e_i = (z_{i-1}, z_i)$, $h_i = z_i - z_{i-1}$, $h = \max_i h_i$. We define the weighted inner product $(w, v)_\rho = \int_0^{z_*} w\bar{v}dz + \rho \int_{z_*}^{z_{\max}} w\bar{v}dz$, and the associated norm $\|\cdot\|_\rho = (\cdot,\cdot)_\rho^{1/2}$.

We shall discretize (1) in depth using two different finite element methods. The first one uses the bilinear form $B(\varphi, \chi) := (\varphi', \chi')_\rho$, for $\varphi, \chi \in Q_h$, where Q_h stand for the complex–valued functions $\chi \in C^0(\bar{I})$, such that $\chi|_{e_j} \in \mathbf{P}_3(e_j)$, the cubic polynomials on e_j, and such that $\chi(0) = \chi(z_{\max}) = 0$. The second one is a nonstandard Galerkin/ finite element formulation due to Baker [3] that does not require continuity of the elements of the finite element space at the interfaces. We let $S_{h,i}$, $i = 1, 2, 3$, stand for the complex–valued functions $\chi \in C^2(\bar{I}_i)$ such that $\chi|_{e_j} \in \mathbf{P}_3(e_j)$, and such that $\chi(0) = 0$ for $\chi \in S_{h,1}$ and $\chi(z_{\max}) = 0$ for $\chi \in S_{h,3}$. The finite element space will be $S_h = S_{h,1} \times S_{h,2} \times S_{h,3}$. On S_h we define for $\gamma > 0$ (sufficiently large for stability) the sesquilinear form

$$B_\gamma(\varphi, \chi) := (\varphi', \chi')_\rho + \varphi'(z_*-)[\bar{\chi}(z_*)] + \left(\bar{\chi}'(z_*-) + \frac{\gamma}{h}[\bar{\chi}(z_*)]\right) [\varphi(z_*)]$$
$$+\rho\varphi'(z_b-)[\bar{\chi}(z_b)] + \rho\left(\bar{\chi}'(z_b-) + \frac{\gamma}{h}[\bar{\chi}(z_b)]\right) [\varphi(z_b)],$$

where $[\psi(z)] = \psi(z+) - \psi(z-)$, for $z = z_*, z_b$.

A continuous–in–range finite element discretization of our problem now consists of seeking $u_h(\cdot, r)$ in $\mathcal{X}_h$ for $0 \le r \le R$, such that for all $\chi \in \mathcal{X}_h$

$$(u_{hr}, \chi)_\rho = (i\mathcal{L}_h(r)u_h + G_h(r), \chi)_\rho, \quad 0 \le r \le R, \quad u_h(\cdot, 0) = u_h^0 \tag{2}$$

where $u_h^0 \in \mathcal{X}_h$ is a suitable approximation to u^0, $\mathcal{L}_h(r) : \mathcal{X}_h \to \mathcal{X}_h$, is defined by

$$(\mathcal{L}_h\varphi, \chi)_\rho = -\alpha\mathcal{B}(\varphi, \chi) + ((\beta(\cdot, r) + i\nu(\cdot, r))\varphi, \chi)_\rho, \qquad \text{for } \varphi, \chi \in \mathcal{X}_h \tag{3}$$

and $(G_h(r), \chi)_\rho = (g(\cdot, r), \chi)_\rho$, for $\chi \in \mathcal{X}_h$, with $(\mathcal{B}, \mathcal{X}_h) = (B, Q_h)$ or $(\mathcal{B}, \mathcal{X}_h) = (B_\gamma, S_h)$. The resulting methods will be referred to in the sequel as the CCUB and SPLN method, respectively, [9].

Equation (2) represents a system of o.d.e.'s for the coefficients of $u_h(r)$ with respect to a finite element basis of $\mathcal{X}_h$. It may be written compactly as

$$u_{hr} = i\mathcal{L}_h(r)u_h + G_h(r), \qquad 0 \le r \le R, \qquad u_h(0) = u_h^0. \tag{4}$$

We discretize (4) in range using the 2–stage Gauss–Legendre implicit Runge–Kutta scheme, [1], given by the constants $a_{11} = a_{22} = 1/4$, $a_{12} = 1/4 - \sqrt{3}/6$, $a_{21} = 1/4 + \sqrt{3}/6$, $\tau_1 = 1/2 - \sqrt{3}/6$, $\tau_2 = 1/2 + \sqrt{3}/6$, $w_1 = w_2 = 1/2$. We seek fully discrete approximations $U^n \in \mathcal{X}_h$ to $u(\cdot, r^n)$ (where $r^n = nk$, $0 \le n \le N$, $R = Nk$) such that $U^0 = u_h^0$, and for $n = 0, \ldots, N-1$ (with $r^{n,j} = r^n + \tau_j k$):

$$U^{n,l} = U^n + k\sum_{j=1}^{2} a_{lj}(i\mathcal{L}_h(r^{n,j})U^{n,j} + G_h(r^{n,j})), \quad l = 1, 2, \tag{5}$$

$$U^{n+1} = U^n + k \sum_{j=1}^{2} w_j(i\mathcal{L}_h(r^{n,j})U^{n,j} + G_h(r^{n,j})). \tag{6}$$

Both schemes are unconditionally stable (conservative in the $\|\cdot\|_\rho$ norm) and of fourth–order of accuracy in range and depth, [1], [9].

3. Effectiveness of the Methods

We compare the relative efficiency of our methods, which are fourth–order accurate in range and depth (this is verified numerically with an extensive experimental study of accuracy and stability [9]), with that of a conservative, second–order Crank–Nicolson finite difference scheme suitably modified at the interfaces, [2], which in the sequel will be referred to as the FDIF method (this is a conservative analog of the IFD(S) method, [10]).

To assess the relative efficiency of the methods, [4], we select a problem with known exact solution and a suitable measure of the error; in this case $E(R) = \|U^N - u(\cdot, R)\|_\rho$. This expression is approximated for small values of k and h by

$$E(R) \cong C_1 h^s + C_2 k^m, \tag{7}$$

(for SPLN and CCUB methods $s = m = 4$, while for FDIF $s = m = 2$) where C_1 and C_2 depend on the particular numerical method used, on R, on the solution $u(z,r)$ of (1), but *not* on k and h. We also assume that C_1 depends *only* on the choice of $\mathcal{X}_h$ and that C_2 depends *only* on the range–stepping scheme. Numerical results for a wide variety of k and h assure the validity of these assumptions. For the computational work W required to achieve the error $E(R)$, a good approximation is $W \cong C_3 NM \cong C_3/kh$, since only banded linear system of order $O(M)$ are solved at each range step. Actually, the CPU time expended for a run of the code on computations that are repeated at every range step, is used as a reasonable measure of W in all the numerical experiments performed. For a given error level $E(R)$ we minimize W with respect to k and h subject to the constraint (7) (as equality). The values of k and h for which W is minimized are given by

$$k_{opt} = \left\{ \frac{E(R)s}{C_2(s+m)} \right\}^{\frac{1}{m}}, \qquad h_{opt} = \left\{ \frac{E(R)m}{C_1(s+m)} \right\}^{\frac{1}{s}}.$$

Then, different methods are compared at each error level on the basis of the measure suggested above, i.e. the associated work $W_{\min}$ obtained for the values k_{opt} and h_{opt}, computed as the CPU time needed to perform the range stepping procedure for the run of the code with $k = k_{opt}$ and $h = h_{opt}$.

As a test problem we took in (1) $z_{\max} = 1$, $z_* = z_b = 1/4$, $R = 1$, $\rho = 8/\pi$, $\alpha = 1$,

$$\beta(z,r) = \begin{cases} -2r^2(\sin 2\pi z)^2, & 0 \le z \le z_* \\ -2r^2(\sin \pi(z+\frac{1}{2}))^2, & z_* \le z \le z_{\max} \end{cases}, \quad 0 \le r \le R,$$

$\nu = |\beta|$, and exact solution

$$u(z,r) = \begin{cases} 8\sqrt{2}z^2 r e^{i\pi r}, & 0 \le z \le z_* \\ \sin(\pi z) r e^{i\pi r}, & z_* \le z \le z_{\max} \end{cases}, \quad 0 \le r \le R,$$

which satisfies the interface conditions at z_*. u^0 and g are determined by the exact solution and the differential equation (1). We summarize the results in Table 1, using a reasonable approximation of the values of C_1 and C_2 emerging from computations, [9]. We conclude that the FDIF method has the advantage at low levels of accuracy (10^{-2}, 10^{-3}). For higher accuracies the high order CCUB and SPLN methods are more efficient as expected, with SPLN being a little better. Analogous results concerning the application of SPLN and CCUB methods to the wide-angle (Claerbout's) parabolic equation can be found in [9].

TABLE 1. Relative efficiency of the CCUB, SPLN and FDIF methods.

	CCUB		SPLN		FDIF	
$E(R)$	k_{opt}	h_{opt}	k_{opt}	h_{opt}	k_{opt}	h_{opt}
	W_{min}		W_{min}		W_{min}	
10^{-2}	.2385	.5657	.2385	.4580	.6953E-1	.1346
	.4999E-1		.4000E-1		.1000E-1	
10^{-3}	.1341	.3181	.1341	.2575	.2199E-1	.4255E-1
	.8000E-1		.6999E-1		.6000E-1	
10^{-4}	.7542E-1	.1789	.7542E-1	.1448	.6953E-2	.1346E-1
	.1800		.1499		.5900	
10^{-5}	.4241E-1	.1006	.4241E-1	.8144E-1	.2199E-2	.4255E-2
	.4399		.3999		.6030E+1	
10^{-6}	.2385E-1	.5657E-1	.2385E-1	.4580E-1	.6953E-3	.1346E-2
	.1239E+1		.1070E+1		.5845E+2	
10^{-7}	.1341E-1	.3181E-1	.1341E-1	.2575E-1	.2199E-3	.4255E-3
	.3489E+1		.2989E+1		.5933E+3	

4. The Variable Interface

Sound propagation in the case of a sloping interface is of major interest and was tested in the following model. Given $R > 0$, we let $0 < s(r) < z_b < z_{\max}$, for $0 \le r \le R$, where $z = s(r)$ is a curve fitting the sloping interface. We seek a complex-valued function $u = u(z, r)$ representing the acoustic field for $0 \le z \le z_{\max}$, $0 \le r \le R$, and satisfying

$$u_r = i\alpha u_{zz} + i[\beta(z,r) + i\nu(z,r)]u, \quad z \in (0, s(r)) \cup (s(r), z_b) \cup (z_b, z_{\max}), \quad 0 \le r \le R. \quad (8)$$

We let $F(s(r), r) = u_z(s(r), r) - \dot{s}(r)[ik_0 u(s(r), r) + u_r(s(r), r)]$. Across the curve $z = s(r)$, u is continuous and satisfies the interface condition $F(s(r)-, r) = \rho F(s(r)+, r)$. We assume that $u(\cdot, r)$ is C^1 across $z = z_b$, satisfies a zero boundary condition at $z = 0$ and $z_{\max}$, and that $u(z, 0) = u^0(z)$, $z \in \bar{I}$. Here α, $\beta(z, r)$, $\nu(z, r)$, ρ and u^0 have their usual definition, as given above in the context of underwater acoustics.

We perform the change of variables $z \to zz_*/s(r)$, where we let $z_* = s(0)$. Then $z = s(r)$ transforms to $z = z_*$ =constant in the new variables, and (8) takes the form

$$u_r = i\alpha(\frac{z_*}{s(r)})^2 u_{zz} + i[\beta(\frac{zs(r)}{z_*}, r) + i\nu(\frac{zs(r)}{z_*}, r)]u + z\frac{\dot{s}(r)}{s(r)}u_z, \quad z \in (0, z_*). \tag{9}$$

We select z_b' and $z_{\max}'$ sufficiently large, and we assume that the second layer (sediment) and the absorbing layer occupy the strips (z_*, z_b') and $(z_b', z_{\max}')$, respectively. We seek then a complex-valued function $u = u(z,r)$, for $0 \le z \le z_{\max}'$, $0 \le r \le R$, satisfying (9) for $z \in (0, z_*) \cup (z_*, z_b') \cup (z_b', z_{\max}')$ and $0 \le r \le R$, where u and u_r are continuous across z_* and the interface condition

$$u_z(z_*-, r) = \rho u_z(z_*+, r) + (1-\rho)\frac{\dot{s}(r)s(r)}{z_*(1+\dot{s}^2(r))}(ik_0 u + u_r), \tag{10}$$

holds, while $u(\cdot, r)$ is C^1 across $z = z_b'$, satisfies a zero boundary condition at $z = 0$ and $z_{\max}'$, and $u(z,0) = u^0(z)$ for $z \in (0, z_{\max}')$.

We derive generalizations of the CCUB and SPLN methods to fully discretize (9). The complication of the methods occurring is due to the fact that the differential operator at the right-hand side of (9) is not hermitian and that the interface condition (10) contains a nonhomogeneous term.

5. Test Problems

5.1. HORIZONTAL INTERFACE

We consider the horizontal interface problem due to H. Bucker of [10], where $z_* = 240$ m, $z_b = 512$ m, $z_{\max} = 1200$ m, $\rho_1 = 1$, $\rho_2 = 2.1$ g/cm^3, $b_1 = b_2 = 0$ db/wavelength, the speed of sound is range-independent, piecewise linear in $[0, z_*]$ with values $c = 1500\ m/sec$ at $z = 0$ and $z = 240$ m, $c = 1498\ m/sec$ at $z = 120$ m and $c = 1505\ m/sec$ in $[z_*, z_{\max}]$, and $c_0 = 1499\ m/sec$. A Gaussian source, of frequency $f = 100$ Hz, is placed at a depth $z_S = 30$ m and the receiver at $z_R = 90$ m. The solution is computed up to $R = 20$ km.

In Fig. 1 we present the transmission loss vs. range graph obtained with IFD(S) and plotted against the corresponding graph obtained by SPLN. We observe that both IFD(S) and SPLN give results that are quite close (and physically correct compared to those of SNAP, [10]). The SPLN code is considerably more economical. For the runs shown in Fig. 1 the IFD(S) code (with $M = 400$, $N = 2500$, an optimized pair) took 40 CPU secs on a MicroVax, while SPLN (with $M = 85$, $N = 205$) required 4 CPU secs.

5.2. VARIABLE INTERFACE

We consider the wedge problem with penetrable lossy bottom of [8], where $s(r) = -r/20 + 200$, $z_b = 1000$ m, $z_{\max} = 4000$ m, $\rho_1 = 1$, $\rho_2 = 1.5$ g/cm^3, $b_1 = 0$, $b_2 = 0.5$ db/wavelength, the speed of sound is $c = 1500\ m/sec$ in $[0, s(r)]$ and $c = 1700\ m/sec$ in $[s(r), z_{\max}]$, and $c_0 = 1500\ m/sec$. A Greene's source, of frequency $f = 25$ Hz, is placed at a depth $z_S = 100$ m and the receiver at $z_R = 30$ m. The solution is computed up to $R = 4$ km. A narrow angle approximation of the acoustic field is given by solving (8). For this problem, we took $z_b' = 1000$ m and $z_{\max}' = 4000$ m, which proved to be a suitable choice.

In Fig. 2 we present propagation loss vs. range graphs obtained with the SPLN code and plotted against the corresponding graph obtained by the IFD(S) code. We observe that the

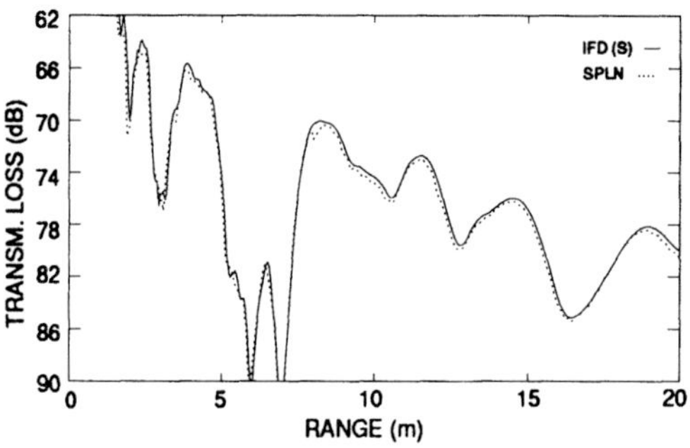

FIG. 1. Horizontal Interface: SPLN vs. IFD(S).

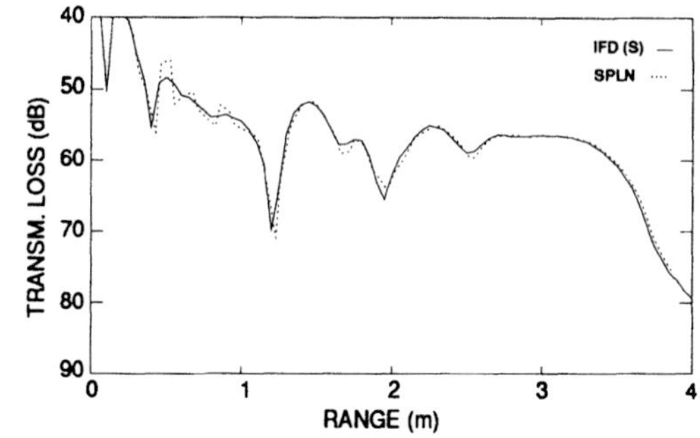

FIG. 2. Variable Interface: SPLN vs. IFD(S).

SPLN results are quite close to those of IFD(S). For the runs shown in Fig. 2 the IFD(S) code (with $M = 2000$, $N = 800$, an optimized pair) took 107 CPU secs on a MicroVax, while SPLN (with $M = 180$, $N = 130$) required 54 CPU secs.

6. Conclusion

The proposed methods are fast, accurate solvers for (1) in environments with horizontal or variable interfaces. We present results from tests on two interesting examples of propagation, one with horizontal interface and another of a wedge with penetrable lossy bottom.

Acknowledgment

This work was partially supported by the European Communities under contract MAS2-CT92-0019 (PRO.MODE).

References

1. Akrivis, G. D. and Dougalis, V. A. (1991) On a class of conservative, highly accurate Galerkin methods for the Schrödinger equation, *RAIRO-MMAN*, **25**, 643–670.
2. Akrivis, G. D. and Dougalis, V. A. (1991) Finite difference discretizations of some initial and boundary value problems with interface, *Math. Comp.*, **56**, 505–522.
3. Baker, G. A. (1973) *Projection methods for boundary value problems for equations of elliptic and parabolic type with discontinuous coefficients*, PhD Thesis, Cornell University.
4. Bona, J. L., Dougalis, V. A. and Karakashian, O. A. (1986) Fully discrete Galerkin methods for the Korteweg–De Vries equation, *Comp. Maths. with Appls.*, **12**, 859–884.
5. Brock, H. K. (1978) *The AESD parabolic equation model*, NORDA TN–12.
6. Collins, M. D. (1989) Applications and time–domain solution of higher order parabolic equations in underwater acoustics, *J. Acoust. Soc. Am.*, **86**, 1097–1102.
7. Greene, R. R. (1984) The rational approximation to the acoustic wave equation with bottom interaction, *J. Acoust. Soc. Am.*, **76**, 1764–1773.
8. Jensen, F. B. and Ferla, M. C. (1990) Numerical solutions of range–dependent benchmark problems, *J. Acoust. Soc. Am.*, **87**, 1499–1510.
9. Kampanis, N. A. (1992) *Galerkin–finite element methods for interface problems in underwater acoustics*, PhD Thesis, University of Crete.
10. Lee, D. and McDaniel, S. T. (1987) Ocean acoustic propagation by finite difference methods, *J. Comp. Maths Applic.*, **14**, 305–423.
11. Tappert, F. D. (1977) The parabolic approximation method, in Keller, J. B. and Papadakis, J. S. (Eds.), *Wave Propagation and Underwater Acoustics*, Lecture Notes in Physics **70** Springer–Verlag, Berlin, 224–287.

DIRECT WAVE PROPAGATION IN THE FREQUENCY DOMAIN VIA THE DIRICHLET-TO-NEUMANN OPERATOR SYMBOL

L. Fishman[a], and M.D. Collins[b]

[a]Applied Mathematical Sciences, Ames Laboratory, Iowa State University, Ames, IA 50011 USA*

[b]Naval Research Laboratory, Washington, D.C. 20375 USA

The wave field splitting/invariant imbedding/phase space approach to direct wave propagation in the frequency domain is outlined. The properties and subsequent applications of the reflection and Dirichlet-to-Neumann operator symbols are briefly illustrated by an exact solution case.

1. THEORY OUTLINE

Recognizing that typical ocean propagation problems are essentially scattering problems in terms of a transition region and transversely inhomogeneous half-spaces, wave field splitting, invariant imbedding, and phase space methods [1-3] reformulate the problem in terms of an operator scattering matrix characteristic of the transition region. For sound propagation in the ocean, the initial frequency-domain modeling is provided by the scalar Helmholtz equation,

$$\partial_x \begin{pmatrix} \phi(\underline{x}) \\ \partial_x \phi(\underline{x}) \end{pmatrix} = \begin{pmatrix} 0 & 1 \\ -\left(\bar{k}^2 K^2(\underline{x}) + \nabla_t^2\right) & 0 \end{pmatrix} \begin{pmatrix} \phi(\underline{x}) \\ \partial_x \phi(\underline{x}) \end{pmatrix}, \tag{1}$$

where $K(\underline{x})$ is the refractive index field, $\bar{k}$ is a reference wave number, and x has been designated as the principal global propagation, or range, direction. Wave field splitting is employed to transform (1) into a representation incorporating the kinematically correct physics of the envisioned scattering experiment. The relevant field quantities then

*This work was supported by grants from the Office of Naval Research (N00014-91-J-1267) and the Joint Electronic Warfare Center of the Joint Chiefs of Staff. The hospitality and support of the S.N. Bose Centre for Basic Sciences in Calcutta, India is also gratefully acknowledged. Finally, a special acknowledgment to the University of Texas System Center for High Performance Computing and Dr. Robert Harkness for computing the figures in this manuscript.

O. Diaschok et al. (eds.), Full Field Inversion Methods in Ocean and Seismo-Acoustics, 27–32.

become the right-traveling ($\phi^+(x,\underline{x}_t)$) and left-traveling ($\phi^-(x,\underline{x}_t)$) wave field components. Invariant imbedding techniques enable the derivation of the scattering (reflection and transmission) operator equations. The phase space analysis transforms the operator equations to equations on well-behaved functions (symbols), in addition to providing the one-way propagation algorithms. The subsequent equations for the reflection and transmission operator symbols are first-order in range, nonlinear (Riccati-like), and, in general, nonlocal. The system allows for well-posed marching of the elliptic scattering problem. The reflection and transmission operator symbol equations are stiff and oscillatory with a subtle (weak and strong) singularity structure. This apparent numerical difficulty is partly overcome by reformulating the calculation in terms of the Dirichlet-to-Neumann operator symbol which generally has a much smoother behavior, and is consequently more easily computed. The physically necessary reflection and transmission operator symbols then follow from a transformation formula. In principle, the transition region can be divided into subregions, allowing for parallel computations and subsequent recombination.

For example, for the limiting case of a transversely homogeneous, two-dimensional environment with the transition region on $x \,\varepsilon\, [a,b]$, the equation for the reflection operator symbol takes the form [3]

$$\begin{aligned}(i/\bar{k})\partial_x \Omega_{\mathbf{R}^+}(x,b;p) &= \Omega_{\boldsymbol{\gamma}}(x,p) + (\Omega_{\boldsymbol{\delta}}(x,p) - \Omega_{\boldsymbol{\alpha}}(x,p))\,\Omega_{\mathbf{R}^+}(x,b;p) \\ &\quad + \Omega_{\boldsymbol{\beta}}(x,p)\,\Omega^2_{\mathbf{R}^+}(x,b;p), \end{aligned} \tag{2}$$

with the initial condition

$$\Omega_{\mathbf{R}^+}(b,b;p) = 0, \tag{3}$$

and where

$$\Omega_{\boldsymbol{\gamma}} = (i/2\bar{k})\left(\frac{K(x)K'(x)}{K^2(x)-p^2}\right), \tag{4}$$

$$\Omega_{\boldsymbol{\beta}} = -\Omega_{\boldsymbol{\gamma}}, \tag{5}$$

and

$$\Omega_{\boldsymbol{\delta}} - \Omega_{\boldsymbol{\alpha}} = 2\left(K^2(x) - p^2\right)^{1/2}. \tag{6}$$

In (4) – (6), an appropriate outgoing (forward) wave radiation condition is understood. The corresponding equation for the Dirichlet-to-Neumann operator symbol is

$$\partial_x \Omega_{\boldsymbol{\Lambda}}(x,b;p) = \Omega^2_{\boldsymbol{\Lambda}}(x,b;p) + \bar{k}^2(K^2(x) - p^2), \tag{7}$$

with the initial condition

$$\Omega_{\boldsymbol{\Lambda}}(b,b;p) = -i\bar{k}\left(K^2(b) - p^2\right)^{1/2}, \tag{8}$$

and where

$$\Omega_{\mathbf{R}^+}(x,b;p) = \frac{i\bar{k}\left(K^2(x)-p^2\right)^{1/2} + \Omega_{\boldsymbol{\Lambda}}(x,b;p)}{i\bar{k}\left(K^2(x)-p^2\right)^{1/2} - \Omega_{\boldsymbol{\Lambda}}(x,b;p)}. \tag{9}$$

The calculation of the wave fields proceeds in the following manner. For the pure scattering problem, assuming a source in the left half-space only, (7) and (8) are solved for $\Omega_{\boldsymbol{\Lambda}}(x,b;p)$ marching from $x=b$ to $x=a$ for each value of p. Only $\Omega_{\boldsymbol{\Lambda}}(a,b;p)$, representing the physical medium, is retained. $\Omega_{\mathbf{R}^+}(a,b;p)$ then follows from (9). The initial value of the reflected wave field is then constructed from

$$\phi^-(a^-,z) = \int_{R^1} dp\, \Omega_{\mathbf{R}^+}(a,b;p)\exp(i\bar{k}pz)\hat{\phi}^+(a^-,p), \tag{10}$$

where $\hat{\phi}^+(a^-,p)$ is the Fourier transform of the incoming right-traveling wave at the transition region boundary. The reflected wave field in the left half-space then follows from standard one-way marching algorithms [1,4]. For the internal transition region wave field, (7) and (8) are again solved for each p value, now retaining all intermediate values of $\Omega_{\boldsymbol{\Lambda}}(x,b;p)$. $\Omega_{\mathbf{R}^+}(x,b;p)$ again follows from (9). The internal wave field is constructed in a layer-stripping fashion employing the right propagation of $\phi^+(x,z)$ in conjunction with the succession of scattering problems generated in the construction of $\Omega_{\mathbf{R}^+}(x,b;p)$. The two-way nature of the elliptic problem in the transition region is accounted for by first marching the Dirichlet-to-Neumann operator symbol (reflection operator symbol) equation from $x=b$ to $x=a$, and then successively calculating the field by marching back from $x=a$ to $x=b$ [3].

2. SYMBOL PROPERTIES

The singularity structure of eqs. (2) and (7) follows from the general theory of Riccati equations. The relationship between Riccati and linear second-order differential equations enables the reflection operator symbol, for example, to be expressed in terms of the Helmholtz equation solutions in the form

$$\Omega_{\mathbf{R}^+}(x,b;p) = \frac{g_1^+(x,p)g_2^+(b,p) - g_2^+(x,p)g_1^+(b,p)}{g_1^-(x,p)g_2^+(b,p) - g_2^-(x,p)g_1^+(b,p)}, \tag{11}$$

where

$$g_i^\pm(x,p) = \pm(i/\bar{k})\left(K^2(x)-p^2\right)^{-1/2}\partial_x\hat{\phi}_i(x,p) + \hat{\phi}_i(x,p), \tag{12}$$

and with the two linearly independent Helmholtz solutions satisfying

$$\left(\partial_x^2 + \bar{k}^2\left(K^2(x)-p^2\right)\right)\hat{\phi}_i(x,p) = 0. \tag{13}$$

From (10) it follows that (11) can be written in the physically insightful alternative form

$$\Omega_{\mathbf{R}^+}(x,b;p) = \frac{\hat{\phi}^-(x,p)}{\hat{\phi}^+(x,p)}, \tag{14}$$

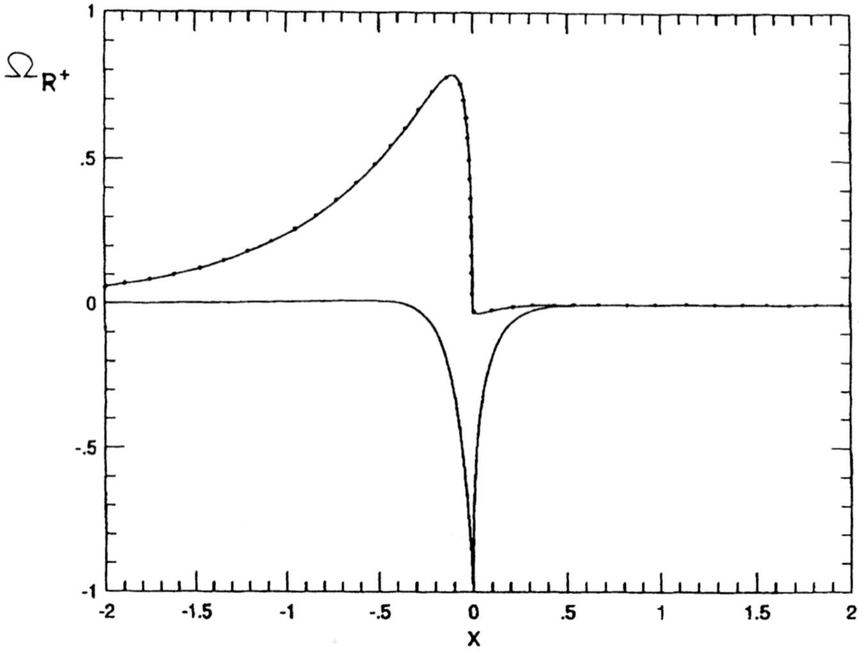

FIG. 1. $\Omega_{\mathbf{R}^+}(x,2;1)$ vs. x for $K^2(x) = 1 + \left(\frac{1}{2}\right)\tanh(5x)$ with $\bar{k} = 1.0$. (——) $\mathrm{Re}\,\Omega_{\mathbf{R}^+}$ and (•-•-•) $\mathrm{Im}\,\Omega_{\mathbf{R}^+}$.

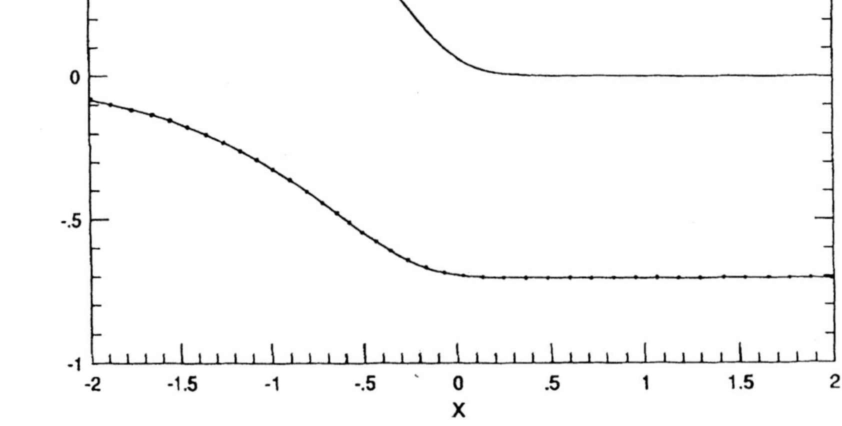

FIG. 2. $\Omega_{\boldsymbol{\Lambda}}(x,2;1)$ vs. x for $K^2(x) = 1 + \left(\frac{1}{2}\right)\tanh(5x)$ with $\bar{k} = 1.0$. (——) $\mathrm{Re}\,\Omega_{\boldsymbol{\Lambda}}$ and (•-•-•) $\mathrm{Im}\,\Omega_{\boldsymbol{\Lambda}}$.

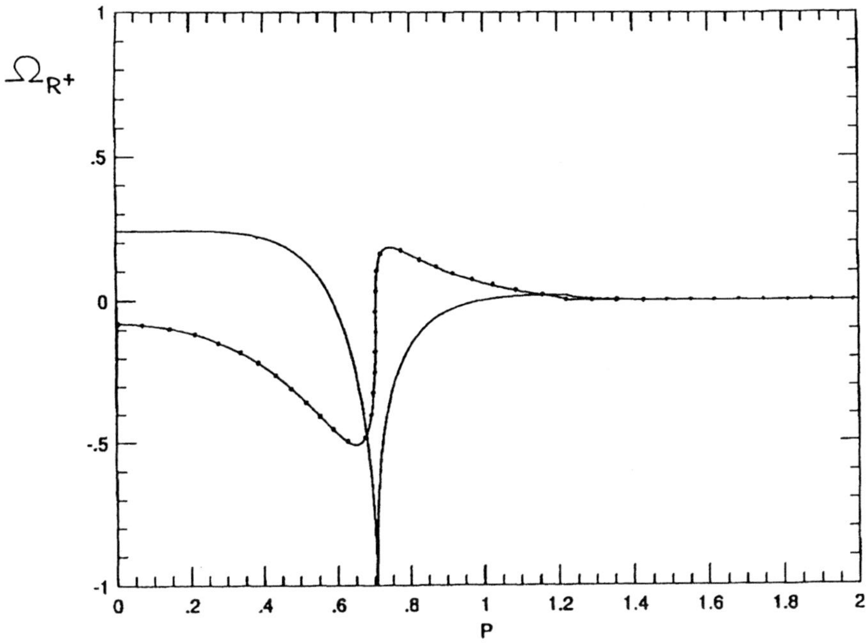

FIG. 3. $\Omega_{\mathbf{R}^+}(-2,2;p)$ vs. p for $K^2(x) = 1 + \left(\frac{1}{2}\right)\tanh(5x)$ with $\bar{k} = 1.0$. (—) $\mathrm{Re}\,\Omega_{\mathbf{R}^+}$ and (•–•–•) $\mathrm{Im}\,\Omega_{\mathbf{R}^+}$.

expressing the reflection operator symbol as the ratio of the left- and right-traveling wave fields for the transverse Fourier components evaluated at the transition region boundary in the invariant imbedding construction. The relationship in (14) can be extended to the case of a general transition region through the solution of the corresponding two-dimensional Helmholtz equation, the identification of the reflection operator kernel, and the subsequent Weyl calculus operator symbol construction [1-3].

$\Omega_{\mathbf{R}^+}(x,b;p)$ can be expressed in closed form whenever (13) has closed-form solutions, enabling a detailed study of the properties of the reflection operator symbol in addition to providing benchmark cases for the numerical algorithms. For example, for the transversely homogeneous model environment $K^2(x) = K_0^2 + A\,\tanh(\rho x) + B\,\mathrm{sech}^2(\rho x)$, which is appropriate for many modeling applications, the required solutions are expressed in terms of hyperbolic functions and Gauss hypergeometric functions with hyperbolic function arguments [5]. The analogous closed-form expressions for $\Omega_{\Lambda}(x,b;p)$ follow from (9).

Some of the properties of $\Omega_{\mathbf{R}^+}(x,b;p)$ and $\Omega_{\Lambda}(x,b;p)$ are readily illustrated by the previously outlined exact solution case. For $p^2 < K_{min}^2$, the transverse Fourier components are propagating, while for $p^2 > K_{max}^2$, the transverse Fourier components are nonpropagating in the model environment. The interesting case is $K_{min}^2 \le p^2 \le K_{max}^2$, where a given transverse Fourier component may be propagating in some sections of the

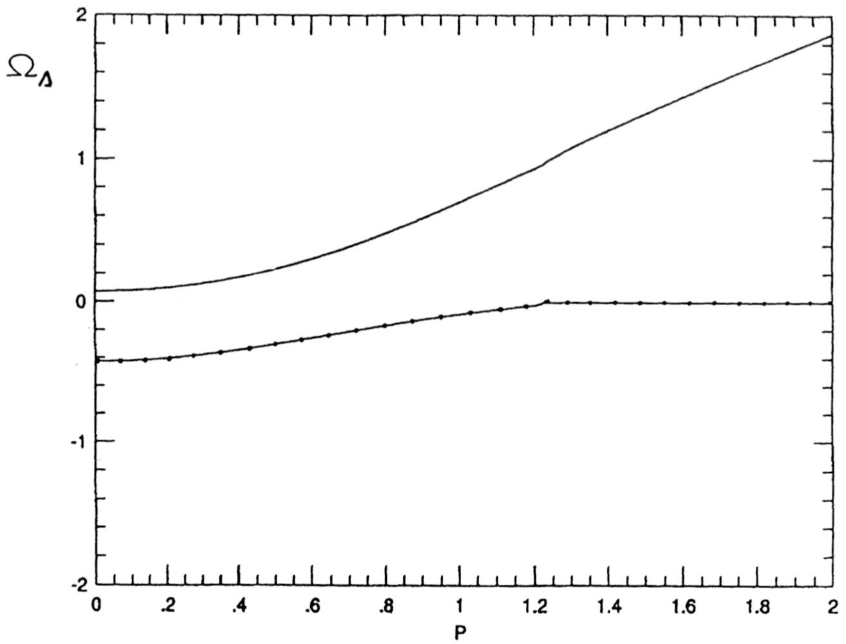

FIG. 4. $\Omega_\Lambda(-2, 2; p)$ vs. p for $K^2(x) = 1 + \left(\frac{1}{2}\right) \tanh(5x)$ with $\bar{k} = 1.0$. (——) $\operatorname{Re} \Omega_\Lambda$ and (•–•–•) $\operatorname{Im} \Omega_\Lambda$.

environment and nonpropagating in others. Figures 1 – 4 illustrate the case $K^2(x) = 1 + \left(\frac{1}{2}\right) \tanh(5x)$ with $\bar{k} = 1.0$ on the interval $[-2, 2]$. This model environment has no trapped modes. Figure 1 demonstrates the (fixed) weak singularity in $\Omega_{\mathbf{R}^+}(x, 2; 1)$, while Figure 2 establishes the correspondingly smooth behavior in $\Omega_\Lambda(x, 2; 1)$. Figures 3 and 4 illustrate analogous results for $\Omega_{\mathbf{R}^+}(-2, 2; p)$ and $\Omega_\Lambda(-2, 2; p)$. A detailed discussion of the singularity structure of $\Omega_{\mathbf{R}^+}(x, b; p)$ and $\Omega_\Lambda(x, b; p)$ will be presented elsewhere. Again, it should be emphasized that while $\Omega_{\mathbf{R}^+}(x, b; p)$ is the desired function, the nonlinear solvers will be applied to the (generally) smoother function $\Omega_\Lambda(x, b; p)$.

REFERENCES

1. L. Fishman, J.J. McCoy, and S.C. Wales, J. Acoust. Soc. Am., 81(5) (1987) 1355.
2. L. Fishman, J. Math. Phys., 33(5) (1992) 1887.
3. L. Fishman, Radio Science, 28(5) (1993) 865.
4. M.D. Collins, J. Acoust. Soc. Am., 93(4) (1993) 1736.
5. N. Rosen and P.M. Morse, Phys. Rev., 42 (1932) 210.

EFFICIENT ACOUSTIC FIELD COMPUTATION FOR ESTIMATING GEOACOUSTIC BOTTOM PARAMETERS USING MATCHED-FIELD INVERSION.

N.G. HENSON, D.E. HANNAY, and P. SCRIMGER
JASCO Research Ltd.,
205 - 9865 W.Saanich Rd.
Sidney, B.C.
CANADA V8L 3S1

S.E. DOSSO
Defence Research Establishment Pacific.
FMO Esquimalt, Victoria, B.C.,
CANADA V0S 1B0

ABSTRACT. Matched-field inversion for source location, environmental properties, or a combination of the two (focalization) is based on comparing acoustic fields measured at an array of receivers to replica fields predicted for a variety of possible source positions and/or environmental parameters using a numerical propagation model. Although this approach is conceptually straightforward and widely applicable, it requires a large number of replica field computations and can be impractically slow if the propagation modeling is not carried out in an efficient manner. This is particularly true for environmental inversion and focalization problems where the expanded parameter search space generally requires a very large number of replica field computations, even if a relatively efficient optimization algorithm such as simulated annealing (SA) is employed. Therefore, to apply these methods to realistic problems, it is important that the propagation modeling be performed as efficiently as possible.

This paper presents a method of inversion using SA optimization and rapid acoustic field modeling based on the use of pre-computed mode functions and wave numbers which are stored in 'look-up' tables. The method is described together with the results of a study which established the optimum size of the grid elements required to achieve accurate propagation modeling and reliable matched-field inversion using this 'look-up' table method. An example is presented for a typical Arctic environment in which several of the unknown geoacoustic parameters are estimated using this efficient inversion method.

1. Introduction

Matched field inversion can become impractically slow when the size of the search space is large. This is often the case when the inversion is performed to obtain estimates of environmental parameters. In these cases, an environmental model $\boldsymbol{m}$ is described by a particular set of values for the environmental parameters which affect the field. Each additional parameter adds an extra dimension to the parameter search space, so the size of the space increases geometrically. A cost function $E(\boldsymbol{m})$ is defined over the parameter space to indicate the degree of mismatch between the replica fields computed for the particular model, and the measured fields. The goal of the inversion is to find the model which minimizes the cost function. The values for the parameters which describe this model are then taken as the estimates for the real parameter values present at the experiment site.

In general, the cost function can have many sub-optimal minima, and there is no direct method for finding the global minimum. A common method for locating the global minimum is the exhaustive search method. This method involves discretizing the parameter search space $\boldsymbol{m}=\{\boldsymbol{m}_i\}$, so that there is a finite number of models. The inversion then involves finding the minimum of the cost function for all realizations of $\boldsymbol{m}$ within the discrete parameter space. A requirement for this method to work is that the discretization of the parameter space be made on a sufficiently fine scale to resolve the global minimum. However, fine discretizations lead to very large search spaces. For example, 5 environmental

O. Diaschok et al. (eds.), Full Field Inversion Methods in Ocean and Seismo-Acoustics, 33–38.

parameters, each with only 10 possible discrete values, lead to a total of 10^5 = 100,000 models. Clearly, the exhaustive search method quickly becomes impractical due to the excessive time required to test all models.

When the number of models is too large for an exhaustive search, more efficient techniques such as simulated annealing (SA) are employed to reduce the number of models tested. SA is based on an analogy with the annealing process of metallurgy, whereby molten metal is cooled slowly so that it crystallizes in a very low energy state. The SA algorithm is an iterative technique which repeatedly perturbs parameter values, and accepts or rejects models based on the change of cost function value. The process is controlled by a parameter T, analogous to temperature, which is slowly reduced (cooling). At any point in the inversion, the algorithm stores an accepted model $\boldsymbol{m}_a$, and the associated value of the cost function $E(\boldsymbol{m}_a)$. A new model $\boldsymbol{m}_n$ is determined by perturbing one or more of the parameters, and the change in the cost function, $\Delta E = E(\boldsymbol{m}_n) - E(\boldsymbol{m}_a)$ is calculated. Perturbations which decrease E are always accepted. Perturbations which increase E are accepted with a probability that is proportional to the Boltzman distribution:

$$P(\Delta E) = \exp(-\Delta E / T) \qquad (1).$$

Because the algorithm allows probabilistic acceptance of increases in cost, it has the ability to escape from sub-optimal or local minima in the cost function. As T decreases, the probability that an increase in cost is accepted becomes smaller. If T is reduced slowly, the algorithm will move out of, and avoid regions in the parameter search space where the cost is high. If, however, T is reduced too quickly, the algorithm can become trapped in sub-optimal minima and the inversion will fail. The initial value of T, and the rate at which it is reduced is referred to as the annealing schedule. The annealing schedule is problem specific, and is usually determined by trial and error after running the algorithm several times on a similar problem.

Since the SA algorithm requires that the replica acoustic fields be computed a large number of times, it is important that this be done as efficiently as possible. If the normal-mode model is used to generate the replica fields, the most time-consuming step is the computation of the wave numbers k_j and mode functions $\phi_j(z_s)$ and $\phi_j(z_h)$ at the source and receiver respectively. The index j, which represents mode number, ranges from 1 to the number of propagating modes M. Once the wave numbers and mode function values are known, computation of the pressure at a receiver $p(z_h)$, due to a source at range r, is relatively fast using the normal mode solution:

$$p(z_h) = \alpha \sum_{j=1}^{M} \phi_j(z_s)\phi_j(z_h) \frac{\exp[ik_j r]}{\sqrt{k_j r}} \qquad (2).$$

where α is a constant.

Dosso et al [1,2] described a method for rapid repetetive replica field computations using 'modal interpolation'. This method is applied by running the normal mode algorithm initially for models on a course grid spanning the entire parameter space. The wave numbers and mode function values at the source and receiver depths for all significant propagating modes at each grid point are saved in 'mode look-up tables'. Once the mode look-up table has been generated, wave numbers and mode function values for any specific set of parameters are found by interpolation from the table values. The field for the model described by the parameters is then computed relatively quickly from the interpolated values using equation (2). An important advantage of this method is that it can be used together with the SA algorithm to provide both rapid prediction of the fields and efficiency in searching the parameter space. In fact, after the mode table has been generated, multiple SA runs can be performed to determine a satisfactory annealing schedule without repeating the modal calculations.

Review of previous work and scope of current work

Ref. [1,2] demonstrated the application of modal interpolation with SA to matched field inversion. Both papers apply these methods to obtain depth and bottom sound speed estimates from synthetic experimental data, for the cases of range independent and range dependent variation of the parameters. The papers assume the field from a harmonic source is measured some distance away by a multi-element vertical array. The inversions were performed for simulated environments in which the geoacoustic properties of the bottom were assumed to be known, but where bottom depth was variable between the source and the receiver. The inversions were performed to estimate the bottom depth at several positions between the source and receiver, and were extended to include bottom sound speed at each of the positions. Mode look up tables were created and consisted of wave numbers and mode function values for a grid in bottom depth - compressional wave speed parameter space. Replica fields for these inversions were generated using the adiabatic normal mode solution due to the range dependent nature of the problems. The cost function used to compare replica fields $p_i(\boldsymbol{m})$ to the simulated data p_i was based on the Bartlett correlator scaled to within (0,1), where 0 represents a perfect match:

$$E(\boldsymbol{m}) = 1 - \frac{\left|\sum_{i=1}^{N} p_i(\boldsymbol{m}) p_i^*\right|^2}{\sum_{i=1}^{N} |p_i(\boldsymbol{m})|^2 \sum_{i=1}^{N} |p_i|^2} \qquad (3).$$

Recently, work has been done by Henson, Hannay, and Scrimger [3] to investigate the accuracy and applicability of using mode look-up tables in matched-field inversion of ocean bottom parameters. Of primary concern in this study was determining the size of the grid elements required for various ocean-bottom parameters to achieve accurate propagation modeling and reliable inversion results. In this study a simple one-layer Arctic environment was considered. The 'base-line' environment included a typical upward refracting sound speed profile in a 400 m water column over a solid bottom. The specific bottom properties were a 1900 m/s compressional sound speed (C_p), a 500 m/s shear speed (C_s), a density (ρ) of 2.0 g/cm^3, a compressional attenuation (α_p) of 0.2 dB/λ and a shear wave attenuation (α_s) of 0.6 dB/λ. The source was located at a fixed depth of 350 m at a range of 40 km from a 20 element vertical hydrophone array with 20 m spacing between phones. The array covered the entire water column, having the first phone at a depth of 20 m and the last phone on the ocean bottom.

The investigation used the normal mode program KRAKENC [4] to generate mode function values and wave numbers for the 'base-line' environment above. Changes in the mode function values and wave numbers were examined when six of the parameters were varied through ranges defined respectively for each parameter. The varied parameters included compressional wave speed, shear wave speed, density, compressional wave attenuation, shear wave attenuation, and water depth. In this work, plots of the mode values versus individual parameters were made to estimate the sensitivity for each parameter.

Figure 1 shows an example of one of these sensitivity plots where the real part of the mode functions are plotted versus the compressional wave speed as it is varied from 1600 m/s to 2200 m/s. The remainder of the parameters were held fixed at their respective 'base-line' model values, with the exception of bottom density which was fixed at 1.90 g/cm^3. The environment as described supported 8 modes as indicated in figure 1. This figure demonstrates the significant variation which occurs in the mode functions when the geoacoustic parameters are varied.

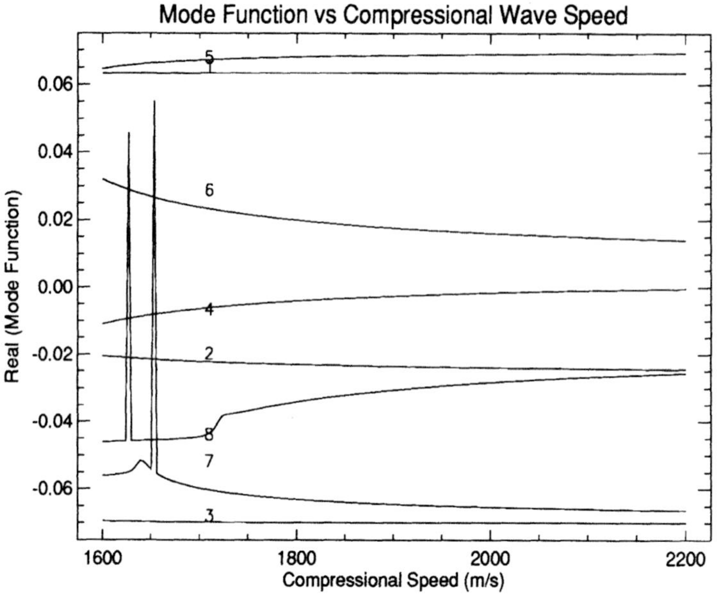

Figure 1: *A plot of the real parts of the eight complex mode functions at 25 Hz; $\phi_j(z=200\ m)$, $(j=1,8)$ as a function of the compressional wave speed. The propagation model is that of a typical Arctic environment with a water depth of 400 m and a single bottom layer with shear speed 500 m/s, density 1.90 g/cm^3, shear wave attenuation 0.6 dB/λ and compressional attenuation 0.2 dB/λ. Note the occurrence of anomalous data points in the curves for modes 7 and 8.*

Two anomalous 'spikes' are noticed on the curves for modes 7 and 8 in figure 1. These spikes result from a simple sign change of the mode function value. (Note that in both cases the peak of the spike is located at a point equal to the negative of the expected function value. Corresponding spikes also occur on the imaginary parts of these respective mode functions). The anomalous data points are not in fact errors because the resulting pressure field is insensitive to simple changes of sign. These points however will obviously cause problems for our interpolation approach. We must therefore preview the mode look-up table values to ensure that non-physical sign changes in the mode values are removed. The previewing is also beneficial because it allows the modeler to confirm that the functions are smooth enough for valid interpolation through the discretized space. Also, erroneous data may be identified during the preview.

In order to take maximum advantage of the 'look-up' table method, it is necessary to represent mode function and wave number plots such as these with as few points as possible and then to employ a strategic interpolation scheme to approximate the function values at all intermediate points. In this work, only linear interpolation was used, however other methods such as polynomial fit interpolation should produce better results for some parameters.

Three parameter demonstration

The inversion procedure described previously was applied to demonstrate its ability to estimate the bottom parameters: C_p, C_s, and ρ. For this inversion, synthetic data was generated with KRAKENC for an environment similar to the 'base-line' environment described previously, but where C_p=1950 m/s, C_s=450 m/s, and ρ=2.1 g/cm^3. The field was sampled by the 20 element vertical array spanning the upward refracting 400 m water column, at a range of 40 km from the source which was at a depth of 350 m. All processing was performed at a frequency of 25 Hz.

A mode table was created to store wave numbers and mode function values for the source and 20 receiver depths on a three dimensional grid consisting of 7 discrete compressional speed values, 5 discrete shear speed values, and 5 discrete density values. The compressional speed values ranged from 1600 m/s to 2200 m/s in 100 m/s steps. Shear speed values ranged from 200 m/s to 600 m/s in 100 m/s steps, and density ranged from 1.5 g/cm^3 to 2.5 g/cm^3 in 0.25 g/cm^3 steps. During the inversion, 2000 iterations of the simulated annealing algorithm were performed. At each iteration, the cost function (3) was evaluated. The initial temperature was set to 1.0 and it was reduced by multiplying by 0.995 per iteration. Figure 2 shows the values of the parameters for the accepted models during each iteration of the inversion.

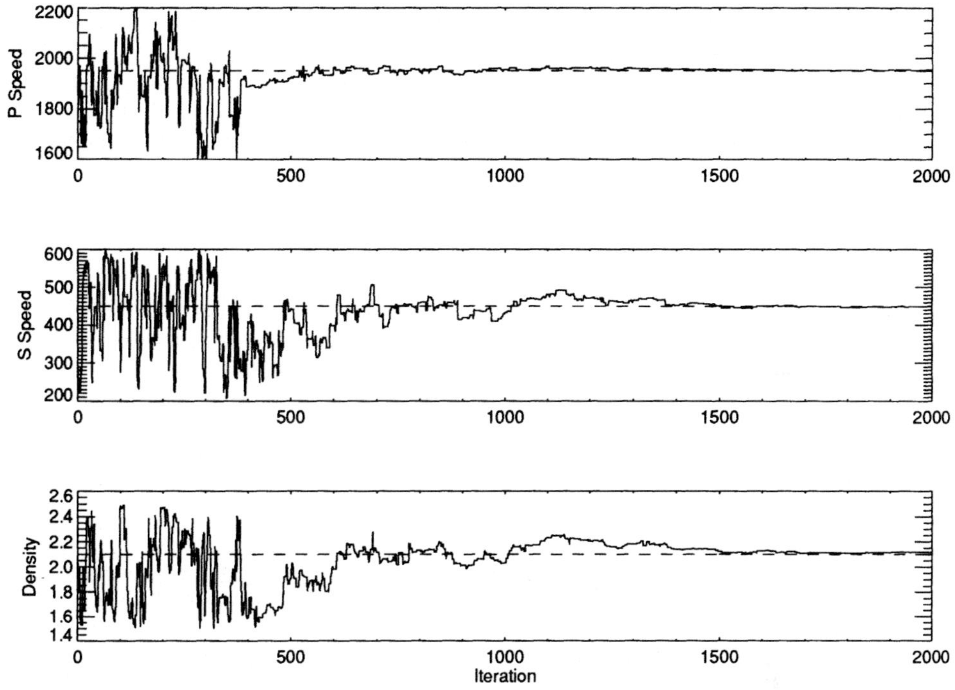

Figure 2: *Accepted parameter values during the Simulated Annealing inversion. Dashed lines indicate synthetic experimental parameter values*

When using SA for minimization of multivariate functions, the parameters for which the cost function is most sensitive tend to 'freeze out' first. This effect is apparent in figure 2 which shows the compressional speed parameter locked at its final value much earlier in the inversion than the shear wave speed and density. It was found that the cost function for this environment is much more sensitive to the compressional wave speed than to shear wave speed and density. Other researchers, e.g. [2,4] have modified the temperature values used for different parameters so that all parameters 'freeze out' at approximately the same time. This was not done in the example shown here.

The values of the parameters at the end of the inversion were: C_p=1954.53 m/s, C_s=450.52 m/s, and ρ=2.12 g/cm^3. The values of the 'experimental' parameters used to generate the synthetic data were purposely placed midway between grid values in all three dimensions to illustrate the case where maximum error in interpolation might occur. It is noted that even for this worst case scenario, very good results were obtained. The errors as a percentage of search range for C_p, C_s, and ρ are 0.75%, 0.13%, and 2.0% respectively. In order to obtain similar accuracies using an exhaustive search without mode tables would require approximately 10^6 runs of the normal mode algorithm. A similar inversion using

SA alone requires 2000 runs of the normal mode algorithm. In our demonstration example using both mode look-up tables and SA, KRAKENC was run only 175 times.

Conclusions

The recently proposed method referred to as mode interpolation was investigated for application to matched field inversion for estimating bottom geoacoustic parameters. This paper presents an overview of more detailed work [3], and a simple demonstration of the method is shown in which three bottom parameters (compressional wave speed, shear wave speed, and density) are inverted from synthetic data. This example shows how substantial time savings can be achieved by greatly reducing the number of times the computationally intensive normal mode algorithm has to be run. Additionally, other significant benefits of mode interpolation are pointed out; such as how the method complements the Simulated Annealing minimization technique which is often used in these types of inversion problems.

References

1. Dosso, S.E.., Ozard, J.M. and Fawcett J.A. (1993) 'Inversion of Acoustic Field Data for Bathymetry and Bottom Sound Speed via Simulated Annealing' in *Acoustic Signal Processing for Ocean Exploration,* Kluwer Academic Publishers, Dordrecht, the Netherlands. pp. 51-56

2. Dosso, S.E.., Yeremy, M.L., Ozard, J.M. and Chapman N.R. (1993) 'Estimation of Ocean-bottom Properties by Matched-Field Inversion of Acoustic Field Data', *IEEE Journal of Ocean Engineering,* Vol. **18**, 232-239.

3. Henson, N.G., Hannay, D.E., and Scrimger, P. (1994) 'Efficient Acoustic Field Computation for Estimating Geoacoustic Bottom Parameters Using Matched Field Inversion', *Defence Research Establishment Pacific Contractors Report No. 94-54.*

4. Kuperman, W.A., Collins M.D., Perkins J.S., and Davis N.R., 'Optimal time-domain beamforming with simulated annealing including application of *a priori* information,' J. Acoust. Soc. Am., vol.88, pp.1802-1810, Oct. 1990

5. Porter, M.B. (1991) 'The KRAKEN Normal Mode Program', SACLANT Undersea Research Centre Memorandum Rep. SM-245

WEAKLY DIVERGENT BUNDLES OF RAYS AND THEIR POSSIBLE USE IN INVERSE METHODS OF OCEAN ACOUSTICS

L.M. BREKHOVSKIKH, V.V. GONCHAROV, V.M. KURTEPOV
Acoustic Wave Propagation Lab., P.P.Shirshov Oceanology Institute
23 Krasikova St., Moscow 117851, Russia

ABSTRACT
Sound signals of high amplitudes were observed at discrete depths up to distances of 3500 km from the source. This effect was explained by the propagation of part of the sound energy along narrow, weakly divergent bundles of rays (WDBR). The structure of WDBRs is conserved at great distances even in the presence of strong inhomogeneities of the ocean - fronts, eddies, etc. The use of WDBRs in acoustic tomography and thermometry of the ocean is discussed.

1. Introduction

We observed in our experiments (1989–90) on long range (up to 3500 km) sound propagation, that there is a concentration of sound at certain depths at all distances. This means that the sound energy is not inevitably spread over all depths and decaying with distance according to the average law of decay [1], but rather concentrates along some bundles of rays, even at very great distances [2–4]. It was shown that the existence of so-called 'Weakly Divergent Bundles of Rays' (WDBR) is the reason for such a concentration. Rays of these bundles leave the source at grazing angles χ, which correspond to the ***extrema*** of function $D(\chi)$ – the ray cycle distance. It appears that the propagation times $T(\chi)$ along rays in WDBRs also concentrate closely near some average value, which is important for the ray acoustic tomography of the ocean. This property of WDBRs follows directly from the relation [4]:

$$dT/du = u^{-1} dD/du. \tag{1}$$

Here $u \equiv c(z') = c(z)/\cos\chi(z)$ is the horizontal phase velocity of the quasi plane wave corresponding to a given ray, whereas z' is the vertical coordinate at the turning point of this ray. Therefore, the functions $T(u)$ and $D(u)$ have ***extrema*** at the same u, i.e., at the same χ. Hence, propagation times along different rays in a WDBR are also close one to another.

One may anticipate that WDBRs can be observed not only in range independent conditions but also under adiabatic conditions (slowly changing waveguide). Our experiments show, however, that WDBRs are observed at great distances (up to 3500 km) even when there is considerable change of the sound channel with distance [2]. Additional proof of this result was obtained by numerical simulations. It appears that though cycle distance D changes considerably with range, some ***extrema*** of the function $D(\chi; r)$are observed almost for the same χ (grazing angle at the source) at all r. This fact ensures the small divergence of the rays with distance.

O. Diaschok et al. (eds.), Full Field Inversion Methods in Ocean and Seismo-Acoustics, 39–44.

Let us consider average energy parameters of the WDBR. We use for this purpose the $z(\chi;r)$ diagram, which specifies the depth z of the ray at a distance r if it leaves the source at the grazing angle χ. For example, if one distributes the total energy of the bundle uniformly over its cross section $\Delta z \simeq \max\{z(\chi;r)\} - \min\{z(\chi;r)\}$, one obtains the squared amplitude (sound intensity) of the bundle:

$$A^2 = I\,R^2(r\cos\chi'\,\cos\chi''\,\Delta z/\Delta\chi)^{-1}, \tag{2}$$

where I is the sound intensity generated by the same source at a distance R in the case of homogeneous medium, χ' and χ'' are grazing angles of some characteristic ray (for example the middle one) at the source and at the distance r, respectively, and $\Delta\chi$ is the angle width of the beam at the source.

The amplitude of a single ray is described also by Eq.(2) if we replace the bundle's cross section $\Delta S = \cos\chi''\,\Delta z$ by the cross section of an infinitely narrow ray tube $dS = \cos\chi\,|dz/d\chi|d\chi$. Introduce the average (over the bundle) derivative $< z' >= \Delta Z/\Delta\chi$ where $\Delta z = \int|z'(\chi;r)|d\chi$ is the sum of quantities $\max\{z(\chi;r)\} - \min\{z(\chi;r)\}$ over all monotonic parts of the $z(\chi;r)$ diagram). Now, we have for the estimate of the average squared amplitude of the ray

$$A_1^2 = IR^2(r\cos\chi'\,\cos\chi''\,\Delta Z/\Delta\chi)^{-1}, \tag{3}$$

which is also the sound intensity measured by a directional antenna resolving different families of rays in the bundles. (More than one family can be present in the bundle due to the existence of caustics for example.) The ratio A/A_1 is approximately equal to the number of such families.

2. Experimental Results

Let us now consider results of the 1989–90 experiments done with the research ships *Academik S. Vavilov* and *Academik Ioffe* (with the source at 1000 m depth, frequency 137 Hz) in the Atlantic (Canarian Abyssal Plain). Acoustic paths were 3500 km long and crossed, at distances of more that 2800 km, the 'tongue' of Mediterranean waters with higher salinity (and, hence, sound velocity). This intrusion caused the formation of a two-channel propagation structure beyond 2800 km. The WDBR existing at smaller distances (χ in the interval $-2.7°$, $2.6°$) suffers considerable structural change and splits into three bundles with χ in the intervals: I $[-2.7°, -2.0°]$, II $[-1.6°, -1.4°]$, III $[2.0°, 2.6°]$. Bundle II (with smaller χ) becomes trapped in the deep, strong channel with the axis around 1500 m depth, and the bundle conserves its weak divergence. Bundles I and III suffer stronger disturbances, being trapped in the upper channel with the axis at 500 m, and their energy is being smeared practically all over the channel. This can be seen in Fig. 1a where the boundaries $\min z(r)$ and $\max z(r)$ are plotted for both bundle II (full lines) and bundle I (dotted lines). Average ray amplitudes vs distance along the same bundles calculated by the use of Eq. (3) are shown in Fig. 1b.

The amplitudes in bundle II are strongly oscillating which is due to the variations in the bundle's cross section (not very visible in Fig. 1a due to the great difference between horizontal and vertical scales). Sound amplitudes in this bundle are 15 dB above the background (formed by all other normally divergent rays, dotted curve) at some distance. Bundle I appears to be more divergent that bundle II and cannot be classified as a WDBR. Ray amplitudes are just 6 dB above the background in this bundle.

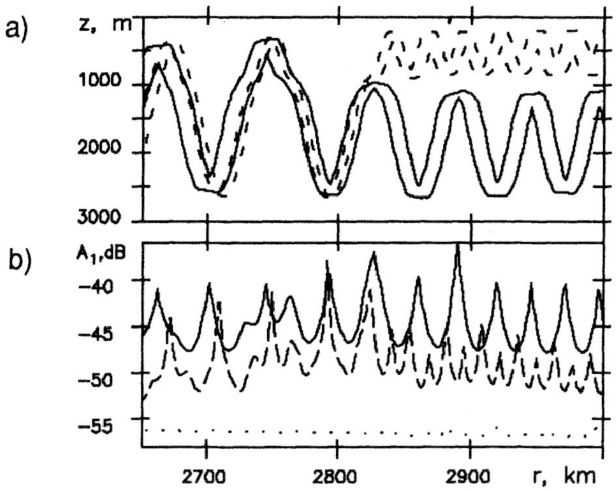

Fig.1. Atlantic experiment. WDBR II (full lines),I (dashed lines); a) boundary rays; b) amplitude along WDBRs.

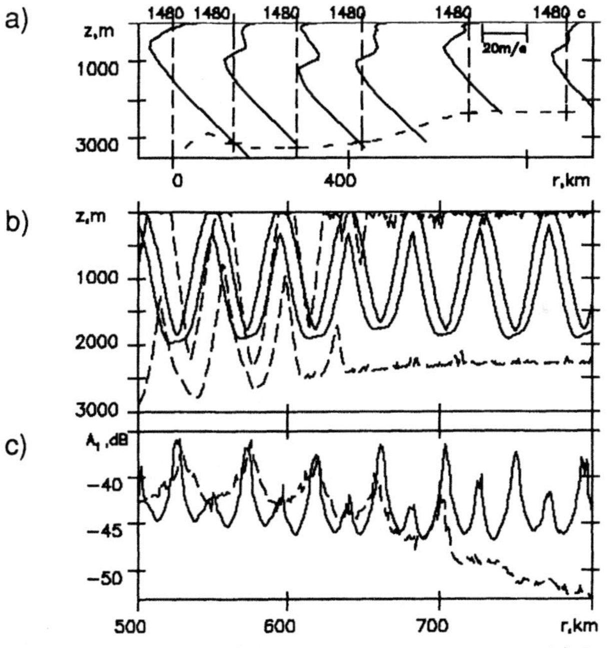

Fig.2. Norwegian sea. a) Environmental conditions; b) boundary rays; c) amplitude along WDBRs. IV(full lines), I (dashed lines).

Horizontal inhomogeneity of the sound channel was strongly present in the case of our experiment in the Norwegian Sea (1990). The largest distance in this experiment was 890 km. The vertical sound velocity profile $c(z)$ at different distances and the general bottom profile can be seen in Fig. 2a. A two-channel profile is seen to be present at middle distances. However, even in this strongly range-dependent conditions of sound propagation, some WDBRs

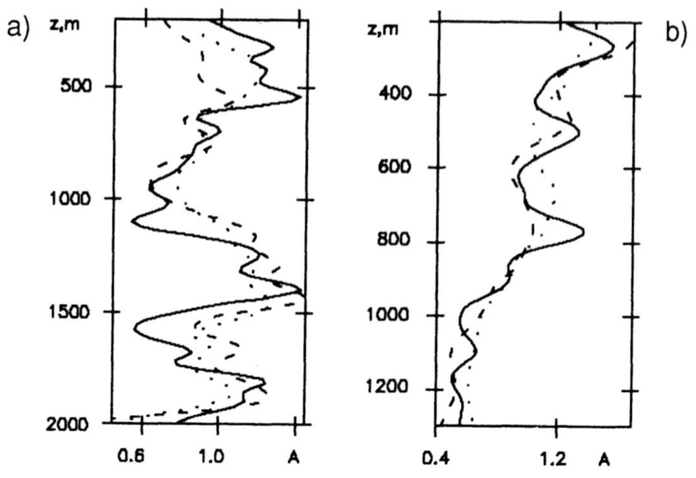

Fig.3. Vertical profiles of the sound amplitudes. Experiment (full lines); WDBRs (dashed lines); all rays (dotted lines) Distances: a) 571 km, b)778 km.

conserve their structure. There are 6 WDBRs corresponding to χ in the intervals: I [$-10°$, $-7.6°$], II [$-7.6°$, $-5.9°$], III (-3.2, -1.7), IV [$1.7°$, $3.1°$], V [$6.4°$, $7.9°$], VI [$7.9°$, $10°$]. In Fig. 2b the boundaries of WDBR IV which retains its structure up to the maximum distance are shown as full lines. The dashed lines indicate bundle I, formed by rays leaving the source at steep angles. This bundle is destroyed at a distance $r \approx 630$ km due to the interaction with the bottom.

The average sound amplitudes A_1 in these bundles are shown in Fig. 2c by the same kind of curves. The amplitude in bundle I drops off beyond a distance of 630 km and remains close to the background level. A similar behaviour was observed for the amplitude of bundle VI, not shown in Fig. 2c. Also not shown is that bundles II and III were considerably affected by the horizontal inhomogeneity of the sound velocity and were only slightly above the background. In contrast to this, the amplitude in bundle IV, though strongly oscillating, was considerably higher (by 17 dB at some distances) than the background level. An analogous behaviour was seen for bundle V (also not shown), and only at distances greater than 800 km does this bundle break up due to interaction with the bottom.

We now turn to a different view of the experimental results for the Norwegian Sea. In Fig. 3a the amplitude is plotted as a function of depth (averaged over 100 m of depth) on a linear scale (full curve) at a distance of 571 km. The same is in Fig. 3b – 778 km. The considerable variations of amplitudes vs depth indicate that WDBRs play an important role. Dotted lines in these figures represent numerically calculated amplitudes taking into account all rays leaving the source in the grazing angle interval [$-11°$, $11°$]. The dashed curve is the result when taking into account only rays forming WDBRs – all six at 571 km

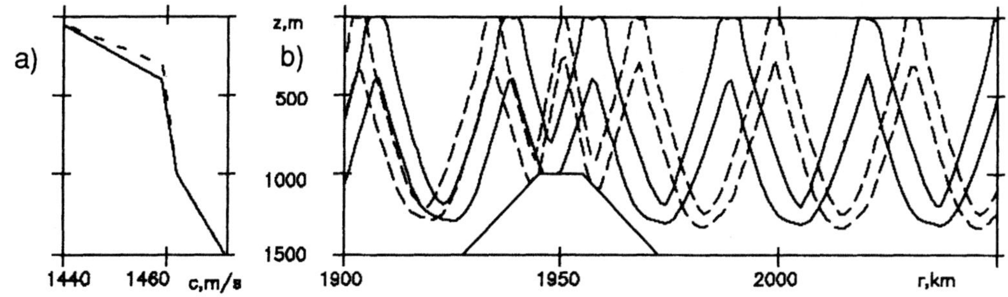

Fig.4. Artic path. a) c(z) profiles before (full line) and after (dashed) the Lomonosov Ridge. b) Boundary rays for bundles II (full line) and I(dashed line).

and the four non-destroyed WDBRs at 778 km. The dashed-curve behaviour is very similar to that of the experimental amplitudes, which means that WDBRs play the main role in the formation of the sound field.

WDBRs may have considerable practical importance. They can be used for effective underwater long-distance sound communication for example. Strongly pronounced inhomogeneity of the sound field both in the horizontal and in the vertical caused by WDBRs increases the effectiveness of the matched field method in solving the inverse problem. Stability of WDBRs in the presence of different kinds of inhomogeneities in the ocean is very important also for Munk-Wunsch ray tomography. Particularly, they can be used as reference rays in 'dynamic' tomography using differences of arrival times of different rays instead of time itself. Also for acoustic thermometry of the ocean is it important to use rays and bundles of rays which are stable under the conditions of strong variability of the medium. Note, however, that working with WDBRs one must use hydrophone arrays or spatial probing of the ocean by a receiver. A single receiver at a fixed position will be ineffective as a rule, because WDBRs can change their position when conditions for sound propagation are changing.

3. Summary and conclusions

In conclusion, let us discuss the possibility of using WDBRs along an acoustic path in the Arctic (Beaufort Sea – Franz Victoria Strait) proposed in the ATOC project for acoustical thermometry [5]. We use two representative $c(z)$ profiles, one before the Lomonosov Ridge and one after (Fig. 4a). Two WDBRs are shown in Fig. 4b leaving the source (at depth 300 m) at grazing angles I $[-7.8°, -7.6°]$ and II $[7.4°, 7.6°]$. The lower turning points of these rays appear to be deeper than the top of the Lomonosov Ridge (at the distance 1950 km in Fig. 4b). This fact influences WDBRs in some measure but does not disturb their general structure. Of course there exist standard WDBRs corresponding to $\chi = 0$, not shown in Fig. 4b. The corresponding rays have short cycle distances and the bundle disintegrates even at moderate distances.

Stability of WDBRs appears to be very important for detecting small, climatic changes of the average temperature of the Arctic Ocean (ATOC project). To show this we have

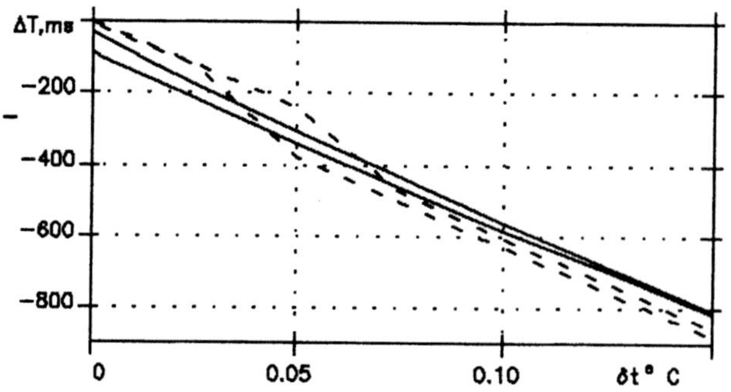

Fig.5. Time delay for the same boundry rays as in Fig.4. vs warming degree.

determined by numerical calculation how arrival times and amplitudes of WDBRs change under the influence of small temperature increases $\Delta t(z)$. We supposed $\Delta t = 0$ at $z = 0$ (ice melting temperature); $\Delta t = \delta t \equiv \max \Delta t$ at $z \equiv h = 30$ m and $\Delta t(z) = \delta t \exp(-|z - h|/H)$, ($H = 5000$ m) at $z > h$, i.e., Δt decreases slowly at great depth. Calculations were done for different $0 \leq \delta t \leq 0.15°$C. According to the calculations, ray amplitudes in WDBRs at a maximum distance of 2900 km at all dt appear to be 10–14 dB higher than the 'background' due to all other of rays, not belonging to WDBRs. The depth of the central ray in a bundle varies between 300 and 1300 m and the vertical cross section of the bundle changes from 40 up to 600 m. Figure 5 shows the decreasing propagation time ΔT for both bundles vs δt, as compared with those at $\Delta t = 0$. The small ΔT difference for bundle I ($-7.8°$, $-7.4°$) at small δt is due to the additional focusing of the rays reflecting at the front slope of the ridge, as seen in Fig. 4b. The considerable difference in ΔT for boundary rays in this bundle for $\delta t > 0.02°$ is due to the broadening of the bundle when it reflects at the rear slope of the ridge and at the plateau adjacent to it. This is also the cause for the decrease in sound amplitude in the bundle by approximately 5 dB.

One additional feature of importance can be seen in Fig. 5. The difference in propagation times of different bundles as well as different rays in the same bundle is considerable (30 ms and more). Therefore different bundles and even different rays in them can in practice be resolved. Also seen in Fig. 5 is the approximately linear relationship $\Delta T/T$ more or less equal to $\Delta c/c$ between the decrease in propagation time ΔT for either bundle and Δc, the increase in sound velocity corresponding to the δt under consideration.

References

1. Brekhovskikh, L.M. and Lysanov, Yu.P. (1991) *Fundamentals of Ocean Acoustics*, Heidelberg, Germany, Springer-Verlag.
2. Brekhovskikh, L.M. et al. (1990) Long range sound propagation experiment in the Canarian Abyssal Plain, the Atlantic, *Acoust. Journal*, **36**, 5, 824–831, (in Russian).
3. Brekhovskikh, L.M. et al. (1992) Acoustical tomography of the ocean and related problems, *In*: Proceedings of the 14th Intern. Congress on Acoustics, Beijing, China, p. P4.
4. Chuprov, S.D. (1989) Selection of modes and rays in underwater sound channels, *In:* Brekhovskikh, L. and Andreeva, I.,*eds.*, Acoustics of the Ocean Medium, Nauka, Moscow, 56–64, (in Russian).
5. Munk, W. (1994) Heard Island and Beyond, *Journal of the Acoustical Society of America*, **95**, 567.

PARALLEL IMPLEMENTATION OF A 3-D RANGE-DEPENDENT RAY MODEL FOR REPLICA FIELD GENERATION

LEWIS DOZIER and PIERRE LALLEMENT
Alliant Techsystems, Inc., Advanced Technology Center
1911 N. Ft. Myer Drive, Suite 601, Arlington, VA 22209, U.S.A.

ABSTRACT. Efficient generation of replica fields at high frequency is essential for acoustic tomography or for full field inversions of small-scale bottom roughness parameters. Such roughness can cause scattering of great significance to high-frequency active sonars operating in shallow water. Three-dimensional effects due to rough sloping bottoms can also be important to such sonars. In this paper we present a ray-theoretic method of computing fully coherent pressure fields, consistently in 2-D or 3-D range dependent environments, which avoids calculations of eigenrays or focusing factors. Also, we present 2-D results from a numerical implementation on a massively parallel computer architecture.

1. INTRODUCTION

The late A. L. Piskarev (1989, 1992) introduced an algorithm which computes the incoherent acoustic intensity averaged locally over a spatial mesh cell. The novelty of his algorithm is that it avoids computation of ray focusing factors which become infinite at caustics. Rather, a dense fan of rays, spread over the entire angular aperture of the source, is traced forward, discretized over a predefined set of spatial mesh cells, from the source to the final range of interest. The rays that pass through any given cell contribute to the averaged intensity for that cell in proportion to their arc lengths within that cell. Remarkably, this method gives a very accurate incoherent intensity without the necessity of computing the angular derivatives associated with focusing factors, and without the numerical problems associated with such factors becoming infinite near caustics. Also, there is no necessity to search for eigenrays, because one run yields a full field computation for the entire domain of interest.

To Piskarev's algorithm, we have added the capability to coherently sum ray contributions and thereby to simulate time series of complex pressure. In addition, we have implemented a 2-D version of our algorithm on a parallel computer architecture, namely, a 128-node Intel Hypercube. Tracing of individual rays is distributed over the nodes. The spatial mesh is then scanned once, and within each cell, the rays are grouped into ray families. Within a family, the rays have a common history (including order) of bottom bounces, surface bounces, and caustic interactions. The total coherent pressure field is obtained as a coherent sum of contributions from each ray family, as will be explained in Sections 2 and 3.

It is crucial to note that whereas the original Piskarev algorithm provides only the incoherent intensity averaged over a cell, our algorithm gives the coherent pressure, which varies from point

O. Diaschok et al. (eds.), Full Field Inversion Methods in Ocean and Seismo-Acoustics, 45–50.

to point within a cell. Thus, our work differs markedly in spirit from that of Piskarev, whose work describes only the slowly varying envelope of the pressure field; our coherent algorithm actually reconstructs its rapid oscillations with surprising accuracy.

2. COHERENT PRESSURE ALGORITHM

Piskarev's original algorithm, as implemented in a computer code he calls DELTA, traces a fan of M rays, spaced densely in vertical and (in the 3-D case) azimuthal angle, out to some specified distance from a point source. The region of the ray trace is divided into spatial cells of a predetermined size, as desired for averaging. The result is a single value, per spatial grid cell, of incoherent averaged intensity as

$$I = \sum_{n=1}^{N} I_n \quad , \tag{1}$$

where N is the total number of rays passing through the grid cell and I_n is the intensity of the nth ray, as defined by Piskarev (1989, 1992), including attenuation. M must be large enough so that the ray tubes represented by the corresponding angular discretization at the source have a cross-section much smaller than the area (or volume, in 3-D) of the grid cell, when they pass through the cell; then N will be large enough that the N rays and associated intensities allow an accurate sum to the total energy in the cell. I_n is proportional to the arc length of the nth ray within the cell, and I_n includes an angular (equivalent to a ray tube area or volume) normalization factor. Thus, if the number of rays in the vertical is doubled, each ray's intensity I_n is halved, so that, assuming N large enough as described above, approximately the same result is obtained for the sum I.

To generalize Eq. (1) to a coherent sum, we sort the N rays into K families; within each family, all rays have the same history, including order, of k_s surface and k_b bottom bounces, and k_c caustic interactions. For each family we define an intensity and a phase as follows: Let rays n_{k_j}, $j = 1, \cdots, N_k$, be the N_k members of family k, so that $N_1 + \cdots + N_K = N$. Then the intensity $\tilde{I}_k$ of the kth family, which does not vary within the grid cell, is taken to be simply

$$\tilde{I}_k = \sum_{j=1}^{N_k} I_{n_{k_j}} \quad . \tag{2}$$

The phase ϕ_k of the kth family at a particular point $\vec{x}$ within the grid cell is obtained from phase changes accumulated according to the family history, and from an estimate of the travel time for the family to that point. That is,

$$\phi_k(\vec{x}) \;=\; \omega t_k(\vec{x}) + \pi k_s + \frac{\pi}{2} k_c + \sum_{j=1}^{k_b} A_{k_j} \quad , \tag{3}$$

where ω is the angular acoustic frequency, and where the travel time t_k and (frequency-dependent) phase shifts A_{k_j} at bottom reflections (computed according to a geophysical bottom, including shear if desired) are associated with the most intense ray in the family. The dependence of t_k on $\vec{x}$ will be described in Section 3. Finally, then, the coherent acoustic pressure is computed as

$$p(\vec{x}) = \sum_{k=1}^{K} \sqrt{\tilde{I}_k}\; e^{i\phi_k(\vec{x})} \tag{4}$$

An alternative view of the situation is that for a given point in a cell, each eigenray to that point is approximated by a locally plane wavefront corresponding to the ray family of which the eigenray is a member. Thus, Eq. (4) may be thought of as an approximation to a coherent sum of eigenrays, but with the advantage of not requiring iteration to find the actual eigenrays.

3. TRAVEL TIME COMPUTATION

To compute $t_k(\vec{x})$, we associate the kth family with a local plane wavefront and compute the time at which the wavefront passes through the point $\vec{x}$. An idealized wavefront in two dimensions, with $\vec{x} = (r,z)$, a point at range r and depth z, is illustrated in Fig. 1. Of course, refraction causes ray bending within the grid cell, so that in general, the rays within a family are neither perfectly straight nor perfectly parallel to each other within a cell. We approximate the wavefront using normals to the most intense ray in the kth family; usually, this would be the ray with the maximum arc length within the cell. First, we implemented an "exact" method where we actually retraced the ray's path exactly within the cell; we then used a root finding algorithm to determine at what point(s) on the ray its normal intersects the point of interest. In most cases only a single point on the ray allows such intersection, and it can be found with quite sufficient accuracy by a simpler and computationally more efficient approach, as illustrated in Fig. 2; the wavefront arrival time at (r,z) is found by linear interpolation (or if necessary, extrapolation) between the ray travel times at the cell entry and exit points, using the normal to a straight line ray path connecting those points. Such an approach clearly does not work, however, if the actual ray suffers a sharp turn (e.g., specular reflection at a boundary) within the cell; in that case, we use a straight line path defined by the ray's cell entry point and entry angle, and extrapolate as illustrated in Fig. 3.

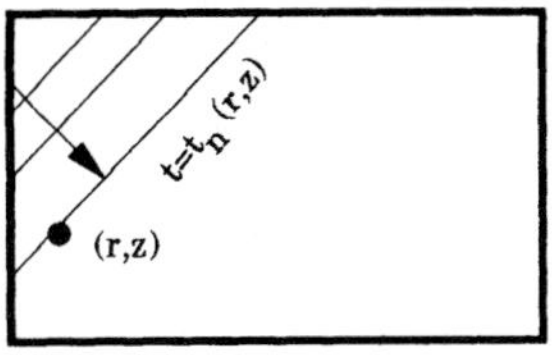

Fig. 1: Wavefront in cell.

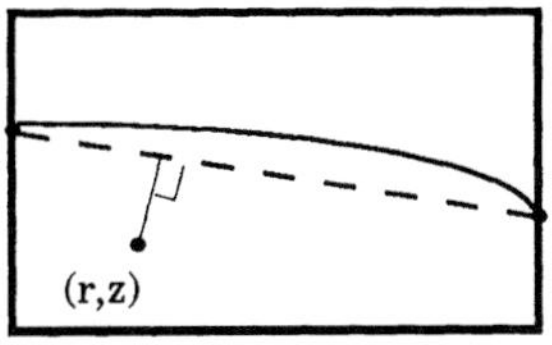

Fig.2: $t_k(r,z)$ interpolation.

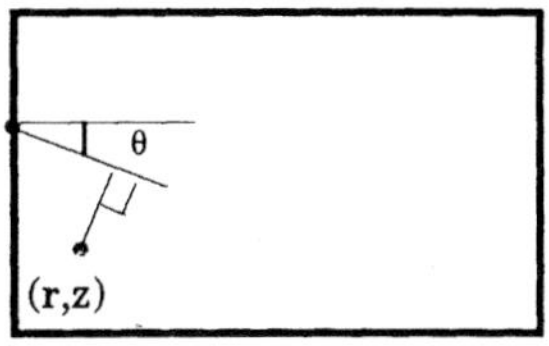

Fig.3: $t_k(r,z)$ extrapolation.

At very high frequency ω, t_k must be computed to a high degree of accuracy, since it appears in the product ωt_k in Eq. (3). Following Piskarev, our code uses splines to smooth the depth dependence of sound speed profiles, and traces rays using fifth order polynomials, so that the trace of a single ray gives a sufficiently accurate travel time for that ray. However, Eq. (3) uses a representative travel time for a *ray family*, not a single ray. As $\vec{x}$ passes from one cell to another, small discontinuities in $t_k(\vec{x})$ will be introduced because the most intense ray in family k in one cell will not be the most intense ray in family k in the next cell. Figure 4 illustrates this situation for a horizontal sequence of points $\vec{x}$ at the tips of the normals to the most intense rays in two adjacent cells. Even though these discontinuities may be small, when multiplied by a large frequency, the total phase error may be non-negligible. More precisely, an accurate evaluation of Eq. (4) requires that phase errors be much less than 2π. For example, at a frequency of 30 kHz, the error δt_k in travel time must satisfy

$$|\delta t_k| << \frac{1}{f} \leq 3\times10^{-5}\ s. \tag{5}$$

Due to volume attenuation at such a frequency, propagation decays to uninteresting levels for travel times at most a few seconds, or ranges at most a few kilometers. At maximum times or ranges of interest, then, Eq. (5) translates into requiring also a *minimum relative accuracy* much less than 10^{-5} at 30 kHz. Since single precision (32-bit) arithmetic is limited to about 7 digits of accuracy, portions of our FORTRAN computer code contain double precision variables. In addition, the cumulative travel time as a ray is traced through consecutive cells is stored in two variables, the first being an integer representing a multiple of 2π, and the second being a floating point residual less than 2π; this prevents truncation errors in travel time calculations from accumulating with range. The ray trace is then able to find the edge of the cell, and the associated travel time, to the accuracy required at such high frequencies.

Once the numerical error has been controlled as described above, there remains an intrinsic error, or discontinuity in phase across cell boundaries, which is related to the cell size. Figure 4 illustrates the rays as straight lines, as though locally perfect plane waves are propagating. In general, the rays are slightly curved, and this curvature would give rise to a small discontinuity even if the arithmetic were of infinite precision. We carried out a numerical experiment to show the relation between cell size and magnitude of error in intensity. A 30 kHz source was placed 1 m below the surface of a medium with an upward refracting linear sound speed profile with an extreme gradient of 1 s^{-1}. Grid cells were 2 m horizontally by 1 m vertically. For a source aperture of ±20° about the horizontal, and using 20000 rays to assure convergence, we found discontinuities of 0.1 dB across vertical cell boundaries and at horizontal ranges on the order of 320 m from the source; the associated errors δt_k were found to be 1.3×10^{-7} s. This example is extreme; for more realistic (smaller) sound speed gradients, the error is even less. Also, it should be noted that these errors do not accumulate systematically with range, but rather are tiny oscillations that are barely visible (if at all) in a transmission loss curve plotted to typical scale.

Note that the cell size in the above example is 20λ vertically by 40λ horizontally (λ=acoustic wavelength). In our high frequency applications so far, we have not needed to use cells smaller than about 10λ by 30λ. The cell dimensions can be thought of as corresponding to intrinsic coherence lengths of the medium, e.g., in the horizontal, a 60-element array with λ/2 spacing.

There is a relation between the cell size, the number of rays required, and the maximum gradient present in the sound speed profile. For smaller cell sizes, more rays must be traced in order to assure that each family is adequately represented within each cell. Thus, it is advantageous to use the largest cell size consistent with the error allowable for an application. However, the stronger the gradient, the smaller the cell size must be to maintain a fixed error, since the local plane wave approximation is not as good.

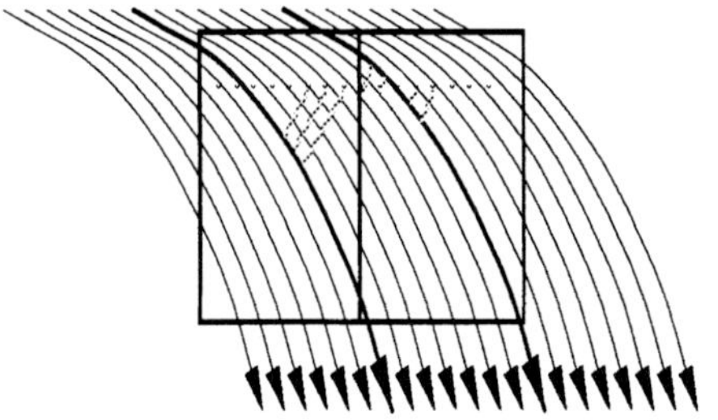

Fig. 4: Travel time discontinuity.

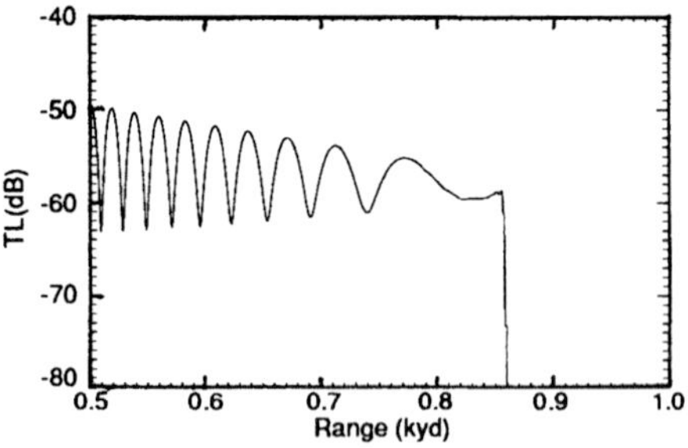

Fig. 5: Porter-Bucker benchmark case.

4. PARALLEL IMPLEMENTATION AND CODE VALIDATION

We have implemented a 2-D version of the coherent pressure algorithm in a parallel computer code that runs on a 128-node Intel Hypercube; we shall refer to this code as "coherent DELTA." This code is organized into two parts. The first, computationally intensive part runs on the Hypercube. It traces a dense fan of rays from the source and organizes the results into a database containing ray family information for cells within declared "areas of interest." A typical run at 35 kHz with 40000 rays on a 500×300 range-depth grid, saving a database in the vicinity of a sloping bottom, takes about 6 minutes. The second part simply reads the database and computes coherent pressure and/or, if desired, incoherent intensity. This part runs easily on a Sun workstation; it takes less than a second on a SPARCstation 2 to compute the coherent pressure at 1000 points.

As a quantitative validation of our code, consider the first case in Section VI of the paper by Porter and Bucker (1987) describing a Gaussian beam tracing algorithm. The profile is strongly downward refracting, so that with both source and receiver at a depth of 66.7 yards, a shadow zone begins at a range of about 850 yards. Figure 5 shows transmission loss given by our algorithm at the benchmark frequency of 2000 Hz. It will be observed that, except for the sharp drop-off at the edge of the shadow zone, our result agrees almost exactly with the reference solution [Figure 3(b) in the Porter-Bucker article], reproducing quite accurately the coherent interference pattern. The sharp drop-off occurs because at present, we have no caustic corrections to account for decay into the shadow zone. These can be added, if desired, much as in the Gaussian beam algorithms.

If one needs, say, a full field coherent intensity plot, the entire (r,z) grid can be declared a single area of interest. Often in applications, however, we are interested in the coherent field only within a subset of the field, e.g., near the location of a receiver. In that case significant storage and computation time can be saved by computing and storing the ray family information only in limited areas. As a striking example, consider a typical bilinear profile and a shallow source;

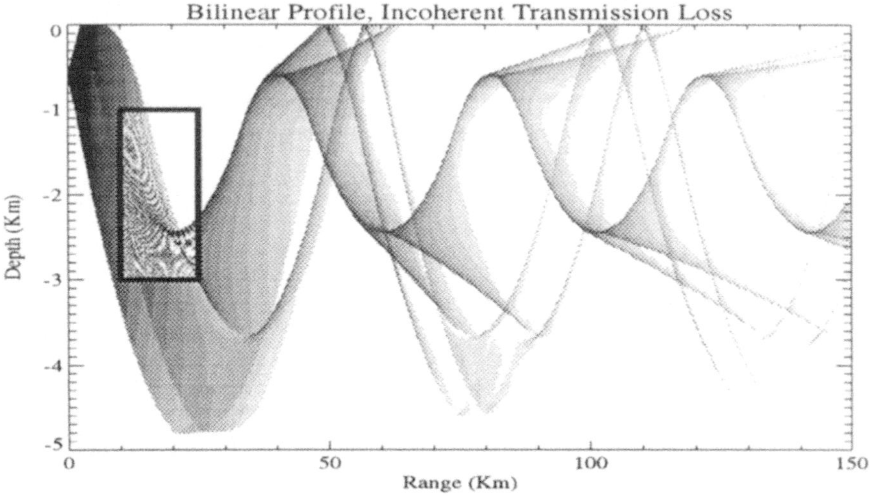

Fig. 6: Coherent DELTA, bilinear profile, sloping bottom.

this case is Run #1 in Piskarev (1992), but with the addition of a gently sloping bottom (depth varies from 4.9 km to 4.3 km). Figure 6 shows a plot of the intensity computed as incoherent, except coherent in a rectangular area of interest around the first cusped caustic.

On the Hypercube, individual ray traces are distributed over parallel processors (nodes). If M rays are to be traced using L nodes, the simplistic approach is to assign roughly M/L rays to each node. However, this approach does a poor job of load balancing among the nodes, since calculation of the steeper rays often terminates much sooner because of accumulated bottom loss (there is an input threshold). A more effective approach is to benchmark the calculation by initially tracing a small sample (5 rays per node) of the rays, spaced evenly over the source aperture. The information gained by observing the computation time of the sample rays is then used to apportion the full ray trace calculation efficiently over the nodes. This approach can save almost a factor of two in run time compared to the simplistic approach described above.

Following the ray trace, the rays are combined into ray families on each node, creating a database in a linked list format, to save storage as compared with a multidimensional array. Then, the local databases on each node are combined into a single database using a binary tree approach.

5. SUMMARY AND CONCLUSIONS

The modular organization of the coherent DELTA code allows it to be used as a propagation engine for a variety of tasks. Currently, it is the workhorse for an Acoustic Time Series Simulator, for which its ray family databases are directly used by echo, noise, and reverberation modules. Arbitrary source directionality is easily included by weighting the ray intensities according to their direction of emission from the source. In the future, we envision using DELTA in inversion schemes, e.g., in which the bottom losses and phase shifts in Eqs. (1) and (3) are varied (according to assumed geophysical bottoms), but without repeating the ray trace.

6. ACKNOWLEDGMENTS. This work was supported by the Advanced Environmental Acoustic Support Program of the U.S. Office of Naval Research (Barry Blumenthal, Program Manager) under contract number N00014-92-C-6009. Also, Henrik Schmidt of the Massachusetts Institute of Technology provided helpful discussions, along with a copy of Piskarev's original (undocumented, largely uncommented, and non-parallel) DELTA computer code that computes incoherent ray sums.

7. REFERENCES

Piskarev, A. L. (1989) "Calculation of the average intensity distributions of sound fields in the ocean", *Sov. Phys. Acoust.* **35**, 418-422.

Piskarev, A. L. (1992) "Soviet Research in Underwater Acoustic Propagation Modeling", EG&G Tech. Report Number 6K1-011A, 31 March 1992 (sponsored by the U.S. Office of Naval Research through Contract Number N00014-91-C-0271).

Porter, M. B. and Bucker, H. P. (1987) "Gaussian beam tracing for computing ocean acoustic fields", *J. Acoust. Soc. Am.* **82**, 1349-1359.

SOME IMPROVEMENT OF THE SHORTEST PATH RAY TRACING ALGORITHM

ZOLTÁN WÉBER
Geophysical Research Group of the Hung. Acad. Sci.
Department of Geophysics, Eötvös University
Ludovika tér 2.
H-1083 Budapest
Hungary

ABSTRACT. A technique for improving the efficiency of the shortest path ray tracing (SPR) algorithm is presented. We provide quantitative measures to determine the accuracy of SPR with varying numbers of connections per node. Our improvement consists of a method that finds more accurate rays in regions of low velocity contrast than that yielded by the original SPR algorithm. This allows the researcher to use fewer connections per node and, as a consequence, to save computer time. In two-dimensional experiments we find that the improved algorithm requires about 80 percent less connections per node saving 70 percent of elapsed time. In three-dimensional tomographic applications these savings should be even greater.

1. Introduction

Ray tracing is the most time-consuming step of nonlinear seismic tomography. Since usually thousands of rays must be traced for a typical tomographic inversion, saving computer time and memory plays an important role in selecting the appropriate forward modelling algorithm. On the other hand, accuracy of both the traveltimes and raypath loci is also very important because the accuracy of tomographic inversions greatly depends on the errors introduced by forward calculations.

Shortest path ray tracing (SPR) is one of the most popular forward methods because it has many potential benefits over other techniques. It works well with block or otherwise discrete velocity models, it is fairly fast when used to trace many rays and it always finds the global minimum traveltime raypath through a network that represents the earth (Moser, 1991). Diffracted raypaths, headwaves and paths to shadow zones can be found correctly. There are also no restrictions to the complexity or the dimensionality of the velocity model. However, when accurate results are needed, the method requires vast memory and intensive calculation. In order to achieve the appropriate accuracy with as little computer time as possible, Fischer and Lees (1993) slightly alter the original SPR algorithm, saving up to 60 percent in time.

In this paper we also propose a method for improving the efficiency of the shortest path ray tracing algorithm. Our approach is simpler than that of Fischer and Lees (1993) and, as a consequence, it is faster and easier to implement. In two-dimensional experiments we manage to save 70 percent in elapsed time. In three-dimensional tomographic applications these savings should be even greater.

O. Diaschok et al. (eds.), Full Field Inversion Methods in Ocean and Seismo-Acoustics, 51–56.

The Method

Like the travelling salesman who wants to find the shortest route from one city to another in order to minimize his time spent on travelling, one can find seismic raypaths by calculating the shortest traveltime paths through a network that represents the earth. The network consists of points. Each point or node is connected, by straight lines, with a restricted number of points in its neighborhood but not with points that lie farther away. These connections are called arcs. The length of an arc represents the traveltime of the seismic wave between the nodes connected by the arc. According to Fermat's principle, the shortest traveltime path between any two nodes along the arcs approximates the seismic raypath between them. Since the shortest path algorithm (Dijkstra, 1959) finds the traveltime from one node to every other node at the same time, SPR is particularly efficient when many rays are required.

One of the possible networks we can use is a rectangular grid of nodes, each node is connected with the nodes in its neighborhood. The seismic velocity or slowness is defined in each block of the grid. This structure facilitates the drawing of contours of equal traveltimes and velocities and the introduction of interfaces.

The accuracy of SPR depends on the number of connections per node (CPN), because the greater the number of CPNs, the greater the number of the possible paths, out of which the shortest one, i.e. the raypath, has to be selected. Unfortunately computer time increases linearly with the number of connections and internal memory, required to store raypath information for later use in an inversion program, can also be very large. On the other hand, using few CPNs produces large inaccuracies. Rays usually follow zig-zag paths that are longer than the straighter true raypaths, increasing the error of traveltimes. The rays themselves are also inaccurate because there are often several equally short paths between two nodes, especially where velocity variations are small. Some of these equally short paths are good approximations of the ray, but most are not, and Dijkstra's (1959) algorithm does not differentiate between them, often choosing a ray not close to the true one. Nevertheless spatial similarity between the traced ray and the true ray is very important for tomographic inversion since traveltime anomalies are back projected along the calculated raypaths.

In order to achieve the appropriate accuracy with as few CPNs as possible, Fischer and Lees (1993) slightly alter the original Dijkstra's (1959) algorithm. If in the course of the SPR procedure they find many raypaths between two nodes with the same traveltime (called 'ties' by the authors), the true ray is approximated by a straight line segment between the two nodes, provided its traveltime is shorter than that of the ties. Our approach is similar to that of Fischer and Lees (1993), but it works on the rays already determined by the *original* Dijkstra's (1959) algorithm. If the number of CPNs is small, SPR usually yields zig-zag raypaths in flat parts of the velocity model, which are far from the true rays. However, these zig-zag rays can be straightened out easily. If a raypath changes its direction at a node, it has a breakpoint at that node. If the elimination of a breakpoint makes the traveltime shorter, the breakpoint is removed. The rays modified in such a way change their directions at significant velocity contrasts, only.

Figure 1 illustrates how the proposed method works. The raypaths produced by the modified SPR with 24 CPNs (*Figure 1c*) are very similar to those calculated by an accurate ray tracing method (*Figure 1a*) and much better than those determined by the original SPR with the same number of CPNs (*Figure 1b*). The zig-zag rays traced by the unmodified algorithm are all straightened out by our simple modification.

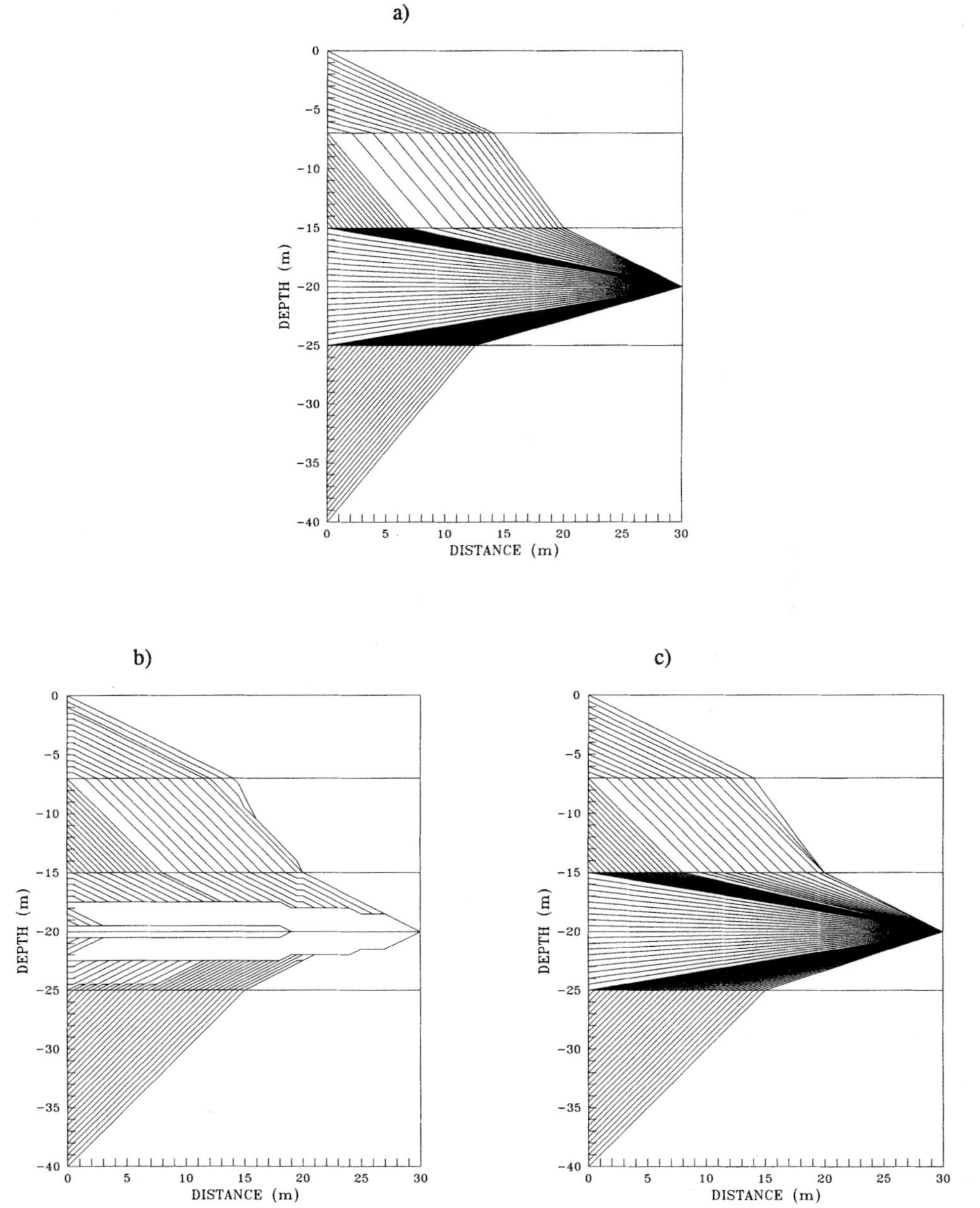

Figure 1: Some raypaths for a simple horizontaly layered model. The velocity interfaces are shown. The velocities are 1500, 1000, 1500 and 1000 m/s from the top to the bottom. The rays are calculated by *a)* an accurate ray tracing algorithm, by *b)* the original SPR with 24 connections per node and by *c)* the modified SPR with 24 connections per node.

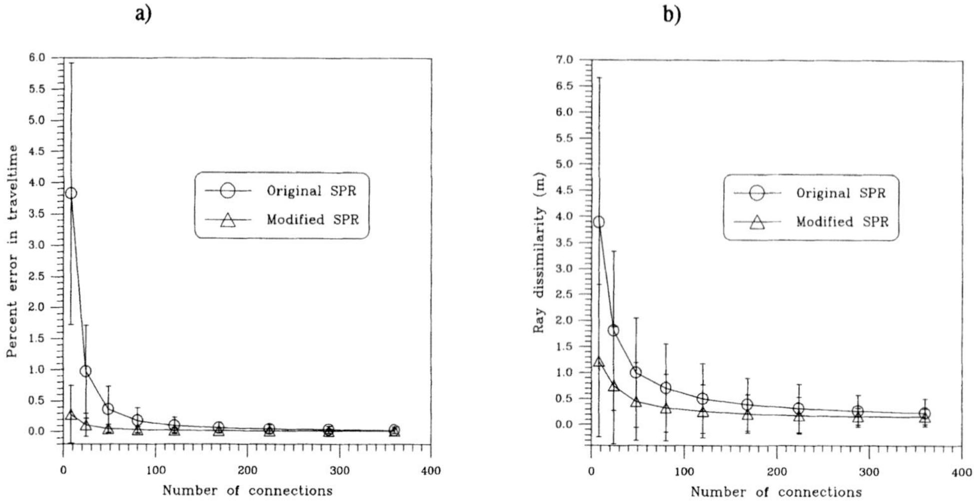

Figure 2: The accuracy of the proposed method: *a)* average errors in travel time and *b)* ray dissimilarities vs. number of connections per node. Vertical bars represent one standard deviation from the mean. The presented method improves rays dramatically for less than 120 connections per node.

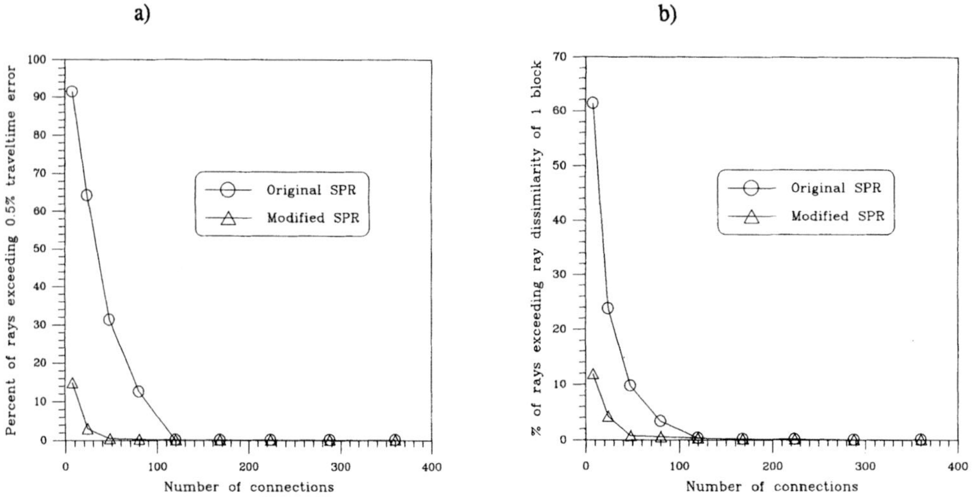

Figure 3: Percent of rays exceeding the maximum error limits of 0.5 percent in traveltime and one block in ray dissimilarity. This figure illustrates how many connections per node are needed to achieve a preset accuracy level.

Experiments

To determine how many CPNs are sufficient to achieve the appropriate accuracy, we computed shortest paths for a horizontaly layered model with five layers. The layers have the same thickness with velocities 800, 900, 1000, 1100 and 1200 m/s from the top to the bottom. We placed 41 seismic sources at the left edge and at the bottom of the model and 41 receivers at the surface and at the right edge of the model. This configuration produces 1681 rays with a broad range of raypath coverage over the model. Reference raypaths and traveltimes are calculated by an accurate ray tracing algorithm.

We compare the traced rays individually to the reference rays by two error measures: percent error in traveltime with respect to the reference and the average distance between the calculated and the reference rays defined as the area enclosed by the two rays devided by the length of the reference ray (ray dissimilarity).

We illustrate averages of the traveltime errors and ray dissimilarities over the 1681 rays (*Figure 2*). Vertical bars represent one standard deviation from the mean. In *Figure 3* we present the percentage of rays exceeding the maximum error limits of 0.5 percent in traveltime and one block in ray dissimilarity. These error limits are usually sufficient for seismic tomography inversion.

We can conclude that our improved algorithm produces improved rays. Without our modification SPR yields long straight line segments in constant velocity areas, but since their directions can be only chosen from a few discrete angles, the full raypaths poorly approximate the true ones (*Figure 1*). Application of our proposed improvement produces straight line segments in homogeneous layers at more arbitrary angles. As a consequence, the length of the rays and ray dissimilarity become smaller, i.e. accuracy increases. Since the number of break-

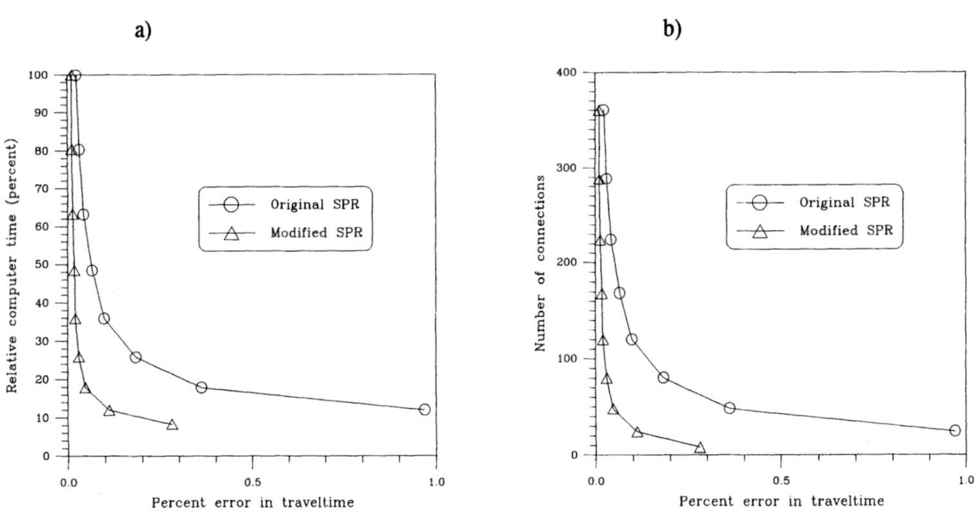

Figure 4: Efficiency of the method: *a)* relative computer time and *b)* number of connections per node vs. percent error in traveltime.

points of the rays also decreases, internal memory, required to store raypath information for later use in an inversion program, is less than for the original SPR. In the example shown in *Figure 1* the number of breakpoints changes from 3252 to 236. Moreover, in contrast with the method proposed by Fischer and Lees (1993), in homogeneous media our procedure *always* finds the exact straight raypaths, *independently* of the number of CPNs (see *Figure 9* of the cited paper).

Figures 2 and 3 show that the modified SPR with 48 CPNs gives more accurate results than the original SPR with 120 CPNs or less. Beyond 120 CPNs average ray dissimilarities change only slightly, while average traveltime errors do not decrease perceptibly. For less than 120 CPNs ray dissimilarities of the traced rays become about one half after the ray modification procedure. *Figure 4* illustrates how computer time and the number of CPNs depend on desired accuracy in traveltime. If average traveltime error must not increase the value of 0.1 percent, our modified SPR algorithm needs about 80 percent less CPNs than the original SPR (24 CPNs instead of 120) and we save about 70 percent in computer time.

Conclusions

This shortest path ray tracing program employed a vast amount of computer time/memory to solve large seismic tomography inversion problems. Efficiency can be increased by decreasing the number of CPNs, but it yields raypaths inaccurate in traveltime and location.

In this paper we proposed a method, by which the number of CPNs, necessary for a given accuracy, can be greatly reduced. For less than 120 CPNs the presented method produces more accurate rays than the original SPR, making it possible to use fewer CPNs and to save about 70 percent in computer time. Additionally, computer memory required to store raypath information for later use can be reduced by up to 90 percent. In contrast with the method proposed by Fischer and Lees (1993), in homogeneous media our procedure *always* finds the exact straight raypaths, *independently* of the number of CPNs. Moreover, it does not change the original Dijkstra's method and, as a consequence, it is faster and easier to implement.

Acknowledgement

The reported investigation was financially supported by the OTKA Bureau (No. F4237). This help is gratefully acknowledged.

References

Dijkstra, E. W. (1959): A note on two problems in connection with graphs, *Numer. Math.*, **1**, 269-271.

Fischer, R. and Lees, M. (1993): Shortest path ray tracing with sparse graphs, *Geophysics*, **58**, 987-996.

Moser, T. J. (1991): Shortest path calculation of seismic rays, *Geophysics*, **56**, 59-67.

A THREE-DIMENSIONAL, BROADBAND, COUPLED NORMAL-MODE SOUND PROPAGATION MODELING APPROACH

CHING-SANG CHIU and JAMES H. MILLER
Naval Postgraduate School
Monterey, CA 93943

WARREN W. DENNER
EOS Research Associates
Monterey, CA 93940

and

JAMES F. LYNCH
Woods Hole Oceanographic Institution
Woods Hole, MA 02543

ABSTRACT A three-dimensional, coupled normal-mode approach for the prediction of pulse propagation in a complex coastal environment is presented. In this broadband approach, the received time-domain sound pressure signal is generated by Fourier synthesis using the product of the source signal spectrum and the source-to-receiver ocean transfer function. For a given frequency, the basic formulation involves decomposing the acoustic pressure into slowly varying complex envelopes that modulate (mode by mode) analytic, rapidly-varying, adiabatic-mode solutions. A coupled set of differential equations governing these complex modal envelopes are derived. Since the formulation allows for sound speed and density to vary three-dimensionally and bathymetry to vary two-dimensionally, it is particularly useful for shallow-water, low-frequency applications including coastal acoustic tomography.

1. Introduction

In a coastal environment, the propagation of low-frequency sound from a projector to a separated receiver is heavily influenced by the variable bathymetry, sediment structure and water-column sound-speed field. These environmental complexities can create significant three-dimensional effects on the transmitted sound, such as horizontal refraction and scattering, which can only be properly modeled by methods containing fully three-dimensional physics. At present, the research on fully 3-D methods is at its infancy. Some pioneering works in this subject area can be found in a recent book edited by Robinson and Lee [8].

The focus of this paper is to discuss a newly developed 3-D method. Specifically, a 3-D, coupled normal-mode approach for the prediction of pulse propagation in a complex coastal environment is presented. In this broadband approach, the received time-domain sound pressure signal is generated by Fourier synthesis using the product of the source signal spectrum and the source-to-receiver ocean transfer function. For a given frequency, the basic formulation involves decomposing the acoustic pressure into slowly varying complex envelopes that modulate (mode by mode) analytic, rapidly-varying, adiabatic-mode solutions. A coupled set of differential equations governing these complex modal envelopes are derived. Since the formulation allows for sound speed and density to vary three-dimensionally and bathymetry

O. Diaschok et al. (eds.), Full Field Inversion Methods in Ocean and Seismo-Acoustics, 57–62.

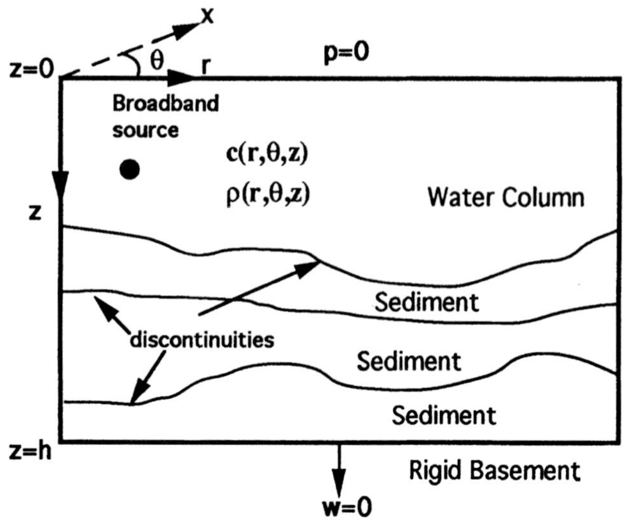

FIGURE 1. An illustration of the complexity of the input environmental model.

to vary two-dimensionally, it is particularly useful for coastal tomography [3, 4, 7] and other active sonar applications.

The modeling approach discussed in this paper has been used to predict the modal arrival structure in the 1992 Barents Sea Coastal Acoustic Tomography Test. The model results as well as a comparison to the data measured by a vertical hydrophone array are presented in a companion paper by the same authors [5].

2. Theoretical Formulation

A sketch of the environmental complexity that can be properly handled by our 3-D coupled normal mode approach is displayed Figure 1. The medium is treated as a waveguide bounded by a flat pressure release surface and a flat rigid basement. Within the layer, both sound speed and density are allowed to be arbitrary functions of the 3-D space. These functions can be discontinuous. Through proper specification of the sound speeds and densities, variable bathymetry and ocean and sediment properties as well as multiple sediment layers can be well modeled.

The time series of sound pressure at a position can be simulated by an inverse Fourier transform of the product of the source signal spectrum $S(f)$ at a unit distance r_o from the source and the source-to-receiver ocean transfer function $P(r, \theta, z; f)$:

$$p(r, \theta, z, t) = \int_{-\infty}^{+\infty} S(f) P(r, \theta, z; f) \exp(-i2\pi f t) \, df \tag{1}$$

In cylindrical coordinates, a position is defined by range r, azimuthal angle θ and depth z. The key to this Fourier synthesis of the time-domain pressure signal is thus the determination of the ocean transfer function at various acoustic frequencies f.

For a point source located at a depth of z_o in a medium with spatially varying sound speed $c(r,\theta,z)$ and density $\rho(r,\theta,z)$, the source-to-receiver ocean transfer function satisfies

$$\nabla^2 P \; - \; \frac{1}{\rho}\nabla\rho \cdot \nabla P \; + \; k^2 P \; = \; -4\pi\frac{r_o}{r}\delta(r)\delta(\theta)\delta(z-z_o) \qquad (2)$$

where $k = 2\pi f/c$ is the acoustic wavenumber.

The coupled mode theory involves casting the solution for P as a linear combination of local normal modes $Z_n(z;r,\theta,f)$ such that

$$P(r,\theta,z;f) \; = \; \sum_n \frac{r_o}{\sqrt{r}} P_n(r,\theta;f) Z_n(z;r,\theta,f) \qquad (3)$$

where the normal modes satisfy

$$\left\{\frac{\partial^2}{\partial z^2} \; - \; \frac{1}{\rho(r,\theta,z)}\frac{\partial\rho}{\partial z}\frac{\partial}{\partial z} \; + \; \left[k(r,\theta,z)^2 \; - \; k_n(r,\theta;f)^2\right]\right\} Z_n(z;r,\theta,f) \; = \; 0 \qquad (4)$$

$$Z_n(0;r,\theta,f) \; = \; 0$$

$$\frac{\partial}{\partial z} Z_n(h;r,\theta,f) \; = \; 0$$

$$\int_0^h \frac{1}{\rho(r,\theta,z)} Z_n(z;r,\theta,f) Z_m(z;r,\theta,f) dz \; = \; \delta_{n,m}$$

In (4), k_n is the horizontal wavenumber (or eigenvalue) of the nth mode and h is total thickness of the waveguide.

The coupled mode equations governing the horizontal structure of the modes, i.e., the modal amplitudes, are obtained by substituting the solution form (3) in the wave equation (2) and applying the conditions in (4). At far field $kr >> 1$, these equations are

$$\left(\frac{\partial^2}{\partial r^2} \; + \; \frac{1}{r^2}\frac{\partial^2}{\partial\theta^2} + {k_n}^2\right) P_n \; = \; -\sum_m \left(\gamma_{mn}\frac{\partial}{\partial r} \; + \; \beta_{mn}\frac{1}{r}\frac{\partial}{\partial\theta} + \; C_{mn}\right) P_m \qquad (5)$$

where

$$\gamma_{mn}\hat{r} \; + \; \beta_{mn}\hat{\theta} \; = \; \int_0^h \frac{2}{\rho} Z_n \nabla_h Z_m dz \; - \; \int_0^h \frac{\nabla_h\rho}{\rho^2} Z_n Z_m dz$$

$$C_{mn} \; = \; \int_0^h \frac{1}{\rho} Z_n \nabla_h^2 Z_m dz \; - \; \int_0^h \frac{\nabla_h\rho}{\rho^2} Z_n \nabla_h Z_m dz$$

are the coupling coefficients with ∇_h^2 and ∇_h representing the horizontal Laplacian and gradient operators, respectively. The neglect of the coupling terms on the right hand side of the equation constitute the so-called adiabatic mode approximation.

To solve the coupled mode equations, we use a two-scale decomposition technique [1] in which the modal amplitudes are expressed as slowly-varying complex envelopes U_n modulating analytic, rapidly-varying, adiabatic-mode solutions , i.e.,

$$P_n \; = \; U_n(r,\theta;f) \; \exp\left[i\varphi_n(r,\theta;f)\right] \qquad (6)$$

$$\varphi_n \; = \; \int_0^r k_n(r',\theta;f)dr'$$

The decomposition is based on the following physical intuition: In the absence of horizontal environmental variations, the coupling coefficients have zero values and the modes propagate independently of each other. In such case and with the neglect attenuation, each modal amplitude P_n is an azimuth-invariant, constant-amplitude sinusoid oscillating rapidly at an acoustic wavenumber of k_n along the radial direction. In the presence of horizontal environmental variations, mode-mode interaction occurs and the envelope of P_n is no longer constant but varies gradually in both the radial and azimuthal directions with the environmental variation length scales. Obviously, it is computationally advantageous to calculate for the slowly varying envelopes instead of the rapidly varying modal amplitudes themselves.

By substituting (6) in (5), the governing equations for the slowly varying envelopes are obtained:

$$\left(\frac{\partial^2}{\partial r^2}+\frac{1}{r^2}\frac{\partial^2}{\partial\theta^2}+i2k_n\frac{\partial}{\partial r}+\frac{F_n}{r}\frac{\partial}{\partial\theta}+E_n\right)U_n = \sum_m\left(G_{mn}\frac{\partial}{\partial r}+\frac{B_{mn}}{r}\frac{\partial}{\partial\theta}+H_{mn}\right)U_m \quad (7)$$

$$E_n = i\frac{\partial k_n}{\partial r} + \frac{i}{r^2}\frac{\partial^2\varphi_n}{\partial\theta^2} - \frac{1}{r^2}\left(\frac{\partial\varphi_n}{\partial\theta}\right)^2$$

$$F_n = \frac{i2}{r}\frac{\partial\varphi_n}{\partial\theta}$$

$$G_{mn}\hat{r} + B_{mn}\hat{\theta} = -\left(\gamma_{mn}\hat{r} + \beta_{mn}\hat{\theta}\right)\exp\left[i\left(\varphi_m - \varphi_n\right)\right]$$

$$H_{mn} = \left(-i\gamma_{mn}k_m - \frac{i\beta_{mn}}{r}\frac{\partial\varphi_m}{\partial\theta} - C_{mn}\right)\exp\left[i\left(\varphi_m - \varphi_n\right)\right]$$

Modal attenuation caused by sea-water and sediment absorption and boundary and volume scattering can be parameterized in these equations by adding imaginary parts to the horizontal wavenumbers. These imaginary parts are referred as the modal attenuation coefficients. For sea-water and sediment absorption only, an integral formula is available for the determination of these coefficients given attenuation profiles and the normal modes [6].

3. Numerical Implementation

To construct the coupled mode solution for the source-to-receiver ocean transfer function, the first step is to solve the equations governing the normal modes (4) at each of the frequency increments within the band of the source signal and at the horizontal locations where the depth profiles of density and sound speed are specified. In our numerical approach, (4) is solved by approximating the first and second derivative terms with fourth-order finite-differences. The fourth-order finite-difference approximations are required to ensure accurate eigenvalue and hence accurate phase and travel time calculations out to a range of approximately 100 km in a shallow water waveguide. The finite-difference approximations cast the eigenvalue-eigenfunction problem (4) into a linear algebra problem which can be easily solved using the QR algorithm and the method of backward substitutions. Once the normal modes are determined, the coupling coefficients are equated using finite difference approximations to obtain the normal mode derivatives and quadrature methods for the evaluation of the integrals. Linear interpolation is then used to make the coupling coefficients continuous in the horizontal coordinates.

The next step is to solve the system of coupled equations governing the modal envelopes, (7). An iterative procedure similar to that introduced by Chiu and Ehret [1] is used. The procedure recasts (7) as

$$\left(\frac{\partial^2}{\partial r^2} + i2k_n\frac{\partial}{\partial r} + E_n\right)U_n^{(k)} - \sum_m \left(G_{mn}\frac{\partial}{\partial r} + H_{mn}\right)U_m^{(k)} = -\left(\frac{1}{r^2}\frac{\partial^2}{\partial\theta^2} + \frac{F_n}{r}\frac{\partial}{\partial\theta}\right)U_n^{(k-1)} + \sum_m B_{mn}\frac{1}{r}\frac{\partial}{\partial\theta}U_m^{(k-1)} \tag{8}$$

where $U_n^{(k)}$ denotes the solution at the kth iteration. At each iteration, the equations are integrated using a fourth- and fifth-order Runge-Kutta algorithm with automatic step size adjustments enabling larger range steps at regions where horizontal environmental variations are mild. A necessary condition for the convergence is that the terms on the right hand side of (8) must be much smaller than the largest term on the left hand side. In an order of magnitude analysis, Chiu and Ehret [2] showed that this is always the case for typical ocean environmental variations. With the neglect of backscattering, the integration at each iteration is started with the following initial conditions:

$$U_n(0,\theta;f) \approx \frac{1}{\rho(0,\theta,z_o)}\sqrt{\frac{2\pi}{k_n(0,\theta;f)}}\exp\left(i\frac{\pi}{4}\right)Z_n(z_o;0,\theta,f) \tag{9}$$

$$\frac{\partial}{\partial r}U_n(0,\theta;f) \approx \frac{1}{\rho(0,\theta,z_o)}\sqrt{2\pi k_n(0,\theta;f)}\exp\left(i\frac{3\pi}{4}\right)Z_n(z_o;0,\theta,f) \tag{10}$$

Finally, the ocean transfer function is built by combining the modal envelopes and the normal modes according to (3) and Fast Fourier Transform (FFT) is used to construct the time-domain sound pressure.

4. Concluding Remarks

In this paper, the formulation of a broadband, coupled normal mode propagation modeling approach is presented. This versatile approach is capable of modeling pulse signal propagation in a complex, three-dimensionally varying ocean environment. Furthermore, at low frequencies and over shallow water, the method is extremely effective because a small number modes is required.

Based on our formulation, a computer model has been written in MATLAB. The model has been used to predict the modal arrival structure in the 1992 Barents Sea Coastal Acoustic Tomography Test. The model results as well as a comparison to the data measured by a vertical hydrophone array are presented in a companion paper by the same authors [5].

Acknowledgments

The first author is grateful to the inversion Conference Organizing Committee (particularly to the Chairman, Orest Diachok) for the invitation to present this paper.

References

1. Chiu, C-S. and L.L. Ehret, "Computation of Sound Propagation in a Three-dimensionally Varying Ocean: A Coupled Normal Mode Approach," in *Computational Acoustics II*, D. Lee, A. Cakmak, and R. Vichnevetsky (editors), North-Holland, 1990.

2. Chiu C-S. and L.L. Ehret, "Three-Dimensional Acoustic Normal Mode Propagation in the Gulf Stream," Chapter 8 of *Oceanography and Acoustics: Coupled Prediction and Propagation Models*, A. Robinson and D. Lee (editors), American Institute of Physics, 1994.
3. Chiu, C-S., J.H. Miller, and J.F. Lynch, "Inverse Techniques for Coastal Acoustic Tomography," in *Environmental Acoustics*, (D. Lee and M. Schultz editors), World Scientific, 1994 (in press).
4. Chiu, C-S., J.H. Miller, R.H. Bourke, J.F. Lynch, and R.D. Muench, "Acoustically-Derived Images of the Barents Sea Polar Front," *EOS Transactions, AGU*, Vol. 75, NO. 3, supplement, p118, 1994.
5. Chiu, C-S., J.H. Miller, W.W. Denner, and J.F. Lynch, "Forward Modeling of the Barents Sea Tomography Vertical Line Array Data and Inversion Highlights," same issue.
6. Frisk, G.V., "*Ocean and Seabed Acoustics*," Prentice Hall, 1993.
7. Miller, J.H., C-S. Chiu, and J.F. Lynch, "Signal Processing for Coastal Acoustic Tomography," in *Environmental Acoustics*, (D. Lee and M. Schultz editors), World Scientific, 1994 (in press).
8. Robinson, A. and D. Lee (editors), "*Oceanography and Acoustics: Coupled Prediction and Propagation Models*," American Institute of Physics, 1994.

DEVELOPMENT OF AN ACOUSTIC FIELD PREDICTOR IN A THREE DIMENSIONAL OCEANIC ENVIRONMENT

F. STURM[a,b], M.C. PELISSIER[a,b] and D. FATTACCIOLI[b]
[a]*Modélisation et Signal / ETMA, Université de Toulon*
BP 132, 83957 La Garde Cedex, France.
[b]*CTSN/Dissuasion Lutte Sous-Marine*
chemin de la gardiole, Le Brusc, 83140 Six-Fours, France.

ABSTRACT. We propose a new model for the prediction of sound propagation in a three-dimensional oceanic environment. It is based on a paraxial approximation of the Helmholtz equation. This model can take into account the range and azimuth dependence of sediment layers and thermal fronts.

1. Introduction

In the presence of eddies or currents, sudden changes with range and azimuth of the sound velocity can occur at certain depths. In such an environment horizontal variations of the sound speed cannot be neglected, leading to the need for three-dimensional models to predict sound propagation in the ocean [1]. One of the main problems is then to take sloping bottoms and sedimental layers into account.

We are currently developing a three-dimensional propagation model based on a modified parabolic equation and assuming no cylindrical symmetry on the geometry of the bottom. The standard parabolic approximation [2], widely used in the underwater acoustics community, is a paraxial approximation of the Helmholtz equation. The envelope of the acoustic pressure is described by a Hankel function. This expresses an outgoing radiation condition at infinity. No backscattering is allowed by the model. Our model keeps the same main features. It is derived from the standard parabolic equation by means of affined mapping which maps the variable physical domain onto a parallelepiped. The characteristics of the bottom geometry are consequently included in the variable coefficients of the modified equation, which therefore contains new derivative terms. This approach thus differs from the one used by most authors, whose numerical models assume cylindrical symmetry of the bottom [3], [4]. A large number of additional references can be found in [5].

Let $Z=Z_{max}(r,\theta)$ be the expression of the bottom in cylindrical coordinates (r,θ,Z) and $\omega(r,\theta)=1/Z_{max}(r,\theta)$. The modified parabolic equation is derived via the change of variable :

$$z = Z\omega(r,\theta) \tag{1}$$

Thus, in the new coordinates, the acoustic field $u(r,\theta,z)$ is the solution of :

$$\frac{\partial u}{\partial r} = A(r,\theta,z)u + B(r,\theta,z)\frac{\partial^2 u}{\partial z \partial \theta} + C(r,\theta,z)\frac{\partial^2 u}{\partial \theta^2} + D(r,\theta,z)\frac{\partial^2 u}{\partial z^2} + E(r,\theta,z)\frac{\partial u}{\partial z} \tag{2}$$

where

O. Diaschok et al. (eds.), Full Field Inversion Methods in Ocean and Seismo-Acoustics, 63–68.

$$A(r,\theta,z)=\frac{ik_0}{2}\left(n^2(r,\theta,\frac{z}{\omega})-1\right) \quad B(r,\theta,z)=\frac{i}{k_0}\frac{z}{\omega r^2}\frac{\partial\omega}{\partial\theta} \quad C(r,\theta,z)=\frac{i}{2k_0 r^2}$$

$$D(r,\theta,z)=\frac{i}{2k_0}\left(\omega^2+\frac{z^2}{r^2\omega^2}\left(\frac{\partial\omega}{\partial\theta}\right)^2\right) \quad E(r,\theta,z)=-\frac{z}{\omega}\frac{\partial\omega}{\partial r}+\frac{i}{2k_0 r^2}\frac{z}{\omega}\frac{\partial^2\omega}{\partial\theta^2}$$

depend on the bottom geometry $\omega(r,\theta)$, the reference wave number $k_0=2\pi f/c_0$ and the (variable) reference refraction index $n(r,\theta,Z)$. f is the frequency of the source, c_0 a reference sound speed and $i=\sqrt{-1}$.

An initial reference condition

$$u(0,\theta,z)=u_0(\theta,z) \quad 0\le z\le 1 \quad 0\le\theta\le 2\pi \tag{3}$$

is given at r=0, and at the surface z=0 the pressure is zero :

$$u(r,\theta,0)=0 \quad r\ge 0 \quad 0\le\theta\le 2\pi \tag{4}$$

2π-periodicity in θ for u and all the data (5)

is moreover assumed.

2. The Bottom Parabolic Model

2.1. INTERFACE CONDITIONS

In the new coordinates, the bottom is flat at z=1. The assumption of a rigid bottom leads to the vanishing of the normal derivative $\partial P/\partial\nu$ of the acoustic pressure P, related to the acoustic field u by :

$$P(r,\theta,z/\omega)=H_0^{(1)}(k_0 r)\,u(r,\theta,z) \tag{6}$$

where $H_0^{(1)}$ denotes the Hankel function of order 0 of the first kind.
At the interface $\xi(r,\theta,Z)=0$ between two sedimental layers, the continuity conditions write :

$$P_1=P_2 \qquad \frac{1}{\rho_1}\frac{\partial P_1}{\partial\nu}=\frac{1}{\rho_2}\frac{\partial P_2}{\partial\nu} \tag{7}$$

(P_1 : pressure in medium 1 ; ρ_1 : density in medium 1 ; ν : normal vector).

But we know [6],[7] that these physical conditions, which are mathematically correct conditions for the wave equation, lead to an ill-posed boundary value problem when the associated partial differential equation is a paraxial approximation of the wave equation. We have, in this situation,to use a consistant paraxial approximation of the normal derivative. In the case of our modified parabolic equation, this approximation of equation $\xi(r,\theta,Z)=0$ [8] may be written as:

$$\frac{\partial P}{\partial\nu}\approx\frac{\partial\xi}{\partial\theta}\frac{1}{r^2}\left(\frac{\partial u}{\partial\theta}+\frac{z}{\omega}\frac{\partial\omega}{\partial\theta}\frac{\partial u}{\partial z}\right)+\omega\frac{\partial\xi}{\partial Z}\frac{\partial u}{\partial z}+ik_0\bar{n}\frac{\partial\xi}{\partial r}u \tag{8}$$

where
$\bar{n}^2(r,\theta,z) = \frac{c_0^2}{\bar{c}^2(r,\theta,z)}\left(1+i\frac{\beta}{27{,}2875}\right) = n^2(r,\theta,z/\omega)$ depends on the sound velocity $\bar{c}(r,\theta,z) = c(r,\theta,z/\omega)$ and the attenuation coefficient β in the corresponding layer (water or sediment).

The interface conditions are thus written as:

$$u_1 = u_2 \quad \text{on} \quad \xi(r,\theta,Z) = 0 \tag{9}$$

$$\frac{1}{\rho_1}\left(a_1(r,\theta,z)u_1 + b(r,\theta,z)\frac{\partial u_1}{\partial\theta} + d(r,\theta,z)\frac{\partial u_1}{\partial z}\right) = \frac{1}{\rho_2}\left(a_2(r,\theta,z)u_2 + b(r,\theta,z)\frac{\partial u_2}{\partial\theta} + d(r,\theta,z)\frac{\partial u_2}{\partial z}\right) \quad \text{on} \quad \xi(r,\theta,Z) = 0 \tag{10}$$

where a_l (l=1,2), b, d depend on the bottom geometry $\omega(r,\theta)$ and the equation of the interface $\xi(r,\theta,z/\omega)=0$, a_l varying also with the reference wave number k_0 and the refraction index of layer l :

$$a_l = ik_0\bar{n}_l\frac{\partial\xi}{\partial r} \qquad b = \frac{1}{r^2}\frac{\partial\xi}{\partial\theta} \qquad d = \frac{1}{r^2}\frac{z}{\omega}\frac{\partial\omega}{\partial\theta}\frac{\partial\xi}{\partial\theta} + \omega\frac{\partial\xi}{\partial Z} \tag{11}$$

On a perfectly reflective bottom, $\xi(r,\theta,Z)=0$, $\omega(r,\theta)Z - 1 = 0$, and u satisfies :

$$ik_0\bar{n}\frac{1}{\omega}\frac{\partial\omega}{\partial r}u + \frac{1}{r^2\omega}\frac{\partial\omega}{\partial\theta}\frac{\partial u}{\partial\theta} + \left(\frac{1}{r^2\omega^2}\left(\frac{\partial\omega}{\partial\theta}\right)^2 + \omega^2\right)\frac{\partial u}{\partial z} = 0 \quad z=1 \quad r\geq 0 \quad 0\leq\theta\leq 2\pi \tag{12}$$

2.2. HOMOTHETIC SEDIMENTAL LAYERS

For the sake of simplicity of the numerical implementation, we add a homotheticity hypothesis : the layers are required to be "parallel" to the rigid bottom and are thus transformed into horizontal planes by the scaling mapping.

The equations of the interface are, in the case of Q sedimental layers : $\omega(r,\theta)Z - S_q = 0$ for q=1,..,Q, where $0 < S_{q-1} < S_q < 1$ for q=2,..,Q; and the corresponding continuity conditions write, for q=1,..,Q :

$$u_{q-1}(r,\theta,S_q) = u_q(r,\theta,S_q) \quad r>0 \quad 0\leq\theta\leq 2\pi \tag{13}$$

$$\frac{1}{\rho_{q-1}}\left(a_{q-1,q}u_{q-1}(r,\theta,S_q) + b_q\frac{\partial u_{q-1}}{\partial\theta}(r,\theta,S_q) + d_q\frac{\partial u_{q-1}}{\partial z}(r,\theta,S_q)\right) = \frac{1}{\rho_q}\left(a_{q,q}u_q(r,\theta,S_q) + b_q\frac{\partial u_q}{\partial\theta}(r,\theta,S_q) + d_q\frac{\partial u_q}{\partial z}(r,\theta,S_q)\right) \quad r>0 \quad 0\leq\theta\leq 2\pi \tag{14}$$

index 0 being relative to the water, and index q to the qth sedimental layer, with :

$$a_{q-1,q} = ik_0\bar{n}_{q-1}(r,\theta,S_q)\frac{S_q}{\omega(r,\theta)}\frac{\partial\omega}{\partial r}(r,\theta) \qquad b_q = \frac{S_q}{r^2\omega(r,\theta)}\frac{\partial\omega}{\partial\theta}(r,\theta)$$

$$a_{q,q} = ik_0\bar{n}_q(r,\theta,S_q)\frac{S_q}{\omega(r,\theta)}\frac{\partial\omega}{\partial r}(r,\theta) \qquad d_q = \frac{S_q^{\;2}}{r^2\omega^2(r,\theta)}\left(\frac{\partial\omega}{\partial\theta}(r,\theta)\right)^2 + \omega^2(r,\theta)$$

3. Numerical Implementation

3.1. DISCRETIZATION OF THE EQUATION

We solve equation (2) with initial condition (3) and boundary conditions (4), (5),(12) using a standard Crank-Nicolson finite differences scheme. The affine mapping has changed the physical domain into a regular computational domain, which is best suited for numerical discretization. We note $u_{i,j}^n = u(n\Delta r, i\Delta\theta, j\Delta z)$ the value of the approximate solution at grid point $r=n\Delta r$, $\theta=i\Delta\theta$ and $z=j\Delta z$.

Let N be the number of grid points in depth and M their number in azimuth. Then, $\Delta z=1/N$. The solution of the discretized equations [8] can be written as the solution of an algebraic linear system $B\,U^{n+1} = F$ where B is a block-tridiagonal matrix of order M*N and F is a vector depending only on previously calculated values of the sound field. Submatrices B1, B2 ,B3 are of order M and almost tridiagonal. Because of the periodicity condition (5), there are terms in the upper right and lower left corner of these matrices. The solution vector is constructed with the following form :

$$U^{n+1} = \left(u_{1,1}^{n+1}, u_{2,1}^{n+1}, \ldots, u_{M,1}^{n+1}; u_{1,2}^{n+1}, u_{2,2}^{n+1}, \ldots, u_{M,2}^{n+1}; \cdots; u_{1,N}^{n+1}, u_{2,N}^{n+1}, \ldots, u_{M,N}^{n+1}\right)^T \qquad (15)$$

$$B = \begin{pmatrix} B2 & B3 & 0 & 0 & \ldots & \ldots & 0 \\ B1 & B2 & B3 & 0 & & & \vdots \\ 0 & B1 & B2 & B3 & & & \vdots \\ \vdots & & \ddots & \ddots & \ddots & & \vdots \\ \vdots & & & B1 & B2 & B3 & 0 \\ \vdots & & & 0 & B1 & B2 & B3 \\ 0 & \ldots & \ldots & 0 & 0 & B1 & B2 \end{pmatrix}$$

3.2. DISCRETE EQUATION ON THE INTERFACE

We first notice that the first continuity condition (13) leads to the continuity of all the r and θ derivatives. We are thus concerned only with the z-derivatives. For the sake of simplicity, we will write $u_{i,j}$ instead of $u_{i,j}^n$ and u_1 (resp. u_2) instead of u_{q-1} (resp. u_q) on the interface.

A Taylor formula in z gives :

$$\left(\frac{\partial^2 u_1}{\partial z^2}\right)_{i,j} \approx \frac{2}{\Delta z}\left(\frac{\partial u_1}{\partial z}\right)_{i,j} - \frac{2}{\Delta z^2}\left(u_{i,j} - u_{i,j-1}\right) \qquad (16)$$

$$\left(\frac{\partial^2 u_2}{\partial z^2}\right)_{i,j} \approx -\frac{2}{\Delta z}\left(\frac{\partial u_2}{\partial z}\right)_{i,j} + \frac{2}{\Delta z^2}\left(u_{i,j+1} - u_{i,j}\right) \qquad (17)$$

A Taylor formula in θ and z gives :

$$\left(\frac{\partial^2 u_1}{\partial z \partial \theta}\right)_{i,j} \approx \frac{1}{\Delta z}\left(\frac{\partial u_1}{\partial z}\right)_{i,j} - \frac{1}{2\Delta z \Delta \theta}\left(u_{i+1,j-1} - u_{i-1,j-1}\right) \tag{18}$$

$$\left(\frac{\partial^2 u_2}{\partial z \partial \theta}\right)_{i,j} \approx -\frac{1}{\Delta z}\left(\frac{\partial u_2}{\partial z}\right)_{i,j} + \frac{1}{2\Delta z \Delta \theta}\left(u_{i+1,j+1} - u_{i-1,j+1}\right) \tag{19}$$

The modified parabolic equations in media 1 and 2 write :

$$\frac{\partial u}{\partial r} = A_l u + B\frac{\partial^2 u_l}{\partial z \partial \theta} + C\frac{\partial^2 u}{\partial \theta^2} + D\frac{\partial^2 u_l}{\partial z^2} + E\frac{\partial u_l}{\partial z} \qquad \text{for } l = 1,2 \tag{20}$$

We replace $\frac{\partial^2 u_1}{\partial z^2}$ (resp. $\frac{\partial^2 u_1}{\partial z \partial \theta}$) in equation (20) for the first medium by its expression (16) (resp. (18)). We do the same for the second medium. Using now the second interface condition (14) we can eliminate $\frac{\partial u_1}{\partial z}$ and $\frac{\partial u_2}{\partial z}$ to obtain the following semi-discrete parabolic equation on the interface :

$$\begin{aligned}\left(\frac{\partial u}{\partial r}\right)_{i,j} = (k_1 + k_2)^{-1}\Bigg\{&\left[k_1\left(A_1 - \frac{2D}{\Delta z^2}\right) + k_2\left(A_2 - \frac{2D}{\Delta z^2}\right) + \frac{k_1 k_2}{d}(\rho_1 a_2 - \rho_2 a_1)\right]u_{i,j} \\ &+ (k_1 + k_2)C\left(\frac{\partial^2 u}{\partial \theta^2}\right)_{i,j} + \frac{B}{2\Delta z \Delta \theta}\left[k_2\left(u_{i+1,j+1} - u_{i-1,j+1}\right) - k_1\left(u_{i+1,j-1} - u_{i-1,j-1}\right)\right] \\ &+ \left[\frac{B}{\Delta z}(k_1 - k_2) + \frac{b k_1 k_2}{d}(\rho_1 - \rho_2)\right]\left(\frac{\partial u}{\partial \theta}\right)_{i,j} + \frac{2D}{\Delta z^2}\left(k_2 u_{i,j+1} + k_1 u_{i,j-1}\right)\Bigg\}\end{aligned} \tag{21}$$

where

$$k_1 = \frac{1}{\rho_1}\left(\frac{2D}{\Delta z} - E\right) \qquad k_2 = \frac{1}{\rho_2}\left(\frac{2D}{\Delta z} + E\right)$$

4. Computational Examples

We propose an example of a 3-D effect due to a brutal discontinuity of the sound velocity. Indeed, the model can take thermal fronts in range and azimuth into account. We consider a water depth varying from 3300 to 3600 meters and a source depth of 300 meters. The axial range runs from 0 to 100 kilometres. The source frequency is 50 Hertz. We calculate propagation losses in water in the presence of the two experimental (South Island) sound velocity profiles plotted in figure 1. We notice that SVP 1 (continuous line in figure 1) leads to a surface sound channel.

Figure 2 shows the propagation losses at receiver depth 300 meters. The sound velocity profile is SVP 1 at the exterior of the thermic fronts. Two thermic fronts are present, with sound velocity profile SVP 2. In the western part, the thermic front appears at range 3 kilometers and is located between the two (fixed) azimuth 225° and 315°. In the eastern part, the thermic front is encountered at range 40 kilometers; its azimuthal boundaries vary between 72° and 30° (northern boundary) and between 130° and 109° (southern boundary).

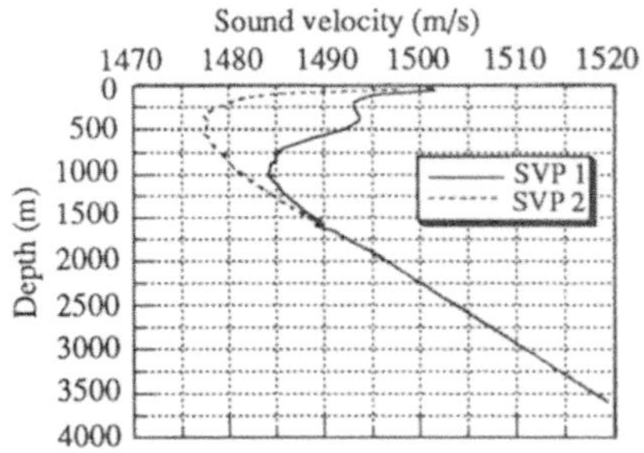

Figure 1 : Sound Velocity Profiles

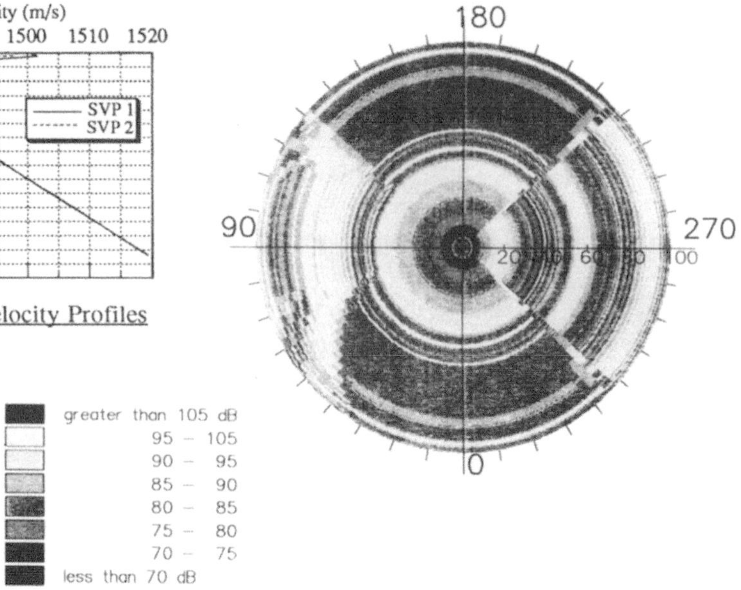

Figure 2 : Propagation losses at receiver depth 300 m

5. Conclusion

We have built a broadly applicable non-symmetric predictor that can take sloping bottoms, various thermal fronts and sediment layers into account. It appears as a 3-D realistic model that permits numerical treatment of sound propagation in general oceanic environments. The next step will be the implementation of a wide-angle equation.

References

1. Lee, D., Scully-Power, P.D. and Botseas G. (1987) A Discussion of Three-Dimensional Wave Propagation Effects, *J. Acoust. Soc. Am. Suppl. 1 82, S44*

2. Tappert, F.D. (1977) The Parabolic Approximation Method, *Wave Propagation and underwater acoustics, edited by J.B. Keller and J.S. Papadakis, Lecture Notes in Physics, Vol. 70 (Springer, New York)*

3. Lee, D., Botseas, G. and Siegmann, W.L. (1992) Examination of the Three-Dimensional Effects using a Propagation Model with Azimuth-Coupling Capability (FOR3D), *J. Acoust. Soc. Am. 91 (6)*, 3192-3202.

4. Collins, M.D. and Chin-Bing, S.A. (1990) A Three-Dimensional Parabolic Equation Model that includes the effects of rough boundaries, *J. Acoust. Soc. Am. 87 (3)*, 1104-1109.

5. Buckingham, M.J. (1992) Ocean -Acoustic Propagation Models, *J. Acoustique*, 223-287

6. Abrahamsson, L. and Kreiss, H.O. (1990) Boundary Conditions for the Parabolic Equation in a Range-Dependent Duct, *J. Acoust. Soc. Am. 87 (6)*, 2438-2441.

7. Bernardi, C. and Pélissier, M.C., in preparation.

8. Pélissier, M.C., Sturm, F. and Fattaccioli, D. (1994) Sound Propagation Modeling using a Modified Parabolic Equation in a 3-D Range-Dependent Oceanic Environment", to be published in the *Proceedings of the 2nd European Conference on Underwater Acoustics, Copenhagen.*

THREE-DIMENSIONAL SOUND PROPAGATION IN AN OCEAN OVERLYING AN ELASTIC BOTTOM

GREGORY J. ORRIS, MICHAEL D. COLLINS
Naval Research Laboratory
Washington, DC 20375, USA

GRANT B. DEANE
Scripps Institution of Oceanography
La Jolla, California 92093, USA

and

MICHAEL B. PORTER
New Jersey Institute of Technology
Newark, New Jersey 07102, USA

ABSTRACT The adiabatic mode parabolic equation is generalized to the case of an ocean overlying an elastic bottom. This three-dimensional model is valid when the medium varies sufficiently gradually with the horizontal coordinates so that both coupling of energy between modes and the azimuthal component of displacement may be neglected. The efficiency of the model is demonstrated by applying it to solve a global-acoustics problem involving diffraction by the Hawaiian Islands.

1. Introduction

There has been relatively little three-dimensional propagation modeling in ocean acoustics [1–3] because three-dimensional calculations have been widely regarded as impractical and unnecessary. Many ocean acoustic propagation problems may be solved accurately with the uncoupled azimuth approximation [4], which is based on neglecting the term in the wave equation that involves azimuthal derivatives. This is an important approximation because it is rarely practical to solve three-dimensional problems. Interest in three-dimensional modeling is currently increasing because of a wide interest in shallow water acoustics.

The adiabatic mode parabolic equation (PE) [5] is a three-dimensional propagation model that is practical for solving many problems of interest. This approach is based on the adiabatic mode solution [6], which arises from the assumption that energy coupling between the modes is negligible, and the parabolic equation method [7,8], which is an efficient approach for solving range-dependent problems in ocean acoustics [9,10]. The adiabatic mode PE has been used to solve global-scale problems at low frequencies [11]. This approach has been generalized to handle the effects of fluid flow and applied to model acoustic propagation from the impact sites of the fragments of Comet Shoemaker-Levy 9 with Jupiter [12].

In this paper, we extend the adiabatic mode PE to handle problems involving elastic ocean bottoms. The PE method was previously extended to elastic media for two-dimensional (range and depth) problems [13–16]. In Section 2, we derive the adiabatic

O. Diaschok et al. (eds.), Full Field Inversion Methods in Ocean and Seismo-Acoustics, 69–75.

mode PE for problems involving an ocean overlying an elastic ocean bottom. The elastic parameters may vary piece-wise continuously in depth. Horizontal variations in the elastic parameters must be sufficiently gradual so that mode coupling and the azimuthal component of displacement may be neglected. In Section 3, we apply the adiabatic mode PE to a global-scale test problem involving diffraction by the Hawaiian Islands. We compare the coupled and uncoupled azimuth solutions to illustrate the importance of three-dimensional effects.

2. The Adiabatic Mode Parabolic Equation

In this section, we describe the adiabatic mode solution for sound propagation in an ocean overlying an elastic bottom. We work in cylindrical coordinates, where z is the depth below the ocean surface, r is the horizontal distance from a time-harmonic point source of circular frequency ω, and θ is the azimuth. The spatially varying parameters are the compressional speed c_p, the shear speed c_s, the density ρ, the compressional attenuation β_p, and the shear attenuation β_s. In the water column, $c_s = \beta_p = \beta_s = 0$ and $\rho = 1$ g/cm^3.

We remove the time-dependent factor $\exp(-i\omega t)$ and the cylindrical spreading factor $r^{-1/2}$ from the dependent variables. We assume that horizontal variations in the elastic parameters are sufficiently gradual so that energy does not couple between modes and that the azimuthal component of displacement is dominated by the other components. Under these assumptions, the following equations of motion are valid in the farfield [16]:

$$(\lambda+2\mu)\frac{\partial^2\Delta}{\partial r^2}+(\lambda+2\mu)\frac{1}{r^2}\frac{\partial^2\Delta}{\partial\theta^2}+\frac{\partial}{\partial z}\left[(\lambda+2\mu)\frac{\partial\Delta}{\partial z}\right]+\rho\omega^2\Delta+ \\ 2\frac{\partial\mu}{\partial z}\frac{\partial^2 w}{\partial r^2}+\omega^2\frac{\partial\rho}{\partial z}w+\frac{\partial}{\partial z}\left(\frac{\partial\lambda}{\partial z}\Delta+2\frac{\partial\mu}{\partial z}\frac{\partial w}{\partial z}\right)=0\,, \tag{1}$$

$$\mu\frac{\partial^2 w}{\partial r^2}+\mu\frac{1}{r^2}\frac{\partial^2 w}{\partial\theta^2}+\mu\frac{\partial^2 w}{\partial z^2}+\rho\omega^2 w+(\lambda+\mu)\frac{\partial\Delta}{\partial z}+\frac{\partial\lambda}{\partial z}\Delta+2\frac{\partial\mu}{\partial z}\frac{\partial w}{\partial z}=0\,, \tag{2}$$

where w is the vertical displacement, Δ is the divergence of the displacement vector, and the complex Lamé parameters λ and μ are defined in [16]. These equations are valid for elastic layers involving piece-wise continuous depth variations in the elastic parameters. The interface conditions described in [16] are applied to handle problems involving fluid layers. The equations of motion are in the form,

$$\frac{\partial^2}{\partial r^2}\begin{pmatrix}\Delta\\ w\end{pmatrix}+\frac{1}{r^2}\frac{\partial^2}{\partial\theta^2}\begin{pmatrix}\Delta\\ w\end{pmatrix}+L^{-1}M\begin{pmatrix}\Delta\\ w\end{pmatrix}=\begin{pmatrix}0\\ 0\end{pmatrix}, \tag{3}$$

where the matrices L and M contain depth operators.

The normal mode representation of the solution of (3) is of the form,

$$\begin{pmatrix}\Delta\\ w\end{pmatrix}=\sum_n\begin{pmatrix}\Delta_n(z;r,\theta)\\ w_n(z;r,\theta)\end{pmatrix}\psi_n(r,\theta)\,, \tag{4}$$

$$L^{-1}M\begin{pmatrix}\Delta_n\\ w_n\end{pmatrix}=k_n^2(r,\theta)\begin{pmatrix}\Delta_n\\ w_n\end{pmatrix}, \tag{5}$$

where (Δ_n, w_n) and k_n^2 are the modes and eigenvalues [17,18] and ψ_n is to be determined. The semicolons in the arguments of Δ_n and w_n indicate gradual variation with r and θ.

Substituting (4) into (3) and applying the assumption that energy coupling between modes may be neglected, we obtain the adiabatic mode wave equation,

$$\frac{\partial^2 \psi_n}{\partial r^2} + \frac{1}{r^2}\frac{\partial^2 \psi_n}{\partial \theta^2} + k_n^2 \psi_n = 0. \tag{6}$$

We factor the operator in (6) into incoming and outgoing operators to obtain

$$\left(\frac{\partial}{\partial r} + i\sqrt{\frac{1}{r^2}\frac{\partial^2}{\partial \theta^2} + k_n^2}\right)\left(\frac{\partial}{\partial r} - i\sqrt{\frac{1}{r^2}\frac{\partial^2}{\partial \theta^2} + k_n^2}\right)\psi_n = 0. \tag{7}$$

Assuming that outgoing energy dominates, we obtain the outgoing adiabatic mode wave equation,

$$\frac{\partial \psi_n}{\partial r} = i\sqrt{\frac{1}{r^2}\frac{\partial^2}{\partial \theta^2} + k_n^2}\psi_n. \tag{8}$$

We rearrange (8) to obtain

$$\frac{\partial \psi_n}{\partial r} = ik_0\sqrt{1+X}\psi_n, \tag{9}$$

$$X = k_0^{-2}\left(\frac{1}{r^2}\frac{\partial^2}{\partial \theta^2} + k_n^2 - k_0^2\right), \tag{10}$$

where k_0 is a representative horizontal wave number. Substituting a rational approximation for the square root in (9), we obtain the adiabatic mode PE,

$$\frac{\partial \psi_n}{\partial r} = ik_0\left(1 + \sum_{j=1}^{m} \frac{a_{j,m}X}{1 + b_{j,m}X}\right)\psi_n. \tag{11}$$

This equation is solved with standard numerical techniques as described in [5]. Choices for the coefficients $a_{j,m}$ and $b_{j,m}$ are given in [16,19].

3. Application to Global Acoustics

There has recently been a great deal of interest in global acoustics [20–22]. To apply the adiabatic mode PE to global-scale problems, we replace the spreading factor $r^{-1/2}$ with $R^{-1/2}$, where

$$R = R_0 \sin\left(\frac{r}{R_0}\right), \tag{12}$$

and R_0 is the radius of the earth. The following generalization of (8) was derived in [5] for this case:

$$\frac{\partial \psi_n}{\partial r} = i\sqrt{\frac{1}{R^2}\frac{\partial^2}{\partial \theta^2} + k_n^2}\psi_n. \tag{13}$$

This equation is solved with the same approach that is used to solve (8).

We apply the adiabatic mode PE to a problem involving a 1-Hz source to the southeast of the Hawaiian Islands at 210° E and 15° N. Since the geographic dependence of the sound speed in the ocean is a minor factor for this problem, we use a single profile that is representative of temperate regions. Since we do not have access to a global data base

Figure 1. Adiabatic mode PE solutions for the first mode at 1 Hz. Azimuthal coupling is included (top) and neglected (bottom). The dynamic range is 100 dB. with red corresponding to high intensity and blue corresponding to low intensity.

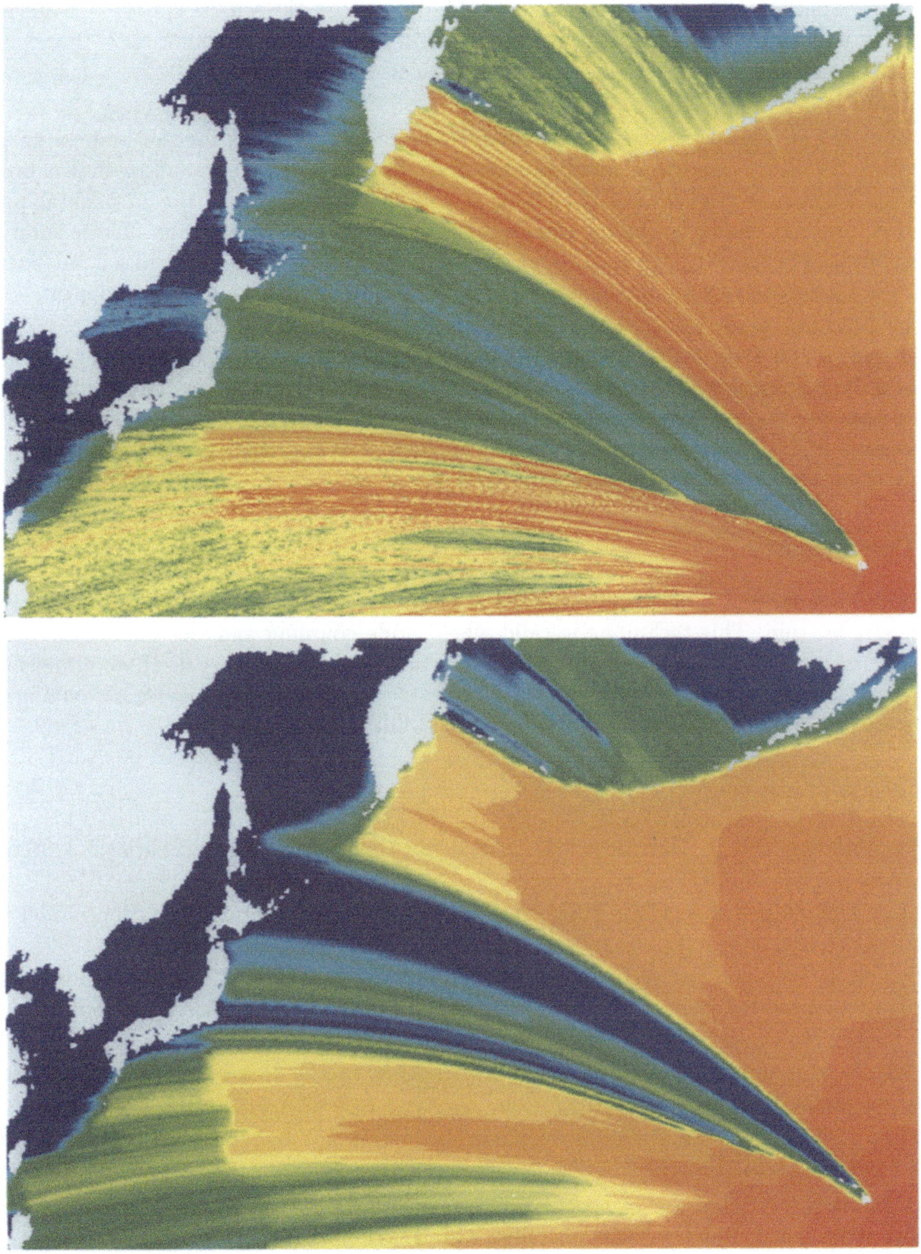

Figure 2. Magnified view of the solutions appearing in Figure 1. Azimuthal coupling is included (top) and neglected (bottom). The dynamic range is 100 dB. with red corresponding to high intensity and blue corresponding to low intensity.

for the parameters of the ocean bottom, we also use constant profiles for these parameters. The wave speeds are linear functions of depth defined by the values $c_p(z_0) = 3400$ m/s, $c_p(z_1) = 6000$ m/s, $c_s(z_0) = 1700$ m/s, and $c_s(z_1) = 3000$ m/s, where $z_0 = 0$ and $z_1 =$ 20 km. The parameters $\rho = 1.5$ g/cm^3, $\beta_p = 0.1$ dB/λ, and $\beta_s = 0.2$ dB/λ are taken to be constants.

Appearing in Figures 1 and 2 are adiabatic mode PE solutions for the first mode that include and neglect azimuthal coupling. The Hawaiian Islands cast a shadow that is broadened by horizontal refraction (an azimuthal coupling effect). Since the horizontal phase speed is larger in shallower water, this is consistent with the ray solution. These solutions are qualitatively similar to the solutions that appear in [11] for the case of a fluid bottom that has a qualitatively similar horizontal phase speed dependence on ocean depth.

The adiabatic mode PE computation required about 5 hours on a Silicon Graphics computer with a 150-MHz MIPS R4400 chip. The computations were done using a three-term rational approximation in the adiabatic mode PE, a range step of 10 km, and 256 azimuthal grid points per degree. The computations were carried out over the entire earth, including the oceans and the continents. The first eigenvalue was obtained as a function of ocean depth using the approach described in [17,18].

4. Conclusions

The adiabatic mode PE has been extended to handle problems involving an ocean overlying an elastic bottom. This technique is valid when mode coupling and the azimuthal component of displacement are negligible and is practical for solving large-scale three-dimensional problems. The efficiency of the model was illustrated by applying it to solve a global acoustics problem involving diffraction around the Hawaiian Islands.

References

[1] H. Weinberg and R. Burridge, "Horizontal ray theory for ocean acoustics," J. Acoust. Soc. Am. 55, 63–79 (1974).

[2] M.D. Collins and S.A. Chin-Bing, "A three-dimensional parabolic equation model that includes the effects of rough boundaries," J. Acoust. Soc. Am. 87, 1104–1109 (1990).

[3] D. Lee, G. Botseas, and W.L. Siegmann, "Examination of three-dimensional effects using a propagation model with azimuth-coupling capability (FOR3D)," J. Acoust. Soc. Am. 91, 3192–3202 (1992).

[4] J.S. Perkins and R.N. Baer, "An approximation to the three-dimensional parabolic-equation method for acoustic propagation," J. Acoust. Soc. Am. 72, 515–522 (1982).

[5] M.D. Collins, "The adiabatic mode parabolic equation," J. Acoust. Soc. Am. 94, 2269–2278 (1993).

[6] A.D. Pierce, "Extension of the method of normal modes to sound propagation in an almost-stratified medium," J. Acoust. Soc. Am. 37, 19–27 (1965).

[7] M.A. Leontovich and V.A. Fock, "Solution of the problem of propagation of electromagnetic waves along the earth's surface by the method of parabolic equation," J. Exp. Theor. Phys. 16, 557–573 (1946).

[8] V.A. Fock, *Electromagnetic Diffraction and Propagation Problems* (Pergamon, New York, 1965), pp. 213–234.

[9] F.D. Tappert, "The Parabolic Approximation Method," in *Wave Propagation and Underwater Acoustics*, edited by J.B. Keller and J.S. Papadakis, Lecture Notes in Physics, Vol. 70 (Springer, New York, 1977).
[10] M.D. Collins, "A split-step Padé solution for the parabolic equation method," J. Acoust. Soc. Am. 93, 1736–1742 (1993).
[11] M.D. Collins, B.E. McDonald, K.D. Heaney, and W.A. Kuperman, "Three-dimensional effects in global acoustics," J. Acoust. Soc. Am. (in press).
[12] M.D. Collins, B.E. McDonald, W.A. Kuperman, and W.L. Siegmann, "Jovian acoustics and Comet Shoemaker-Levy 9," J. Acoust. Soc. Am. (in press).
[13] R.R. Greene, "A high-angle one-way wave equation for seismic wave propagation along rough and sloping interfaces," J. Acoust. Soc. Am. 77, 1991–1998 (1985).
[14] M.D. Collins, "A higher-order parabolic equation for wave propagation in an ocean overlying an elastic bottom," J. Acoust. Soc. Am. 86, 1459–1464 (1989).
[15] B.T.R. Wetton and G.H. Brooke, "One-way wave equations for seismoacoustic propagation in elastic waveguides," J. Acoust. Soc. Am. 87, 624–632 (1990).
[16] M.D. Collins, "Higher-order parabolic approximations for accurate and stable elastic parabolic equations with application to interface wave propagation," J. Acoust. Soc. Am. 89, 1050–1057 (1991).
[17] M.B. Porter and E.L. Reiss, "A numerical method for bottom interacting ocean acoustic normal modes," J. Acoust. Soc. Am. 77, 1760–1767 (1985).
[18] M. Porter, "The KRAKEN normal mode program," SACLANTCEN Memorandum SM-245 (SACLANT Undersea Research Centre, La Spezia, Italy, 1991).
[19] A. Bamberger, B. Engquist, L. Halpern, and P. Joly, "Higher order paraxial wave equation approximations in heterogeneous media," SIAM J. Appl. Math. 48, 129–154 (1988).
[20] A.B. Baggeroer and W. Munk, "The Heard Island Feasibility Test," *Physics Today* 45(9), 22–30 (1992).
[21] K.D. Heaney, B.E. McDonald, and W.A. Kuperman, "Perth-Bermuda sound propagation (1960): Adiabatic mode interpretation," J. Acoust. Soc. Am. 90, 2586–2594 (1991).
[22] B.E. McDonald, M.D. Collins, W.A. Kuperman, and K.D. Heaney, "Comparison of data and model predictions for Heard Island acoustic transmissions," J. Acoust. Soc. Am. 96, 2357–2370 (1994).

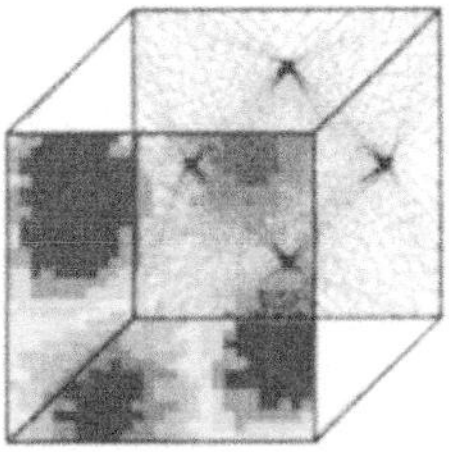

section 2

Signal Processing and Global Search Strategies

PARAMETER ESTIMATION THEORY BOUNDS AND THE ACCURACY OF FULL FIELD INVERSIONS

ARTHUR B. BAGGEROER AND HENRIK SCHMIDT
Massachusetts Institute of Technology
Cambridge, MA USA 02139

ABSTRACT The Cramer-Rao parameter estimation bounds are derived to evaluate limits on the attainable accuracy for full field inversions. These bounds are applied to several inversion scenarios: i) matched field tomography for both narrowband and broadband signals, ii) broadband, full field tomography for coherent signals and iii) noise tomography. In the case of adiabatic propagation with a mode resolving receiver array these bounds specify the performance limits for full field inversions by both matched field and ocean acoustic tomography and they can be expressed in terms of the phase and group slownesses of the propagation and the mode functions at the source.

1. Introduction

Parameter estimation bounds have long been an important topic in the signal processing literature and have led to fundamental insights into algorithm performance; as such they provide a very useful paradigm for understanding and comparing the performance of full field inversion methods. They indicate the coupling among the propagation physics, environmental uncertainties, array geometries, signal structure and data statistics. Most importantly, they provide benchmarks to which the performance of *ad hoc*, but usually simpler, inversion algorithms can be compared.

Full field inversion in ocean and seismo acoustics is a problem in parameter estimation. Tomography concerns estimating environmental parameters such as speeds, gradients, densities, boundary roughness or coefficients of an orthogonal expansion while source localization concerns estimating the position and/or velocities of sources. Often tomography and source localization are embedded in the same problem. More generally, the parameters can change in time which leads to a filtering problem. Nevertheless, all of these are problems in parameter estimation.

The literature on inversion methods is extensive, see for example the bibliographies in references [1, 2, 3] for applications in geophysical inverse methods, ocean acoustic tomography and matched field processing. Performance analyses for specific inversion methods have been done and an extensive number of simulations have demonstrated the accuracies of methods; however, relatively little has appeared on evaluating fundamental limits on performance limits. Parameter estimation bounds, such as the Cramer-Rao, address this from the perspective of determining attainable accuracy regardless of inversion method or algorithm employed.

O. Diaschok et al. (eds.), Full Field Inversion Methods in Ocean and Seismo-Acoustics, 79–84.

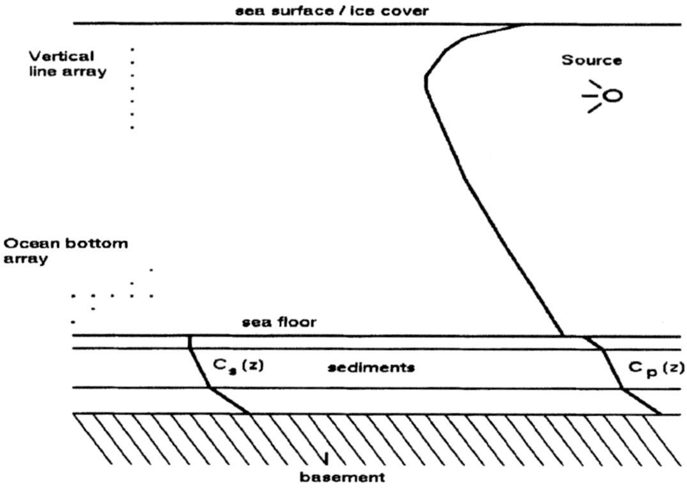

FIGURE 1. Model for the seismo/acoustic environment and receiver array.

2. Model and notation

We consider a model for the seismo/acoustic environment and receiver array indicated in Figure 1. While the figure suggests a range independent environment and hydrophone sensors, these are not inherent restrictions.

The complex envelope of the received signal is given by[1]

$$\mathbf{R}(f) = \tilde{b}(f)S_s(f)\mathbf{G}(f,\mathbf{a}) + \mathbf{N}(f,\mathbf{a}), \quad f \in \Delta W; \tag{1}$$

where

$\mathbf{a}$	is a vector of the unknown parameters;
$\mathbf{G}(f,\mathbf{a})$	is a vector of Green's function for the propagation to the receiver array;
$S_s(f)$	is the Fourier transform of a coherent source signal;
$\tilde{b}(f)$	is a random process incorporating amplitude and phase variability;
$\mathbf{N}(f,\mathbf{a})$	is a stationary noise vector with spectral covariance matrix, $\mathbf{K}_n(f,\mathbf{a})$;
ΔW	is the frequency band occupied by the signal.

The vector $\mathbf{a}$ can include both source parameters, *e.g.* location and velocity, such as done for matched field localization and environmental parameters, *e.g.* speeds, gradients, densities, orthogonal expansion coefficients, as done for matched field tomography.

Several categories of inverse problems fit within this model depending upon the assumptions made about the quantities defined above.

1. Matched field tomography: For the usual matched field localization and tomography models the source is a stationary random process; therefore, incoherent across a frequency band. For this we set $S_s(f) = 1$ and let $\tilde{b}(f)$ have the power spectral density equal of the source, $S_b(f)$. Usually the spectral covariance matrix, $\mathbf{K}_n$ is assumed not

[1]Complex envelopes are used to represent narrowband signals and processes. † is the complex transpose.

to depend upon **a**. In most, but not all, of the literature to date the formulations have been for a single frequency.

2. Ocean acoustic tomography: For full field ocean acoustic tomography the source is coherent across its frequency band. For this $\tilde{b}(f)$ is a single scalar random variable with variance σ_b^2 and $S_s(f)$ is the source signal, *e.g.* an M sequence or an FM sweep. It is usually assumed in this case as well that the noise does not depend upon **a**. In practice, ocean acoustic tomography exploits just travel times of the rays or modes, or primarily the phase information in the signal, and not signal amplitudes; as such it is not exactly a full field approach.
3. Noise tomography: Finally, one can formulate "noise tomography" for estimating environmental parameters. Here, there is no signal, *i.e.* $S(f) = 0$ and one exploits the parametric dependence of spatial structure of the noise covariance.

3. Cramer-Rao bounds

The Cramer-Rao bound is the most extensively used bound in the signal processing literature. It provides a lower bound on the variance of the parameter estimates, $\hat{a}(\mathbf{R})$, based upon the signal structure, *i.e.* the Green's function, and the noise covariance, $\mathbf{K}_n$. For Gaussian signals and noise it leads to results similar to those for minimizing mean square errors. The lower bound on the variance of any unbiased estimator is given by

$$\sigma^2_{\hat{a}(\mathbf{R})} \geq \left[\mathbf{J}^{-1}\right]_{i,i}, \tag{2}$$

where $\mathbf{J}$ is the so called information matrix. For Gaussian signals the elements of this matrix are given by [4] [2]

$$\mathbf{J}_{i,j} = Tr\left[\mathbf{K}_r^{-1}(\mathbf{a})\frac{\partial \mathbf{K}_r(\mathbf{a})}{\partial A_i}\mathbf{K}_r^{-1}(\mathbf{a})\frac{\partial \mathbf{K}_r(\mathbf{a})}{\partial A_j}\right], \tag{3}$$

where Tr is the trace operator and $\mathbf{K}_r(\mathbf{a})$ is the covariance of the total observation vector. In order to give the Cramer Rao bounds for the three categories indicated above several quantities need to be defined. These are:

$$\begin{aligned}
d^2(f,\mathbf{a}) &= \mathbf{G}^\dagger(f,\mathbf{a})\mathbf{K}_n^{-1}(f)\mathbf{G}(f,\mathbf{a}); & d^2(\mathbf{a}) &= \int_{\Delta W} |S_s(f)|^2 d^2(f,\mathbf{a})df;\\
l_i(f,\mathbf{a}) &= \mathbf{G}^\dagger(f,\mathbf{a})\mathbf{K}_n^{-1}(f)\tfrac{\partial \mathbf{G}(f,\mathbf{a})}{\partial A_i}; & l_i(\mathbf{a}) &= \int_{\Delta W} |S_s(f)|^2 l_i(f,\mathbf{a})df;\\
l_{i,j}(f,\mathbf{a}) &= \tfrac{\partial \mathbf{G}^\dagger(f,\mathbf{a})}{\partial A_i}\mathbf{K}_n^{-1}(f)\tfrac{\partial \mathbf{G}(f,\mathbf{a})}{\partial A_j}; & l_{i,j}(\mathbf{a}) &= \int_{\Delta W} |S_s(f)|^2 l_{i,j}(f,\mathbf{a})df.\\
\gamma(f,\mathbf{a})) &= \tfrac{2}{1+S_b(f)d^2(f,\mathbf{a})} & \gamma(\mathbf{a}) &= \tfrac{2}{1+\sigma_b^2 d^2(\mathbf{a})}
\end{aligned} \tag{4}$$

All these terms have physical interpretations. The terms $d^2(f,\mathbf{a})$ and $d^2(\mathbf{a})$ are narrowband and broadband signal to noise ratios (SNR) for the Green's function referenced in the space of the additive noise; they are generalizations of array gain and output SNR expressions. The terms $l_i(f,\mathbf{a})$ and $l_i(\mathbf{a})$ are narrowband and broadband measures of the mean of the

[2] If the parameter vector is modeled as a random vector with a probability distribution, there is an additional term incorporating this *a priori* information.

parameter sensitivity in the same space; $l_{i,j}(f,\mathbf{a})$ and $l_{i,j}(\mathbf{a})$ are narrowband and broadband measures of the convexity of the parameter sensitivity. For example, in classical radar/sonar parameter estimation of range and doppler the broadband quantities $l_i(\mathbf{a})$ and $l_{i,j}(\mathbf{a})$ lead to the well known bounds on range and doppler in terms of derivatives of the radar/sonar ambiguity function.[4] The terms $\gamma(f,\mathbf{a})$ and $\gamma(\mathbf{a})$ are close to unity at low SNR's and become important at high SNR's when phase can be estimated by the receiver. Similar results appear for the sonar/radar range/doppler estimation problem; they are the result of using a complex gain for the signal component.

Implementing the bound involves evaluating the Green's function, $\mathbf{G}(f,\mathbf{a})$, and the gradients, $\frac{\partial \mathbf{G}(f,\mathbf{a})}{\partial a_i}$. The function itself can be found using SAFARI/OASES for range independent problems while adiabatic normal modes can be used when appropriate.[5] The gradients are far more difficult. They can be evaluated directly by perturbing each parameter such as done in the companion paper; alternatively, one can use analytic expressions for the separated Green's function, $G(z,z_s|f,k)$.[6, 7]

This matrix provides important insights into the inversion problems. First, an eigenvector decomposition of $\mathbf{J}$ describes the coupling among the estimation errors of the components of the parameter vector $\mathbf{a}$. This is implemented in the companion paper. [6] Second, one can determine the effective number of degrees of freedom, N_{eff} in the estimate from the eigenvalues of $\mathbf{J}$ by

$$N_{eff} = \frac{(\sum_i \lambda_i(\mathbf{a}))^2}{\sum_i \lambda_i^2(\mathbf{a})} \tag{5}$$

where $\lambda_i(\mathbf{a})$ are the eigenvalues of the matrix with elements

$$\rho_{i,j} = [\mathbf{J}^{-1}]_{i,j}\sqrt{J_{i,i}J_{j,j}}. \tag{6}$$

3.1. MATCHED FIELD TOMOGRAPHY

Matched field tomography assumes a stationary random process with power density spectrum $S_b(f)$; consequently, the processing is incoherent with respect to frequency. Additionally, it is also usually assumed that the noise does not depend upon the parameter vector $\mathbf{a}$; since the noise propagates in the same waveguide as the signal, this is an approximation for matched field tomography; however, the signal dependence is usually dominant. For matched field localization and tomography the elements of the information matrix are given by [3]

$$\begin{aligned} \mathbf{J}_{i,j} &= \int_{\Delta W} df\ S_b^2(f)\gamma(f,\mathbf{a})\,[\ Re\ (d^2(f,\mathbf{a})l_{i,j}(f,\mathbf{a}) - l_i^\dagger(f,\mathbf{a})l_j(f,\mathbf{a})) \\ &+ \gamma(f,\mathbf{a})(Re(l_i(f,\mathbf{a}))\ Re(l_j(f,\mathbf{a}))\ )\ ] \end{aligned} \tag{7}$$

3.2. OCEAN ACOUSTIC TOMOGRAPHY

In full field ocean acoustic tomography the source is coherent and $b(f)$ is a single random variable with variance σ_b^2. It also assumes that the additive noise does not depend upon the

[3] Re = real part.

parameter vector **a**. For ocean acoustic tomography the elements of the information matrix are given by

$$\begin{aligned}\mathbf{J}_{i,j} &= \sigma_b^4\gamma(\mathbf{a})[\,Re\,(d^2(\mathbf{a})l_{i,j}(\mathbf{a}) - l_i^{\dagger}(\mathbf{a})l_j(\mathbf{a})) \\ &+ \gamma(\mathbf{a})\,(Re(l_i(\mathbf{a}))\,Re(l_j(\mathbf{a})\,)\,]\end{aligned} \tag{8}$$

In both Eqs. 4 and 5 the terms on the first line can be associated with the convexity of the "ambiguity function" while the second relates to changes in the received signal to noise ratio due to changes **a**.

These results can be simplifed to find bounds on environmental parameters which can be expressed in terms of the phase and group slownesses. We assume i) an array which can resolve all the propagating modes with a variance σ_i^2, and ii) narrowband signals with energy E_s and centered at f_o with a mean square bandwidth $(\Delta f)^2$

$$E_s = \int_{\Delta W} |S_s(f)|^2 df, \quad f_o = \frac{\int_{\Delta W} f|S_s(f)|^2 df}{E_s}, \quad (\Delta f)^2 = \frac{\int_{\Delta W}(f - f_o)^2|S_s(f)|^2 df}{E_s}$$

The results are for a range invariant channel; however, extension to adiabatic channels are direct. The extension to broadband signals is also straightforward. The first assumption is the most important since it leads to results which may be overly optimistic. It also does not include a large class of bottom interaction problems where it is difficult to sample the modes within the bottom.

We define the following quantities:

$$\alpha_i = \frac{|\Psi_i(z_s, f_o, \mathbf{a})|^2}{\rho_s\sigma_i^2 c_o p_{p_i}(f_o, \mathbf{a})}, \quad C(z_s, f_o, \mathbf{a}) = \sum_{i=1}^{N_p}\alpha_i, \quad P_i = \frac{\alpha_i}{C(z_s, f_o, \mathbf{a})}. \tag{9}$$

where z_s is the source depth, ρ_s is the density at the source depth, $p_{p_i}(f_o, \mathbf{a})$ and $p_{g_i}(f_o, \mathbf{a})$ are the phase and group slownesses of the *ith* mode, and c_o is a reference sound speed. P_i has all the properties of a discrete probability distribution. We use the notation $\overline{x}$ to represent an average over this distribution, or

$$\overline{x} \equiv \sum_{i=1}^{N_p} x_i P_i.$$

The quantity $\sigma_b^2 d^2(\mathbf{a})$ is an output SNR and is given by

$$\sigma_b^2 d^2(\mathbf{a}) = \frac{C(z_s, f_o, \mathbf{a})}{R/\lambda_o} E_s \tag{10}$$

and the elements of the information matrix, $\mathbf{J}_{i,j}$ are given by

$$\begin{aligned}\mathbf{J}_{i,j} = \; & (\sigma_b^2 d^2(\mathbf{a})R)^2\gamma(\mathbf{a}) \\ \times \; & \overline{\left[f_o^2((\frac{\partial p_{p_i}(f_o,\mathbf{a})}{\partial a_i} - \overline{\frac{\partial p_{p_i}(f_o,\mathbf{a})}{\partial a_i}})(\frac{\partial p_{p_i}(f_o,\mathbf{a})}{\partial a_j} - \overline{\frac{\partial p_{p_i}(f_o,\mathbf{a})}{\partial a_j}})) + (\Delta f)^2(\frac{\partial p_{g_i}(f_o,\mathbf{a})}{\partial a_i}\frac{\partial p_{g_i}(f_o,\mathbf{a})}{\partial a_j}) \right]}\end{aligned} \tag{11}$$

This equation has several useful interpretations. First, at high signal to noise ratios (SNR) the elments are proportional to the SNR, or $\sigma_b^2 d^2(\mathbf{a})$, which in turn scales inversely as the range measured in wavelengths. The terms involving the phase and group slownesses present an interesting tradeoff and indicate the relative performance of matched field tomography *versus* ocean acoustic tomography. Both involve second moment like terms; The phase slowness term is amplified by the center frequency; however, it is a variance like term with the mean shift in the phase slowness removed. This is intuitively satisfying since matched field tomography depends upon modal phase interactions and mean shifts do not contribute. The group slowness term is amplified by the signal bandwidth and has the form of mean square shift. Again this is intuitively satisfying since high bandwidth leads to accurate group delay estimates and the mean square measure suggests that a large spread in modal arrival changes has a high sensitivity to parameter changes. Finally, if one is in the high SNR regime, the performance increases with range. The signal level is decreasing with the cylindrical spreading of $1/\sqrt{R}$; this is balanced by the term R scaling the change of the phase and group delays. This extends in range until the output SNR is less than unity. The usefulness of the Cramer Rao bound in this region is often in question: Note that this dependence is very different for the source localization problem.

3.3. NOISE TOMOGRAPHY

Several authors have suggested tomographic inversions based upon parametric dependencies in noise field spectral covariance. The Cramer Rao bounds for this can be derived directly from the general result for term of the information matrix, **J**, by omitting the signal term.

4. Summary

We have derived a unified theory for Cramer-Rao bounds which is applicable to a large number of full field inversion problems.

References

1. Menke, W., *Geophysical Data Analysis: Discrete Inverse Theory, Revised Edition*, Academic Press, Inc., 1985
2. Worcester, P.F., Cornuelle, B.D., and Spindel, R.C., "A review of ocean acoustic tomography: 1987 - 1990," *Reviews of Geophysics Supplement*, pp. 557 - 570, April 1991
3. Baggeroer, A.B., Kuperman, W.A., and Mikhalevsky, P.N., "An overview of matched field processing in ocean acoustics," *IEEE Journal of Oceanic Engineering*, 18(4), pp. 401 - 424, October 1993
4. van Trees, H.L., *Detection, Estimation and Modulation Theory, Part III*, John Wiley and Sons, 1972
5. Jensen, F.B., Kuperman, W.A., Porter, M.B., Schmidt, H., *Computational Ocean Acoustics*, American Inst. of Physics Press, New York, 1994
6. Schmidt, H. and Baggeroer, A.B.,"Physics-imposed resolution and robustness issues in seismo-acoustic parameter inversion," NATO Conference on Full Field Inversion Methods in Ocean and Seismic Acoustics (this meeting)
7. Borodin, V.A., personal communication

PHYSICS-IMPOSED RESOLUTION AND ROBUSTNESS ISSUES IN SEISMO-ACOUSTIC PARAMETER INVERSION

HENRIK SCHMIDT and ARTHUR B. BAGGEROER
Massachusetts Institute of Technology
Cambridge, MA 02139, USA

ABSTRACT Adaptive optimization schemes have become important tools for rapid matched field environment inversion. In spite of several successful applications to both synthetic and real data it has been found that these algorithms often fail to converge to the correct solution even for relatively simple environments. We use the theoretical Cramer-Rao resolution bounds to demonstrate that this lack of robustness is a result of the adaptive use of the local parameter sensitivity to adjust the search space. Thus, the waveguide physics yields highly inhomogeneous and coupled resolution properties within even a limited parameter space, which often forces adaptive optimization schemes to converge on a wrong local minimum of the cost function.

1. Introduction

The use of matched field inversion for environmental parameters generally requires simultaneous estimation of so many parameters that full mapping of the ambiguity or cost function is prohibitive for computational reasons. Therefore a combination with some form of random, global search algorithm is required. In this regard optimization schemes such as *simulated annealing* [1] and *genetic algorithms* [2] have been successfully applied to both synthetic and real data. Nevertheless, others have found that direct use of such schemes are not automatically ensured success [3].

Such lack of robustness is associated with the modifications of the original algorithms, introduced to accelerate convergence. An example is *adaptive simulated annealing* (ASA) of which successful use has been found to require significant manual tuning based on a solid understanding of the fundamental waveguide physics [3].

We here use the theoretical Cramer-Rao bounds to analyze the fundamental parameter resolution properties of the waveguide inversion problem and discuss their effect on the robustness of algorithms such as ASA. In addition, we discuss how the Cramer-Rao bounds can be used *a priori* to improve the robustness for a particular environment inversion problem and, ultimately, as an integral part of new, improved adaptive optimization algorithms.

2. Adaptive Optimization Schemes

The fundamental idea behind the adaptive optimization schemes is to adjust dynamically the search interval for each parameter based on the local resolution. Thus, once a strong minimum of the cost function is found, the search interval is narrowed in accordance with the local sensitivity of the field to parameter perturbations. This procedure often fails

O. Diaschok et al. (eds.), Full Field Inversion Methods in Ocean and Seismo-Acoustics, 85–90.

for the environmental inversion problem [3], primarily due to the parameter coupling and resolution properties imposed by the complex waveguide physics. In the following we use the Cramer-Rao resolution bounds to analyze and explain this phenomenon.

3. Cramer-Rao Bounds

The resolution issue in parameter estimation concerns the accuracy of the estimates, i.e. the local errors that occur at high signal to noise ratios when the correct global maximum of the ambiguity function has been identified. Commonly, this is expressed in terms of the mean square error of the estimates. The mean square error of a particular estimator can be found *a posteriori* for some inversion schemes, e.g. genetic algorithms [2], but in general is difficult to determine. One often can find limits on the performance of any estimator and there are a number of bounds in the literature; the most extensively employed are the Cramer-Rao lower bounds [4, 5]. These bounds are expressed in terms of the inverse of the Fisher information matrix [6],

$$J_{ij}(\mathbf{a}_T) = \operatorname{Tr}\left[\mathbf{K}^{-1}(\mathbf{a}_T)\frac{\partial \mathbf{K}(\mathbf{a}_T)}{\partial a_{Ti}}\mathbf{K}^{-1}(\mathbf{a}_T)\frac{\partial \mathbf{K}(\mathbf{a}_T)}{\partial a_{Tj}}\right]. \quad (1)$$

where $\mathbf{a}_T$ is a vector representing the true values of the parameters, e.g. sound speed, depth and density, and $\mathbf{K}$ is the covariance matrix for the acoustic field on the receiving array. The lower bounds on the estimate are expressed in terms of the elements of the inverse of the Fisher information matrix, $R_{ij} = J_{ij}^{-1}$, which we refer to as the *resolution matrix.* The mean square error of any unbiased estimator of the parameters can be shown to be bounded from below by the corresponding main diagonal term of the of the resolution matrix, or

$$E\left[(\hat{a}_i - a_{Ti})^2\right] \geq R_{ii}(\mathbf{a}_T), \quad (2)$$

where $\hat{\mathbf{a}}$ represents the vector of parameter estimates. Note that the estimated accuracy depends on the actual value of *all* parameters.

It is the resolution represented by Eq. (2) which is used by many adaptive optimization schemes to re-adjust the parameter search space. This procedure ignores, however, the coupling of the parameters, i.e. the uncertainty in associating a change in the acoustic field with a change in one particular parameter. This coupling is represented by the off-diagonal terms in the Cramer-Rao resolution matrix, and as will be discussed below such coupling may severely affect the convergence performance of adaptive schemes.

4. Resolution and Robustness Issues

In the following we will use the Cramer-Rao bounds for a simple, canonical inversion problem to demonstrate the complexity of the associated parameter resolution and coupling properties, and discuss the effect on the robustness of adaptive optimization schemes.

4.1. CANONICAL INVERSION PROBLEM

For the analysis we choose the MFP benchmark problem shown in Fig. 1 [7]. In spite of its simplicity this environment displays several of the most important parameter resolution and coupling properties affecting the inversion robustness. The sound speed at the seabed and in the subbottom as well as the subbottom depth is sought through inversion of the

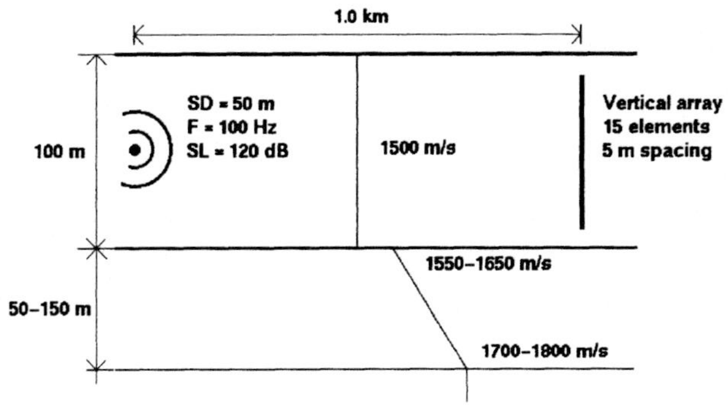

FIGURE 1. Matched field inversion for bottom sound speeds in shallow water environment using vertical line array.

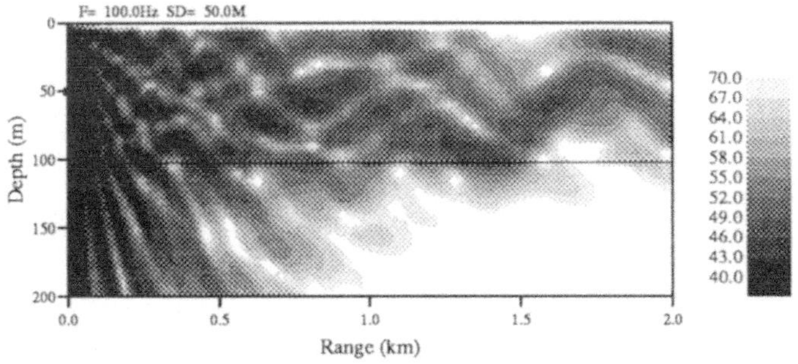

FIGURE 2. Transmission loss contours for 100 Hz point source at depth 50 m in shallow water environment.

acoustic field on a 15-element vertical array at a range of 1 km from a 100 Hz source at mid depth. The search intervals for the parameters are shown in Fig. 1. Contours of the transmission loss for the mean value of the parameters are shown in Fig. 2. The depth-averaged transmission loss at the range of the array is 55 dB, and the ambient noise is assumed to be white, with a spectral level of 50 dB. With a source strength of 120 dB, this translates into an average SNR of 15 dB on the array.

4.2. RESOLUTION VARIABILITY

For the subbottom depth 200 m, Fig. 3 shows contours of the resolution in the (a) seabed and (b) subbottom speed as a function of the actual value of the two unknown parameters. It is clear that the resolution is highly inhomogeneous, in particular for the subbottom speed, ranging from better than 4 m/s at the lower right corner of the graph, to more than 20 m/s in the upper left corner. The physical reason for this is pretty obvious. The upper left corner corresponds to a weak gradient, yielding insignificant refraction of bottom penetrating sound back to the array. In contrast, the strong gradients represented by the lower right corner yield strong refraction, resulting in a higher sensitivity, and therefore better resolution, to the subbottom speed.

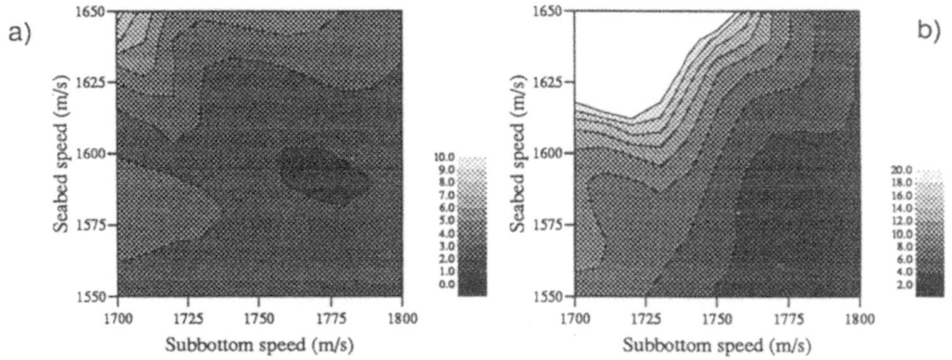

FIGURE 3. Contours in m/s of resolution in (a) seabed speed and (b) subbottom speed, mapped over the parameter space.

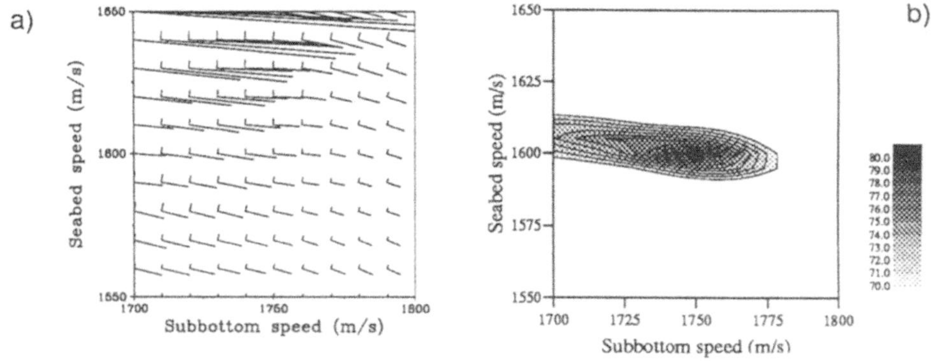

FIGURE 4. Cramer-Rao bounds and ambiguity function for seabed and subbottom sound speed inversion. (a) Eigenvectors of resolution matrix plotted over parameter search space. (b) Maximum likelihood ambiguity function.

Figure 3 represents the diagonal elements of the resolution matrix, R_{ii} and does not reflect the coupled resolution represented by the off-diagonal terms. To illustrate the coupling, we compute the eigenvectors of R_{ij} which represent the main axes of the *error ellipsoid* and map them over the parameter space, as shown in Fig. 4a. Here, the angles of the eigenvectors relative to the parameter axes represent the coupling between the parameters. There is little coupling between these two parameters, and the length of the eigenvectors is consistent with the resolutions shown in Fig. 3. Figure 4b shows the associated ambiguity surface obtained with a maximum-likelihood beamformer for the true parameters being 1600 m/s for the seabed speed and 1750 m/s for the subbottom speed. The shape of the ambiguity mainlobe is consistent with the eigenvectors of the resolution matrix, and it is interesting to observe the asymmetric elongation of the main lobe in the direction of the larger resolution bounds.

The implication of such resolution inhomogeneity for adaptive schemes is as follows: Assume that the search algorithm finds a low cost for speeds in the lower right corner of Fig. 3, corresponding to a sharp gradient. Because of the high resolution it will then narrow down the sound speed search space significantly. Now, assume that the true depth of the subbottom is much smaller, and that the true speeds are somewhere in the upper left corner of the plots in Fig. 3; then, because of the narrow search space, the optimization algorithm

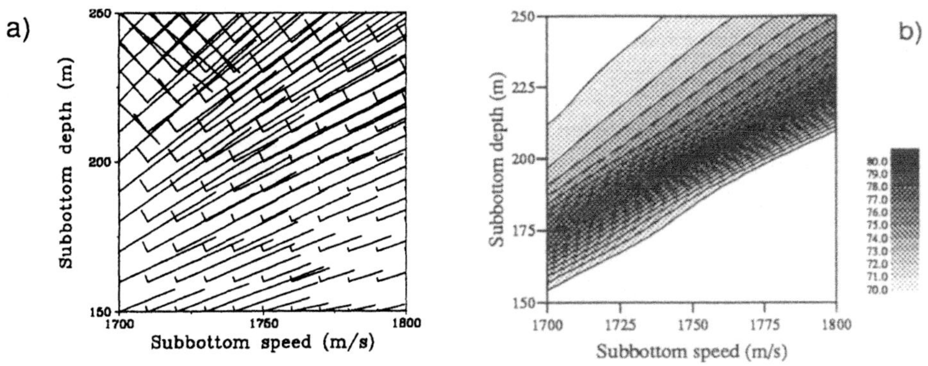

FIGURE 5. Cramer-Rao bounds and ambiguity function for subbottom depth and sound speed inversion. (a) Eigenvectors of resolution matrix plotted over parameter search space. (b) Maximum likelihood ambiguity function.

has very little chance of converging to the right solution. This scenario assumes that the sound speed gradient is the dominant factor and, in fact, the Cramer-Rao bounds provide direct evidence for this as discussed in the next section.

4.3. PARAMETER COUPLING

We next address the resolution and coupling associated with the estimation of the subbottom depth and sound speed, assuming the seabed speed is known to be 1600 m/s. Figure 5a shows the direction and magnitude of the eigenvectors of the resolution matrix. The angle of the eigenvectors is close to 45° indicating strong coupling between these two parameters. Figure 5b shows the associated ambiguity surface obtained with a maximum-likelihood beamformer for the true parameters being a depth of 200 m and sound speed 1750 m/s. Again, the shape of the ambiguity mainlobe is consistent with the eigenvectors of the resolution matrix. It is easily verified that the "ridge" represents parameter combinations yielding a constant sound speed gradient. This indicates that the field is much more sensitive to the sound speed gradient than to the actual values of speed and depth, which makes sense physically.

This is an example of a parameterization with strong coupling in combination with a high aspect ratio for the error ellipse. In such cases the diagonal terms of the resolution matrix are small, consistent with the sharpness of the ridge in Fig. 5b when cut along the individual parameter axes. An adaptive algorithm ignoring the coupling may therefore get stuck anywhere along the ridge, failing to converge to the correct solution. The Cramer-Rao bounds could be used *a priori* for selecting optimal, uncoupled parameterizations. As discussed in the following, this is not always possible, however.

4.4. COUPLING VARIABILITY

Figure 6 shows the eigenvectors of the resolution matrix and the ambiguity function for the seabed speed and the subbottom depth. The Cramer-Rao bounds show strong coupling along the diagonal from upper left to lower right, with decreasing coupling towards the other corners. This is an example of *coupling variability* within the parameter space. Such behavior reduces the possibility of selecting an optimal, uncoupled parameterization *a priori*. The

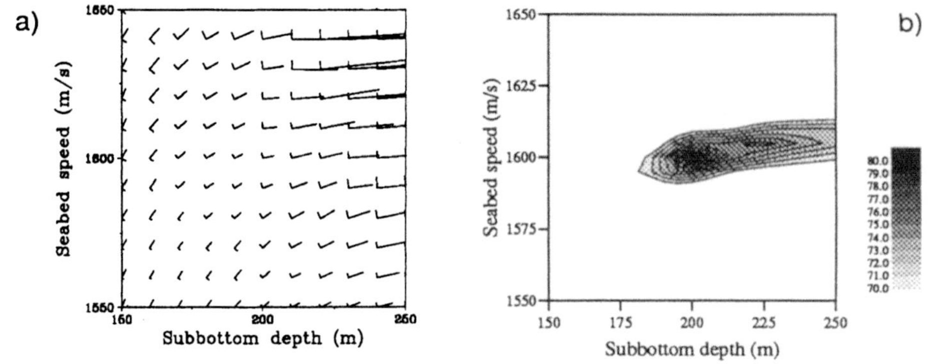

FIGURE 6. Cramer-Rao bounds and ambiguity function for subbottom depth and seabed sound speed inversion. (a) Eigenvectors of resolution matrix plotted over parameter search space. (b) Maximum likelihood ambiguity function.

curved shape of the ambiguity function in Fig. 6b is clearly consistent with the coupling variability.

5. Conclusion

We have used the Cramer-Rao bounds to demonstrate the subtleness of the environmental inversion problem. The complex waveguide physics introduces strong variability of the resolution, and coupling between the parameters, factors severely affecting the robustness of adaptive optimization schemes. It has been demonstrated how the Cramer-Rao bounds can be used to *a priori* analysis of a particular inversion scenario with the objective of selecting an optimal and robust parameterization. Because of inhomogeneous coupling properties this is often not possible and ultimately one should develop new adaptive schemes based on not only the individual parameter resolutions, but also the coupling terms. Such adaptive use of the full resolution matrix would be consistent with the use of the local gradient in linear inversion schemes. This work was supported by ONR, Arctic Program Office.

References

1. Collins, M.D., Kuperman, W.A and Schmidt, H. (1992) Non-linear inversion for ocean-bottom properties, *J. Acoust. Soc. Am.*, **92**(5), 2770–2783.
2. Gerstoft, P. (1994) Global inversion using genetic algorithms, *J. Acoust. Soc. Am.*, **95**(2), 770–782.
3. Siedenburg, C., Lehtomaki, N., Arvelo, J., Rao, K. and Schmidt, H. (1994) Iterative full-field inversion using simulated annealing, *This volume.*
4. Van Trees, H.L. (1970) *Detection, Estimation and Modulation Theory, Part III*, John Wiley and Sons, New York.
5. Baggeroer, A.B. and Schmidt, H. (1994) Parameter estimation theory bounds and the accuracy of full field inversions. *This Volume.*
6. Baggeroer, A.B., Kuperman, W.A. and Schmidt, H. (1988) Matched field processing: Source localization in correlated noise as an optimum parameter estimation problem, *J. Acoust. Soc. Am.*, **82**(2), 571–587.
7. Porter, M.B. and Tolstoy, A. (1994) The matched-field processing benchmark problems, *J. Comp. Acoust.*, (To appear).

STATISTICAL APPROACH TO OCEAN ACOUSTIC TOMOGRAPHY, CRAMER-RAO BOUNDS FOR ACCURACY OF SOUND-SPEED FIELD RECONSTRUCTION

V.V. BORODIN and G.R. MINASIAN
N.N. Andreyev Acoustics Institute, 117036, Shvernik str. 4, Moscow, Russia

ABSTRACT This paper describes a statistical approach to ocean acoustic tomography. Deterministic, quasi-deterministic and random Gaussian signals in Gaussian noise background are examined. The maximum-likelihood approach is used to estimate the sound-speed. An expression of the Green's function variational derivative with respect to sound speed is applied to find the maximum-likelihood estimate. Fisher information matrices and the Cramer-Rao lower bounds for the various signal models are derived.

1. Introduction

It is assumed that the sound field in the ocean is generated by M sources located at the points $X_1, \ldots, X_M$ emitting signals $s,(t), \ldots, S_M(t)$. We will consider three cases of the transmitted signal statistics. In the first case, $s_1(t)$ is a known deterministic signal. In the second case, the sources transmit a quasi-deterministric signal and in the third case, the sources transmit zero-mean Gaussian noise with covariance matrix $K_S(i, j, t-t')$. N omnidirectional receivers are located at points $y_1, \ldots, y_N$. The noise field is assumed to be a stationary zero-mean Gaussian field with cross-spectral density matrix $K_N(s, r, \omega)$. The observation model is

$$P(y_S, t) = P_N(y_S, t) + \sum_{j=1}^{M} \gamma(y_S, x_j, t) s_j(t). \qquad \begin{array}{l} S = 1, \ldots, N \\ j = 1, \ldots, M \;, \\ t \in [O, T] \end{array} \tag{1}$$

where $\gamma(y_S, x_j, t)$ is the Green's function of the oceanic waveguide. The observation model in the frequency domain is

$$P(y_S, \omega_n) = P_N(y_S, \omega_n) + \sum_{j=1}^{M} G(y_S, x_j, \omega_n) s_j(\omega_n). \qquad \omega_n = 2\pi n/T, \tag{2}$$

where $P(y_S, \omega_n)$, $P_N(y_S, \omega_n)$, $G(y_S, x_j, \omega_n)$, and $s_j(\omega_n)$ denote complex spectra of the appropriate quantities.

The Green's function $G(y, x, \omega)$ satisfies the Helmholtz equation

$$(\Delta + \frac{\omega^2}{c^2(y)}) G(y, x, \omega) = \delta(y - x) \tag{3}$$

O. Diaschok et al. (eds.), Full Field Inversion Methods in Ocean and Seismo-Acoustics, 91–95.

with the appropriate, boundary conditions at the surface and the bottom, and the conditions of cancellation at infinity.

The sound speed is assumed to be

$$c(x) = c_0(z) + \tilde{c}(x),$$

where $c_0(z)$ is a reference known sound speed profile depending only on depth, and $\tilde{c}(x)$ is an unknown perturbation depending, in general on three coordinates. Tomography is aimed at estimating the field $\tilde{c}(x)$. The perturbation field can be written as an expansion in basis functions $\varphi_\alpha(x)$

$$\tilde{c}(x) = \sum_\alpha C_\alpha \varphi_\alpha(x). \tag{4}$$

The tomography problem is then reduced to the estimation of the unknown coefficients C_α. These functions may be eigenfunctions of the equations of motion, empirical orthogonal functions or any other convenient set of functions.

2. Maximum likelihood equations

The optimum decision statistics for the tomographical problem is the log-likelihood ratio. The log-likelihood function of observations for a deterministic signal is

$$L \sim \sum_{\omega>0} \langle K_N^{-1}(\omega)(P(\omega) - \sum_{j=1}^{M} G(x_j,\omega)s_j(\omega)), (P(\omega) - \sum_{j=1}^{M} G(x_j,\omega)s_j(\omega)) \rangle, \tag{5}$$

where broken brackets denote the inner product $\langle u, v\rangle = \sum_{s=1}^{N} u_s v_s^*$, N is the number of receivers, $K_N(\omega) = K_N(s,r,\omega)$ is the noise cross-spectral density matrix, $P(\omega)$ and $G(x_j,\omega)$ are vectors of the observed and modeled signals. For a signal with unknown phase the optimum statistic is

$$L \sim \sum_{\omega>0}\Big(\langle K_N^{-1}(\omega)P(\omega), P(\omega)\Big\rangle + \sum_{i,j}\Big\langle K_N^{-1}(\omega)G(x_j,\omega), G(x_i,\omega)\Big\rangle s_j^*(\omega)s_i(\omega)\Big)$$
$$-2|\sum_{\omega>0}\sum_i \Big\langle K_N^{-1}(\omega)P(\omega), G(x_i,\omega)\Big\rangle s_i(\omega)|. \tag{6}$$

If the phase and amplitude of the transmitted signal are unknown, then we form the joint maximum likelihood estimation (MLE) of amplitude and phase

$$L \sim \sum_{\omega>0}\langle K_N^{-1}(\omega)P(\omega), P(\omega)\rangle - \frac{|\sum_{\omega>0}\langle K_N^{-1}(\omega)P(\omega), G(x_i,\omega)\rangle s_i(\omega)|^2}{\langle K_N^{-1}(\omega)G(x_j,\omega), G(x_i,\omega)\rangle s_j^*(\omega)s_i(\omega)}. \tag{7}$$

For Gaussian signals the log-likelihood ratio has the form

$$L = \sum_{\omega>0}\Big(\langle (K_N^{-1}(\omega) - K_{\Sigma}^{-1}(\omega)P(\omega), P(\omega)\rangle + \ln\det(K_N(\omega)K_{\Sigma}^{-1}(\omega))\Big) \tag{8}$$

where $K_{\Sigma}(\omega) = K_N(\omega) + \sum_{i,j} G(x_s, x_j, \omega)G^*(x_r, x_i, \omega)æ_{ij}(\omega)$.

In the maximum likelihood method we need to differentiate the expressions Eqs.(5–8) with respect to C_α and to find zeroes of the derivative

$$\frac{\partial L}{\partial C_\alpha} = 0. \tag{9}$$

Only the Green's function depends on coefficients C_α by expressing $\partial G(x_j, \omega)/\partial C_\alpha$ in terms of the variational derivative

$$\frac{\partial G(x, x_0, \omega)}{\partial \bar{c}(x')} = \frac{2\omega}{c^3(x)} G(x, x', \omega) G(x', x_0, \omega), \tag{10}$$

we obtain

$$(\frac{\partial L}{\partial C_\alpha}) = \sum_{i=1}^{M}\sum_{j=1}^{M}\int \frac{d^3x'}{c(x)} \sum_{\omega>0} \omega^2 \langle K_N^{-1}(\omega) G(x_i, \omega), \mathcal{F}(\omega, x', x_j)\rangle \varphi_\alpha(x') s_i^*(\omega) s_j(\omega)$$

$$-\sum_{j=1}^{M}\int \frac{d^3x'}{c(x)} \sum_{\omega>0} \omega^2 \langle K_N^{-1}(\omega) P(\omega), \mathcal{F}(\omega, x', x_j)\rangle \varphi_\alpha(x') s_j(\omega) = 0, \tag{11}$$

where $\mathcal{F}(\,)$ is defined as

$$\mathcal{F}(\omega, x, x', x_0) = \frac{2\omega^2}{c^2(x)} G(x, x', \omega) G(x', x_0, \omega).$$

Differentiating the expressions (6) and (7) with respect to C_α, we obtain the equations of MLE for these coefficients

$$2\mathrm{Re}\int \frac{\varphi_\alpha(x)}{c(x)} \Big(\sum_{\omega>0}\sum_{i,j} \Big\langle K_N^{-1}(\omega) G(x_j, \omega), \mathcal{F}(\omega, x, x_i) \Big\rangle s_j^*(\omega) s_i(\omega) - $$

$$- \frac{\sum_{\omega>0}\sum_i \langle K_N^{-1}(\omega) P(\omega), \mathcal{F}(\omega, x, x_i)\rangle s_i(\omega)}{|\sum_{\omega>0}\sum_i \langle K_N^{-1}(\omega) P(\omega), G(x_i, \omega)\rangle s_i(\omega)|} \Big) d^3x = 0 \tag{12}$$

and

$$2\mathrm{Re}\int \frac{\varphi_\alpha(x)}{c(x)} \Big(\frac{|\sum_{\omega>0}\sum_i \langle K_N^{-1}(\omega) P(\omega), G(x_i, \omega)\rangle s_i(\omega)|^2}{\sum_{i,j} \langle K_N^{-1}(\omega) G(x_j, \omega), G(x_i, \omega)\rangle s_j^*(\omega) s_i(\omega)}$$

$$\times \sum_{\omega>0}\sum_{i,j} \langle K_N^{-1}(\omega) G(x_j, \omega), \mathcal{F}(\omega, x, x_i)\rangle s_j^*(\omega) s_i(\omega)$$

$$- \sum_{\omega>0} \langle K_N^{-1}(\omega) P(\omega), \mathcal{F}(\omega, x, x_i)\rangle s_i(\omega) \Big) = 0. \tag{13}$$

For a Gaussian signal, using the expression for a variational derivative of the covariance operator $K_S(x)$

$$\frac{\partial K_S(\omega)}{\partial c(x)} = \frac{1}{c(x)} [K_S(x) + K_S^{+}(x)],$$

we obtain the maximum likelihood equations

$$\int \sum_{w>0} \Big(\langle ((K_{\Sigma}^{-1}(\omega)(K_S(x) + K_S^{+}(x))K_{\Sigma}^{-1}(\omega))P(\omega), P(\omega)\rangle \Big) -$$

$$Sp(K_{\Sigma}^{-1}(\omega)(K_S(\omega,x) + K_S^{+}(\omega,x)))\Big) \frac{\varphi_\alpha(x)}{c(x)} \mathrm{d}^3x = 0, \tag{14}$$

where $K_S(x) = \sum_{i,j} |\mathcal{F}(\omega, x, x_i) >< G(\omega, x_j)|\, æ_{i,j}(\omega)$.

Solving this non-linear set of equations, we obtain the estimates of C_α.

3. Cramer-Rao bound

It is well known that the MLE is asymptotically unbiased and efficient. The covariance matrix of the MLE errors is asymptotically equal to the inverse of the Fisher information matrix (FIM), which in the case of a deterministic signal has the form

$$I_{\alpha\beta} = E[\frac{\partial L}{\partial C_\alpha}\frac{\partial L}{\partial C_\beta}]$$

$$= \sum_{i=1}^{M}\sum_{j=1}^{M} \int\int \frac{\mathrm{d}^3x\mathrm{d}^3x'}{c(x)c(x')} \sum_{\omega>0} \langle K_N^{-1}\mathcal{F}(\omega,x,x_i), \mathcal{F}(\omega,x',x_j)\rangle \varphi_\alpha(x)\varphi_\beta(x')s_i^*(\omega)s_j(\omega). \tag{15}$$

Denoting

$$I(x,x') = \sum_{i=1}^{M}\sum_{j=1}^{M}\sum_{\omega>0} \langle K_N^{-1}\mathcal{F}(\omega,x,x_i), \mathcal{F}(\omega,x',x_j)\rangle s_i^*(\omega)s_j(\omega) \tag{16}$$

the FIM can be written as

$$I_{\alpha\beta} = \int\int \frac{\varphi_\alpha(x)\varphi_\beta(x')}{c(x)c(x')} I(x,x')\mathrm{d}^3x\mathrm{d}^3x,$$

$I(x, x')$ is the correlation beween signals scattering from inhomogeneities x and x', and at the same time, $I(x, x')$ coincides with the ambiguity function in the problem of point scatterer localization in a bistatic active sonar scenario.

In the case of signal with unknown phase the FIM can be divided into blocks: $I_{\alpha\beta}$ is a FIM of estimates of the coefficients C_α. The non-diagonal block $I_{\alpha\varphi}$ and $I_{\varphi\alpha}$ contains the cross-information of the coefficients C_α and phase ϕ. The diagonal block $I_{\varphi\varphi}$ represents the information quanity of the phase estimate. Inverting this block – matrix we obtain a new FIM $I_{\alpha\beta}$ for estimates of $C\alpha$ for the case of a signal with unknown phase

$$\hat{I}_{\alpha\beta} = I_{\alpha\beta} - I_{\alpha\varphi}I_{\varphi\varphi}^{-1}I_{\varphi\beta} \tag{17}$$

If a transmitting system has seveal sources emitting different signals with phases ϕ_m, FIM has the form

$$\hat{I}_{\alpha\beta} = I_{\alpha\beta} - I_{\alpha\varphi_m}I_{\varphi_m\varphi_n}^{-1}I_{\alpha_n\beta} \tag{18}$$

where $I_{\varphi_m\varphi_n}$ is the FIM of the phase estimates, $I_{\alpha\varphi}$ contains the cross-information of the estimates C_α and φ_m. The expression for FIM of the phase estimate is

$$I_{\varphi_m\varphi_n} = \sum_{\omega}\sum_{i_m}\sum_{j_m} \langle K_N^{-1}G(x_{i_m}), G(x_{j_n})\rangle s_{i_n}^*(\omega)s_{j_m}(\omega). \tag{19}$$

and the expression for the cross-information is

$$I_{\alpha\varphi_n} = i\sum_{\omega}\sum_{\substack{n\\m}}\sum_{i_m}\sum_{i_n}\langle K_N^{-1}\mathcal{F}(\omega, x, x_{i_n}), G(x_{i_m})\rangle s_{i_n}^*(\omega)s_{i_m}(\omega)\varphi_\alpha(x)\mathrm{d}^3x$$

where i_m is the number of sources in the the array. $I_{\varphi_m\varphi_n}$ represents the cross-correlations between signals emitted by the mth and nth arrays. If the arrays emit orthogonal signals, the matrix $I_{\varphi_m\varphi_n}$ becomes diagonal. $I_{\alpha\varphi_m}$ represents the cross-correlation between signals scattered by an inhomogeneity located at point X_α and the direct signal emitted by the mth source. Thus the kernel of the Fisher information operator is

$$I(x,x') = \sum_{\omega>0}\sum_{i,j}\langle K_N^{-1}\mathcal{F}(\omega, x, x_i), \mathcal{F}(\omega, x', x_j)\rangle s_i^*(\omega)s_j(\omega) -$$
$$\frac{\sum_{\omega>0}\sum_{i,j}\langle K_N^{-1}\mathcal{F}(\omega, x, x_i), G(x_j)\rangle s_i^*(\omega)s_j(\omega)\sum_{\omega>0}\sum_{i,j}\langle K_N^{-1}G(x_j), \mathcal{F}(\omega, x', x_i)\rangle s_i(\omega)s_j^*(\omega)}{\sum_{\omega>0}\sum_{i,j}\langle K_N^{-1}G(x_i), G(x_j)\rangle s_i^*(\omega)s_j(\omega)}. \tag{20}$$

For a Gaussian signal model FIM has the form

$$I_{\alpha\beta} = \int\int \varphi_\alpha(x)\varphi_\beta(x')\sum_{\omega>0} Sp\Big\{K_\Sigma^{-1}(\omega)\frac{\delta K_S}{\delta c(x)}K_\Sigma^{-1}(\omega)\frac{\delta K_S}{\delta c(x')}\Big\}\mathrm{d}^3x\mathrm{d}^3x'.$$

Using the expression for $\delta K_S(x)/\delta c(x)$ we obtain

$$\begin{aligned}I_{\alpha\beta} = \int\int \frac{\varphi_\alpha(x)\varphi(x')}{c(x)c(x')}\sum_{\omega>0} 2\omega^4 \mathrm{Re} Sp\sum_{ijlm}\Big(&\langle K_N^{-1}(\omega)G(x_i), \mathcal{F}(\omega, x, x_l)\rangle\\ &\times\langle K_N^{-1}(\omega)G(x_j), \mathcal{F}(\omega, x', x_m)\rangle + \langle K_N^{-1}(\omega)\mathcal{F}(\omega, x, x_l)\mathcal{F}(\omega, x', x_i)\rangle\\ &\times\langle K_N^{-1}(\omega)G(x_j), G(x_m)\rangle\Big)æ_{lj}(\omega)æ_{lm}(\omega)\mathrm{d}^3x\mathrm{d}^3x'.\end{aligned} \tag{21}$$

The quantity $\langle K_N^{-1}(\omega)\mathcal{F}(\omega, x, x_l)\mathcal{F}(\omega, x', x_i)\rangle$ represents the correlation of the field scattered by inhomogeneities located at the points x and x'. The quantity $\langle K_N^{-1}(\omega)G(x_j), \mathcal{F}(\omega, x', x_m)\rangle$ is the correlation of the direct field with the field scattered by inhomogeneities located at the point x.

STATISTICAL FORMULATION OF RAY, INTERFERENCE AND MODE ACOUSTIC TOMOGRAPHY AND ITS POTENTIAL ACCURACY EVALUATION

V.V. BORODIN and G.R. MINASIAN
N.N. Andreyev Acoustics Institute,
117036, Shvernik str. 4, Moscow, Russia

ABSTRACT This work is a continuation of previous work [1]. The cases of tomography of the global, large-scale and small-scale inhomogeneities are considered. Fisher information matrices of the estimates of travel times of rays (modes) are derived. The simulation results of potential accuracy of sound-speed field reconstruction are presented.

1. Global inhomogeneities

Analysis conducted in the present work shows that, on the scale of the inhomogeneities we want to restore, the wave approach naturally leads to all other forms of the tomographical problem. We first consider global inhomogeneities, the sizes of which exceed a ray cycle for a given mean sound-speed profile. In this case, Green's function is conveniently presented in the form of the sum of adiabatical normal modes

$$G(x,x_0)=\sum_n g_n(r,r_0)u_n(r,z)u_n(r_0,z_0), \tag{1}$$

where $u_n(r,z)$ are the local normal modes, and $g_n(r,r_0)$ is calculated by the method of "vertical modes and horizontal rays",

$$g_n(r,r_0)=e^{i\int_{r_0}^{r}k_n(r')\mathrm{d}r'}V_n(l,\partial) \tag{2}$$

where $k_n(r)$ is the local modal wave number, and the functions $V_n(l,\partial)$ satisfy the appropriate parabolic equations. Substituting (2) into (1) and using the WKB approximation for $u_n(r,z)$ and rejecting in the integral the fast-oscillating terms connected with cross-mode interference, we obtain

$$\frac{\partial G(x_1,x_o)}{\partial C_\alpha}=i\omega\sum_n g_n(r_1,r_0)u_n(r_1,z_1)u_n(r_0,z_0)\int_{z_n}\frac{\varphi_\alpha(x)\mathrm{d}S_n}{c^2(x)}, \tag{3}$$

where the functions $\varphi_\alpha(x)$ characterize the spatial size of the inhomogeneities we want to restore. Note that we would achieve this result by differentiating the phases of normal modes with respect to C_α,

$$\frac{\partial}{\partial C_\alpha}\int_{r_0}^{r_1}k_n(r)\mathrm{d}l=-\omega\int_{r_0}^{r_1}\frac{\varphi_\alpha(x)\mathrm{d}S_n}{c^2(x)}.$$

O. Diaschok et al. (eds.), Full Field Inversion Methods in Ocean and Seismo-Acoustics, 97–101.

Thus, it is shown that the main contribution to the derivative of Green's function with respect to C_α for global inhomogeneities introduces a phase or temporal structure (for a non-tone transmitted signal) of adiabatical normal modes but not an amplitude structure. Hence, to define the global inhomogeneities, it is possible without loss of information to measure only times of arrival of the normal modes for reconstruction of the sound speed.

Assume the signal bandwidth is large enough that the separate normal modes are resolved in time, and at the same time narrow enough so that it is possible to neglect a dependence on frequency of trajectories of Brulluen's rays. Then, the expression for FIM is

$$I_{\alpha\beta} = \sum_n i_n \int_{z_n} \frac{\varphi_\alpha(x)\mathrm{d}S}{c^2(x)} \int_{z_n} \frac{\varphi_\beta(x)\mathrm{d}S}{c^2(x)}, \tag{4}$$

where

$$i_n = \sum_{\omega>0} \sum_{a,r} \sum_{i,j} K_N^{-1}(\omega, s, r) G_n(x_s, x_i, \omega) G_n^*(x_r, x_j, \omega) \omega^2 s_i^*(\omega) s_j(\omega)$$

and $G_n(x_s, x_i, \omega) = g_n(r_s, r_i) u_n(r_s, z_s) u_n(r_i, z_i)$.

The quantities i_n represent the Fisher information of the travel-time estimates of the normal modes.

For a stationary Gaussian signal we will use the formula [1,(21)] for FIM. Having substituted into it expression (3), we obtain

$$I_{\alpha\beta} = \sum_{\omega>0} \sum_{n,m} R_n(\omega) R_m(\omega) \omega^2 [\int_{z_n} \frac{\varphi_\alpha(x)\mathrm{d}S}{c^2(x)} - \int_{z_m} \frac{\varphi_\alpha(x)\mathrm{d}S}{c^2(x)}] \times$$
$$\times [\int_{z_n} \frac{\varphi_\beta(x)\mathrm{d}S}{c^2(x)} - \int_{z_m} \frac{\varphi_\beta(x)\mathrm{d}S}{c^2(x)}], \tag{5}$$

where $R_n(\omega) = \sum_{s,r} \sum_{i,j} K_N^{-1}(\omega, s, r) G_n(x_s, x_i, \omega) G_n^*(x_r, x_j, \omega) æ_{ij}(\omega)$ is the signal-to-noise ratio (SNR) for nth normal mode, and $\sum_{\omega>0} R_n(\omega) R_m(\omega)^2 \omega = i_{nm}$ is the Fisher information of the estimate of the delay between signals traveling on the nth and mth normal modes.

2. Large-scale inhomogeneities

Now, we will examine the case of large-scale inhomogeneities such that the typical scale of functions $\varphi_\alpha(x)$ are much larger than the Fresnel's zones. In this case, the integral in the expression [1(10)] for $\partial G(y_s, x_i, \omega)/\partial C_\alpha$ can be calculated by the method of stationary phase,

$$\frac{\partial G(y_s, x_i, \omega)}{\partial C_\alpha} = \frac{\omega}{2i} \sum_{\mu=1}^{M} A_\mu(y_s, x_i) e^{-i\omega t_\mu(y_s, x_i)} \int_{z_\mu} \frac{\varphi_\alpha(x)}{c^2(x)} \mathrm{d}S, \tag{6}$$

where z_μ is the trajectory of the direct ray connecting y_s and x_i, A_μ the amplitude of the signal and t_μ the travel time of the signal of the μth ray. Substituting this expression into [1,(15)] for FIM for a deterministic signal, we obtain

$$I_{\alpha\beta} = \sum_{n,m} \sum_{\mu mn} i_{\mu nm} \int_{z_\mu} \frac{\varphi_\alpha(x)}{c^2(x)} \mathrm{d}S \int_{z_\mu} \frac{\varphi_\beta(x)}{c^2(x)} \mathrm{d}S. \tag{7}$$

Here, $i_{\mu nm} = \sum_{\omega>0} R_{\mu nm}(\omega)\omega^2$ is the Fisher information of the estimates of times of arrival of rays. $R_{\mu nm}$ is the SNR at the appropriate correlation peak. We assume that sizes of receiving and transmitting arrays are less than a Fresnel zone. Here, m and n are the number of arrays, μnm the μth ray connecting the mth and nth arrays and z_μ its trajectory.

For a signal with unknown initial phase, the expression for FIM [1, (17–19)] can be deduced to

$$\hat{I}_{\alpha\beta} = I_{\alpha\beta} - R\sum_{\omega}\sum_{\mu(m,n)} r_{\mu nm}(\omega)\omega \int_{z_\mu} \frac{\varphi_\alpha(x)}{c^2(x)}\mathrm{d}S \sum_{\omega}\sum_{\mu(m,n)} r_{\mu nm}(\omega)\omega \int_{z_\mu} \frac{\varphi_\beta(x)}{c^2(x)}\mathrm{d}S, \tag{8}$$

where $R = \sum_{\mu(m,n)}\sum_\omega R_{\mu nm}(\omega)$ is the total SNR, and $r_{\mu nm}(\omega) = R_{\mu nm}(\omega)/R$.

Finally, for a Gaussian signal, FIM has the form

$$I_{\alpha\beta} = \sum_{n,m}\sum_{\mu mn}\sum_{\nu_{mn}}\sum_{\omega} R_{\mu nm}(\omega)R_{\nu mn}(\omega)\omega^2\left[\int_{z_\mu} \frac{\varphi_\alpha(x)}{c^2(x)}\mathrm{d}S - \int_{z_\nu} \frac{\varphi_\alpha(x)}{c^2(x)}\mathrm{d}S\right]\times$$
$$\times\left[\int_{z_\mu} \frac{\varphi_\beta(x)}{c^2(x)}\mathrm{d}S - \int_{z_\nu} \frac{\varphi_\beta(x)}{c^2(x)}\mathrm{d}S\right]. \tag{9}$$

Note that the integrals $\int_{z_\mu} \frac{\varphi_\beta(x)}{c^2(x)}\mathrm{d}S$ are derivations of $\partial t_\mu/\partial c_\alpha$, and $i_{\mu nm}$, $\sum_\omega R_{\mu nm}(\omega)R_{\nu nm}(\omega)\omega^2$ define variances of the estimates of travel time and travel time difference, $i_{\mu nm} = D^{-1}_{\mu nm}$;$\sum_\omega R_{\mu nm}(\omega)R_{\nu nm}(\omega)\omega^2 = D^{-1}_{\mu\nu nm}$.

Thus, expression (9) can be rewritten as

$$I_{\alpha\beta} = \sum_{n,m}\sum_{\mu mn} D^{-1}_{\mu mn}\frac{\partial t_\mu}{\partial c_\alpha}\frac{\partial t_\mu}{\partial c_\beta} : I_{\alpha\beta} = \sum_{n,m}\sum_{\mu mn} D^{-1}_{\mu\nu mn}\frac{\partial(t_\mu - t_\nu)}{\partial c_\alpha}\frac{\partial(t_\mu - t\nu)}{\partial c_\beta}. \tag{10}$$

In such a form, expressions for FIM have a clear physical meaning. According to expression (10), the only parameter in the sound field which carries information about the characteristics of the medium is either signal travel time (for a deterministic signal) or travel time difference (for a random stationary signal). FIM expresses the accuracies of the estimates of these quantities through their derivative with respect to c_α.

3. Small-scale inhomogeneities

Next, we examine the case of smaller-scale inhomogeneities where the size of variability of the function $\varphi_\alpha(x)$ is less than Fresnel's zones of the appropriate ray congruentions. A kernel $I(x, x')$ can be expressed, as we have already shown [1], through spatial-frequency correlation of fields scattering by inhomogeneities located at the points x and x'. If there is a developed multipath, this correlation has a sharp maximum when $x = x'$. The sharpness of the maximum is defined by the number of multipaths. Indeed, the more complex the interference structure of the field at the receiving point, the more sensitive is the field to small deviations from the initial positions of the scatterers. In the vicinity of the maximum, $I(x, x')$ can be approximated by the Gaussian function

$$I(x, x') \simeq I(x, x)e^{-\frac{1}{2}\langle D^{-1}(x)(x-x'),(x-x')\rangle} \tag{11}$$

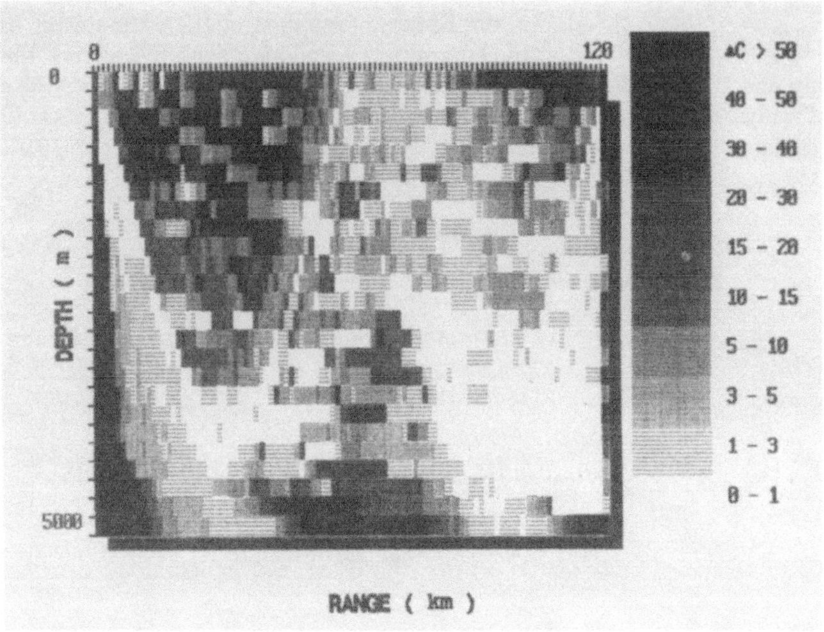

Figure 1. Range separation

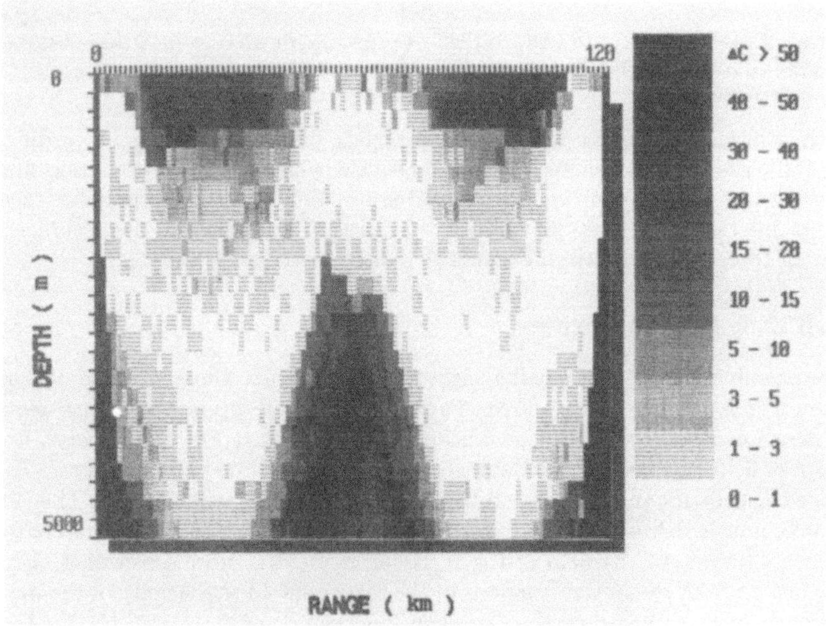

Figure 2. Depth separation

The matrix $D^{-1}(x)$ is

$$D^{-1}(x) = \left.\frac{\partial^2 lnI(x,x')}{\partial x \partial x'}\right|_{x=x'}. \tag{12}$$

$D^{-1}(x)$ is similar to a covariance matrix MLE of point scatterer coordinates in active sonar. In our case, $D(x)$ defines a resolving power or the amount of detail with which we can restore the sound speed. The resolution depends on transmitter/receiver arrays configuration, on the transmitter and receiver mutual location, and on a bandwidth of the transmitting signal. The larger the array size and the wider the bandwidth, the higher the resolution.

Now, choosing a simplest bases $\varphi_\alpha(x)$ in the form of indicative functions of a rectangular parallelipiped, we obtain an expression for the information matrix of the coefficients C_α. The coefficients C_α are the values of the sound speed inside the α-th cell.
If we choose a cell size essentially exceeding the resolving power of the sonar, then the non-diagonal elements of $I_{\alpha\beta}$ become zero, and the expression for the diagonal elements can be reduced to the form

$$\begin{aligned} I_{\alpha\alpha} &= \int_{V_\alpha}\int_{V_\alpha} \frac{I(x,x)}{c^2(x_0)} e^{-\frac{1}{2}\langle D^{-1}(x)(x-x'),(x-x')\rangle} \mathrm{d}^3 x \mathrm{d}^3(x-x') \\ &= \frac{I(x_\alpha,x_\alpha)V_\alpha(2\pi)^3}{c^2(x_\alpha)}. \end{aligned} \tag{13}$$

Then, the variance of the sound speed estimate is

$$D_c(x_\alpha) = \frac{c^2(x_\alpha)}{I(x_\alpha, x_\alpha V_\alpha(2\pi)^{3/2}\sqrt{\det D(x_\alpha)}}. \tag{14}$$

The quantity $\sqrt{\det D(x_\alpha)}$ represents a volume of the ellipsoid of errors for an estimate of the point scatterer coordinates.

4. Simulation results

The unperturbed sound-speed profile was chosen as the one existing in oceans with the sound channel axis at a depth of 1.5 km. The bottom depth was 5.2 km. Receiving and transmitting arrays were vertical linear arrays 5 m long. The frequency for the deterministic signal was 3 kHz, the pressure at 1 m was 10^5 Pa, the pulse duration is 10 s. Four pairs of receiver/transmitter array systems were used. The distance between each receiver/transmitter pair was 120 km. In this case the sound field has a clear convergence zone structure. Figures 1 and 2 display a standard deviation ΔC of sound-speed estimates in m/s. In Fig. 1 we used 4 transmitter/receiver pairs separated in range by 3 km (transmitter depth: 200 m; receiver depth: 150 m). In Fig. 2, we used 4 transmitter/receiver pairs separated in depth: at 10 m, 400 m, 900 m and 1200 m. As shown in the figures, the sound speed has a clear convergence zone structure. The sound speed along the ray congruention connecting transmitter and receiver was restored well, with $\Delta C \sim$ 1–3 m/s. But, away from the ray congruention, there are large errors and the sound speed cannot be restored.

References

1. Borodin, V.V. and Minasian, G.R. (1994) "Statistical approach to ocean acoustic tomography, Cramer-Rao bounds for accuracy of sound-speed field reconstruction", (in this issue).

PRECISION AND SENSITIVITY OF COHERENT AND INCOHERENT INVERSION METHODS: A COMPARATIVE STUDY

M. JOÃO RENDAS
I3S/CNRS URA 1376, Université de Nice Sophia Antipolis
*Ave. Albert Einstein, Les Lucioles 1, 06560 Valbonne, France**

GEORGES BIENVENU
THOMSON SINTRA ASM
BP. 157, 06903 Sophia Antipolis Cedex, France.

and

JOSÉ M. F. MOURA
Dep. Elect. & Comp. Eng., Carnegie Mellon University
5000 Forbes Ave., Pittsburgh, USA.

ABSTRACT This paper presents a comparative study of the precision and sensitivity of coherent and incoherent inversion methods to modeling errors. It is based on the use of performance analysis tools: (i) the Cramer Rao lower bound, that establishes a bound on the precision of parametric estimates; (ii) a generalization of the classical ambiguity function that assesses the global performance under incorrect modeling assumptions, e.g. introduction of biases. Results showing the behaviour of the two methods are presented for specific examples.

1. Introduction

Acoustic tomography systems exploit the structure of the underwater propagation channel to estimate oceanographic parameters of interest. The accuracy of the propagation model used is crucial to the performance of these systems. In this paper, we consider two kinds of methods. *(i) Fully coherent* methods, that use directly the signals captured on the sensors of the receiving array: the so-called "matched field" approach. These methods are certainly the most powerful, taking advantage of the whole information contained in the acoustic field as sampled by the sensors. *(ii)* Methods that use only a set of selected features of the acoustic field, such as the time delays between the different paths that propagate from the source to the receiving points. These methods are *a priori* less performing than the matched field method, they are often used, however, since they are expected to be less sensitive to propagation modeling mismatches.

This paper addresses this question, comparing the **sensitivity to modeling errors** of the **two approaches** to inversion problems. The sensitivity of the first method is assessed by applying directly the ambiguity function defined in [4]. To study the second method, we use large sample results to derive the statistical characteristics of the estimated features.

The paper is organized in the following way. Section 2 states the problem and defines notation and nomenclature. Section 3 briefly decribes the performance analysis tools used

*on leave from Instituto Superior Técnico, Universidade Técnica de Lisboa, Portugal, under an ECE Fellowship (Human Capital and Mobility).

O. Diaschok et al. (eds.), Full Field Inversion Methods in Ocean and Seismo-Acoustics, 103–108.

and section 4 defines the statistical distribution of the estimated inter-path delays used by the incoherent method. Finally, Section 5 applies the expressions previously developed to specific examples of ocean tomography, comparing the sensitivity of the two methods.

2. Problem Statement

The paper addresses the sensitivity study in the framework of a controlled acoustic experiment in the ocean. The signal emitted by a source at a *known* position is observed at a distant array of *known* geometry and localization, during a time interval T. The source signal $s(t), t \in T$, is usually known, providing in this way absolute time reference. Here, we address the case where the signal $s(t)$ is not exactly known at the receiver side, but is rather a pseudo-random Gaussian series, of *known* spectral characteristics, $S_s(\omega)$.

The transfer function of the ocean between source and the outputs of the receiving array is denoted throughout the paper by $\mathcal{H}^0[s(t)]$. This operator combines the effect of the propagation channel and of the receiving operator. To make our analysis concrete, we assume that the spatial variation of the sound speed induces multipath propagation, i.e., that $\mathcal{H}^0[s(t)]$ can be decomposed in the following way:

$$\mathcal{H}^0[s(t)] = \sum_{p=1}^{P^0} \mathcal{H}_p^0[s(t)], \tag{1}$$

where $\mathcal{H}_p^0[s(t)]$ denotes the transfer function along path p. We further assume that the medium is non-dispersive (linear), and that the signal received through each path is an attenuated and delayed version of the original signal, i.e.: $\mathcal{H}_p^0[s(t)] = a_p^0 s(t - \tau_p^0)$.

The signal radiated by the source is observed at the receiver side superimposed to a zero mean stationary Gaussian observation noise with spectral density matrix $\mathbf{S}_n(\omega)$. For simplicity, we assume that the noise is spatially white, i.e., $\mathbf{S}_n(\omega) = S_n(\omega)I_K$, where $S_n(\omega)$ is a real valued function, and I_K stands for the identity matrix of order K.

We denote by $r(t)$ the vector that collects the signals received at the K receiving sensors at time t. With the Gaussian hypothesis on the source signal and observation noise, $r(t)$ is stationary Gaussian, with zero mean and spectral density matrix given by:

$$\mathbf{S}_r(\omega) = S_s(\omega)h^0(\omega)h^{0^H}(\omega) + \mathbf{S}_n(\omega) \tag{2}$$

where $h^0(\omega)$ is the K-dimensional complex *resultant vector* combining the P^0 received paths. We denote the mathematical model used at receiver by $\mathcal{H}_\gamma[s(t)]$, explicitly showing its dependency on the desired ocean parameters γ. We assume that $\mathcal{H}_\gamma[s(t)]$ follows the same general form as $\mathcal{H}^0[s(t)]$.

3. Sensitivity Measure

The sensitivity analysis presented in this paper is based on a definition of ambiguity recently proposed [4]. In this section, we briefly present the definition of the function of interest to this study, referring the interested reader to [4], where a detailed presentation of the tool is given.

Statistically motivated parametric estimation methods are based on the knowledge of a family of conditional pdf's, that describes the dependency of the observed data distribution on the parameter of interest: $\mathcal{G}_\gamma = \{p(r|\gamma), \gamma \in \Gamma\}$. In the case of the *fully coherent methods*

r must be identified with the data collected at the sensors, while for the *incoherent methods* r is the output of a pre-processing estimation step.

When $\mathcal{G}_\gamma$ perfectly reflects the behaviour of the physical environment, the ambiguity function defined in [4] can be used to predict the global performance of the inversion mechanism, identifying possible large errors or configurations of poor observability of the parameter γ. Motivated by the close relationship between Maximum Likelihood estimators and the Kullback-Leibler directed divergence between probability density functions (pdf's) for exponential distributions, it defines ambiguity between two values $\gamma_0, \gamma \in \Gamma$ as a normalized version of the Kullback-Leibler distance between the corresponding elements of $\mathcal{G}_\gamma$.

In this study, we are interested in quantifying the impact of imperfect world knowledge, i.e., the fact that the true world's behaviour is described by a given family of distributions $\mathcal{G}_\gamma^0 = \{p^0(r|\gamma)\}$, and a distinct model is used in the fitting operation done at the receiver site: $\mathcal{G}_\gamma$. The observed data r is governed by a single member of $\mathcal{G}_\gamma^0$, that we denote by $p^0 = p^0(r|\gamma_0)$. The ability to correctly predict γ_0 using r under model mismatch can be analyzed studying a normalized version of the Kullback-Leibler distance between p^0 and the elements of $\mathcal{G}_\gamma$:

$$\mathcal{S}(\gamma) \triangleq 1 - \frac{I(p^0 : p(r|\gamma))}{I_{sup}(p^0)} \tag{3}$$

In this equation $I(p:q) = \mathrm{E}_p\{\ln(p/q)\}$ is the Kullback-Leibler directed divergence between pdf's p and q, where $\mathrm{E}_p\{\cdot\}$ is expectation with respect to the pdf p, and $I_{sup}(p^0)$ is an upper bound on $I(p^0 : p(r|\gamma))$.

The normalization in (3) is suitable to analyze the performance of a single method. However, when comparing several methods, as in this paper, it is more convenient to work directly with the Kullback directed divergence $I(p^0 : p(r|\gamma)$. In this way, we can not only compare the global performance of the different approaches, but also their local performance, using the relation between the gradient of the Kullback divergence and the Fisher information matrix (see [1]). Ideally, $I(p^0 : p(r|\gamma))$ is zero for $\gamma = \gamma_0$, and has large values for all $\gamma \neq \gamma_0$. Modeling errors induce systematic biases in the estimation procedure, that are flagged by the fact that $I(p^0 : p(r|\gamma)$ has its minimum at an erroneous value $\gamma^* \neq \gamma_0$.

3.1. INCOHERENT METHOD

For the incoherent inversion method, the input to the fitting operation are the inter-path delay estimates. Large observation interval assumptions justify the use of a Gaussian distribution to characterize these differential delay estimates: $p^0 = \mathcal{N}(\tau^0, \Sigma^0)$, where τ^0 is a vector collecting the $P^0 - 1$ true inter-path delays measured at a reference sensor, and Σ^0 is the variance of the delay estimates, predicted by the Cramer Rao lower bound.

The model used at the receiver in the fitting operation corresponds to applying the same large observation interval hypothesis, but using the wrong ocean/array model: $p(\hat{\tau}|\gamma) = \mathcal{N}(\tau(\gamma), \Sigma(\gamma))$. In Section 4 we give expressions for Σ^0, and $\Sigma(\gamma)$. With the previous assumptions, the Kullback directed divergence between densities p^0 and $p(\hat{\tau}|\gamma)$ is:

$$\mathrm{I}(\mathrm{p}^0 : \mathrm{p}(\hat{\tau}|\gamma)) = \frac{1}{2}\ln\frac{|\Sigma(\gamma)|}{|\Sigma^0|} + \frac{1}{2}\mathrm{tr}\left[\Sigma^0\Sigma(\gamma)^{-1} - I\right] + \frac{1}{2}\left(\tau^0 - \tau(\gamma)\right)^{\mathrm{T}}\Sigma(\gamma)^{-1}\left(\tau^0 - \tau(\gamma)\right). \tag{4}$$

3.2. FULLY COHERENT

The *fully coherent method* requires use of the Kullback directed divergence between the data records themselves. Invoking a large observation interval assumption, we use the asymptotic characterization of the directed divergence between stationary Gaussian processes derived in [2], that gives $I(p^0 : p(r|\gamma)$ in terms of the spectral densities of the observations:

$$\bar{\mathrm{I}}(\mathrm{p}^0 : \mathrm{p}(\mathrm{r}|\gamma)) = \lim_{\mathrm{T}\to\infty} \frac{1}{\mathrm{T}} \mathrm{I}(\mathrm{p}^0 : \mathrm{p}(\mathrm{r}|\gamma)) = \frac{1}{2} \int \left[\mathrm{tr}\left[S^0(\lambda) S_\gamma(\lambda)^{-1}\right] - \mathrm{K} - \ln \frac{|\mathrm{S}^0(\lambda)|}{|\mathrm{S}_\gamma(\lambda)|} \right] d\lambda. \quad (5)$$

4. Statistical Characterization of Selected Features

We present in this section the expressions of Σ^0 and $\Sigma(\gamma)$. In [3] we derived the following expression for the Fisher information matrix for the estimation of parameters from observations following a multipath model:

$$[\mathcal{F}_\alpha]_{ij} = \frac{N}{2\pi} \int \left\{ K_1 \mathrm{Re} \left\{ \frac{\partial h}{\partial \alpha_i}^H P_{h^\perp} \frac{\partial h}{\partial \alpha_j} \right\} + K_2 \mathrm{Re} \left\{ \frac{\partial h}{\partial \alpha_i}^H h \right\} \mathrm{Re} \left\{ \{ h^H \frac{\partial h}{\partial \alpha_j} \} \right\} d\omega \quad (6)$$

where N is the number of snapshots used, $P_{h^\perp}$ denotes the orthogonal projection matrix in the orthogonal complement of the resultant vector h, and the following definitions hold:

$$K_1 = \frac{S_s^2(\omega)}{S_n(\omega)E} \|h\|^2, \qquad K_2 = \frac{S_s^2(\omega)}{E^2}, \qquad E = S_n(\omega) + S_s(\omega)\|h\|^2. \quad (7)$$

Besides the inter-path delays (τ), other parameters enter the multipath model, describing the wavefront shape across the receiving array (δ). The Cramer Rao lower bound on the inter-path delays is obtained by computing the inverse of corresponding block of the partition of the complete Fisher information matrix. Let $\alpha = [\tau \quad \delta]$, and $\mathcal{F}_\tau, \mathcal{F}_\delta$, and $\mathcal{F}_{\tau\delta}, \mathcal{F}_{\delta\tau}$ be the upper and lower diagonal, and the off-diagonal blocks of $\mathcal{F}_\alpha$, respectively. Then, the Cramer Rao lower bound on the inter-path delays is

$$CRB(\tau) = \left[\mathcal{F}_\tau - \mathcal{F}_{\tau\delta} \mathcal{F}_\delta^{-1} \mathcal{F}_{\delta\tau} \right]^{-1}. \quad (8)$$

In the paper we consider two different ways of modeling the wavefront shape: *(i)* no *a priori* knowledge, i.e., all the inter-path delays defining the shape of the each wavefront accross the array are estimated along with the inter-path delays; *(ii)* planar wavefront approximation, in which case a single parameter defines the orientation of the wavefront accross the array. The two approaches are compared in the simulations presented.

5. Examples

The model used at receiver assumes that the ocean is horizontally stratified, with two distinct homogeneous layers. In the superficial layer (up to depth y_{duct}), the sound speed decreases linearly with depth (with rate g_0), increasing linearly from y_{duct} until the ocean bottom. The ocean boundaries are perfectly flat, with reflection coefficients depending on the grazing angle. The duct and ocean depths, and the sound speed gradient in the deep layer (g_1) are perfectly known, as well as all other parameters relating to source (position, spectrum) and receiver array (geometry, localization, gain).

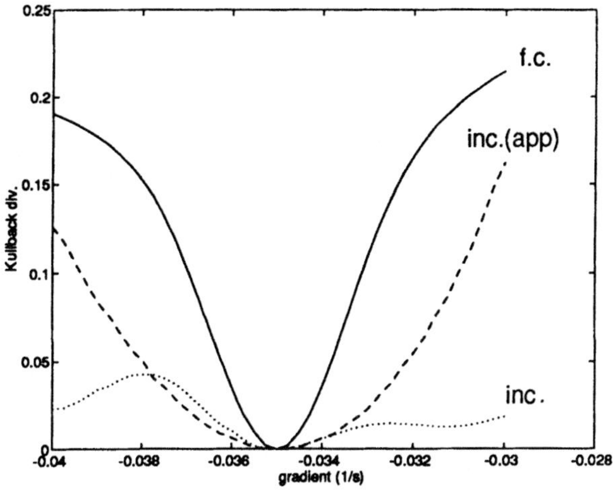

FIGURE 1. Kullback divergence for correct model.

Figure 1 shows the Kullback divergence when the propagation model used at receiver perfectly matches the oceans' transfer function. For this example, the distance between the source and the receiving array is 6 Km, located at the same depth (200 m). The solid curve (f.c.) corresponds to the fully coherent and the dashed (inc. (app)) and dotted (inc.) curves to the incoherent method, under the two different wavefront shape models (dashed line for the planar wavefront approximation, and dotted line for the full parametrization). Using the relation between Kullback divergence and Cramer Rao bound, we can conclude that, as expected, the matched field approach yields the best local performance (larger curvature of the Kullback divergence at the true parameter value), followed by the incoherent method with planar wavefront approximation. The need to estimate a large number of (somewhat redundant) parameters in the detailed wavefront method explains its poor performance. In the following plots, only the plane wave approximation is considered. Except for the incoherent method with no a priori knowledge of the wavefront shape, no secondary lobes are present, and thus no large systematic errors are to be expected.

Figure 2 shows (solid lines) the result of using wrong values for the receiving array's depth *(a)* and for the velocity at the surface *(b)*. To facilitate comparison, we also plot (dashed lines) the curves for no model mismatch. Note that the curvature of the plots around their peak values is only slightly increased, suggesting that local behavior is not significantly affected by model mismatch. Receiver mismatch results in a bias in the gradient estimate. Sensitivity to knowledge of array immersion is larger than to the value of the sound speed at the duct: while a small error in antenna immersion (190 m instead of the actual value of 200 m) induces a considerable systematic bias, a corresponding velocity error (1470 m/s instead of 1480) results in a negligible error.

Figure 3 shows the Kullback divergence for a larger array-source separation (50 Km). Solid plots correspond to a wrong value of the sound speed gradient in the lower layer ($g_1 = .0165s^{-1}$, instead of .016). The dashed lines show the Kullback divergence for no model mismatch. This figure suggests that the matched field approach is less ambiguous than the incoherent method, the sensitivity of both approaches to model errors being comparable.

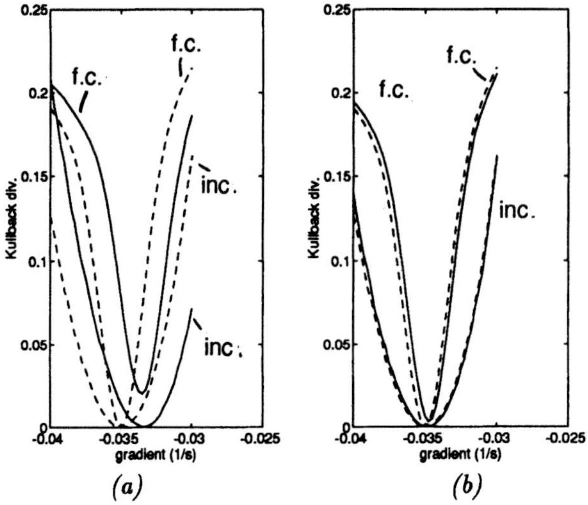

FIGURE 2. Kullback divergence *(a)* wrong array depth; *(b)* wrong velocity at duct.

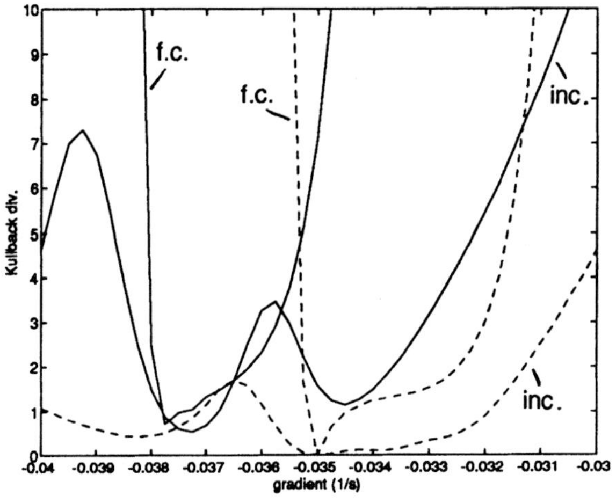

FIGURE 3. Kullback divergence for wrong lower gradient (distant source).

References

1. Solomon Kullback. *Information Theory and Statistics.* Peter Smith, 1978.
2. M. S. Pinsker. *Information and Information Stability of Random Variables and Processes.* Holden-Day, Inc., 1964.
3. M. João D. Rendas and José M. F. Moura. Cramer-Rao bounds for source location systems in multipath environments. *IEEE Trans. on Acoustic, Speech and Signal Processing*, Vol. 39, No. 12:2593–2610, Dec. 1991.
4. M. João D. Rendas and José M. F. Moura. Ambiguity analysis in source location. *Submitted for publication*, March 1994, *also* Research Report no. 94-23, I3S, CNRS-URA 1376, Université de Nice Sophia Antipolis, France.

A SENSITIVITY STUDY FOR FULL-FIELD INVERSION OF GEO-ACOUSTIC DATA WITH A TOWED ARRAY IN SHALLOW WATER

S.M. JESUS
UCEH - University of Algarve
Campus de Gambelas, PT-8000 Faro, Portugal

ABSTRACT Inversion of acoustic data for estimating bottom acoustic parameters has been the subject of a considerable number of studies. Usually, signals are received on vertical arrays of sensors and transmitted from sound sources being towed away from the array location in order to form a syntethic aperture array. That configuration is greatly dependent on the knowledge of the source-receiver distance which is, in practice, relatively difficult to measure with the required precision. Also, since the vertical array is generally moored, or slowly drifting, the area that can be surveyed with such a method is limited to a tenth of a mile in shallow water. Changing area requires the recovery and redeployment of the whole system. This paper explores the possibility of using an horizontal array and a sound source simultaneously towed by a single ship where the source-receiver distance is constant. It has been shown that sensitivity to sound speed variations is higher on the first bottom layers and it increases with array length. Density and attenuation (both compressional and shear) have in general little influence on the acoustic field structure and are therefore difficult to estimate. Increasing the signal frequency bandwidth by incoherent module averaging has no significant influence on sensitivity. Mismatch cases, mainly those related to array/source relative position, showed that deviations of more than $\lambda/2$ in range and $\lambda/5$ in depth may give erroneous *extremum* location and therefore biased final estimates.

1. Introduction

Estimating ocean bottom morphological characteristics in coastal shallow waters is of considerable economical and scientific interest. Sound propagation in shallow water is known to be dominated by a strong signal interaction with the medium boundaries. In principle, comparison of the received acoustic field with that predicted by a suitable propagation model would allow for estimating bottom and surface physical properties. Estimation of bottom environmental parameters from the acoustic data received on an array of sensors is known to be an ill-conditioned inversion problem for which an analytical solution is unknown. Ill-conditioning strongly depends on the non-linearity of the function to be inverted and on the dimension of the parameter space. Brute force inversion, by extensive forward modeling exploration of the whole search space, has been widely used on matched-field processing for source localization, after the pioneering work of Bucker [1]. Using this approach for geo-acoustic data inversion would be computationally prohibitive due to the high dimension of the parameter space to be searched. Therefore, the alternative taken by several authors, combines a matched-field type of technique (its output is a multidimensional ambiguity surface), with a powerful search algorithm that, in principle, allows a quick convergence to the *extremum* of that surface [2] [3] [4]. Simply speaking, the matched field technique provides a parameter dependent cost function that the search algorithm attempts to minimize. In

O. Diaschok et al. (eds.), Full Field Inversion Methods in Ocean and Seismo-Acoustics, 109–114.

other words, the treatment of the inverse problem has two aspects: one is the computing of the cost function and the other is the search algorithm.

The present study concentrates on the cost function system dependence and on its operational characterization. Previous studies used either vertical arrays or sound sources being towed away from the receiver location in order to create a synthetic aperture and resolve mode arrivals. In principle, a shipborne only system would allow easier deployment and lower cost for surveillance of large areas. In our study it is assumed that the ship is towing both the source and the array, such that the source-receiver range is constant. To obtain an idea of the expected performance of the system and draw some conclusions on its operation, this study presents the cost function sensitivity to variations of: array length (spacing and number of sensors), source depth, receiver depth, source range, sensor noise, source frequency and frequency band. The canonical case consists of a 64 hydrophone - 4 m spacing towed array at 100 m depth and a harmonic 100 Hz source also at 100 m depth and at 200 m range.

2. Theoretical background

2.1. THE DATA MODEL

The deterministic sound pressure at the receiver location r_l, z_l is modeled as the solution of the wave equation for a narrowband point source exciting a horizontally stratified range-independent environment:

$$p_l(\omega_k, r_l, z_l, z_0; \gamma) = \int_0^\infty g(\kappa, \omega_k; \theta_l, \gamma, z_0) J_0(\kappa r_l) \kappa d\kappa, \tag{1}$$

where l denotes the lth array sensor, ω_k is the kth frequency bin, z_0 is the source depth and γ is a vector containing all the pertinent environmental parameters under estimation. Thus, at time snapshot n, the L sensor array received acoustic pressure, can be modeled as a multivariate complex normally distributed random variable

$$\mathbf{y}_n(\omega_k, \gamma_T) = b_n(\omega_k)\mathbf{p}(\omega_k, \gamma_T) + \boldsymbol{\epsilon}_n(\omega_k) \qquad k = 1, \ldots, K \tag{2}$$

where $\boldsymbol{\epsilon}$ is the sensor noise assumed to be zero mean and uncorrelated both in time and from sensor to sensor. The scalar b_n is a complex random variable that accounts for the non-deterministic amplitude variation at the receiver due to the environmental inhomogenities and fluctuations that are not included in the sensor noise. Subscript $_T$ denotes the true environmental parameter value under estimation.

2.2. THE BROADBAND CONVENTIONAL MATCHED-FILTER

Given model (1)-(2), the problem is to detect, at each single frequency, a known signal $\mathbf{p}(\omega_k, \gamma)$ in white noise, for which the optimal processor is the matched-filter. Let

$$\Phi_{\rm CMF}(\omega_k, \gamma) = |\mathbf{y}(\omega_k, \gamma_T)^H \mathbf{p}(\omega_k, \gamma)|^2 \qquad \gamma \in \Gamma \tag{3}$$

be the matched-filter output based on model replica prediction for search parameter γ with Γ denoting the whole environmental parameter search space. Thus, the final broadband optimal detector of a single known signal in white noise will be

$$\hat{\gamma}_T = \arg\max_{\gamma} \frac{1}{K} \sum_{k=1}^{K} |\sigma_s^2(\omega_k)|^2 \Phi_{\rm CMF}(\omega_k, \gamma) \tag{4}$$

where $\sigma_s^2(\omega_k)$ is the source power at frequency ω_k.

2.3. CORRELATION OF DIRECTIONAL DATA

A common problem encountered when analysing geo-acoustic data is the superposition of the direct path source arrival with the bottom reflected data of interest. A possibility to separate those arrivals is by analysing the data in the wavenumber space domain and by filter out the direct path arrivals. For an horizontal array the arrivals associated with the steepest vertical angles, which are those that have a stronger interaction with the bottom, correspond to those arriving closer to broadside. That approach implies a transformation of the acoustic data from the hydrophone space to the wavenumber space which may be expressed by the Green's function. In practice, since the acoustic pressure is a discrete function defined over a finite array aperture, it implies that an estimate of the predicted Green's function can be given by

$$\hat{g}_p(k_j,\omega,\gamma) = \frac{e^{i\pi/4}}{\sqrt{2\pi k_j}} \sum_{l=1}^{L} p_l(\omega,\gamma) e^{-ik_j r_l} \sqrt{r_l}, \qquad j = 1,\ldots,N_w \tag{5}$$

where the discretization over the wavenumber space has been arbitrarily performed over N_w equally spaced points in $[0, 2\pi/d]$, d being the array sensor spacing (assumed constant). A one-to-one mapping from the wavenumber to the bearing space may be performed using $k_j = (2\pi f/c)\cos(\theta_j + \pi/2)$ for $\theta_j \in [-90^\circ, +90^\circ]$. With that definition -90° direction is aft (towards the source) and +90° is end fire. A similar expression to (5) may be used for the received data Green's function at time-snapshot n giving $\hat{g}_{y,n}(k_j,\omega,\gamma_T)$. Obviously, when computing (5) there is a windowing spatial effect that actually reduces the array resolution power. Based on (5) and once a given bearing sector $\theta_j \in [\theta_l, \theta_h]$ has been selected, a possible cost function can be defined as

$$\Phi_{\text{WS-CMF}}(\omega_k,\gamma) = |\sum_{j=j_l}^{j_h} \hat{g}_y^*(\theta_j,\omega_k,\gamma_T)\hat{g}_p(\theta_j,\omega_k,\gamma)|^2 \tag{6}$$

and an incoherent broadband function can be defined as the average of (6) over the required frequency band as

$$\Phi_{\text{WS-CMF}}(\gamma) = \frac{1}{k_h - k_l + 1} \sum_{k=k_l}^{k_h} \Phi_{\text{WS-CMF}}(\omega_k,\gamma)\,. \tag{7}$$

Expression (6) has been written as time independent for a question of generality, where in practice it is time-snapshot dependent. In practice time averaging is often performed over a "reasonable" time window where the acoustic field is assumed to be stationary.

2.4. BROADBAND MAXIMUM-LIKELIHOOD MATCHED-FIELD

Given data model (1)-(2) the random complex scalar $b_n(\omega_k)$ is assumed random $N(0,\sigma_b^2)$. The sensor noise $\boldsymbol{\epsilon}_n(\omega_k)$ is assumed also $N(\mathbf{0},\sigma_\epsilon^2\mathbf{I})$. With these assumptions, at time snapshot n, the observation vector is therefore $N(b_n(\omega_k)\mathbf{p}(\omega_k,\gamma_T),\sigma_\epsilon^2\mathbf{I})$. At this point one desires to determine the "best estimator" of γ_T given the set of broadband vectors $\mathbf{Y}^n =$

$[\mathbf{y}_n^T(\omega_1), \mathbf{y}_n^T(\omega_2), \ldots, \mathbf{y}_n^T(\omega_K)]^T$. If the time interval T, used for calculating each individual Fourier transform, is such that $T >> \tau_0$, where τ_0 is the correlation time of the most coherent signal or noise, the vector $\mathbf{Y}^n$ has a near-block-diagonal covariance matrix because the Fourier coefficients at different frequencies are asymptotically uncorrelated. Independently from the nature of the assumed data model the log-likelihood function is given by

$$L(Y) = -\sum_{n=1}^{N}[(\mathbf{Y}^n - \boldsymbol{\mu}^n)^H \mathbf{R}_n^{-1}(\mathbf{Y}^n - \boldsymbol{\mu}^n) + \log\det(\pi \mathbf{R}_n)] \quad (8)$$

where ^{H} denotes Hilbert transpose, $\boldsymbol{\mu}^n = E[\mathbf{Y}^n]$ and $\mathbf{R}_n = \mathrm{COV}[\mathbf{Y}^n]$ which is, with the assumptions above a diagonal matrix. Inserting those two quantities into (8) one easily obtains

$$L(Y) = -\sum_{n=1}^{N}\sum_{k=1}^{K}\{\frac{1}{\sigma_\epsilon^2(\omega_k)} \parallel \mathbf{y}_n(\omega_k) - \mathbf{b}_n(\omega_k)\mathbf{p}(\omega_k, \gamma_T) \parallel^2 - L\log[\pi\sigma_\epsilon^2(\omega_k)]\} \quad (9)$$

Since $\mathbf{p}(\omega_k, \gamma_T)$ is given by (1), for the search parameter γ as $\mathbf{p}(\omega_k, \gamma)$, the only unknown is the random component $\mathbf{b}_n(\omega_k)$. An usual assumption is to introduce a least-squares estimate of the signal, which in that case is given simply by the projection of $\mathbf{y}_n(\omega_k)$ onto the vector $\mathbf{p}(\omega_k, \gamma)$ given by

$$\hat{\mathbf{b}}_n(\omega_k) = \frac{\mathbf{p}(\omega_k, \gamma)^H \mathbf{y}_n(\omega_k)}{\parallel \mathbf{p}(\omega_k, \gamma) \parallel^2} \quad (10)$$

Thus, introducing (10) into (9) one gets the estimator expression

$$\hat{\gamma}_T = \arg\min_{\gamma} \sum_{n=1}^{N}\sum_{k=1}^{K}\{\frac{1}{\sigma_\epsilon^2(\omega_k)} \parallel \mathbf{y}_n(\omega_k) - \frac{\mathbf{p}(\omega_k, \gamma)^H \mathbf{y}_n(\omega_k)}{\parallel \mathbf{p}(\omega_k, \gamma) \parallel^2} \mathbf{p}(\omega_k, \gamma) \parallel^2\} \quad (11)$$

and after some straight forward manipulations and noting that the only n snapshot dependent quantities are the observation vectors $\mathbf{y}_n(\omega_k)$ one can rewrite (11) as

$$\hat{\gamma}_T = \arg\min_{\gamma} \sum_{k=1}^{K} \frac{K}{\sigma_\epsilon^2(\omega_k)} \mathrm{tr}\{[\mathbf{I} - \frac{\mathbf{p}(\omega_k, \gamma)\mathbf{p}(\omega_k, \gamma)^H}{\parallel \mathbf{p}(\omega_k, \gamma) \parallel^2}]\hat{\mathbf{R}}(\omega_k)\} \quad (12)$$

where the matrix $\hat{\mathbf{R}}(\omega_k)$ is the data sample covariance matrix estimate at frequency ω_k given by the time snapshot average of the data outer products. The only unknown quantity is the noise power over the required frequency band, $\sigma_\epsilon^2(\omega_k); k = 1, \ldots, K$ that can be assumed constant for white noise.

3. Simulation results

The simulation environment is shown on table 1. A computer code based on SAFARI, FIPP [5] has been developed to implement the cost functions defined above and necessary looping for all the environmental search parameters. The system scenario includes a 64 hydrophone array at 4 m spacing with a 100 Hz sound source both at 100 m depth and at 200 m range. Testing was performed with the conventional matched-filter approach and in one case a comparison was made with the other cost functions.

TABLE 1. Canonical environment

Depth(m)	**P vel.**(m/s)	**S vel.**(m/s)	**P att.**(dB/λ)	**S att.**(dB/λ)	**Dens.** (g/cm^3)
0.0	1500	0.0	0.0	0.0	1
140	1550	130	0.1	1.7	1.49
145	1700	350	0.8	2.0	1.88
150	2500	900	0.01	0.01	2.4

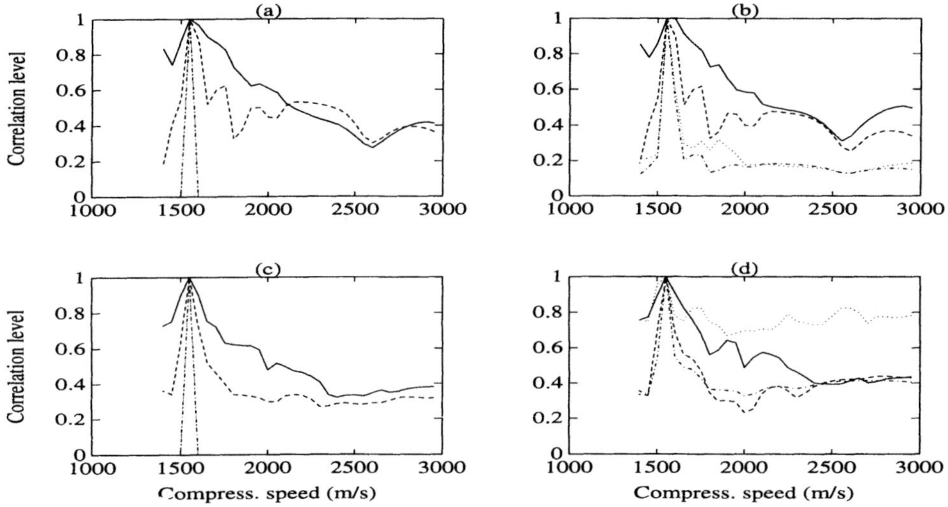

Figure 1: Correlation level versus compressional speed of first sediment layer for CMF (- - dashed), WS-CMF (– solid), LMS (... dotted) and ML (-.-. dash-dot). Narrowband 100 Hz in (a) without noise and (b) SNR = 10 dB; broadband 85 to 115 Hz in (c) without noise and in (d) with SNR = 10 dB.

Array aperture: the results show that varying array aperture from 63 m up to 2016 m increases the sensitivity to bottom parameters, mainly to the shallower sediment layers. Attenuation sensitivity is of the order of 10^{-3}. *Frequency*: varying source frequency between 25 and 200 Hz does change the angle of incidence and signal penetration into the bottom. An increase of frequency improves the sensitivity to compressional parameter variation while for deeper layers lower frequencies give better results. Shear parameters showed a higher sensitivity at lower frequencies that provided also smoother curves (less *minima/maxima*). *Bandwidth*: varying the source bandwidth between 2 and 60 Hz showed that there is no increase in sensitivity with, however, a higher smoothness of the cost function behaviour. *Source-receiver positions*: changing relative source-receiver depth does change the sensitivity curve according to the higher or lower transfer of energy between source and receiver that is depending on the mode excitation *vs* depth. Changing source receiver range has a similar behaviour depending on the modal interference pattern *vs* range. In both cases placing the source and the array at high energy transfer locations does improve the sensitivity. *System parameters mismatch*: higher sensitivity to depth than to range mismatch was observed. The accuracy to which sensor depth should be known has to be better than a $\lambda/5$ while an accuracy of $\lambda/2$ will be enough for sensor range. *Signal-to-noise ratio*: the narrowband and broadband performance of the algorithms was compared and the results are shown in figure 1. For reference the well known least mean squares (LMS) cost function has also been calculated and is shown in figure 1. Since the working signal to noise ratio (SNR) will be relatively high, only the SNR=∞, in (a) and (c), and 10 dB, in (b) and (d), are shown. In both cases the maximum likelihood (ML) estimate provided the best results together with the MLS estimator for SNR=∞ and with the conventional matched-filter (CMF) for SNR=10 dB. At lower SNR (not shown) CMF provided the highest performance. There was only a slight increase in performance with increasing bandwidth.

4. Conclusion

It was demontrated that cost function sensitivity to sound speed variations is higher to the bottom layers and it increases with array length. An increased sensitivity is generally accompanied by a cost function with non-monotonic behavior creating local *minima* and making it problematic to reach the global minimum. Density and attenuation (both compressional and shear) have in general little influence on the acoustic field structure and are therefore difficult to estimate. Increasing the signal frequency bandwidth by incoherent module averaging has no significant influence on sensitivity. A cost function relaying on the conventional matched filter has shown low sensitivity to sensor noise and has been extended to match directional data from bottom arrivals at several frequencies. A technique for providing a maximum likelihood broadband estimate of the peak location has been derived and showed a discrete performance at high SNR. Mismatch cases, mainly those related to array/source relative position, showed that deviations of more than $\lambda/2$ may give erroneous *extremum* location and therefore biased final estimates.

Acknowledgment: this research was supported by MAST 2 program of the European Community under contract MAS2-CT920022.

References

1. H.P. Bucker, "Use of calculated sound fields and matched-field detection to locate sound sources inshallow water", J. Acoust. Soc. Am. **59**, 368-373 (1976).
2. M.D. Collins and W.A. Kuperman, "Non-linear inversion for ocean-bottom properties", J. Acoust. Soc. Am. **92**(5), 2770-2783 (1992).
3. A. Turgut, "Simulated annealing and genetic algorithms in shear modulus inversion of shallow-water sediments", J. Acoust Soc. Am. **93**, (4) Pt.2, 2aAO15 (1993).
4. P. Gerstoft, "Inversion of seismoacoustic data using genetic algorithms and *a posteriori* probability distributions", J. Acoust. Soc. Am. **95** (2), 770-782 (1994).
5. H. Schmidt "SAFARI: Seismo-Acoustic Fast Field Algorithm for Range Independent Environments", La Spezia, Italy, SACLANTCEN SR-113 (1988).

OPTIMIZING ENVIRONMENTAL PARAMETRIZATION AND EXPERIMENTAL DESIGN FOR SHALLOW WATER SOUND SPEED INVERSION

NICHOLAS C. MAKRIS, JOHN S. PERKINS, SCOTT P. HECKEL*
Naval Research Laboratory
Washington, D.C. 20375, USA

JOSKO CATIPOVIC
Woods Hole Oceanographic Institution
Woods Hole, MA 02543, USA

ABSTRACT The effect of environmental parametrization and experimental design on 3-D sound-speed estimation by full-field inversion of shallow-water acoustic data is investigated. A theoretical lower bound on parameter estimation error is used to compare various environmental representations. These include an Empirical Orthogonal Function (EOF), discrete cell, and physics-based internal-wave representation.

1. Introduction

Recently there has been a great deal of interest in 3-D ocean sound speed estimation by full-field inversion of acoustic measurements, a process which has also been referred to as Matched Field Tomography [1]. Most modelling and experimental work to date has been done for deep water where bottom interaction is negligible. Empirical Orthogonal Functions (EOFs) have typically been used to represent the ocean sound speed [1][2]. If they are properly chosen, a small number of EOFs can very accurately represent 3-D structure. However, this requires a priori knowledge of the environment, which is usually based upon physical oceanography. It has been suggested that a Cartesian grid of cells may be a more reasonable representation for a highly variable and unpredictable shallow water environment. This is because there are far fewer constraining assumptions in the cell representation. But a large number of cells may be necessary for an accurate representation. In terms of parameter estimation, we show that there are no inherent flaws with either of these representations. We find a similar result for a physics-based internal wave parametrization of fine-scale structure.

In our illustrative examples, we assume that the upper layer of the water column undergoes temporal variations arising from a combination of atmospheric, oceanographic and coastal stresses and gradients in pressure and temperature. The range-dependent bathymetry and geo-acoustic properties of the lower water column and sediment are assumed to be invariant and known. This partitioning of the environment into variable and deterministic regions is shown in Figure 1. The upper layer is then parametrized by the schemes previously discussed. Each scheme is quantitatively appraised via a theoretical lower bound on parameter estimation error for some practical measurement geometries.

*Planning Systems Inc., Mclean VA. 22102, USA

O. Diaschok et al. (eds.), Full Field Inversion Methods in Ocean and Seismo-Acoustics, 115–120.

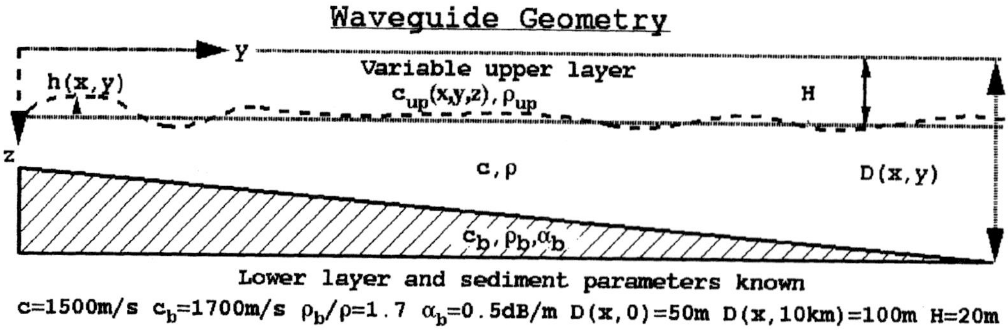

Figure 1.

2. Mesoscale Sound Speed Parametrizations: EOF versus Cell

The same unknown upper layer is used for both the EOF and cell representations. This is shown in Figure 2 over a 10 km (x) by 10 km (y) by 20 m (z) block. A crude physical motivation for this choice is that vertical mixing of the water mass will be more thorough in the shallower bathymetry near the shore. A temperature-driven downward-refracting profile is likely to appear in deeper water further from the coast where the upper layer is more distinct. Over forcing time scales short enough for the Coriolis force to be negligible, no variations are anticipated along-shore (in the x direction).

To obtain an EOF expansion, we first subtract the mean sound speed in the upper layer $\bar{c}_{\mathrm{up}}$ from the variable upper-layer sound speed $c_{\mathrm{up}}(x,y,z)$ to arrive at $c'_{\mathrm{up}}(x,y,z) = c_{\mathrm{up}}(x,y,z) - \bar{c}_{\mathrm{up}}$. We expand the upper ocean into distinct basis vectors for the Cartesian coordinate system [3],

$$c'_{\mathrm{up}}(x,y,z) = \sum_{p,q,r} a_{pqr} L_{pr}(x) G_{qr}(y) F_r(z). \tag{1}$$

This representation is physically meaningful because it maintains independent statistics for x (along-shore), y (seaward), and z (depth) upper layer variations. The entire upper layer shown in Figure 2a can be represented very accurately with only two nonzero EOF coefficients, a_{111} and a_{121}; ie to less than 0.01 m/s maximum error at any point.

Alternatively, the upper layer sound speed $c_{\mathrm{up}}(x,y,z)$ can be parametrized by the mean vertical slope $\bar{s}_j$ for each of a set of horizontal cells j=1,2,3..., with known intercept c=1500 m/s at $z=H$=20 m. Clearly, satisfaction of the spatial Nyquist Criterion would require thousands of cells to obtain the same precision in representing the 3-D sound speed as is obtained by two EOF terms. However, it is common knowledge that the more parameters M necessary to represent the environment adequately, the more independent measurements are required for a properly constrained inversion. For a single source, we equate the maximum number of independent equations available in the inversion with the number of acoustic sensors (hydrophones) N. Then the inversion can be represented as a set of nonlinear equations, $p_i = f_i(\chi_1, \chi_2, \chi_3 ..., \chi_M)$, where p_i is the complex pressure measured at hydrophone i=1,2,3...N, and χ_j are the parameters to be estimated. In an attempt to avoid an under-determined inversion, we seek parametrizations and sensor deployments such that the number of hydrophones N is greater than or equal to the number of parameters M to be estimated. For example, in the EOF formulation, the M=3 parameters are $\chi_1 = a_{111}$, $\chi_2 = a_{121}$ and $\chi_3 = \bar{c}_{\text{up}}$. In the cell formulation the parameters are $\chi_j = \bar{s}_j$ for j=1,2,3...M, where we have chosen M=3 for comparison with the EOF representation.

We use the Cramer-Rao bound from Estimation and Detection Theory [4] to compare the minimum estimation error of alternative parametrizations. For example, the Cramer-Rao upper bound on the resolution of parameter χ_i is

$$E\left[(\hat{\chi}_i - \chi_i)^2\right] \geq \left[J^{-1}\right]_{ii} . \tag{2}$$

The left hand side of Eq. (2) gives the minimum variance obtainable in an unbiased measurement of χ_i, where $\hat{\chi}_i$ is the estimated value. For complex pressure measurements on an acoustic array, the Fisher Information Matrix $\boldsymbol{J}$ has been shown to have elements [5]

$$J_{ij} = \boldsymbol{Tr}\left(\boldsymbol{C}^{-1}(\boldsymbol{\chi}) \frac{\partial \boldsymbol{C}(\boldsymbol{\chi})}{\partial \chi_i} \boldsymbol{C}^{-1}(\boldsymbol{\chi}) \frac{\partial \boldsymbol{C}(\boldsymbol{\chi})}{\partial \chi_j} \right), \tag{3}$$

where $\boldsymbol{C}(\boldsymbol{\chi})$ is the covariance matrix of the field measured by the array ($C_{ij} = < p_i p_j^* >$). This form of the Cramer-Rao bound has typically been used in source localization [5].

We consider the following measurement geometry. An f=500 Hz monopole source is placed at z=H/2=10 m and x=y=0 m, and an N-element λ-spaced vertical receiving array is placed at x=0 m, y=10 km with shallowest sensor at z=1 m. Using the environment in Figure 1 with the upper layer shown in Figure 2a, the covariance at the array is computed for a 2-D range-depth ocean cross section using the parabolic equation method [6]. The optimal resolution of EOF coefficients, and mean sound speed, as a function of SNR is shown in Figure 3a for an N=4 element array and in Figure 3b for an N=32 element array. (The SNR is defined as Tr($\boldsymbol{C}_S$)/(Tr($\boldsymbol{C}_N$)) where $\boldsymbol{C}_S$ is the signal covariance with no noise and $\boldsymbol{C}$=$\boldsymbol{C}_S$+$\boldsymbol{C}_N$ is the total covariance. The noise covariance $\boldsymbol{C}_N$ is approximated with the Kuperman and Ingenito formulation [7] for a range-independent waveguide of 200 m depth with the same geo-acoustics parameters found in the lower layer and sediment.) For each array, EOF coefficient bounds may be converted to minimum sound speed errors in the water column by multiplying the standard deviations $\sigma_i = \sqrt{E\left[(\hat{\chi}_i - \chi_i)^2\right]}$ by the corresponding EOF composite for the upper layer $L_{11}(x)G_{i1}(y)F_1(z)$ shown in Figure 2b,

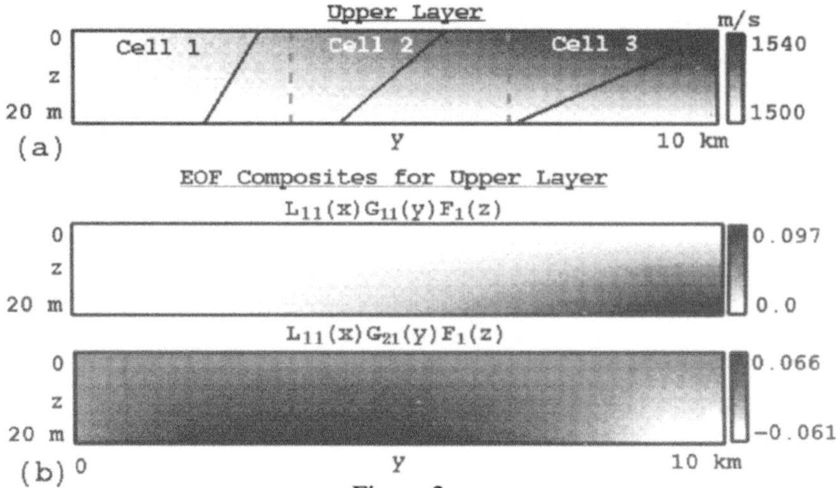

Figure 2.

where i=1,2. Clearly, the minimum estimation error (less than 2.0 m/s for either EOF term) is within acceptable bounds for the 4-element array at high SNR, but quickly decays as SNR approaches and becomes less than unity. As anticipated, order of magnitude increases in parameter resolution, for high and low SNR, are obtained with the N=32 element array.

The fact that a 2-D slice experiment can resolve the entire 3-D upper layer is a consequence of the constraining assumptions of the EOF parametrization. Even if two EOF coefficients could adequately represent the ocean, a greater number would be necessary to represent all plausible oceans, otherwise inversion is unnecessary. But without sufficient a priori information, inadequate restriction on the number of plausible EOF coefficients could lead to an apparently but not actually under-determined system. This is not critical for the present analysis because redundant or irrelevant parameters simply have large resolution bounds because they are not properly constrained. They do not affect the resolution of properly constrained parameters. Similarly, relevant parameters have large error bounds when the constraining equations are under determined.

Resolution bounds for the cell parametrization of the upper layer are shown in Figure 3c as a function of SNR for the N=4 element array. Clearly, the minimum error is within acceptable bounds at high SNR for cells 2 and 3, but quickly decays as SNR approaches and becomes less than unity. Cell 1 has poor resolution even at high SNR where σ_1 is equal to the maximum upper layer slope of 2.0 1/s. This is because the mean slope is increasingly less significant in representing the sound speed as the cell number decreases (see Figure 2a). As a consequence, the minimum estimation error increases with decreasing cell number. Although it is not shown, order of magnitude increases in the resolution of all cells are obtained with the N=32 element array. For example at SNR=100, all minimum error bounds are less than 0.05 1/s, and at SNR=1 they are less than 0.35 1/s. This implies that there is nothing inherently flawed in a cellular representation of mean slope. However, it does suggest that the Cramer-Rao bound can be used to optimize experimental design by taking advantage of a priori knowledge to insure that unknown parameters are properly constrained. For example, we now know that N=4 sensors give poor resolution for small slopes, while N=32 sensors can give more than adequate resolution, in the above scenario.

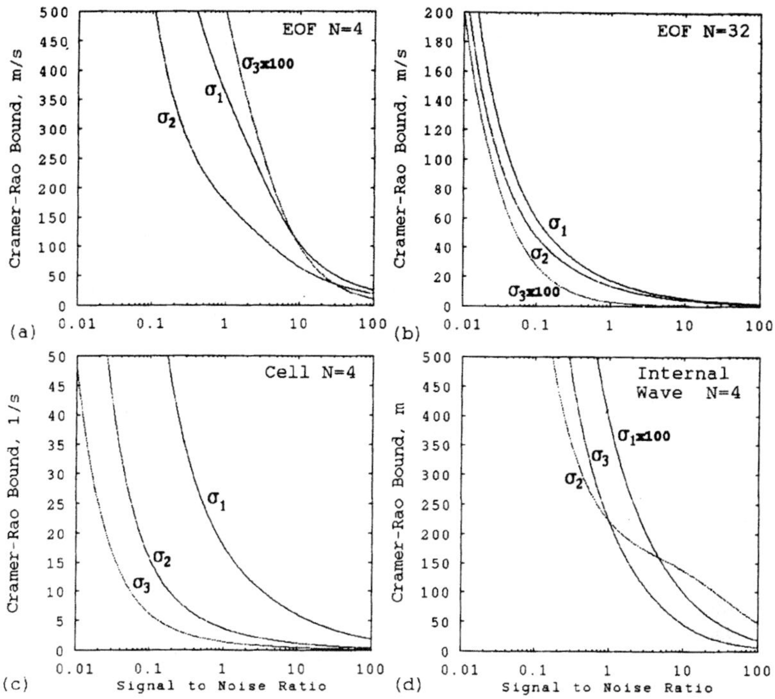

Figure 3.

3. Fine-Scale Sound Speed Parametrization: Internal Waves

Fine-scale variations in sound-speed structure may also have a significant effect upon acoustic propagation. These are typically due to such phenomena as internal waves. For example, we consider the two layer ocean of Figure 1. We assume that a long internal wave of wavelength λ_o^{int}=2 km and amplitude η_o=2 m propagates along the interface between the upper and lower layer. The upper layer is assumed to have a constant sound speed 1520 m/s which is greater than that of the lower layer by Δc=20 m/s. The propagation speed of the internal wave can be characterized by the shallow water, reduced gravity expression

$$c_{\text{int}} = \sqrt{\frac{\Delta\rho}{\rho} gH}\,, \tag{4}$$

which is non-dispersive and is valid for $H \ll D$. Typical density differences $\Delta\rho$ between the upper and lower layer are small enough to make $\Delta\rho / \rho \approx 10^{-3}$. Therefore, such long internal waves are essentially stationary over the period of a typical 10 km acoustic measurement, ie $c_{\text{int}} \approx 0.4$m/s. Both EOF and cell representations would require a large number of parameters to precisely represent such internal waves. A more economical parametrization of the waveguide sound speed structure would be a Fourier Series of long internal wave displacements at the upper-lower layer boundary,

$$h(x,y) = \sum_{n,m} \eta(k_x^{\text{int}}(n), k_y^{\text{int}}(m)) e^{i(k_x^{\text{int}}(n)x + k_y^{\text{int}}(m)y)}, \tag{5}$$

where $k^{\text{int}} = \sqrt{(k_x^{\text{int}})^2 + (k_y^{\text{int}})^2}$. In the case of a single planewave, the amplitude $\chi_1 = \eta_o = |\eta|$, phase $\chi_2 = \theta_o = \tan^{-1}\{\text{Im}(\eta) / \text{Re}(\eta)\}$, wavelength $\chi_3 = \lambda_o^{\text{int}} = 2\pi / k_o^{\text{int}}$, and propagation direction $\chi_4 = \varphi_o = \tan^{-1}(k_{oy}^{\text{int}} / k_{ox}^{\text{int}})$ of the boundary disturbance would be a sufficient parametrization. However, horizontal aperture is necessary to resolve the propagation direction. To maintain use of a single vertical array, as in the examples of the previous section, we assume that the wave travels off-shore, along the y axis. Therefore, only three unknown variables remain in this physics-based parametrization. The resolution of these parameters is shown in Figure 3d for the same source and geometry as in Figures 3a-c. Once again, the minimum estimation error is within acceptable bounds at high SNR, but quickly decays as SNR approaches and becomes less than unity. Although it is not shown, a substantial increase in resolution is obtained with the N=32 element array. For example, at SNR=0.1 we obtain σ_1=0.9 m, σ_2=71 m and σ_3= 46 m respectively, for parameters χ_1, χ_2, χ_3.

4. Conclusion

Minimum estimation error is within acceptable bounds for either EOF, cell or internal wave parametrizations in high signal to noise ratio (SNR), implying that the choice of parametrization is not limited by the theoretical bound. Rather, it is dependent upon the number of independent measurements, physical constraints, degrees of freedom and the spatial resolution specified. In lower SNR, more independent measurements are necessary to maintain the same minimum estimation error, implying that the theoretical bound may be useful in optimizing experimental design by taking advantage of a priori physical information.

References

1. A. Tolstoy, O. Diachok, L.N. Frazer (1991) "Acoustic tomography via matched field processing," *J. Acoust. Soc. Am.*, **89**, 1119-1127.
2. M.D. Collins and W.A. Kuperman (1991) "Focalization: Environmental focusing and source localization," *J. Acoust. Soc. Am.*, **90**, 1410-1423.
3. Personal communication with Colin Shen, Naval Research Laboratory, Wash. DC.
4. H.L. Van Trees (1970) *Detection, Estimation and Modulation Theory, Part III*, John Wiley and Sons, New York.
5. A.B. Baggeroer, W.A. Kuperman and H. Schmidt (1988) "Matched field processing: Source localization in correlated noise as an optimum parameter estimation problem," *J. Acoust. Soc. Am.* **83**, 571-587.
6. M.D. Collins (1992) 'A self-starter for the parabolic equation method,' *J. Acoust. Soc. Am.*, **92**, 2069-2074.
7. W.A. Kuperman and F. Ingenito (1980) "Spatial correlation of surface noise in a stratified ocean," *J. Acoust. Soc. Am.* **67**, 1988-1996.

ITERATIVE FULL-FIELD INVERSION USING SIMULATED ANNEALING

CLINTON SIEDENBURG and NORM LEHTOMAKI
Alliant Techsystems, Inc., MSW, Mukilteo, Washington 98275
JUAN ARVELO and KODALI RAO
Alliant Techsystems, Inc., ATC, Arlington, Virginia 22209
HENRIK SCHMIDT
Massachusetts Institute of Technology, Cambridge, Massachusetts 02139

ABSTRACT. Inversion for bottom properties at low frequency in shallow water environments can be a very difficult problem especially when little *a priori* information is available. This paper explores the sensitivity of the inversion process with respect to source aperture, frequency, bandwidth, receive aperture, shear wave speed, and geoacoustic model parameter selection. The exploitation of these sensitivity factors is illustrated by inverting three distinct environments.

1. Introduction

In addition to physical parameters of the environment, it is found that inversion for bottom properties can be very sensitive to receiver configuration, placement, orientation, and operating frequency. At least eleven parameters are required to specify the simple three-layer inversion problems discussed here when the water column properties are completely known. The ambiguity surface for this eleven-dimensional parameter space is extremely complicated due in part to the many local minima that obscure the search for the global minimum. A robust inversion scheme depends not so much on the tools employed as it does upon an understanding of the complexities of the underlying physics. The goal in this work is to begin to develop a robust inversion technique based upon observed physical principles that will unravel this complexity. We use a fast-field propagation model (OASES[1]) and a nonlinear optimization routine (Adaptive Simulated Annealing[2]) as the inversion tool set to search for the true geoacoustic model. The reduced reliance upon *a priori* information and the focus on exploiting the physics to reduce a complex inversion problem into a sequence of easier inversions distinguishes this work from the earlier work performed by Collins/Kuperman/Schmidt[3].

2. Inversion Approach

Six parameters are used to specify the surface sediment layer while five are used for the basement. The surficial sediment layer is specified by the thickness, density, and the choice of four parameters describing the dilatational and shear wave characteristics. The basement layer is similarly specified except the thickness parameter has no meaning for the half space.

The least squares single frequency cost function used here is $C = 1 - |u_{x_i}^H \cdot u_{p_i}|^2$ where u_{x_i} is the normalized measured pressure vector of the receive elements and u_{p_i} is the normalized predicted pressure vector at frequency, f_i. For a single omni source, the cost function for a broadband pulse is a simply weighted sum over frequency of CW cost functions.

O. Diaschok et al. (eds.), *Full Field Inversion Methods in Ocean and Seismo-Acoustics*, 121–126.

3. Sensitivity to Model Parameters

Figure 1 details the three environments from which these concepts have been deduced. Each environment is a three-layer problem having a water column above a sediment layer which is over a basement. Environment 1 is an example of a high shear speed environment with an iso-velocity water column. Environment 2 is a slightly upward refracting, low shear speed, very good propagating environment. Environment 3 is a severely downward refracting environment with very poor propagation characteristics. A realistic SNR of 30dB enables matches of the cost function to within 0.001. Ambiguity plots of the cost function are created by plotting $-log(C)$ where values above 3 (30dB) indicate a good complex pressure field match.

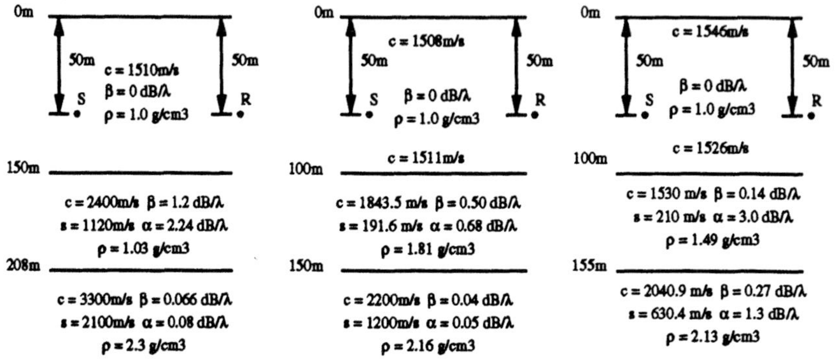

Figure 1. Geoacoustic models for Environments 1, 2, and 3.

3.1 PARAMETER SENSITIVITY AS A FUNCTION OF FREQUENCY

The sensitivity of the pressure field to the geoacoustic parameters below the surface sediment layer decreases with increasing frequency. This is due to the high sediment attenuation allowing little penetration of energy into the basement. A three-layer problem as modeled here, can be reduced to a two-layer problem by increasing frequency allowing five parameters of the top layer to be estimated separately from the basement parameters. This greatly simplifies the inversion problem. For Environments 1 and 2, 240Hz is sufficiently high to isolate the surface sediment layer. The highly bottom interactive environment, Environment 3, requires 960Hz.

3.2 PARAMETER SENSITIVITY AS A FUNCTION OF KR

Pressure field sensitivity is influenced by the model parameters as well as the receive aperture placement and extent. Long apertures and/or rapid pressure field variations increase parameter sensitivity. The term "KR product" is used to parameterize this concept which is directly analogous to the "BT product" of time series analysis. The time extent, T, is directly analogous to the spatial extent of the receive aperture, R. Analogous to time series plots, rapid changes in the transmission loss plots indicate high spatial frequency bandwidth, K. High KR products, as do high BT products, indicate a lot of information and so sensitivity is high.

When the search volume is large, simulated annealing may have difficulty converging. Although a cost surface that resembles the indicator function (unity everywhere except at the true parameter values where it is zero) yields the most accurate estimate of the parameters, simulated annealing would be required to correctly and simultaneously guess all the values. This is an extremely unlikely event. When well widths are a larger fraction of the search bounds, simulated annealing more easily converges to the true solution but with less accuracy. Aperture extent and position controls the well widths of the sensitivity surface. When no *a priori* information is available (requiring widest search bounds), the suggested inversion technique starts at a place where the KR product is low, reduces the bounds by a thresholding algorithm to be discussed later, and iterates with a larger KR product to improve accuracy.

The following figures show that sensitivity increases with KR. Figure 2 shows four sensitivity curves for the real part of the Lame′ parameter μ (related to shear wave) in the surface layer of Environment 1. The aperture begins at 10km (K is constant) and maintains an inter-element spacing of 4m. The curve with the widest peak width at the $-log(C) = 3$ line is an aperture of 12 elements. The aperture doubles for each curve, ending at 96 elements for the last. There is roughly a halving of the widths (or doubling of the sensitivity) for each doubling of the aperture length. Figure 3 shows a comparison for the density of a fixed 767m aperture (R is constant) placed first at 10km (solid) where K is low and then at 500m (dashed) where K is high. The sensitivity plots show a remarkable increase in sensitivity.

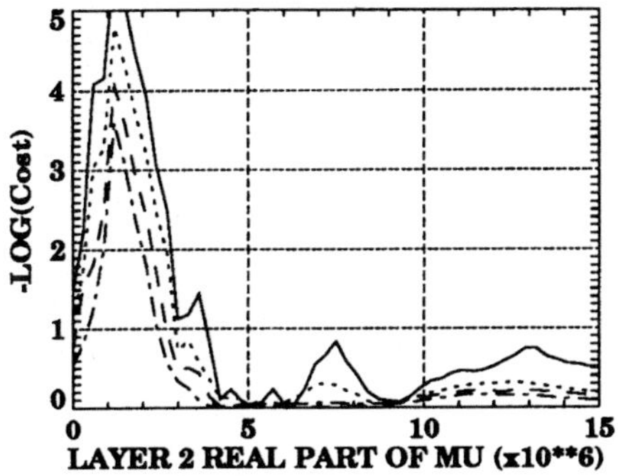

Figure 2. Sensitivity as a function of R.

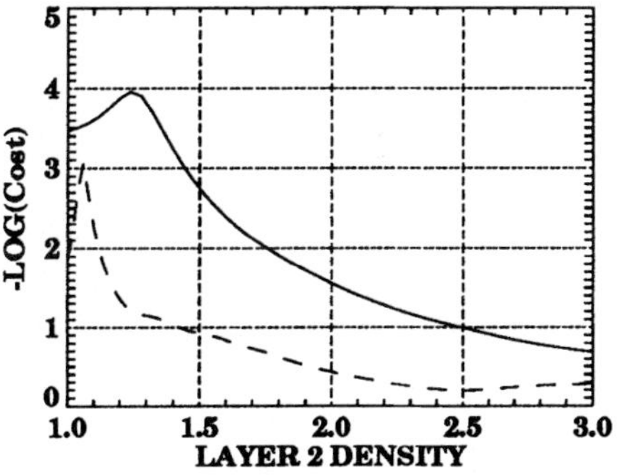

Figure 3. Sensitivity as a function of K.

3.3 INVERSION STRATEGY AS A FUNCTION OF SHEAR WAVE SPEED

The shear wave speed being either high or low provides a further reduction of two more dimensions in the initial stages of the optimization process. Environment 1 exemplifies a high shear wave speed problem. Sensitivity plots show there is no sensitivity to Lame′ parameter λ and extreme sensitivity to μ. This leaves only the surface layer density and complex parameter μ to be estimated for Environment 1 at 240Hz with the aperture placed at 10km. The parameter λ can be estimated accurately at a closer range of 500m (higher KR).

Environment 3 is a low shear speed problem in which shear can be initially ignored. Transmission plots show that both shear speed, s, and density, ρ, seem to modify the overall attenuation but leaves the shape of the curve the same. Both level and structure of the transmission loss curves show great sensitivity to dilatational speed.

3.4 IMPLICATIONS OF ACOUSTICALLY COUPLED MODEL PARAMETER SETS

The small number of physically meaningful parameters that can be chosen to model the acoustic properties of a layer are ρ, λ, μ, $\sqrt{(\lambda+2\mu)/\rho}$, and $\sqrt{\mu/\rho}$. The latter two are the complex dilatational and shear speeds, respectively. Although ρ, λ, and μ are statistically uncoupled, they are not necessarily acoustically uncoupled. This means ambiguity plots may have ridges that lie on a diagonal. Suppose that a sharp ridge lies on a diagonal and simulated annealing converges to a point along the diagonal away from the true values. One technique for progressively reducing search bounds is to run one-dimensional sensitivity plots at the non-optimal convergence point, threshold the sensitivity curve at $-log(C) = 2$, and set the new search bounds to the minimum and maximum intersection points. Doing this at a point up or down the sharp ridge would result in bounds that no longer contain the true values. Ambiguity ridges oriented either vertically or horizontally do not give rise to this pitfall. The desire is to pick a set of parameters that are acoustically uncoupled.

Environment 2 exemplifies the choice of parameters either resulting in coupled or uncoupled ambiguity plots. The ambiguity plot for the basement properties at 60Hz for ρ-μ is shown in Figure 4. Figure 5 shows the ρ-*s* ambiguity plot that has horizontal contours eliminating acoustic coupling where the $-log(C)$ surface is high.

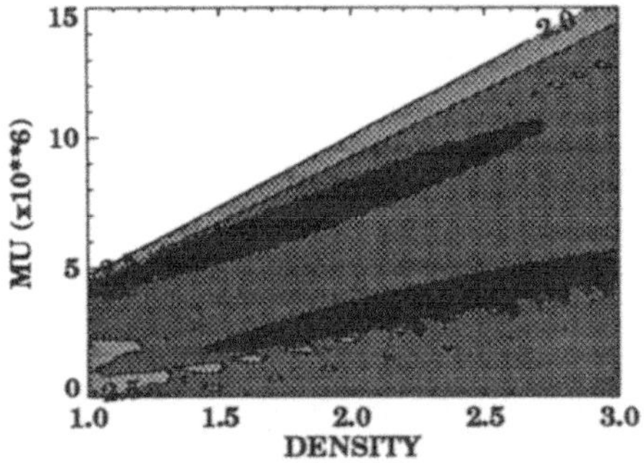

Figure 4. Coupled ambiguity plot.

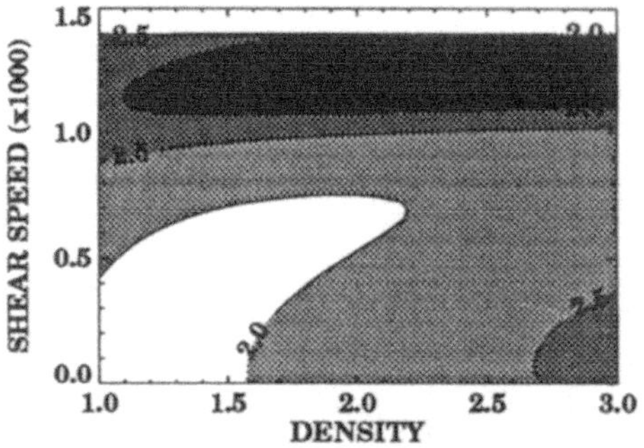

Figure 5. Uncoupled ambiguity plot.

3.5 HLA VERSUS VLA APERTURES

A surprising result is that little difference is found between the sensitivities of a vertical line array or horizontal line array of similar length and spacing in Environment 1. Figure 6 compares the sensitivity plots of surface sediment density for the HLA (solid) and VLA (dashed) cases at 240Hz and 2km. The VLA uses a 50-element aperture with a 3m inter-element spacing. The HLA uses a 48-element aperture with a 4m spacing. Although not shown here, the same effect is true at much closer ranges, lower frequencies, and for other parameters.

3.6 BROADBAND VERSUS CW EXCITATION

A broadband pulse generally yields some increase in sensitivity to the model parameters over CW pulses due to the addition of information from the other frequencies. This increase in sensitivity is obtained at a penalty of greatly increased CPU loading due to calculating the channel transfer function at each frequency of the modelled broadband pulse. As an example, in Environment 1 the sensitivity for the CW case is the same as that of the wideband case for the surface sediment parameter ρ using a 240Hz center frequency with 240Hz of bandwidth. A 30Hz center frequency pulse with 30Hz of bandwidth yields fewer ambiguous peaks and some increase in sensitivity. It should be noted that broadband excitation does not preclude the ability to perform the inversion at a single frequency so long as the spectral content of the received signal can be decomposed into narrow bands of energy around the frequencies of interest.

4. Inversion Examples

4.1 ENVIRONMENT 1

The inversion is performed in a number of steps in which simulated annealing is first run where KR is low to converge to a solution that matches to 10^{-3}. A set of one-dimensional sensitivity plots is run at the point of convergence. The bounds are reduced by taking the minimum and maximum parameter values resulting from the intersection of the line $-log(C) = constant$ and the sensitivity curve. The inversion process is started at a high frequency where the basement parameters are of no importance and at a far range where there is little sensitivity to the dilatational wave characteristic. By first moving the aperture closer to the source and then lowering frequency, the eleven-parameter problem was sequentially inverted with little *a priori* information. The final parameter estimates are shown along with the transmission loss for both the true values (solid) and estimated values (dashed) in Figure 7.

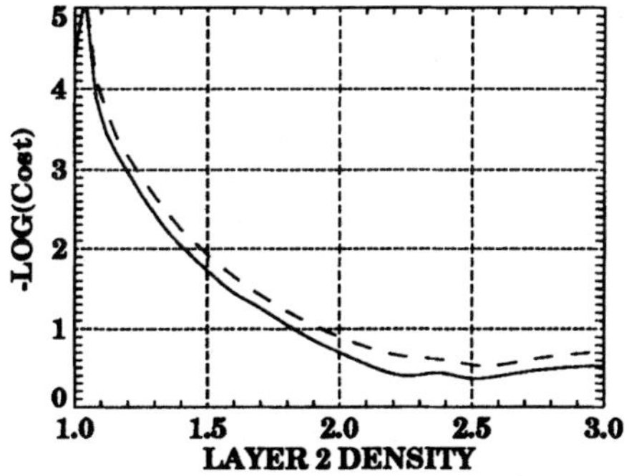

Figure 6. HLA versus VLA at 240Hz.

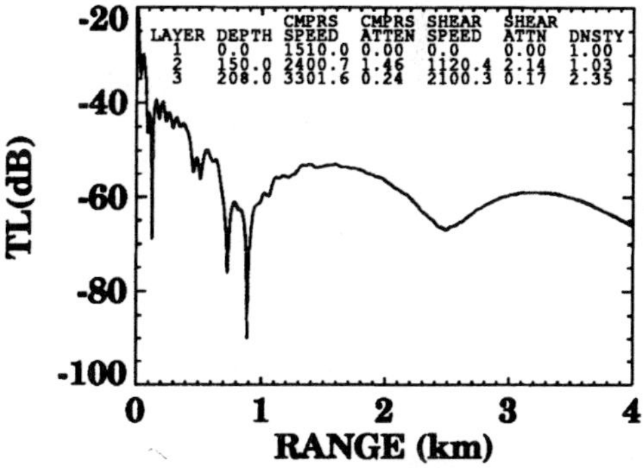

Figure 7. Environment 1 inversion.

4.2 ENVIRONMENT 2

Although 240Hz is sufficiently high to gain immunity to the effect of the basement parameters, the interference among the many strong modes causes a high KR product so that convergence of simulated annealing is difficult. By employing a 21-element VLA projector shaded with a Hanning function, most of the modes giving rise to the rapidly varying transmission loss function are removed. The aperture is placed at 10km and convergence of simulated annealing occurs. Figure 8 shows that the true (solid) and estimated (dashed) transmission loss curves are nearly indistinguishable although the estimated values indicate significant percentage error.

4.3 ENVIRONMENT 3

Due to the highly downward refracting environment, 960Hz is used to isolate the surface sediment properties from the basement. To start the inversion, shear wave is ignored. A relatively low-KR region is observed by the receive aperture beginning at 3.0km. Simulated annealing is run until it appears that the lowest cost is obtained. A sensitivity analysis is performed and tight bounds are set on the density and dilatational wave parameters. The bounds are opened wide for the shear wave properties and simulated annealing is run again. The resulting transmission loss is shown in Figure 9 along with the parameter match.

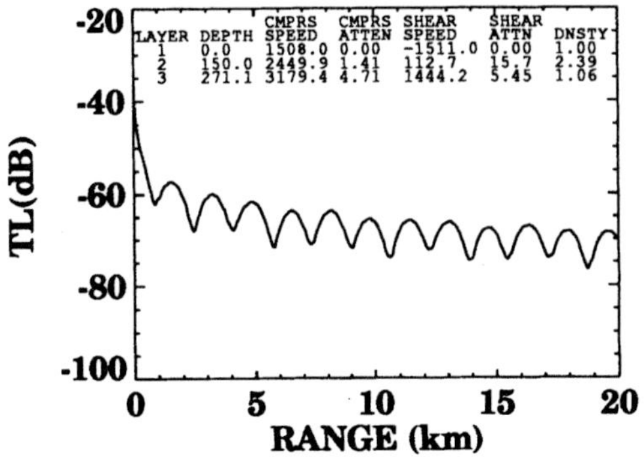

Figure 8. Environment 2 inversion.

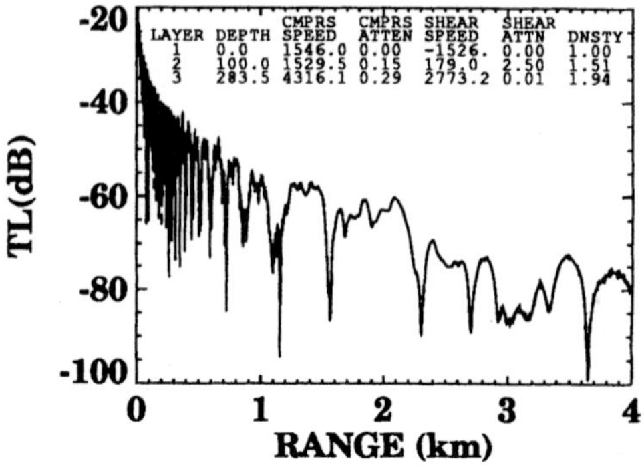

Figure 9. Environment 3 inversion.

5. Conclusions

Inversion of bottom properties with simulated annealing is a complicated problem. This problem, however, can be broken down into a number of solvable steps by taking advantage of the observations made of the physics involved. These observations lead to: 1) increasing frequency to isolate the surface sediment parameters, 2) placing the receive aperture at low and high KR positions to select the correct sensitivity level for simulated annealing, 3) modifying the inversion strategy by assumptions in shear wave speed, and 4) choosing an uncoupled set of parameters for proper search bound reduction. Use of horizontal and vertical receive apertures are comparable. A CW instead of broadband inversion analysis is sufficient. Broadband pulses may be decomposed into the desired frequencies for CW analysis. Example inversions using these principles are shown. A judicious use of these principles may yield a robust methodology for inversion of shallow water environments with little *a priori* information.

6. Acknowledgements

This work was supported by the Advanced Environmental Acoustic Support Program of the U.S. Office of Naval Research, under contract number N00014-92-C-6009.

7. References

1) H. Schmidt, "OASES Version 1.6", Department of Ocean Engineering, Massachusetts Institute of Technology.

2) L. Ingber, "Adaptive Simulated Annealing (ASA)," [ftp.caltch.edu: /pub/ingber/asa.Z], Lester Ingber Research, McLean VA (1993).

3) M.D. Collins, W.A. Kuperman and H. Schmidt, "Nonlinear inversion of ocean-bottom properties," J. Acoust. Soc. Am. 92, 2770-2783 (1992).

A COMPARATIVE STUDY OF TWO WAVE-THEORETIC INVERSION SCHEMES IN OCEAN ACOUSTIC TOMOGRAPHY

G.ATHANASSOULIS[2,1] J.PAPADAKIS[1,3] E.SKARSOULIS[1,2] M.TAROUDAKIS[1,3]

[1] *Foundation for Research and Technology-Hellas, Inst. of Applied & Computational Mathematics, P.O.Box 1527, 711 10 Heraklion, Crete, GREECE*

[2] *National Technical University of Athens, Dept. of Naval Architect. & Marine Engng P.O.Box 64070, 157 10 Zografos, Athens, GREECE*

[3] *University of Crete, Dept. of Mathematics, 714 09 Heraklion, Crete, GREECE*

ABSTRACT A comparative study of two wave-theoretic schemes for ocean-tomography inversion is presented. The first scheme is based on the notion of modal arrivals (modal inversion), while the second scheme is based on the recently introduced notion of wave arrivals (peak inversion). A case has been chosen for which both schemes are applicable. For modal inversion an iterative scheme has been used, while peak inversion is carried out as a single-step procedure. The retrieval of the actual sound speed profile is satisfactory for both schemes, with peak inversion giving slightly better results.

1. Introduction

Ocean acoustic tomography, introduced by Munk and Wunsch [1] as a tool for monitoring the sea environment, has been traditionally implemented using ray-theoretic modelling. In this context, arrival times are modelled as travel times along eigenrays connecting source and receiver. Their variations are associated with variations of the sound-speed profile in the water column, forming the basis of the corresponding ray-inversion scheme [1]. This approach, however, suffers from restrictions posed by ray separability conditions, as well as by water-boundary interferences, not easily considered in ray acoustics, especially when ray tomography is applied to shallow water conditions. Furthermore, the applicability of ray-inversion techniques to low frequencies is questionable, since ray theory is in fact a high-frequency asymptotic method.

An alternative approach, also proposed by Munk and Wunsch [1], is modal tomography [2], [3], based on normal-mode theory. Now the peaks of the received acoustic signal are associated with arrivals of specific propagating modes. Under certain conditions, modal arrivals can be identified in the arrival pattern, finally leading to an inversion scheme, usually called modal-inversion scheme. Nevertheless, modal arrivals are not always identifiable, and this is the major restriction to the applicability of this scheme. In any case, this approach has some advantages over the traditional ray-theoretic approach, since bottom and surface effects can be taken into account and there is no restriction for low-frequency cases.

Recently another approach has been developed [4] that does not require the interpretation/identification of actual arrivals (i.e. observed arrivals during the experiment) as ray or modal ones. Consequently, the resulting inversion scheme is free from any restrictions posed by the conditions of mode or ray resolvability. The major issues of this approach are: (i) the modelling of arrival times as the time instants corresponding to the significant peaks

O. Diaschok et al. (eds.), Full Field Inversion Methods in Ocean and Seismo-Acoustics, 127–132.

of the arrival pattern (wave arrivals), and (ii) the calculation of the perturbation of the so-defined arrival times analytically by means of an appropriate Fréchet derivative. These two steps permit to formulate a novel (perturbative) inversion scheme, which will be called in the sequel *peak-inversion scheme.*

This paper presents a comparative study of the modal and peak inversion schemes, including also a short presentation of the theoretical and numerical background of the two methods. Although the presentation here is restricted to the range-independent case, both methods can also be applied to range-dependent environments.

2. The ocean sound channel

In the framework of linear acoustics, the ocean–sound channel intervening between an acoustic source (S) and a receiver (R) can be considered as a linear system with transfer function $H_{SR}(\omega)$, which is essentially the Green function of the studied environment. If the source excitation function is $X_S(\omega)$, the signal at the receiver in the frequency domain is a complex function $X_R(\omega)$ given by

$$X_R(\omega) = H_{SR}(\omega) \cdot X_S(\omega). \tag{1}$$

By taking the inverse Fourier transform of (1), we can express the received signal in the time domain as:

$$x_R(t) = \frac{1}{2\pi} \int_{-\infty}^{+\infty} H_{SR}(\omega) X_S(\omega) e^{j\omega t} d\omega \ . \tag{2}$$

The amplitude $a(t) = |x_R(t)|$ is called *arrival pattern.* If the emitted signal $x_S(t)$ is a pulsed signal, then $a(t)$ consists, in general, of a sequence of peaks. These peaks are the *actual arrivals.* Their temporal locations, τ_i, $i = 1, 2 \ldots, I$, (arrival times) are the most stable and well–defined characteristics of the arrival pattern, and consist a basic set of observables in tomographic experiments. Perturbations in the sound–speed field cause perturbations to the arrival times. The latter perturbations $\delta\tau_i$, $i = 1, 2 \ldots, I$, are used in travel–time ocean acoustic tomography for the retrieval of the corresponding sound–speed perturbations.

The above notion describes the essence of an ocean-tomography setting, although some specific technical issues have not been included herewith. E.g., in practice, the cross–correlation of the received signal with a replica of the emitted signal is considered, instead of the received signal itself [5], [6].

3. Modal arrivals

Under certain conditions, the acoustic energy propagating via a specific mode forms a distinguishable peak in the arrival pattern. In this case actual arrivals can be associated with separate arrivals of specific modes of propagation, called *modal arrivals.* Conditions of mode resolvability involve the signal bandwidth and the source-receiver distance and they are also related to the modal group velocity at which modal energy propagates [3]. The group velocity corresponding to the n-th mode is defined as

$$v_{g,n} = \frac{\partial \omega}{\partial k_n}\Big|_{\omega_0}, \tag{3}$$

where ω_0 is the central frequency of the signal. The associated modal travel time is

$$\tau_n = \frac{R}{v_{g,n}}. \tag{4}$$

A range-independent environment will be considered in the subsequent analysis and the sound-speed profile $c(z)$ will be represented by means of a series expansion of the form

$$c(z) = c_0(z) + \sum_\ell \theta_\ell \cdot f_\ell(z), \tag{5}$$

where $c_0(z)$ represents a reference sound–speed profile, $f_\ell(z)$ constitute a set of sound–speed modes, and θ_ℓ are the corresponding sound–speed modal parameters. In order to indicate the parametric dependence of the sound–speed field on $\vec{\theta}$ we shall write $c(z;\vec{\theta})$, where $\vec{\theta}$ is the vector of modal parameters.

Since the modal group velocity is associated with the eigenvalues of the pressure expansion, the perturbations of the modal travel time, due to some perturbation of the sound speed with respect to some background environment ($\vec{\theta} = \vec{\theta}_b$), can be easily expressed in terms of the eigenvalue perturbations [3]. The final expression is:

$$\delta\tau_n = \sum_\ell \delta\theta_\ell \; R \int_0^\infty \frac{\partial Q_n(z;\vec{\theta}_b)}{\partial\omega}\Big|_{\omega_0} f_\ell(z)dz = \sum_\ell \delta\theta_\ell A_{n\ell}, \tag{6}$$

where $\delta\vec{\theta} = \vec{\theta} - \vec{\theta}_b$ is the deviation of actual condition from the background one, $Q_n(z;\vec{\theta}_b)$ is a known kernel calculated at the background environment [7], and $A_{n\ell}$ is the corresponding influence matrix. This formula is used in modal inversion for the recovery of the sound speed disturbance $\delta c(z)$, from known (measured) perturbations $\delta\tau_n$ in modal travel times.

4. Wave arrivals

Wave arrivals are defined as the significant local maxima of the arrival pattern, and they consist a direct theoretical counterpart to actual (experimental) arrivals [4]. The arrival times τ_i, corresponding to the background conditions ($\vec{\theta} = \vec{\theta}_b$), satisfy the equation

$$a'(\tau_i;\vec{\theta}_b) = 0, \tag{7}$$

where $a(t;c(z)) = a(t;\vec{\theta})$ is the amplitude of the pressure at receiver's location (arrival pattern), and a prime denotes differentiation with respect to time t. The perturbed arrival times $\tau_i + \delta\tau_i$, corresponding to the actual sound speed field, considered as a perturbation of the background profile ($\vec{\theta} = \vec{\theta}_b + \delta\vec{\theta}$), will then satisfy the equation:

$$a'(\tau_i + \delta\tau_i;\vec{\theta}_b + \delta\vec{\theta}) = 0. \tag{8}$$

A first-order Taylor expansion of this equation, in connection with eq. (7), gives the formula:

$$\delta\tau_i = -\sum_\ell \delta\theta_\ell \frac{\partial a'(\tau_i;\vec{\theta}_b)/\partial\theta_\ell}{a''(\tau_i;\vec{\theta}_b)} = \sum_\ell \delta\theta_\ell B_{i\ell}. \tag{9}$$

The perturbation formula (9) is not subject to any restrictive assumptions concerning the acoustic signal, the environment through which the sound propagates, or the propagation model that is to be used. For the calculation of the influence coefficients $B_{i\ell}$ the knowledge of

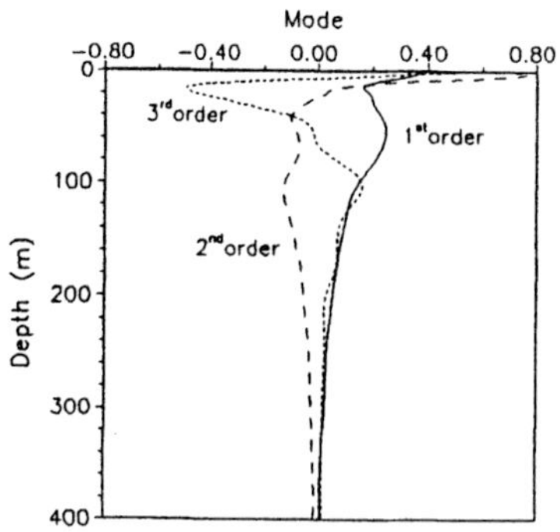

FIGURE 1. The sound speed modes of the test environment

the transfer function $H_{SR}(\omega)$ and its derivatives $\partial H_{SR}(\omega)/\partial f_\ell$ is required. Time derivatives appearing in (9) can be circumvented by using elementary properties of Fourier transform. Analytic expressions for $H_{SR}(\omega)$ and $\partial H_{SR}(\omega)/\partial f_\ell$ can be derived, by using e.g. normal mode theory [4], although any other model for solving the direct problem can be used.

The perturbation formula (9) comprises a system of linear equations for the perturbations $\delta\vec{\theta}$ of modal coefficients, and can be subsequently used as an inversion system for the determination of $\delta\vec{\theta}$ from known (measured) arrival-time perturbations $\delta\tau_i$.

5. Inversion

The perturbation formulas (6) and (9) can be written in matrix form:

$$\{\delta\tau\} = [\mathbf{G}]\{\delta\theta\}, \tag{10}$$

where $\{\delta\tau\}$ is an M-dimensional vector of the travel-time perturbations, $[\mathbf{G}]$ is an $M \times L$ matrix containing the influence coefficients (calculated at the background environment), and $\{\delta\theta\}$ is an L-dimensional vector containing the sound–speed modal-parameter perturbations $\delta\theta_\ell$. Usually, $M \geq L$ and the system (10) (inversion system) is to be solved by applying singular value decomposition.

As an application example a range-independent environment is considered consisting of a 400m-deep water column over a semi-infinite bottom. In order to concentrate on the acoustic energy propagation through the water, an artificially absorbing bottom [4] is considered. The sound-speed profile is described by a linear reference profile $c_0(z)$, ranging from 1494 m/sec at the surface to 1509 m/sec at 400 m depth, and three sound-speed modes $f_\ell(z)$, $\ell = 1, 2, 3$, presented in figure 1. For the "actual" condition the following values for the coefficients θ_ℓ are assumed: $\vec{\theta} = (-44.713, 44.438, -14.892)$. In order to recover the actual sound speed profile, we start from a background profile determined by $\vec{\theta}_b = (-19.21, 27.84, -11.10)$. A gaussian pulse of central frequency 200Hz and bandwidth 100Hz is used to simulate the tomographic signal. It is considered that the source and the receiver are placed at the same depth of 100m and at a range of 100km. Figure 2 shows the arrival pattern for the background environment (dotted line) and also for the actual environment (solid line). The most prominent peaks of the background pattern, modelled as wave arrivals, are consecutively numbered from 1 to 11; all these peaks can be also identified as modal arrivals except for peak No 5.

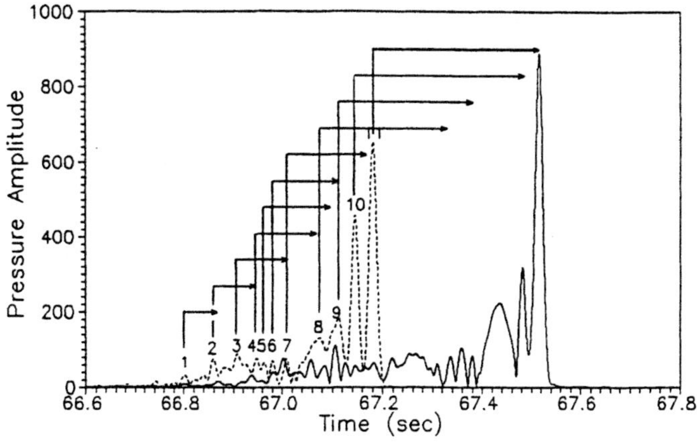

FIGURE 2. The arrival patterns of the background (dashed) and the actual environment (solid)

Figure 2 also shows, in the form of horizontal arrows, the predicted arrival displacements using the wave-arrival approach (eq. (9)). It is seen that the prediction is very successful for the first six peaks (No 1 to No 6) and also for the last two peaks (No 10 and No 11), while for the three intermediate ones (No 7 to No 9) there seems to be no one-to-one correspondence between background and perturbed conditions.

Thus, the arrival pattern can be divided into three parts: (i) the early arrivals, which are subject to moderate form changes (apart from shifting), (ii) the late arrivals, which also undergo minor form changes, and (iii) the intermediate part which is subject to severe form changes (including stretching and generation of new peaks) from background to perturbed conditions. For the early and late parts of the arrival pattern the perturbation formula (9) works very satisfactorily, while in the intermediate part the prediction is rather poor.

As a first step, the three most significant peaks (No 9, 10, 11 in figure 2) are considered for the inversion. From the modal point of view these peaks correspond to modes 4, 1, and 2, respectively. Figure 3 presents the results obtained by the two inversion methods. Modal-inversion results have been obtained from an iterative scheme [7], after ten iterations. On the other hand, the results from peak inversion have been obtained in a single step, as the solution of the inversion system (10). It is clear that the profile of the actual environment is estimated with acceptable accuracy by both methods. The method based on the wave arrivals gives better results and the sound speed profile is almost perfectly recovered. In order to examine the sensitivity of the peak-inversion scheme, a further application has been carried out by using 8 wave arrivals, the first six (No 1-6) and the late two (No 10, 11), i.e. excluding the intermediate part of the arrival pattern. This resulted in an even better approximation of the actual sound-speed profile.

6. Conclusions

A comparative study between two wave-theoretic inversion schemes (modal and peak inversion) has been presented. The modal approach offers some insight in the structure of the arrival pattern and results in a relatively simple scheme; nevertheless its applicability critically depends upon mode resolvability, and, still, an experienced user is required for the identification of modal arrivals in the background and actual arrival pattern. Mode

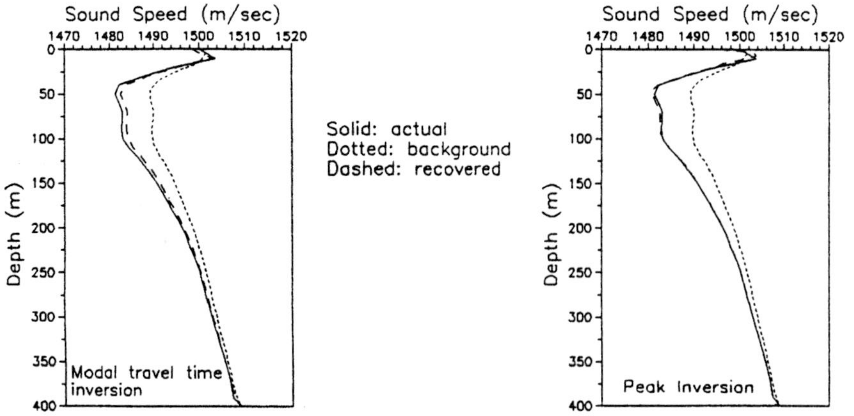

FIGURE 3. The actual, background and recovered sound speed profiles

resolvability condition limits the applicability of modal inversion mainly to shallow water and low frequencies.

On the other hand, wave arrivals are free from any restrictive assumptions such as ray or modal resolvability, and they may apply for tomographic signals of arbitrary bandwidth. The numerical efficiency of the perturbation formula (9) for the prediction of arrival displacements is impressive. The applicability of (9) is free of any kind of arrival-identification procedure, and thus the method can be used for continuous tracking of arrivals during tomographic experiments; furthermore the method can be automated to a great extent. However, the use of full-wave representation for the arrival pattern results in a computationally heavier scheme than the modal one.

Acknowledgements

This work was partly supported by the European Union, under contract MAST-0008 C(MB) (THETIS). We would like also to thank Miss G.Tzompanopoulou for her computational assistance.

References

1. Munk, W.H. and Wunsch, C. (1979) Ocean acoustic tomography: A scheme for large scale monitoring, *Deep-Sea Res.* **26A**, 123–161.
2. Munk, W. and Wunsch, C. (1983) Ocean acoustic tomography: Rays and modes, *Rev. Geophys. Space Phys.* **21**, 777-793.
3. Shang, E.C. (1989) Ocean acoustic tomography based on adiabatic mode theory, *J. Acoust. Soc. Am.* **85**, 1531–1537.
4. Athanassoulis, G.A. and Skarsoulis, E.K. (1994) Arrival-time perturbations of broadband tomographic signals due to sound-speed disturbances. A wave-theoretic approach, (submitted for publication).
5. Spindel, R.C. (1979) An underwater acoustic pulse compression system, *IEEE Trans. Acoust. Speech Signal Proc.* **ASSP-27**, 723–738.
6. Metzger, Jr.K. (1983) *Signal processing equipment and techniques for use in measuring ocean acoustic multipath structures*, Ph.D. Thesis, Univ. of Michigan, Ann Arbor.
7. Taroudakis, M.I. (1994) A comparison of the modal-phase and modal-travel time approaches for ocean acoustic tomography, in Proc. 2nd European Conference on Underwater Acoustics. Copenhagen, Denmark.

DIRECT INVERSE ESTIMATION FOR A FLUCTUATING SHALLOW WATER ENVIRONMENT

R.M. OBA
Naval Research Laboratory
Stennis Space Center, MS 39529-5004
USA

ABSTRACT Consider a shallow, strongly stratified ocean with sediment modelled by a fluid half space bottom. In the radially symmetric case, P.A. Deift and D.C. Stickler in [1] proposed a direct, non-perturbative inverse method for finding a sound speed profile in perfect stratification. They restrict themselves to cases without propagating modes and require that the sound speed in the half space converge rapidly to some constant. Their model uses a continuous wave point source. The method first makes integral transforms of information from horizontal acoustic data. Then techniques arising from one dimensional inverse quantum scattering are used to find depth dependence. The data needed for recovery is the normal derivative of the pressure in the entire top surface. Even in a perfectly stratified environment, this type of data is likely to be contaminated or to be available only from estimates. Thus one needs to look at inversion data arising from averages of actual propagation with environmental fluctuation that is dominated by stratification. Since this type of averaged information may not satisfy the appropriate technical conditions of non-averaged data, it must be related the allowable inversion data. The inversion technique can then be applied, and the resulting sound speed profile can then be compared in numerical experiment to the profile ensembles which gave rise to it. The reintroduction of several propagating modes can be explored through methods of P.A. Deift and E. Trubowitz [2], if good estimates of the vertical eigenvalues can be made. The direct inversion method and inverse quantum scattering methods also allow ensembles of sounds speeds to provide a typical or averaged sound speed profile that will have some of the characteristics of the average propagation.

1. Introduction

This analysis outlined here is directed toward answering the question of what relation a single velocity profile has to the average pressure field of an ensemble of profiles, and specifically how to pick the single profile. The format here will start with quick overview of the determistic forward problem and an introduction of the various symbols. Then we procede to an overview of the Stickler-Deift [1] method. They used integral transforms to rewrite the inverse problem for a stratified medium (ocean plus sediments plus sediment with a pressure release surface driven by a continuous wave source at depth z_0) into problem of inverse scattering from Schroedinger quantum theory. Then some aspects of a stochastic model and average solutions over varied velocity profiles is considered. This will lead to a brief analysis of the type of adjustments required to obtain invertible data.

2. Forward Problem

In this section and the next, the notation of [1] will be adhered to as closely as is possible. Consider the standard ocean acoustic propagation problem that depends only on r and z in

O. Diaschok et al. (eds.), Full Field Inversion Methods in Ocean and Seismo-Acoustics, 133–138.

a cylindrical coordinate system with its origin at $z = 0$ and with the positive z-axis passing through the point source located at $z = z_0$. The boundary conditions include a pressure release surface pressure release surface at $z = 0$ and outward going radiation condition as $z \to \infty$ and $r \to \infty$. This analysis requires a continuous wave source at depth z_0 in $e^{i\omega t}$ with angular time frequency ω, which is small enough that there are no propagating modes. Thus, this is a low frequency method, unless facts from the propagating modes are reintroduced (as in [2]). Stickler and Deift take sound speed $c(z)$ to be depth dependent only, and constant density is assumed for all z and r. Finally $c(z) \to c_\infty$ will be small in a sense of an integral norm. The pressure field $u(x)$ in r and z in a cylindrical coordinate system satisfies

$$\frac{1}{r}\frac{\partial}{\partial r}\left(r\frac{\partial u}{\partial r}\right) + \frac{\partial^2 u}{\partial z^2} + k_\infty^2 u = -\frac{\delta(r)\delta(z-z_0)}{2\pi r} \tag{1}$$

with $u = 0$ on the top surface $z = 0$, and $u(r, z)$, outward going as $r \to \infty$. The condition on $c(z) \to c_\infty$ as $z \to \infty$ is as follows: the index of refraction is $n(z) = \frac{c_\infty}{c(z)}$, and define $k_\infty = \frac{\omega}{c_\infty}$, the wave number as $z \to \infty$. The variation of n from unity (times k_∞) is $q(z) = k_\infty^2[1 - n(z)]$. Also $k_\infty^2 - k^2 = \beta^2$. Take q to have finite integral

$$\int_0^\infty |q(z)|z \, dz < 1, \tag{2}$$

where ω is taken small enough. For small ω (2) will hold automatically, if $c(z) \equiv c_\infty$ for $z > z_s$ large enough in some constant velocity sub-sediment starting at z_s. This inequality implies there are no propagating modes [4]. In order to have sufficient convergence properties, it is desirable to have $c(z)$ (and so $q(z)$) be twice continuously differentiable except at water/sediment transitions. The pressure field, $u(r, z)$, has the following representation

$$u(r,z) = \int_0^\infty v(\beta,z)J_0(\beta r)\beta \; d\beta \,, \quad v(\beta,z) = \int_0^\infty u(r,z)J_0(\beta r)r \; dr. \tag{3}$$

where $J_0(x)$ is the Bessel function of the first kind, and the function $v(\beta, z)$ is the Hankel transform of u. This separation of variables shows that $v(\beta, z)$ satisfies

$$\frac{\partial^2 v}{\partial z^2} + [k_\infty^2 n^2 - \beta^2]v = -\frac{\delta(z-z_0)}{2\pi} \tag{4}$$

with $v(\beta, z) = 0$ when $z = 0$ and $v(\beta, z)$, outward going as $z \to \infty$.

Consider the operator $-\frac{d^2}{dz^2} + q$ on $0 < z < \infty$. The Jost solutions $f(z, k)$ solve the Schroedinger equation

$$-\frac{d^2 f}{dz^2}(z,k) + q(z)f(z,k) = k^2 f(z,k), \tag{5}$$

with radiation condition $f(z, k) \sim e^{ikz}$, as $z \to \infty$. (The $\sim$ is $=$ in constant velocity sub-sediment.) The Dirichlet solution $y(z, k)$ of the Schroedinger equation (5) satisfies $y(0, k) = 0$ and $y'(0, k) = 1$. The function $v(\beta, z)$ has the Green's function form expressed in terms of the f and y as

$$v(\beta,z) = \begin{cases} \frac{y(z,k)f(z_0,k)}{2\pi f(0,k)}, & z \le z_0 \\ \\ \frac{y(z_0,k)f(z,k)}{2\pi f(0,k)}, & z \ge z_0 \end{cases} \tag{6}$$

For k real, y can be expressed $y(z,k) = \frac{f(z,k)f(0,-k)-f(z,-k)f(0,k)}{2ik}$. Finally, the scattering operator

$$S(k) = \frac{f(0,-k)}{f(0,k)}, \tag{7}$$

quantifies how and incoming (upward) wave $f(z,-k)$ is converted to an outgoing (downward) wave $f(z,k)$ at the surface. $S(k)$ will contain only phase information. Differentiation shows that the discriminant $\frac{df}{dz}(x,k)f(x,-k) - f(x,k)\frac{df}{dz}(x,-k) = 2ik$ for k real. Since there are no propagating modes, the operator $-\frac{d^2}{dz^2}+q$ on $0 < z < \infty$ has no proper eigenvalues[2]. In this case, knowledge of the scattering operator $S(k)$ is sufficient to determine $c(z)$. In [1] another adjustment to ω is suggested instead of this smallness assumption. This is unlikely to be simultaneously satisfied by all sound speeds in an ensemble, so the smallness condition will be required.

3. Inverse Problem

This inversion technique will be rather delicate, and Stickler [3] has proposed a more robust alternative. This method does provide framework for the inverse methods of [6] and [7] in the case of normal modes (in finite depth models). Inversion begins with physical measurement of the normal derivative of the pressure field at the pressure release surface:

$$g(r) = \frac{\partial u}{\partial z}(r,z)\bigg|_{z=0}.$$

This measurement can be used to determine

$$h(k) = \frac{\partial v}{\partial z}(\beta,z)\bigg|_{z=0}$$

as follows: for $0 \le \beta \le k_\infty$, Hankel transform as in 3 to find

$$h(k) = v(\beta,0) = \int_0^\infty g(r)J_0(\beta(k)r)r\,dr$$

directly for $0 \le k \le k_\infty$. For $\beta = i\eta$, $\eta \ge 0$, the Hardy space of square integrable functions analytic in the upper half plane is needed to rewrite the Mellin transform H of h along a contour for which the integral converges. Take the Hankel transform of g

$$G(\eta) = \int_0^\infty g(r)rJ_0\left(r\sqrt{k_\infty^2+\eta^2}\right)dr;$$

and then Mellin transform the result

$$H(t) = \frac{e^{-\pi t/2+i\pi/4}}{\sqrt{2\pi}}\int_0^\infty G(\eta)\eta^{it-1/2}\,d\eta.$$

Now inverse Mellin transform to the appropriate contour for $\beta = i\eta$, $\eta \ge 0$ to compute [8]

$$h(k) = h(i\eta) = \frac{1}{\sqrt{2\pi}}\int_{-\infty}^{\infty} H(t)k^{-it-1/2}\,dt$$

for $k_\infty \le k \le \infty$. These transforms to find $h(k)$ will require some care, as it will be necessary to find the zeros of $\mathrm{Im}h(k)$. The positive zeros of $\mathrm{Im}h(k)$ partially determine $|f(0,k)|$.

Reconstruction if $f(0,k)$ from $|f(0,k)|$ follows by standard inverse scattering method. First argue that initial value $y(0,k) = 0$ for all k, so that the zeros (k^2) of $y(z_0,k)$ are the eigenvalues of the self-adjoint operator $-\frac{d^2}{dz^2} + q$, on $0 \le z \le z0$ of the Sturm-Liouville problem with Dirichlet boundary conditions. These eigenvalues $k^2 = \mu_j^2 > 0$, since the operator $-\frac{d^2}{dz^2} + q$ on $0 \le z < \infty$ has no negative eigenvalues corresponding to square integrable eigenfunctions. The zeros of $y(z_0,k)$ are

$$\mu_j^2 = \frac{\pi^2 j^2 + \delta_j}{z_0^2}, \quad \text{where } \delta_j = \delta_j(z_0)$$

for $j \ge 1$. The δ_j are bounded as $j \to \infty$, but need not converge to 0. Size estimates for $y(z_0,k)$ as an entire function, imply it has an infinite product expansion [5]

$$y(z_0,k) = z_0 \prod_{j=1}^{\infty} \frac{\mu_j^2 - k^2}{(\pi j z_0^{-1})^2}, \quad y(z_0,k) \sim \frac{\sin kz_0}{k} \quad \text{as } k \to \infty;$$

if $q \equiv 0$ then $\sim$ can be replaced by $=$. In terms of computational approximation, take $\delta_j \equiv 0$ for $j > N$ so that

$$y(z_0,k) \approx \prod_{j=1}^{N} \frac{\mu_j^2 - k^2}{(\pi j z_0^{-1})^2 - k^2} \frac{\sin kz_0}{k}$$

For k real $f(0,-k) = \bar{f}(0,k)$ (where $\bar{z}$ indicates complex conjugate of z), it follows that

$$\frac{2iky(z_0,k)}{|f(0,k)|^2} = 2i\mathrm{Im}\frac{f(z_0,k)}{f(0,k)} = 4\pi i\mathrm{Im}h(k).$$

Hence $h(k)$ determines the positive zeros and hence all the zeros of $y(z_0,k)$. Rewrite this to find

$$|f(0,k)|^2 = \frac{ky(z_0,k)}{2\pi\,\mathrm{Im}h(k)}$$

Suppose (see [7]) $f(0,k)$ can be written as

$$f(0,k) = \exp\left(\int_0^\infty \gamma(t)e^{ikt}dt\right), \quad \text{then} \quad \log|f(0,k)| = \int_0^\infty \gamma(t)\cos kt\, dt.$$

The reciprocal cosine Fourier transform gives

$$\gamma(t) = \int_0^\infty \log|f(0,k)|\cos kt\, dk$$

which with (7) give the half-line scattering coefficient

$$S(k) = \frac{\bar{f}(0,k)}{f(0,k)} = \exp\left(-2i\int_0^\infty \gamma(t)\sin kt\, dt\right) = \exp\left(\frac{4i}{\pi}\int_0^\infty \frac{\log|f(0,k)|}{\kappa^2 - k^2}d\kappa\right),$$

which gives $S(k)$ explicitly formulted in terms of $|f(0,k)|$.

$S(k)$ will determine $(kR)^-$. Physically $R(k)$ is the reflection coefficient for $q(z)$ for the whole line problem created by setting $q(z) = 0$ for $z < 0$. Introduce the projection ϕ^- for a square integrable function ϕ onto the Hardy space of square integrable functions in the lower-half k plane can be calculated from the forward Fourier transform

$$\hat{\phi}(\lambda) = \frac{1}{\pi}\int_{-\infty}^{\infty} \phi(k)e^{2ik\lambda}dk$$

followed by the inverse Fourier transform from the *positive* half-line only

$$\phi^{-}(k) = \int_{0}^{\infty} \hat{\phi}(\lambda)e^{-2ik\lambda}d\lambda \, .$$

One must find $(kR)^{-}$, the projection of $kR(k)$ onto that Hardy space, as it is needed in the trace formula that follows. It has been shown [2] that

$$(kR)^{-} = k[1 - S(k)]^{-} \, .$$

This provides an algorithm to find $(kR)^{-}$ from physically measured data.

In the case that the Schroedinger Equation for y on $0 \leq z < \infty$ has no square integrable eigenfunction, $q(z)$ can be computed by solving simultaneously the trace formula [2]

$$q(z) = \int_{-\infty}^{\infty} (kR)^{-} f^2(z,k)dk \, .$$

One may combine this into (5) to find an integro-differential equation which may be solved by iteration which converges to a solution $f(z,k) \sim e^{ikz}$ as $z \to \infty$.[2] This process is substantially simplified if $q(z) \equiv 0$ in a constant velocity sub-sediment in $z > z_s$. Pick $z_m < z_s$ depending upon $(kR)^{-}$ which will permit convergence. On the interval $z_m < z < z_s$ set

$$f_0(z,k) = e^{izk} \, .$$

For the iterative step set

$$q_j(z) = \int_{-\infty}^{\infty} (kR)^{-} f_j^2(z,k)dk.$$

Then let $f_{j+1}(z,k)$ be the solution of

$$-\frac{d^2 f_{j+1}}{dz^2}(z,k) + q_j(z) f_{j+1}(z,k) = k^2 f_{j+1}(z,k)$$

with initial value of $f(z_s,k) = e^{ikz}$, and solving the equation from right to left on the interval $z_m \leq z \leq z_s$. Once the iterations have found $f(z_m,k)$, $\frac{df}{dz}(z_m,k)$, and $q(z)$ on $z_m \leq z \leq z_s$ for all k, pick $z_{m'} < z_m$, so that the scheme will again converge. Iterate as before on $z_{m'} \leq z \leq z_m$. This process will eventually yield $q(z)$ on $0 \leq z \leq z_s$ and hence the velocity profile $c(z)$ in the range of z dependence.

4. Some Aspects of Ensemble Averages

This method cannot be used where there is uncertainty in the sub-sediment velocity because of the asymptotics. Source depth uncertainty can be treated. A simple example is outlined for definiteness. Take z_0, z_s (a fluid half space sediment) to be fixed, and consider the Pekeris type model [9] with uncertainty in the upper layer:

$$q(z) = \begin{cases} q_0 + \epsilon\tau & z \leq z_s \\ 0 & z > z_s \end{cases} \, .$$

where $q_0 < 0$ is small enough that there are no proper eigenfunctions, $\epsilon < |q_0|$, and τ is the random paramenter for the ensemble, say uniformly distributed on $-1 \le \tau \le 1$. Set $\kappa_\tau = \sqrt{k^2 - q_0 - \epsilon\tau}$; Then $y(z,k) = \kappa_\tau^{-1} \sin \kappa_\tau z$ for $z \le z_s$ and

$$f(z,k) = e^{ikz_s} \left[\cos \kappa_\tau (z - z_s) + \frac{ik}{\kappa_\tau} \sin \kappa_\tau (z - z_s) \right] .$$

By differentiating (6)

$$h(k) = \frac{dv}{dz}(0,k) = \frac{[\kappa_\tau \cos \kappa_\tau (z_s - z_0) - ik \sin \kappa_\tau (z_s - z_0)]}{2\pi [\kappa_\tau \cos \kappa_\tau z_s - ik \sin \kappa_\tau z_s]},$$

which may be averaged with respect to τ to obtain the initial data for the ensemble average intial data for inversion. Unity minus scattering coefficient can be written

$$1 - S(k) = 1 - \frac{\kappa_\tau \cos \kappa_\tau z_s + ik \sin \kappa_\tau z_s}{\kappa_\tau \cos \kappa_\tau z_s - ik \sin \kappa_\tau z_s} e^{-2ikz_s} = -\frac{ik(\kappa_\tau - k) \sin 2\kappa_\tau z_s}{k^2 - q_0 \cos^2 \kappa_\tau z_s} .$$

Finally estimate the smallness of ω. Let n_0 and c_0 correspond to the water layer so that $q_0 = k_\infty^2 [1 - n_0^2] = \frac{\omega^2}{c_\infty^2} \left[1 - \frac{c_\infty^2}{c_0^2} \right]$. Using (2) find

$$1 > \int_0^{z_s} q_0(z) z \, dz = \frac{z_s^2 \omega^2 [c_\infty^2 - c_0^2]}{2 c_0^2 c_\infty^2},$$

which implies $\omega < \frac{\sqrt{2}\, c_0 c_\infty}{z_s \sqrt{c_\infty^2 - c_0^2}}$ in this case. Finer estimates are available from analysis of the formula for $y(z_s, k)$, which requires $-q_0 + \epsilon < \frac{\pi^2}{4z_s^2}$.

Acknowledgements

This work was supported by the by Office of Naval Research and Naval Research Laboratory. NRL contribution NRL/PP/7181–93–0060. Approved for public release; distribution is unlimited.

References

1. Deift, P.A. and Stickler, D.C. (1981), "Inverse problem for a stratified ocean and bottom," *J. Acoust. Soc. Am.* **70**, 1723–1727.
2. Deift, P.A. and Trubowitz, E. (1979), "Inverse scattering on the line," *Comm. Pure Appl. Math.* **XXXII**, 121–251.
3. Stickler, D.C. (1983), "Inverse scatteing in a stratified medium," *J. Acoust. Soc. Am.* **74**, 994–1005.
4. Coddington, E.A. and Levinson, N. (1955), *Theory of Ordinary Differential Equations.* McGraw–Hill Inc. New York.
5. Trubowitz, E. (1977), "The inverse problem for periodic potentials," *Comm. Pure Appl. Math.* **XXX**, 321–337.
6. Gel'fand, I.M. and Levitan, B.M. (1951), "On the determination of a differential equation from its spectral function," *Izv. Akad. Nauk SSSR* **15**, 309–360 ; (1955) *Am. Math. Soc. Transl.* **1**, 253–304.
7. Faddeyev, L.D. (1959), "The inverse problem in the quantum theory of scattering," *Uspekhi Mat. Nauk* **14**, 5–119; (1963) *J. Math. Phys.* **4**, 72–104.
8. Van Winter,C. (1971), "Fredholm equations on a Hilbert space of analytic functions," *Trans. Am. Math. Soc.* **162**, 103–139.
9. Pekeris, C.L. (1946), "Theory of propagation of sound in a half-space of variable sound velocity under conditions of formation of shadow zone," *J. Acoust. Soc. Am.* **18**, 295–315.

NULL SPACE REDUCTION
IN THE LINEARIZED TOMOGRAPHIC INVERSION

A. VESNAVER
OGS, Department of Geophysics of the Lithosphere
P.O. Box 2011, 34016 Trieste, Italy

ABSTRACT The null space is the origin of the non-uniqueness of solutions of linearized tomographic inversion. Usually its negative effects are dealt by introducing a priori information by means of constraints, covariance matrices and damping factors. In this paper a singular values analysis is introduced which allows to identify the null space causes: uncrossed pixels and linearly dependent rays. Both can be controlled or eliminated modifying properly the tomographic grid, accordingly to the estimated local reliability. Some synthetic examples are presented with a velocity distribution which is typical of the sea water, and the relationship between the acquisition geometry, the pixel distribution and the quality of results is discussed.

1. THE UNIQUENESS PROBLEM

Tomographic inversion of seismic travel times may be linearized dividing the earth into pixels, where the propagation velocity is assumed to be constant. In this case, the vector $\vec{t}$ of the recorded travel times can be related by a linear function to the vector $\vec{u}$ of the pixels slowness:

$$\vec{t} = A\vec{u} \; . \tag{1}$$

A is the tomographic matrix, whose elements a_{ij} are the path lengths of the i-th ray in the j-th pixel. Each row of A is associated to a ray, and each column to a pixel. System (1) is usually either under- or over-determined, and a least squares solution is sought; unfortunately, we rarely get a unique and reliable solution because it is rank deficient. In fact, the so-called null space exists, constituted by vectors $\vec{u}_o$ being solution of the associated homogeneous system:

$$A\vec{u}_o = 0 \; . \tag{2}$$

Any linear combination of vectors $\vec{u}_o$ with a solution of (1) still satisfies system (1), and therefore the number of possible solutions is infinite in this case.

Singular value decomposition (SVD) allows to express matrix A by the following product:

$$A = UWV^T \; , \tag{3}$$

where: $U^TU = I$, $V^TV = VV^T = I$, and I is the identity matrix. The elements w_j of the diagonal matrix W are the singular values of A. The columns of the matrix V corresponding to null singular values constitute an ortonormal basis of the null space, whilst the columns of U corresponding to non-null singular values are an ortonormal base of the range (see e.g. Press et al., 1989).

O. Diaschok et al. (eds.), Full Field Inversion Methods in Ocean and Seismo-Acoustics, 139–145.

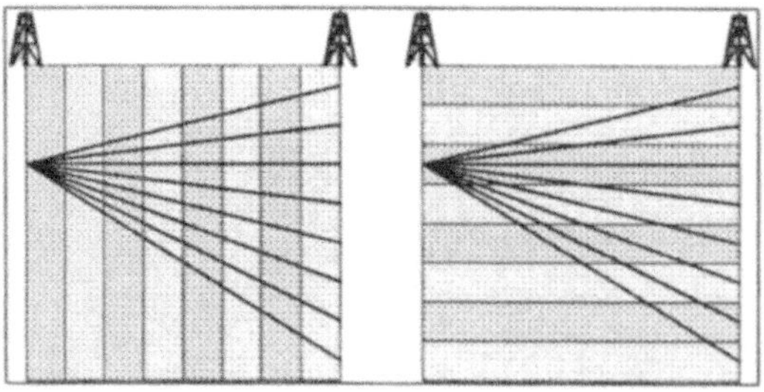

Fig.1

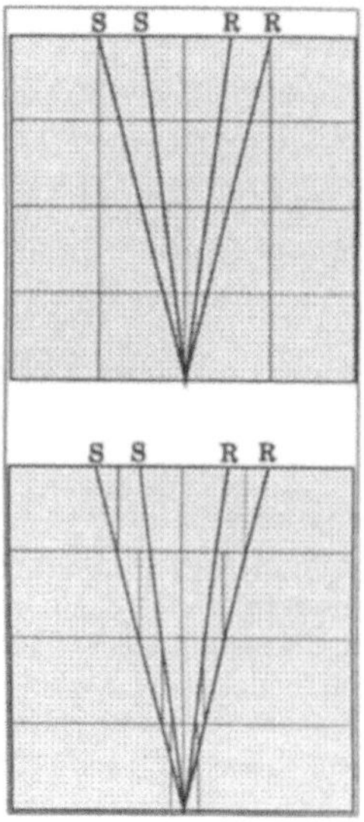

Fig.2

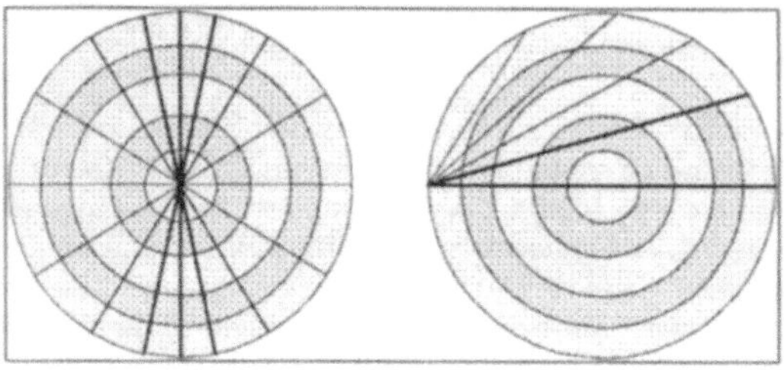

Fig.3

From the practical point of view, the inversion is obstacled not only by the null singular values, but also by those which are too small, because they often produce data overflows or underflows. Therefore, we have to introduce the generalised concept of quasi-null singular values, i.e. those whose ratio w_i/w_{max} with respect to the largest singular value w_{max} is smaller than a chosen threshold ρ, depending on the available computational precision and the noise in the data. This procedure may also be viewed as a generalisation of the concept of condition number. It is straightforward then to define the quasi-null space as that spanned by the columns of V associated with the quasi-null singular values.

There are two categories of vectors in the quasi-null space base: vectors associated to (quasi) void pixels, and vectors associated to (quasi) linearly dependent rows (i.e. rays) of the matrix A. The first group may be found automatically building the reliability vector $\vec{r}$ of the quasi-null space, whose components r_i are given by:

$$r_i = r_i(\rho) = 1 - \sum v_{ij}^2 \; . \qquad (4)$$

The summation is carried out over the index j of the quasi-null singular values. The indexes i for which r_i is practically equal to 1 indicate the quasi-void pixels (i.e. crossed by too short ray paths, or not crossed at all). The indexes of the remaining quasi-null singular values identify the ill-conditioned component of the range, i.e. the columns of the matrix U obtained by quasi-linearly dependent rays.

The reliability vector $\vec{r}(\rho)$ is a measure of the information provided by the experimental data to the inversion solution: it is 1 in the model zones well constrained by rays, it is close to zero where instabilities occur, and zero in the uncrossed pixels, where any velocity is acceptable. The related ρ parameter must be chosen according to the available data quality and computational precision.

After the ambiguity origin identification, it can be easily removed modifying the tomographic grid, adapting the pixels shape to the acquisition geometry and to the preliminary velocity estimates. For example, we see in Fig.1 that only rotating the pixels orientation we can transform a tomographic inversion from undetermined (left) to well-conditioned (right).

In fact, the rays on the left are all linearly dependent, because of the Thales theorem, since they are crossing a set of parallel lines (the vertical pixels boundaries); this does not happen on the right, where a resolvable system is obtained.

2. THE ANGULAR COVERAGE PROBLEM

It is well known that inversion instabilities occur in practice when a limited angular coverage is available, due to the acquisition geometry and to the irregular ray density produced by the spatial velocity variations. This phenomenon is easily explained by the concepts just introduced: nearly parallel rays may produce quasi-null singular values, which may be eliminated modifying the tomographic grid (Fig.2).

Fig.3 allows to remark this statement: on the left infinite rays crossing five shell-shaped pixels at any angle are unable to get a unique solution, since all rays passing through the pixels centre are linearly dependent; on the right, five rays are sufficient to solve the problem.

TABLE 1.

Layer number	Layer bottom (m)	Layer velocity (km/s)
1	25	1.534
2	50	1.524
3	75	1.534
4	100	1.509
5	150	1.511
6	200	1.511
7	400	1.511
8	800	1.517

3. SOME EXAMPLES

Fig.4 shows a simple model composed by eight homogeneous layers, each divided into two pixels. Transmitted and reflected waves were modelled, assuming that only the deepest interface is reflecting, using a minimum time ray tracing technique (Vesnaver, 1993). The velocity distribution simulates the average water structure in the Sicily Channel during the spring and summer (Boehm et al., 1994), reported in Table 1. The spatial coordinates are expressed in km, the velocity in km/s. An acquisition geometry was initially tested which is typical for marine seismic data. A source and a cable with 96 receivers where set at a depth of 11 m. The distance between the source and the nearest receiver is 200 m, the receiver interval is 25 m.

Fig.5 displays the velocity field estimated adopting the acquisition geometry in Fig.4. The dual tomography was used as the inversion algorithm (Carrion, 1991). We see that only the first layer is correctly resolved, with the main contribution of the transmitted waves. The poor quality estimate is indicated also by the corresponding reliability vector (Fig.6), using a ρ value equal to 10: this means that only the singular values of the same order of the maximum were considered in the summation (4).

Fig.4 Synthetic Model

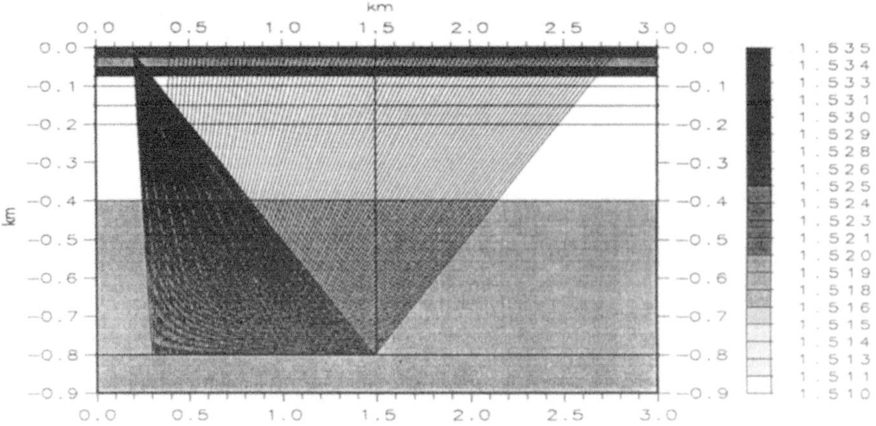

Fig.5 Tomographic Inversion

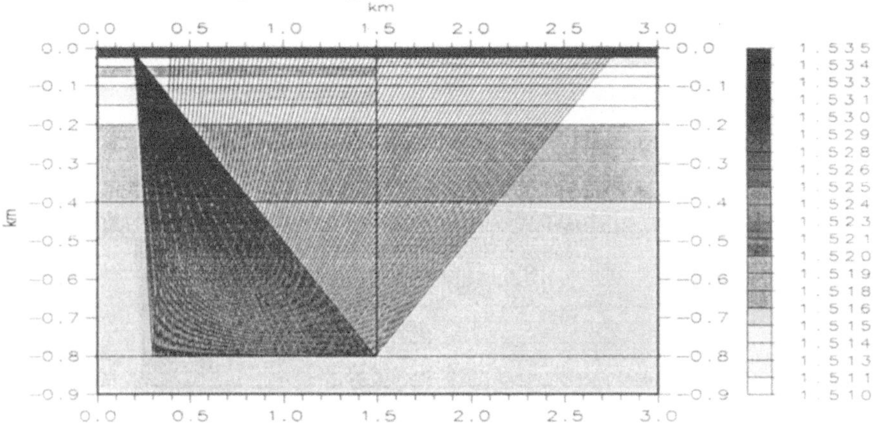

Fig.6 Local Reliability

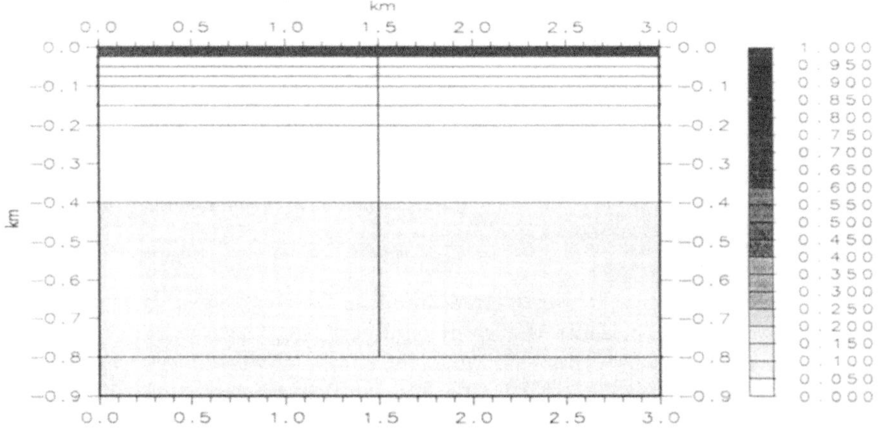

Fig.7 Tomographic Inversion

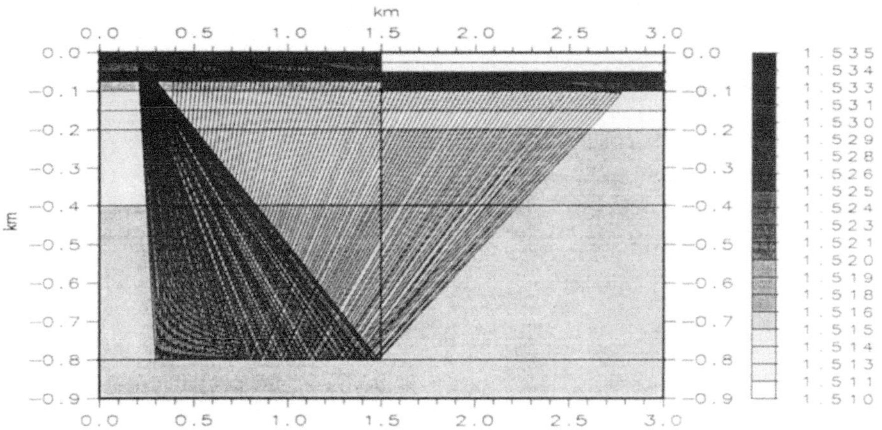

Fig.8 Local Reliability

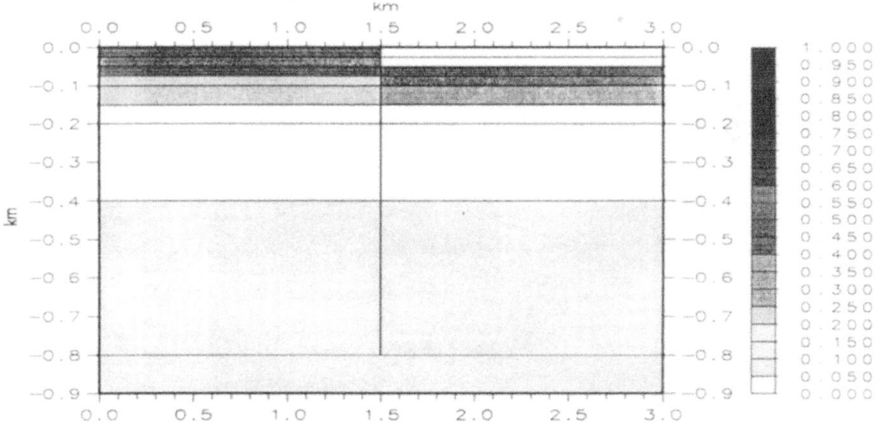

Fig.9 Tomographic Inversion

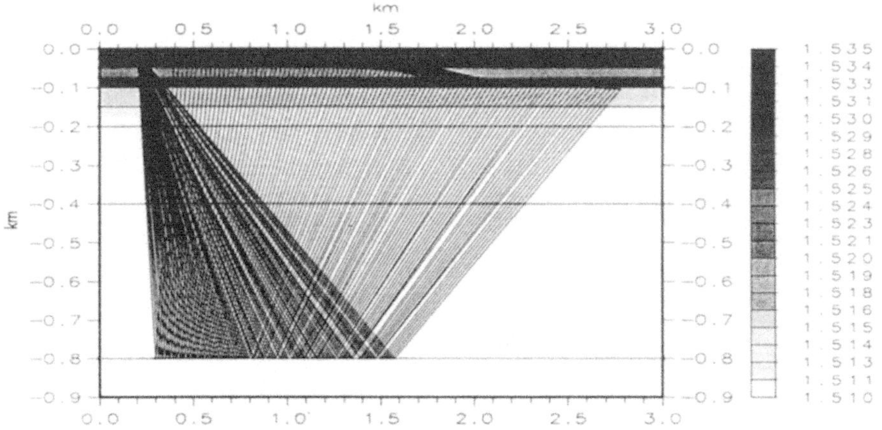

Fig. 10 Local Reliability

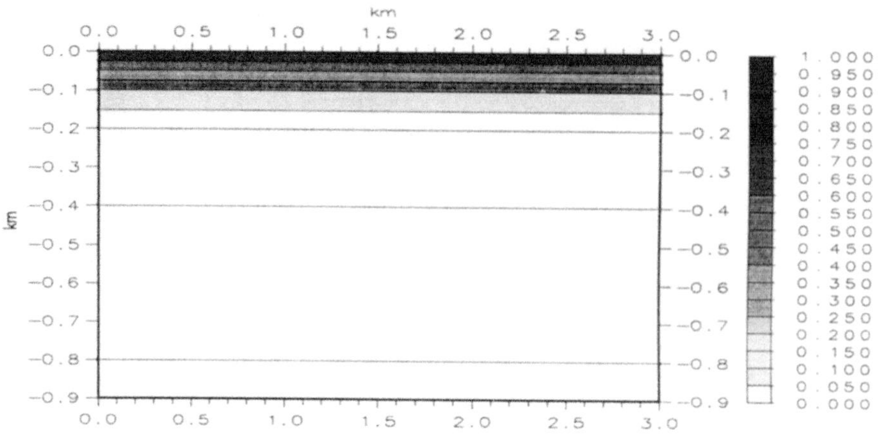

Fig. 11 Tomographic Inversion

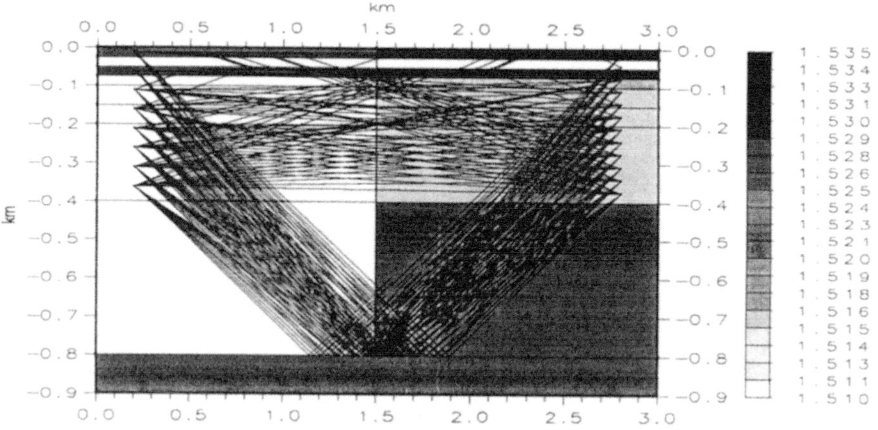

Fig. 12 Local Reliability

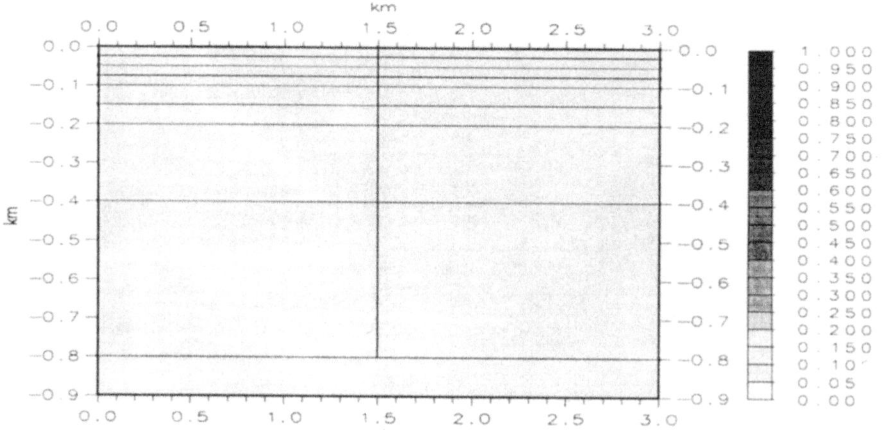

If the seismic cable is dipping (which is undesirable for conventional seismic acquisition), we improve the inversion results (Fig.7) as well as its reliability (Fig.8), except in the uppermost pixels on the right, which are no more covered by rays. On the other hand, merging together the pixels of the same layer (Fig.9), the high velocity layers at surface are better identified. The estimated reliability (Fig.10) confirms that a more stable and accurate solution is obtained in this case.

A better result may be obtained if a totally different acquisition geometry is adopted (Fig.11), which resembles one typical of reservoir geophysics (cross-well). In this case, the rays cover the model in such a way that rays are much more independent, and the image quality obtained is much improved. The estimated reliability (Fig.12) is the best obtained with respect to all other acquisition geometries because it is flat, i.e., the distribution of linearly independent rays is well balanced in space for that model discretization. This means that the cross-well geometry is the best choice for a model where the velocity distribution is horizontally layered.

4. CONCLUSIONS

Singular values decomposition of the tomographic matrix identifies its null and quasi-null space and may provide practical hints to remove the inversion instabilities and ambiguities, including those due to a limited angular coverage. Modifying the pixels grid we may adapt it to the available acquisition geometry and data quality. In this way we can reach an reasonable compromise between the local resolution and the desired reliability.

References

1. Boehm, G., Crise, A. and Vesnaver, A. (1994) Travel time inversion of marine seismic data for ocean sound speed reconstruction, *Proceedings of the NATO Conference on full field inversion methods in ocean and seismic acoustics*, Kluwer, The Netherlands, (this monograph).
2. Carrion, P. (1991) Dual tomography for imaging complex structures, *Geophysics*, **56**, 1395–1404.
3. Press, W.H., Flannery, B.P., Teukolsky, S.A. and Vetterling, W.T. (1989) Numerical recipes, *Cambridge University Press*, United Kingdom.
4. Vesnaver, A. (1993) Minimum time ray tracing of reflected, refracted and diffracted waves in irregular grids, *Expanded Abstracts of 55th EAEG Meeting*, P059.

NON-LINEAR INVERSION OF SEISMIC DATA BY SUCCESSIVE APPROXIMATION OF MODEL PARAMETER PROBABILITY DISTRIBUTION FUNCTIONS

W. T. Wood
Naval Research Laboratory, Code 7432, Stennis Space Center, MS, 39529
M. K. SEN
The University of Texas Institute for Geophysics, 8701 Mopac Blvd., Austin, TX, 78759

ABSTRACT. Inversion of seismic data from a 1-D elastic earth is accomplished by choosing models randomly from a distribution which is the reciprocal of the objective function. Minima in the objective function correspond to peaks in distribution, thereby enhancing the probability that more models will be chosen from this area. Results show good recovery of dominant model parameters values and their uncertainties.

1. Introduction

Recently several methods of non-linear inversion have proven successful for inverting (i.e. recovering elastic parameter profiles from) 1-D seismic data, a potentially highly non-linear problem. Simulated annealing (Sen and Stoffa, 1991) uses an optimization scheme based on the annealing or freezing of a material into its lowest possible energy state. Genetic algorithms (Stoffa and Sen, 1991) are based on the evolution of a population of interbreeding individuals into a race ideally suited for its environment. Both simulated annealing and genetic algorithms can be adjusted for faster convergence with run time control parameters, but this adjustment requires several experimental runs. A conceptually simple alternative to these methods which requires minimal user supplied control parameters is presented here.

To state the problem of nonlinear inversion as a global optimization problem we first define the forward problem,

$$\mathbf{d}_{obs} = \mathbf{f}(\mathbf{m}_{true}) + \mathbf{n}_{ran}$$

where $\mathbf{m}$ is a vector of model parameter values, $\mathbf{f}$ is an operator which mimics a physical process (in our case the propagation of elastic waves) sufficiently well to yield a synthetic data set $\mathbf{d}$, and $\mathbf{n}_{ran}$ is a vector of random noise. We assume that if $\mathbf{m} = \mathbf{m}_{true}$ (the "true" earth model) then $\mathbf{d} = \mathbf{d}_{obs}$ (the observed data set). Let e be some error function of $\mathbf{d}_{obs}$ and $\mathbf{d}_{syn} = \mathbf{f}(\mathbf{m}_i)$ such that

$$\begin{aligned} e \;=\; e(\mathbf{d}_{obs},\mathbf{d}_{syn}) \;&=\; 0 \quad \text{when} \quad \mathbf{d}_{obs} = \mathbf{d}_{syn} \\ &>\; 0 \quad \text{when} \quad \mathbf{d}_{obs} \neq \mathbf{d}_{syn}. \end{aligned}$$

We assume that when the minimum value of e is found, $\mathbf{m}_i = \mathbf{m}_{true}$ and the problem is

O. Diaschok et al. (eds.), Full Field Inversion Methods in Ocean and Seismo-Acoustics, 147–152.

solved, (this will not be strictly true in the presence of noise). To find the minimum of $e(\mathbf{d}_{obs},\mathbf{d}_{syn})$ we probe the model space for $\mathbf{m}_{true}$.

2. Method

The optimization technique discussed here is based on the assumption that areas of model space which yield low objective function values are more likely to contain the global minimum than those areas which yield high objective function values, i.e. the probability of a model parameter value being correct is inversely proportional to the error function value. We therefore want to probe the valleys in the error function more than the peaks. This is done by letting the reciprocal of the error function evaluations at each iteration form a probability density function from which to chose the next model parameter estimate. Let

$$P(\mathbf{m}_i : \mathbf{d}_{obs}) \quad = e^{-1}$$

where P is a non-normalized probability density function from which the next model will be drawn. Using non-uniform PDFs in the sampling of model space has been employed by Rothman (1986) and also by Styblinski and Tang (1990). Although this concept is simple, the practical implementation is somewhat more complex.

The process is started with a number of trial models chosen randomly from a distribution which is uniform over a user defined range and zero elsewhere. This is done to avoid coding the algorithm to create a PDF based on one point, or very few points. This number should be large enough to get a rough estimate of the PDF. In this study it was chosen to be 40.

During the course of the inversion a single model parameter value may be associted with high or low error, depending on the rest of the parameter values in that model. A plot of error vs. value for a particular model parameter will in general be multi-valued. Because we are interested only in the PDF peaks the low P (high error) points are not used. Throughout the inversion only a limited number of the best models are kept (here n_{kept} was set to 40) and it is the envelope formed by these best models which we take as the marginal PDF (PDF of one parameter while the others are fixed) for the next model choice. The stiffness of the envelope is important with regard to the resolution of the method and is discussed later.

Once the marginal PDF for a particular model parameter has been determined, a random value must be chosen. To map a uniform distribution (conveniently generated by RAN3.FOR, Press et al., 1989, for example) to the non-uniform distribution desired here we use the normalized cumulative probability density function (CDF), a numerical integral of the PDF. Using the uniform random number as a value for y on the CDF curve, the corresponding x value on the curve becomes the next choice for the current model parameter. This process is repeated for each model parameter.

As the number of trials increases, the error function, and therefore the PDF becomes better defined, i.e. peaked at the correct model value. A tall narrow peak in the PDF leads to a step-like appearance in the CDF indicating that almost all choices for y (the uniformly distributed random number) will give nearly the same value for x (model value). Thus the areas of model space with low error are probed more densely. However, at no point does the PDF reach zero, so there is always a finite probability that the global minimum will be found.

Crucial to the proper function of this algorithm is a reasonable stiffness of the envelope (smoothness of the resulting PDF). A PDF which is too smooth results in poor resolution and many poor models chosen. An overly rough PDF includes those points which have nearly the correct model value but low P. Upon integration this detracts from the area of the peak, again decreasing the resolution. The stiffness of the envelope is therefore adjusted so that the number of envelope points is about the same as n_{kept}. If there are significantly more than n_{kept} envelope points then the envelope becomes too rough. If there are significantly fewer then the detail of the shape of the PDF is obscured.

The effect of this is to increase resolution as the range of best models shrinks. This is similar to the convolutional smoothing of Styblinski and Tang (1990) who start with a smooth unimodal approximation to the multi-modal objective function and decrease the smoothness as the inversion progresses. The number of models kept, n_{kept}, was chosen to be the same as the number of peaks that are expected in the PDF, so that these peaks can be portrayed by the PDF envelope.

Another attractive feature of this algorithm is that the marginal PDF for each model parameter at each iteration is indicative of the sensitivity of the data to that model parameter, (Stoffa and Sen 1991). Assuming that all models were drawn from the latest PDF these models form a distribution whose variance or standrd deviation can be used as a measure of uncertainty in the result. This allows for not just an estimate of the answer but an estimate of its uncertainty as well.

3. Application to Seismic (Scattered Wave) Data

Applying this algorithm to seismic data requires first choosing a method of generating synthetic seismograms, and a method of comparing seismograms. Here a synthetic shot gather is generated for a 1-D elastic earth by ray tracing from the source to each receiver, and using the Zoepritz equations (Aki and Richards, 1980) at each interface to compute reflection coefficients. The source and receivers are assumed to lie in an acoustic medium (water) although loss due to mode conversion in the sediments is included. Each of 7 traces computed contained 1024 samples at a sample rate of 0.00032 s and a frequency range of 250 to 650 Hz.

To compare seismograms we use simply the sum of the absolute values of the differences in the seismograms (the L1 norm),

$$e = \sum_{i=1}^{n_{dat}} |d^i_{obs} - d^i_{syn}|$$

where n_{dat} is the number of traces times the number of samples per trace.

The graphs to the top and right in Figure 1a show the first 10000 evaluations of P vs. model parameter value for thickness and v_p (acoustic velocity) respectively. Note the strong peak at the correct velocity and arrival time, (v_p = 1.53km/s, t = 0.45km) as well as smaller peaks corresponding to error function minima. Figure 1b shows the first 1000 points of v_p vs. t for layer 1. Note the distinct concentration of points near the global minimum, although the entire slice of model space is being sampled. Figure 1c shows points (9000 to 10000) from the same slice. At this point in the inversion the points are concentrated on the peaks of the marginal PDFs, and very strongly concentrated around the global minimum. However there are still points distributed throughout the slice, showing that even though a very low minimum has been found, these distant portions of

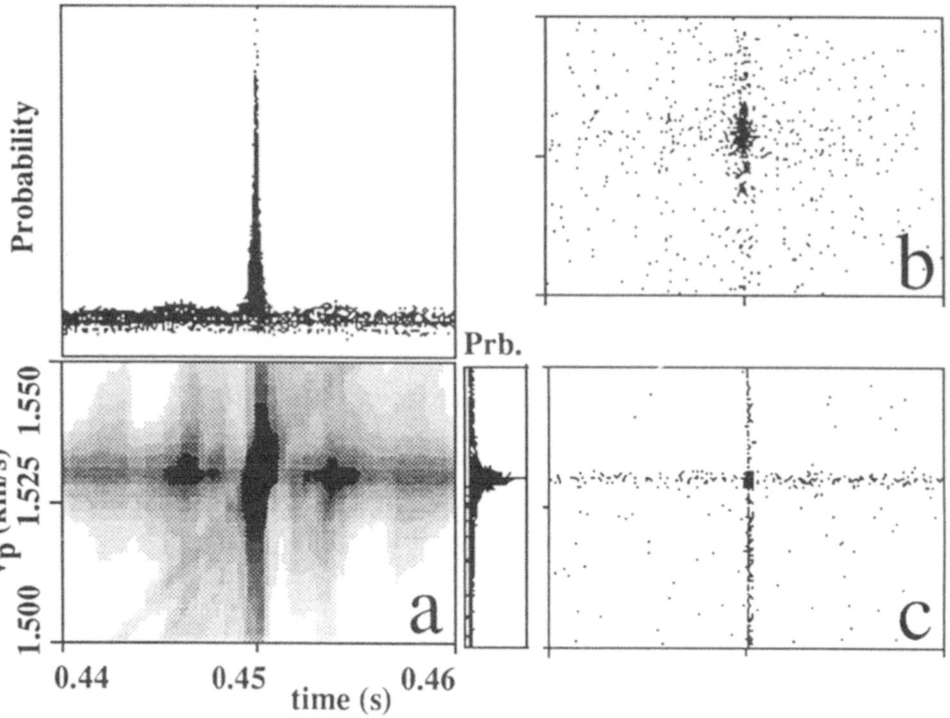

Figure 1. a) A slice of the objective function showing the arrival time vs. v_p in layer 1, (darker is lower). Note the strong local minima along v_p= 1.53, (the correct answer) and smaller minima to the lower left and upper right. The graphs to the top and right show the reciprocal of the objective function for each of the first 10000 trials (normalized to yield probability) vs. time and v_p respectively. Objective function minima correspond to peaks in the PDF. b) and c) show the locations of trials 1 to 1000, and 9000 to 10000 respectively. Note the tendency to probe the minima more intensley as the number of trials increases.

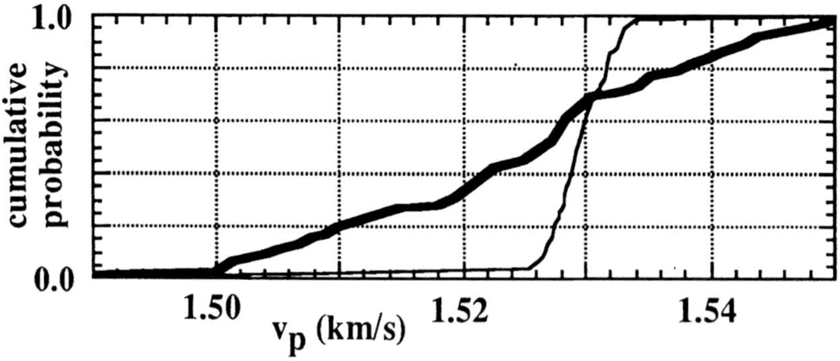

Figure 2. The CDF used to compute estimates for v_p of layer 1 at iteration 200 (bold) and 1000 (fine). Models are more likely to be chosen from the steeper portions of the curve. Note the convergence to v_p = 1.53, the correct value.

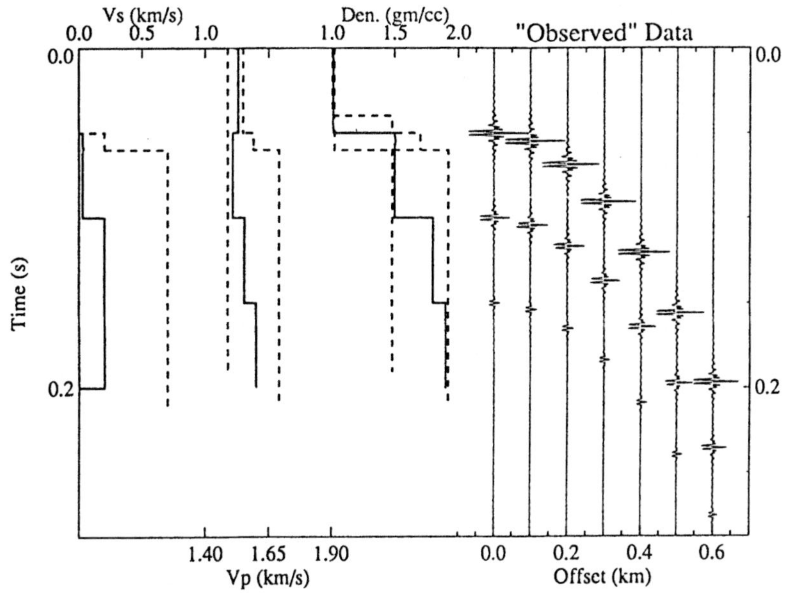

Figure 3. The true model (solid line) is shown to the left and the corresponding synthetic data ($\mathbf{d}_{obs}$) is shown to the right. The dashed lines represent boundaries beyond which no models are tried. Note that arrival time is also a model parameter.

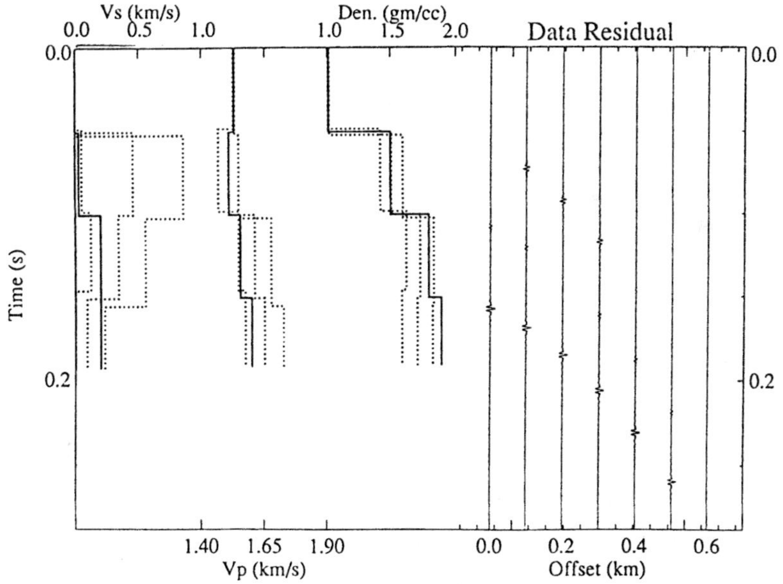

Figure 4. The true model (solid line) is compared to the best estimate +/- 2σ (dots). The difference between $\mathbf{d}_{obs}$ and the best $\mathbf{d}_{syn}$ is shown at right. Note the good recovery of arrival v_p, and density especially in the upper layers. The small data residual indicates good recovery of pertinent model parameters.

model space have not been excluded.

The convergence shown in Figure 1 can also be seen in Figure 2. This is a plot of two CDFs of the v_p for the first layer after 200 trials (bold) and 1000 trials (fine). After 200 trials a random y value will map to a wide range of v_p values, (~1.5 to 1.55 km/s) indicating a broad PDF peak. After 1000 trials however, the likely range has narrowed substantially (~1.528 to 1.534 km/s) about the correct value of 1.53 km/s. As the number of trials increases this and the other CDF curves steepen about the most probable model parameter values.

Also important in the inversion is the choice of model parameter bounds within which to conduct the search. Figure 3 shows the true model (solid) and starting bounds to the left (dashed), and the "observed" data to the right, (data and model have been advanced by 0.4 s). The bounds used here are wide enough to include all reasonable models of soft sediment. The bounds in the layer arrival times were set at 0.01 above and below the correct value. Note that this is several times the dominant wavelength of the wavelet, making the problem significantly non-linear.

4. Results

The result of this inversion ($\mathbf{m}_{best}$) stopped arbitrarily after 20000 trials is shown in Figure 4. The three dotted lines for each profile represent the value estimate $\pm 2\sigma$ where σ is the standard deviation about the best model for the group of n_{kept} models at the last iteration. The solid line represents $\mathbf{m}_{true}$ while the difference seismogram between $\mathbf{d}_{obs}$ and $\mathbf{d}_{syn} = \mathbf{f}(\mathbf{m}_{best})$ is shown to the right, at the same scale as the seismogram in Figure 3. Note small data residual indicating recovery of the most important parameters. Comparing $\mathbf{m}_{best}$ to $\mathbf{m}_{true}$ we see the largest uncertainties (least sensitivity) for v_s, (shear velocity). Also indicated is an increase in uncertainty with depth as expected.

5. Conclusions

The algorithm presented here shows promise for the inversion of non-linear problems in general and seismic data inversion in particular for several reasons. There is always a non-zero probability of finding the global minimum, the conceptual simplicity leads to a minimum of user supplied run time control parameters, and the resulting map of the error function provides a measure of sensitivity of data to model parameters.

References

1. Aki, K. and Richards, P. G. (1980) *Quantitative Seismology*. W. H. Freeman and Company, New York.
2. Press, W. H., Flannery, B. P., Teukolsky, S. A., and W. T. Vetterling, W. T., (1989), *Numerical Recipes*, Cambridge University Press, New York.
3. Rothman, D. H., (1986) Automatic estimation of large residual statics corrections, *Geophysics*, 51, 332-346.
4. Sen, M. K. and Stoffa, P. L. (1991) Nonlinear one-dimensional seismic waveform inversion using simulated annealing, *Geophysics*, 56, 1624-1638.
5. Stoffa, P. L. and Sen, M. K. (1991) Nonlinear multiparameter optimization using genetic algorithms: Inversion of plane wave seismograms, *Geophysics*, 56, 1794-1810.
6. Styblinski, M. A. and Tang, T. S. (1990) Experimants in Nonconvex Optimization: Stochastic Approximation with Function Smoothing and Simulated Annealing, *Neural Networks*, 3, 467-483.

EXPERIMENTAL CONFIRMATION OF FOCALIZATION

ÅGE KRISTENSEN, A. CAITI, F. INGENITO, MICHAEL D. MAX
SACLANT Undersea Research Centre
La Spezia, Italy

JONATHAN M. BERKSON, MICHAEL D. COLLINS, LAURIE T. FIALKOWSKI, NICHOLAS C. MAKRIS, B. EDWARD MCDONALD, JOHN S. PERKINS
Naval Research Laboratory
Washington, DC 20375, USA

and

W.A. KUPERMAN
Scripps Institution of Oceanography
La Jolla, California 92093, USA

ABSTRACT Focalization is implemented in a form suitable for processing field data and applied to data from a shallow-water ocean environment. The results indicate that focalization is a viable signal-processing technique.

1. Introduction

Matched-field processing (MFP) techniques are designed to solve the inverse problem of determining the location of an acoustic source using data from an array of receivers [1,2]. Under ideal conditions, the acoustic data is compared with solutions of the wave equation (the replica fields) to determine the source location that corresponds to the best agreement with the data. Since departure from ideal conditions tends to be the rule rather than the exception, MFP has been an active area of research for nearly two decades [3].

Environmental mismatch is a serious problem that frequently arises in applications of MFP. If the environmental parameters are not known accurately, the replica fields will not be accurate, and MFP techniques may fail [4–7]. A straightforward but usually impractical approach for overcoming mismatch is to measure the environmental parameters accurately. This approach usually ranges from very difficult to hopeless because the regions of interest are usually very large relative to a representative wavelength. Even small regions can be difficult to characterize when bottom interaction is significant because the bottom parameters are difficult to obtain.

An effective approach for overcoming relatively weak mismatch is to develop processors that are relatively insensitive to mismatch [8,9]. When the mismatch is too great for these approaches to handle, it is necessary to include both source and environmental parameters (and possibly other parameters, such as corrections to the locations of the receivers) in the search space. Focalization involves adjusting the environmental parameters until the source location comes into focus in the ambiguity surface [10]. The focalization simulations presented in [10] illustrate a parameter hierarchy in which source parameters tend to outrank environmental parameters. It is possible in many cases to determine the location of

O. Diaschok et al. (eds.), Full Field Inversion Methods in Ocean and Seismo-Acoustics, 153–158.

the source without determining the environmental parameters. This can be a significant advantage if the main goal is to determine the location of the source.

In this paper, we describe an implementation of focalization that is suitable for processing field data and present experimental results that demonstrate that focalization is a viable signal-processing technique. The implementation involves the multi-valued (MVB) Bartlett processor [11], which is used to extract the source signal from noise. Simulated annealing [12–14] is used to search the parameter space, which consists of the range and depth of the source and the sound speed and attenuation of the ocean bottom. Although the structure of the ocean bottom ordinarily calls for a larger set of parameters, two parameters seem to be adequate for the focalization problem we considered, apparently due to the parameter hierarchy. During the focalization parameter search, the estimate for the source location alternates between four points but is attracted to the true source location. The data was taken in the first of a sequence of three shallow-water experiments that were carried out in the Straits of Sicily (February 1992), off the coast of Plymouth, England (August 1992), and off the coast of Majorca, Spain (March 1993).

2. Implementation of Focalization

Although the simulations presented in [10] indicate that focalization should be an effective signal-processing technique, they are based on matching quantities that would be relatively difficult to extract from data. In this section, we describe an implementation of focalization that is suitable for processing data.

A focalization problem is defined by specifying a parameter space and a cost function. The parameter space is partitioned into the source subspace, which includes the source coordinates, and the focusing subspace, which may include environmental, receiver, and other parameters. The cost function provides a measure of the agreement between the data and the replica field. The solution of the focalization problem is the point in the parameter space that corresponds to the global minimum of the cost function. Since the dimension of the focalization parameter space may be high, it is important to use an effective optimization technique, exploit the parameter hierarchy, and suppress degradations due to noise.

We have implemented focalization using the MVB processor [11] as the cost function and a combination of simulated annealing and exhaustive search to locate the global minimum of the cost function. The MVB processor is a technique for handling interfering sources and low signal-to-noise ratio that is based on the eigenvectors of the covariance matrix, which is the standard input data to MFP. The MVB processor exploits the fact that signals from discrete sources tend to be partitioned into different eigenvectors and ambient noise tends to be distributed among the eigenvectors [15,16]. The MVB processor can surgically extract signals from moderately noisy data. When knowledge of the noise covariance is available, the noise-canceling processor [17] should be useful for localizing a source buried in noise in an uncertain environment.

MFP problems are usually solved by exhaustively searching the range-depth cost function. This approach is not practical for focalization problems that involve high-dimensional parameter spaces. We apply simulated annealing to search the focusing subspace. To avoid difficulties associated with the extremely multi-modal nature of the range-depth cost function, we exhaustively search the source coordinates for every sample of the focusing parameters. This approach was used in [10] and is practical because the source subspace is of low dimension. This is often the only sensible approach because it is generally necessary to

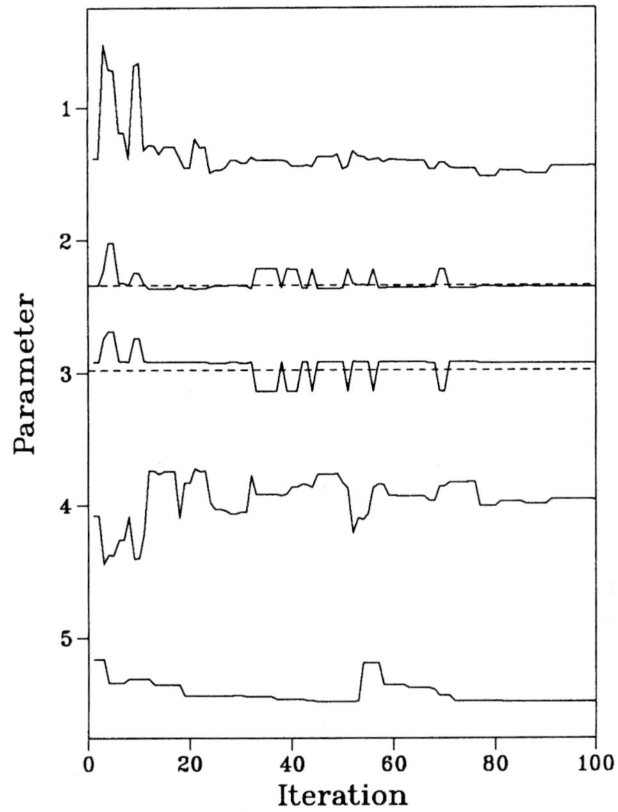

FIGURE 1. Convergence of the parameter search for a focalization calculation involving data from the Straits of Sicily. The parameters are (1) the cost function, (2) source range, (3) source depth, (4) sediment sound speed, and (5) sediment attenuation. The dashed lines for the source parameters correspond to the actual source coordinates.

sample the source subspace densely to compute replica fields for range-dependent problems.

Other implementations of focalization would also be effective. We selected simulated annealing because the parameter space may be of high dimension and the cost function may have local minima. Other nonlinear optimization techniques, such as genetic algorithms, would also be effective. A genetic algorithm has been used to solve nonlinear inversion problems in ocean acoustics [18]. High-resolution processors are not effective for focalization because they are sensitive to mismatch [8]. We selected the MVB processor because it is a low-resolution processor that suppresses noise. The multi-constraint processor [8] is another low-resolution processor that should be effective for focalization because it suppresses ambiguities.

3. Application to Data

In this section, we apply focalization to 125-Hz data that was taken in the Straits of Sicily on a 15-element vertical array (which was actually a subarray of a 64-element array) that

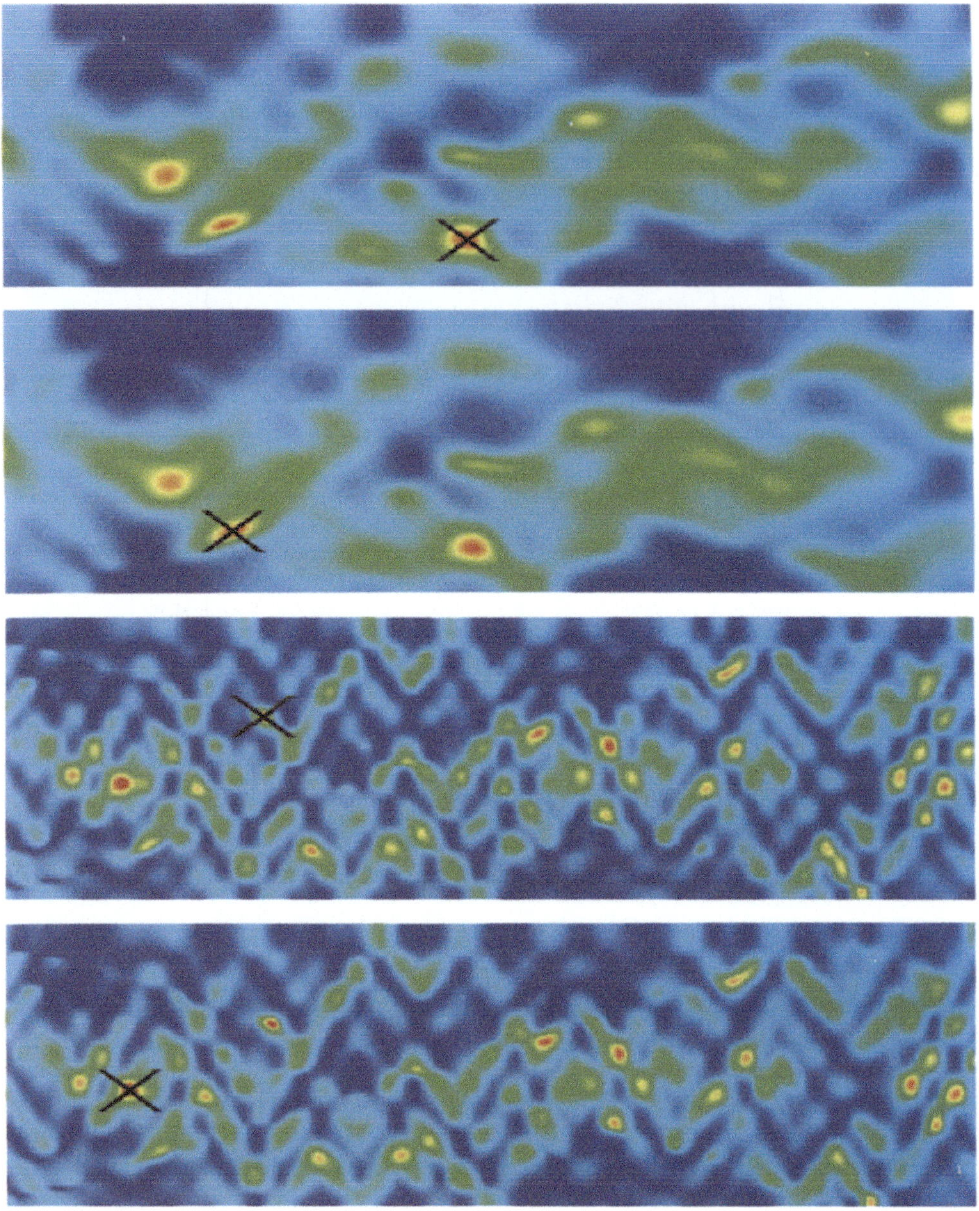

FIGURE 2. Some of the ambiguity surfaces encountered during the parameter search. From top to bottom corresponds to iteration numbers 5, 10, 40, and 100. Red corresponds to the most likely source locations. Blue corresponds to the least likely source locations. The crosses mark the peaks in the ambiguity surfaces, which span the water column and the ranges 3.5 to 7 km from the array.

was moored on the bottom in 143 m of water at 37°25.77′ N and 12°21.63′ E. The top receiver was 44.5 m below the ocean surface and the receiver spacing was 4 m. The source was suspended 75 m below the ocean surface from the R.V. Alliance, which was allowed to drift away from the array.

We constructed the covariance matrix for data corresponding to a source range of 4 km. The first eigenvector was found to contain more than 95 percent of the energy and to vary regularly as the source range increased. These are indications of high-quality data and high signal-to-noise ratio. Although the sediment is believed to be relatively complicated in this region, we used a very simple parametrization in order to exploit the parameter hierarchy. We model the sediment as a homogeneous half space in which the sound speed and attenuation are the only unknowns. We assume a density of 1.95 g/cm^3 based on sediment samples obtained during the experiment.

In the parameter search, we constructed the MVB ambiguity surfaces over the entire water column and over 3.5 to 7 km in range from the array. The focalization parameter search is illustrated in Figure 1. The energy decreases in the typical irregular fashion of simulated annealing. The source parameters sample several points but are definitely attracted to a point close to the true source location. The source range is recovered accurately, and the error in the source depth estimate of 82 m is less than a wavelength. The environmental parameters were attracted to values near 1650 m/s and 0.2 dB/λ. Due to the parameter hierarchy, these values may differ significantly from the true values despite the fact that the source was localized. We searched for these parameters over the ranges of 1550 to 1750 m/s and 0.2 to 0.5 dB/λ.

Some of the MVB ambiguity surfaces appear in Figure 2. The parameter search attempts to bring the source into focus at three false locations before settling on the correct location. The ambiguity surfaces corresponding to iterations 5 and 10 contain relatively little structure because the sound speed parameter is less than 1600 m/s. The ambiguity surfaces corresponding to iterations 40 and 100 contain more structure because the sound speed parameter is greater than 1650 m/s.

4. Conclusion

We have implemented focalization in a form that is suitable for processing field data using the MVB processor as the cost function and simulated annealing to search the parameter space for the global minimum of the cost function. We have experimentally confirmed focalization as a viable signal-processing technique.

5. Acknowledgments

This work was supported by the Office of Naval Research under grant number N00014-89-D-0041.

References

[1] H.P. Bucker, "Use of calculated sound fields and matched-field detection to locate sound sources in shallow water," J. Acoust. Soc. Am. 59, 368–373 (1976).

[2] A.B. Baggeroer, W.A. Kuperman, and H. Schmidt, "Matched field processing: Source localization in correlated noise as an optimum parameter estimation problem," J. Acoust. Soc. Am. 83, 571–587 (1988).

[3] A.B. Baggeroer, W.A. Kuperman, and P.N. Mikhalevsky, "An overview of matched field methods in ocean acoustics," IEEE J. Ocean. Eng. 18, 401–424 (1993).
[4] D.R. Del Balzo, C. Feuillade, and M.M. Rowe, "Effects of water-depth mismatch on matched-field localization in shallow water," J. Acoust. Soc. Am. 83, 2180–2185 (1988).
[5] A. Tolstoy, "Sensitivity of matched field processing to sound-speed profile mismatch for vertical arrays in a deep water Pacific environment," J. Acoust. Soc. Am. 85, 2394–2404 (1989).
[6] D.F. Gingras, "Methods for predicting the sensitivity of matched-field processors to mismatch," J. Acoust. Soc. Am. 86, 1940–1949 (1989).
[7] R.M. Hamson and R.M. Heitmeyer, "Environmental and system effects on source localization in shallow water by the matched-field processing of a vertical array," J. Acoust. Soc. Am. 86, 1950–1959 (1989).
[8] H. Schmidt, A.B. Baggeroer, W.A. Kuperman, and E.K. Scheer, "Environmentally tolerant beamforming for high-resolution matched field processing: Deterministic mismatch," J. Acoust. Soc. Am. 88, 1851–1862 (1990).
[9] A.M. Richardson and L.W. Nolte, "*A posteriori* probability source localization in an uncertain sound speed, deep ocean environment," J. Acoust. Soc. Am. 89, 2280–2284 (1991).
[10] M.D. Collins and W.A. Kuperman, "Focalization: Environmental focusing and source localization," J. Acoust. Soc. Am. 90, 1410–1422 (1991).
[11] M.D. Collins, L.T. Fialkowski, W.A. Kuperman, and J.S. Perkins, "The multi-valued Bartlett processor and source tracking," J. Acoust. Soc. Am. (in press).
[12] N. Metropolis, A.W. Rosenbluth, M.N. Rosenbluth, A.H. Teller, and E. Teller, "Equations of state calculations by fast computing machines," J. Chem. Phys. 21, 1087-1091 (1953).
[13] S. Kirkpatrick, C.D. Gellatt, and M.P. Vecchi, "Optimization by simulated annealing," Science 220, 671–680 (1983).
[14] H. Szu and R. Hartley, "Fast simulated annealing," Phys. Let. 122, 157–162 (1987).
[15] R.O. Schmidt, "A signal subspace approach to multiple emitter location and spectral estimation," Ph.D. dissertation, Stanford Univ., Stanford, California (1981).
[16] D.H. Johnson, "The application of spectral estimation methods to bearing estimation problems," Proc. IEEE 70, 1018-1028 (1982).
[17] M.D. Collins, N.C. Makris, and L.T. Fialkowski, "Noise cancellation and source localization," J. Acoust. Soc. Am. 96, 1773–1776 (1994).
[18] P. Gerstoft, "Inversion of seismoacoustic data using genetic algorithms and *a posteriori* probability distributions," J. Acoust. Soc. Am. 95, 770–782 (1994).

ESTIMATION OF BOTTOM PARAMETERS FROM REAL DATA BY GENETIC ALGORITHMS

M. LAMBERT[1], P. GERSTOFT, A. CAITI[2] and R. AMBJØRNSEN
SACLANT Undersea Research Centre, Viale San Bartolomeo 400, 19138 La Spezia, Italy

ABSTRACT An inversion scheme for estimation of the physical parameters of marine sediments from the pressure field measured in the water column is presented. It is based on a global optimization technique called Genetic Algorithms. In order to constrain the solution, regularization is introduced via the use of shape functions and the data observations are weighted according to their importance. Real transmission loss data in very shallow water have here been successfully inverted. An analysis of the uncertainties of the results is carried out to assess the reliability of the inversion. It is found that for this site and frequency only the first few wavelengths of the sediment have influence on the pressure field in the water. [1] [2]

1. Introduction

Our work aims at retrieving the physical characteristics of the ocean floor from its response to an acoustic source placed above it. In this paper the inversion is posed as an optimization problem, which is solved by a directed Monte Carlo search using Genetic Algorithms (GA) and *a posteriori* probabilities [1–3].

The inversion problem is stated as an optimization problem. The goal is to find the model vector $\mathbf{m}$ that minimizes or maximizes a 'fitness' function ϕ. Its formulation depends both on the problem at hand and on the measured data available (in our case, the amplitude of the pressure field versus range). We choose a quadratic deviation:

$$\phi(\mathbf{m}) = \mathbf{e}^T\mathbf{W}\mathbf{W}\mathbf{e}\ , \qquad \mathbf{e} = \mathbf{d}_{obs} - \mathbf{d}_{cal}(\mathbf{m})\ . \tag{1}$$

$\mathbf{d}_{obs}$ and $\mathbf{d}_{cal}$ are normalized unit vectors containing the observed and calculated amplitude of the pressure field, $\mathbf{m}$ is the model vector containing the physical parameters, $\mathbf{W}$ is a diagonal matrix containing the weighting for each observation point. In the present application, in order to give more importance to the near-field, we weight the points by $\mathbf{W} = \mathrm{Diag}\left[r_1^{-1/2}, r_2^{-1/2}, \ldots\right]$, where r is the distance from the source to the receiver. The scaling of the data vectors allows us to work only with the shape of the pressure field. The functional behavior is not necessarily regular and its minimization is obtained by GA.

2. Genetic algorithms and a posteriori statistics

GA is based on a random walk in the search space where a probability rule is used to help the search to proceed. GA can be used with a cooling temperature to combine advantages of

[1]Present address: Laboratoire des Signaux et Systems, Supelec, 91192 Gif/Yvette, France
[2]Present address: DIST, Universtity of Genova, Via Opere Pia 11a, Genova, Italy

O. Diaschok et al. (eds.), Full Field Inversion Methods in Ocean and Seismo-Acoustics, 159–164.

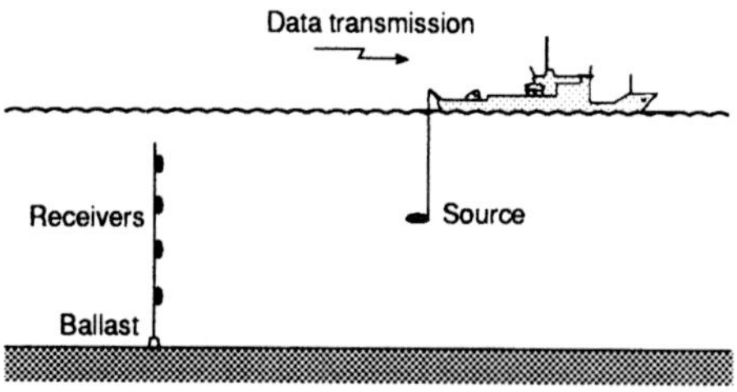

FIGURE 1. Measurement configuration.

Simulated Annealing (SA) with those of GA. In GA, the system we work with is discretized into M parameters using a model vector $\mathbf{m}$ called a chromosome. Each parameter, $j = 1,\ldots,M$ can take 2^{n_j} values between its lower and its upper bounds according to an *a priori* distribution. GA essentially consists of performing the evolution of a population of q model vectors in the search space using three operations: selection, crossover and mutation. More details can be found in [2,3].

During the optimization procedure, all samples of the search space are stored and used to estimate the *a posteriori* probabilities which are a measurement of how much we trust the solution. For a system of N parameters, the result is a N-dimensional space. This is displayed by the marginal probability density function. The samples are ordered according to their energy and the probability distribution is scaled using the Boltzmann distribution,

$$\sigma(\mathbf{m}^k) = \frac{\exp\left[-\phi(\mathbf{m}^k/T)\right]}{\sum_{l=1}^{N_{obs}} \exp\left[-\phi(\mathbf{m}^l/T)\right]}. \tag{2}$$

The temperature T is taken equal to the average of the best 50 samples (by experience a good choice). A marginal probability distribution can be calculated for each parameter [2].

3. Regularization

Preliminary experience with synthetic data [4] showed the necessity of regularization to have *a priori* well-behaved solutions. Regularization is introduced via shape functions:

$$\mathbf{m} = \sum_{i}^{M_s} \mu_i \mathbf{h}_i, \tag{3}$$

where $\mathbf{h}_i$ is the ith shape function and μ_i is the weight associated with the shape function. The advantages of shape functions are:

- to constrain the solution to belong to a certain class of expected profiles.
- to describe the variation of the parameters with fewer coefficients and in so doing to reduce the number of unknowns.
- to link correlated parameters.

Regularization by use of shape functions, can decrease the correlation between parameters and this can improve the inversion results [3].

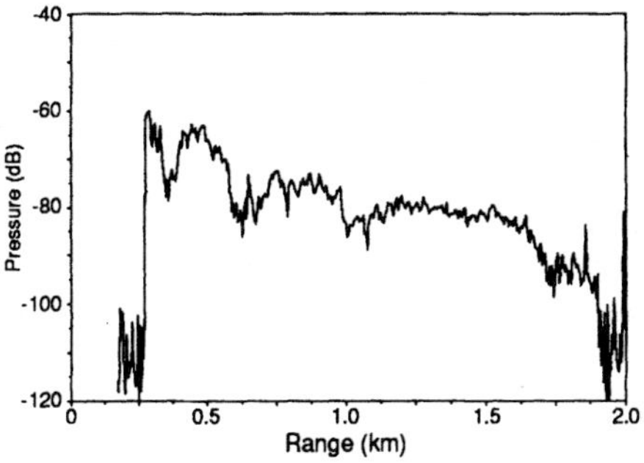

FIGURE 2. Range averaged signal for a receiver at 5 m depth.

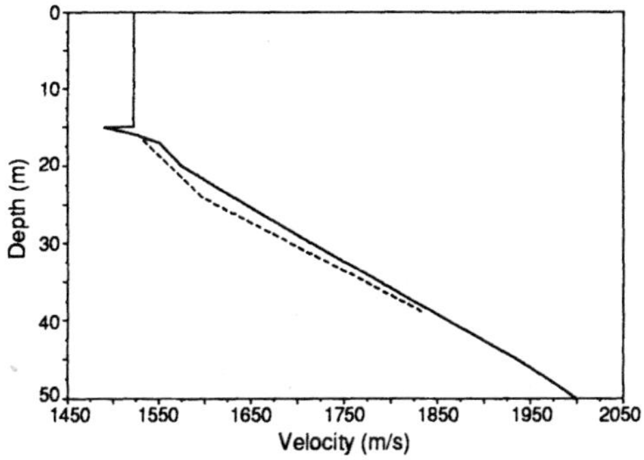

FIGURE 3. P-velocity profiles. The solid line is the broadband estimated profile, the dotted line, the inverted estimate by SNAP.

4. Inversion of real data

The experiment TELLARO took place in the Gulf of La Spezia in June 1992. The measurement configuration is sketched in Fig. 1: four receivers are fixed to the bottom at several depths (2, 5, 10 and 15 m) and a 330 Hz source is towed at a fixed depth of 7 m. The recording made with the 5 m-deep receiver is chosen because of its low noise compared to the others. But the data are still very noisy and we are obliged to smooth them by averaging the signal from six neighboring range values (Fig. 2).

From CTD measurements we have precise information on the water velocity profile and the water depth, and it is reasonable to assume a constant water velocity (1523 m/s). The water depth was 16 m. Since the data are very noisy especially at long range, we use the signal between ranges 270 and 1600 m. In order to decrease the influence of the long range the data are weighted by $1/\sqrt{r}$ where r is the distance between the source and the receiver.

Earlier, a seismic broadband analysis at the Tellaro site was carried out to estimate the bottom velocity profile. The S-wave velocity was estimated to about 100 m/s [5] and therefore, its influence on the transmission loss can be neglected. The P-velocity profile was estimated with the Herglotz-Wiechert analysis of the refracted arrival and provided the broadband estimated velocity profile in Fig. 3. This profile is used for reference but it is not necessarily the correct one. As it is a broadband experiment it probably has a deeper penetration depth than the single frequency measurement.

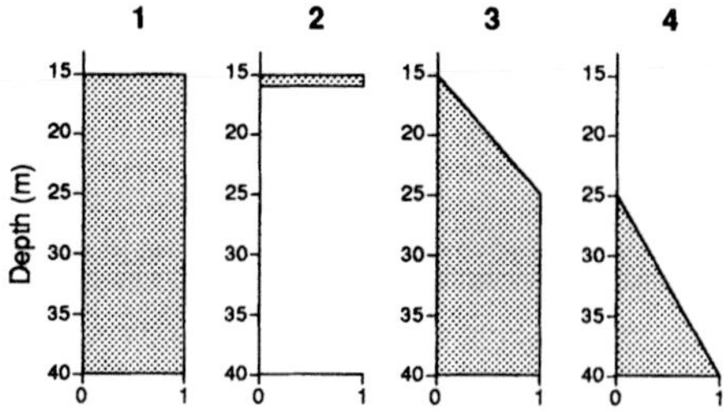

FIGURE 4. Shape functions for the speed and attenuation in the sediment.

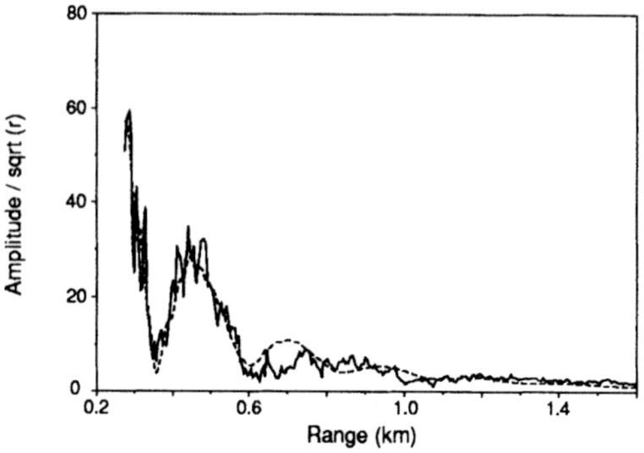

FIGURE 5. Pressure field at 330 Hz for the parameters in Fig. 3. The dashed line is the inverted result, while the solid line is the data.

The GA parameters we use for the inversion are: population size q (64), reproduction size f (0.5), crossover rate p_x (0.8), mutation rate p_x (0.05). All the known parameters are put at their measurement values. The studies of GA with synthetic data [4] have shown that the influence of the density on long range propagation is not important and is fixed to 1.7 g/cm^3. To describe the sediment we use ten layers of different thicknesses. Velocity and attenuation values in each layer have to be retrieved.

The first forward model we used with GA was SNAP which is a normal mode approach [6]. In this model the sediment and the basement are assumed to have constant attenuations. The result for the P-velocity profile is presented Fig. 3. The number of unknowns is six (four shape functions, plus sediment and basement P-attenuation values). The four shape functions used are depicted in Fig. 4. Fig. 5 shows the measured and the calculated pressure fields obtained with the P- velocity profile presented Fig. 3. We observe a good agreement at short ranges. The *a posteriori* probabilities obtained are presented in Fig. 6; we observe a lot of small spikes without really good location of the solution (this is different of from our experience with synthetic data [5]). For attenuation values we have problems because we obtain the maximum value allowed, which indicates that the interval is not large enough, or, more likely, that an incomplete parametrization of the environment is used. In general it is difficult to choose a 'correct' parametrization of the environment, since we are only using a few of the parameters describing the environment and there is little guidance on how to select them.

The second forward model we use is SAFARI/OASES [7]. It permits a varying P-attenuation in the sediment. For the P-velocity profile in each layer we have two possibilities:

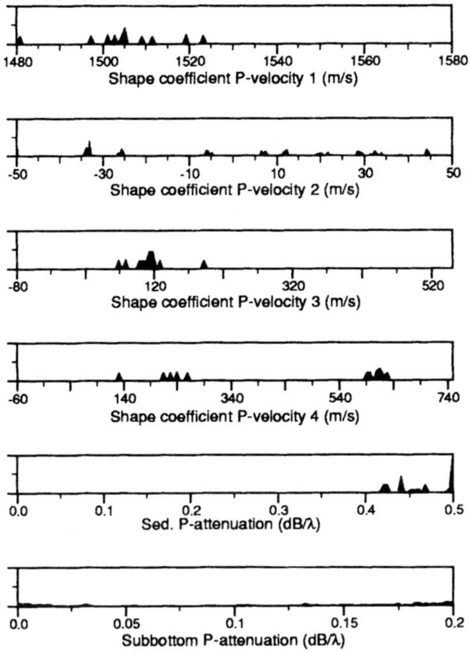

FIGURE 6. Probability distributions for the four shape functions, the P-attenuations in the sediment and in the subbottom.

either a constant speed or a linear variation in $1/c(z)^2$ with depth (to first order approximation this velocity variation is a linear profile with depth). Both options are tested. P-velocity and P-attenuation profiles are described using four shape functions (same as with SNAP) and results are presented Fig. 7. We observe that the first seven meters of the P-velocity profile is well determined which corresponds to a penetration depth of about 3 wavelengths. The P-attenuation is not as well-determined, especially at the sediment surface where the value found (0.5 dB$/\lambda$) is the maximum allowed in the search space. This means that we do not have enough loss in the system. The pressure field corresponding to the linear P-velocity profile is presented in Fig. 8 showing that the varying P-attenuation allows to better fit the small oscillations at short range.

5. Conclusions

A global inversion technique for the estimation of parameters in a marine environment from the measurement of the pressure field is considered, and genetic algorithms are shown to be very efficient in selecting the appropriate parameter values. The *a posteriori* probabilities help us in evaluating the solution. Regularization is introduced via shape function to constrain the solution and to reduce the number of unknown parameters and the observed data is weighted according to their assumed quality.

For the Tellaro test site it was found that for a frequency of 330 Hz only the first few meters of the sediment are important for the wave propagation, and therefore information about the deeper layers can only be retrieved with considerable uncertainty. The standard

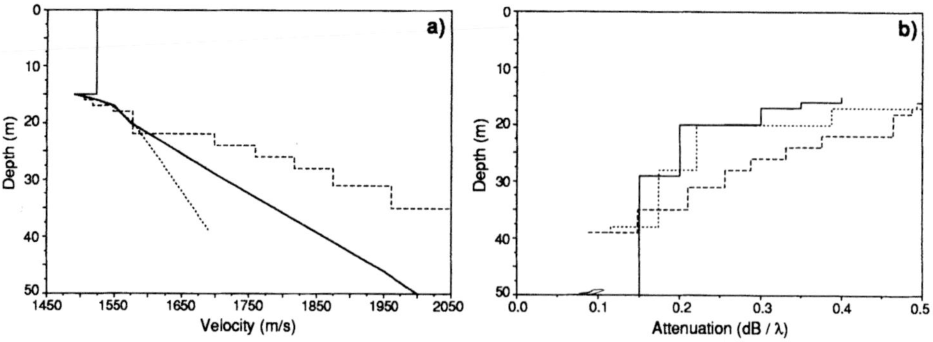

FIGURE 7. P-velocity (a) and P-attenuation (b) profiles in the case where the P-velocity value in each layer is constant (dashed line) and a straight line (dotted line). The broadband estimate is the solid line.

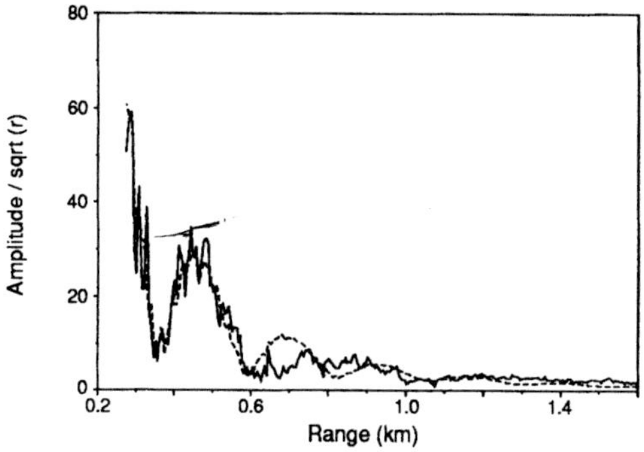

FIGURE 8. Pressure field at 330 Hz for the parameters in Fig. 7. The dashed line is the inverted result, while the solid line is the data.

propagation models OASES and SNAP have been used as forward models for the inversion. While OASES is more general, SNAP was found to be faster and sufficiently accurate for this inversion.

References

1. Stoffa, P.L., and Sen, M.K. (1991), "Nonlinear multiparameter optimization using genetic algorithms: inversion of plane-wave seismograms," *Geophysics* **56**, 1794-1810.
2. Gerstoft, P. (1994), "Inversion of seismoacoustic data using genetic algorithms and *a posteriori* probability distributions," *J. Acoust. Soc. Am.* **95**(2), 770-782.
3. Gerstoft, P. (1994), "Global inversion by genetic algorithms for both source position and environmental parameters," *Journal of Computational Acoustics*, to appear.
4. Lambert, M. (1994), "Inversion of seismo-acoustic real data using genetic algorithms," SM-276, SACLANT Undersea Research Centre, La Spezia, Italy.
5. Caiti, A., Akal T. and Stoll R.D. (1993), "Shear wave velocity in seafloor sediments by inversion of interface wave dispersion data," SR-205, SACLANT Undersea Research Centre, La Spezia, Italy.
6. Jensen, F.B. and Ferla M.C. (1979), "SNAP: the Saclantcen Normal-mode Acoustic Propagation Model," SM-121, SACLANT Undersea Research Centre, La Spezia, Italy.
7. Schmidt, H. (1987), "SAFARI: Seismo-acoustic fast field algorithm for range independent environments. User's guide," SR-113, SACLANT Undersea Research Centre, La Spezia, Italy.

MATCHED FIELD INVERSION FOR GEOACOUSTIC PROPERTIES OF YOUNG OCEANIC CRUST

N. R. CHAPMAN and K. S. OZARD
Defence Research Establishment Pacific
FMO Victoria, B.C. Canada
V0S 1B0

ABSTRACT A geoacoustic profile for young oceanic crust has been estimated from experimental data using matched field inversion with simulated annealing. The effect of errors in experimental geometry was investigated, and the inversion performance was significantly improved by including geometrical parameters such as range in the simulated annealing search algorithm.

Introduction

Matched field (MF) processing has been applied to a wide variety of inverse problems in underwater acoustics [1]. Geoacoustic properties of the ocean bottom have been estimated by MF inversion using acoustic field data for specific experimental geometries [2,3]. The inversion proceeds by specifying a set of geoacoustic models, and then searching the model parameter space to obtain the best correlation between measured and modelled acoustic fields. The practical application of the inversion depends in part on having an efficient method for searching the multi-dimensional space of ocean bottom models. In this paper, simulated annealing is used as the search algorithm in an investigation using MF inversion of vertical line array (VLA) data to determine geoacoustic profiles for young oceanic crust. The inversion is capable of estimating both the compressional and shear properties of the basalt.

Successful inversion requires that the experimental geometry be accurately known, however, there is usually some degree of uncertainty in the measurements of the variables that define the experimental arrangement. In general, the inversion performance is sensitive to mismatch in the range, water depth, and array tilt. The effect of errors in the experimental geometry on the model parameter estimation was investigated in a simulation study using a range-independent ocean bottom consisting of a fluid sediment layer over an elastic basement. The relatively high sensitivity to geometrical mismatch indicated that the experimental uncertainties could be tolerated in the inversion by including variables such as the range and depth as parameters in the search process. This approach is an application of the concept of 'focalization' [4].

In the remainder of the paper, the method for MF inversion is described, and the results of the simulation study on the effect of geometrical errors are presented. Following this, the MF inversion method, including focalization, is applied to experimental data that were obtained at a deep water site near the Juan de Fuca Ridge, and the results are summarized.

O. Diaschok et al. (eds.), Full Field Inversion Methods in Ocean and Seismo-Acoustics, 165–170.

Matched Field Inversion

The components of the matched field inversion method include a geoacoustic model, a propagation model for calculating the acoustic field at the receiver, a cost function based on a matched field processor for assessing specific models, and an efficient search algorithm for searching the model parameter space. For this work, the geoacoustic model consisted of three layers; a water layer, a homogeneous fluid sediment layer, and an underlying homogeneous elastic upper crust. The sediment layer was described by the thickness, h, the density, ρ, and the sound speed, c, and attenuation, α; the upper crust was descibed by the density and the sound speed and attenuation of the compressional (p) and shear (s) waves propagating in the layer (Fig. 1). The form of this model is a realistic approximation to the layer 2A upper crust environment for the experimental site [5].

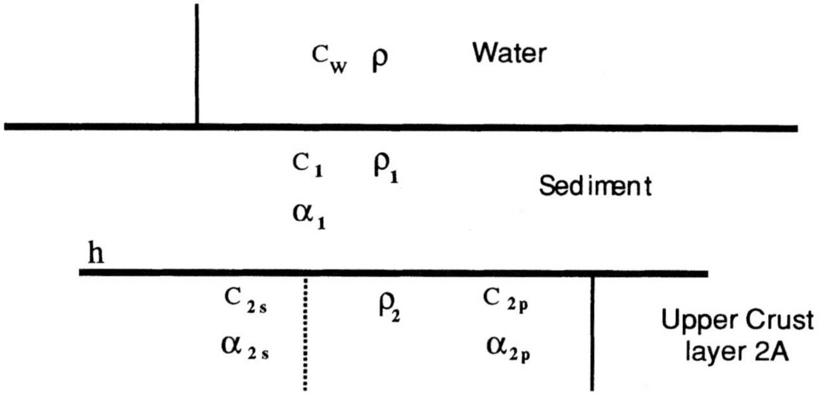

Fig. 1. Geoacoustic model for the upper crust at the experimental site.

The cost function, $E(m)$, is based on the Bartlett matched field processor which describes the correlation between the measured and modelled acoustic fields. It is given by

$$E(m) = p^*(m)Rp(m) \tag{1}$$

where R is the cross-spectral matrix and $p(m)$ is the modelled field calculated for a specific model, m. Although the Bartlett processor has low resolution and high sidelobes, it is robust under conditions of noise and mismatch. The modelled fields were calculated using the normal mode model KRAKENC [6]; this model uses a complex root finder to locate the discrete complex wavenumbers corresponding to the modes of an elastic waveguide.

To estimate the model parameter values, simulated annealing was used to search the model parameter space for the optimum solution. The search was initiated with random values for the parameters chosen from the imposed bounds; the inversion then proceeded through a series of iterations involving random perturbations of the model parameters, as a control parameter analogous to the temperature was reduced. Perturbations that decreased the cost function were accepted unconditionally. Those that increased the cost function were accepted conditionally, with probability of acceptance given by the Boltzmann distribution

$$P(\Delta E_i) = \exp(-\Delta E / s_i T) \tag{2}$$

where s_i is a scale factor for the ith parameter and T is the control parameter. The scale factor accounts for the different sensitivities of the model parameters, so that all parameters converge at about the same rate. The temperature was reduced according to the schedule

$$T_{j+1} = \gamma^{j+1} T_0, \text{ with } \gamma \leq 1, \text{ and } j = 0,1,... \tag{3}$$

The parameter perturbations were selected according to

$$m_i = m_i + \xi^3 \delta_i \tag{4}$$

where ξ is a random number from a uniform distribution on [-1,1], and δ_i is the maximum perturbation for the ith parameter.

Simulations

The simulations were carried out for a VLA consisting of 16 hydrophones spaced 45 m apart, with the top element at a depth of 200 m. The array was located 2 km from a 10-Hz CW sound source at a depth of 46 m. The geoacoustic profile was designed to model the ocean environment at the experimental site. The ocean bottom consisted of two homogeneous layers: a sediment layer that was 75 m thick, with density 1.8 g/cm^3, sound speed 1600 m/s, and attenuation 0.2 dB/λ; and an elastic half-space upper crust with density 2.4 g/cm^3, p-wave speed 2700 m/s, s-wave speed 1250 m/s and both attenuations 0.1 dB/λ. The sound speed profile for the water column was a profile measured during the experiment, which was truncated at a depth of 1000 m. For this environment, there were 14 modes at 10 Hz.

Initially a study was carried out to determine the sensitivity of the Bartlett processor to the model parameters. Each parameter was tested separately, while all the others were fixed at their true values. The sediment layer thickness and the upper crust p-wave speed were the most sensitive, followed by the sediment sound speed and the upper crust s-wave speed. The least sensitive parameters were the densities and attenuations for both layers; the attenuations were fixed at their true values for all the inversions. Similar tests indicated that the sensitivities for geometrical parameters such as the range, ocean depth and array depth were equivalent to those for the most sensitive geoacoustic parameters. The results for an inversion under ideal conditions of no mismatch are shown in Table 1. The final correlation after 2500 iterations was 0.9961.

The effect of range errors was investigated in inversions for which the modelled fields were calculated at array locations that were shifted by 1/3, 2/3 and 4/3 wavelengths from the true range. The results are shown in Table 2. Using the final MF correlation and the error in the estimate for the most sensitive parameter as a measure of the effect, it is evident that both the correlation and the accuracy are significantly decreased. The high sensitivity to range error indicated that the uncertainty could be tolerated in the inversion by including the range as a search parameter. In this approach, the inversion simultaneously localizes the source and estimates the optimum parameters of the waveguide model [4]. A simulation was carried out to search the range over a 50-m range mismatch, and the results are shown in the last column of Table 2. The range was estimated to within 3 per cent of the error, and the final MF correlation and the errors in the model parameter estimates were comparable to those for the ideal case.

Table 1. MF inversion estimates for the ideal case of no mismatch.

m_i	True Value	Bounds	s_i	Estimate
h (m)	75	(50 - 100)	1.0	77.8
c_1 (m/s)	1600	(1500 - 1700	10.0	1596
c_{2p} (m/s)	2700	(2500 - 2900)	1.0	2658
c_{2s} (m/s)	1250	(1050 - 1450)	10.0	1285
ρ_1 (g/cm^3)	1.8	(1.6 - 2.0)	100	1.8
ρ_2 (g/cm^3)	2.4	(2.1 - 2.7)	100	2.2

Table 2. Effect of range error on MF inversion estimates.

m_i	1/3 λ	2/3 λ	4/3 λ	Range Search
Correlation	0.9527	0.8476	0.6495	0.9977
h (m)	67.0	59.3	56.6	78.0
c_1 (m/s)	1608	1530	1700	1580
c_{2p} (m/s)	2751	2893	2895	2766
c_{2s} (m/s)	1097	1264	1119	1201
ρ_1 (g/cm^3)	1.6	1.7	1.6	1.8
ρ_2 (g/cm^3)	2.2	2.4	2.2	2.2
Range (km)	-	-	-	2.00

Experimental Data Inversion

The data were obtained in an experiment carried out using two ships at a site near the Juan de Fuca Ridge in the Northeast Pacific. One ship monitored a 16-element, 675-m VLA that was suspended at a depth of 413 m, and the other ship towed a 10-Hz CW sound source at a depth of 46.3 m in an arc at a radius of 5.5 km from the VLA. The range was determined from GPS navigation measurements that were recorded on each ship, and the ocean depth was measured using a 3.5-kHz profiler on the source ship. The VLA data were processed using a 4K FFT to obtain the spectral components at 10 Hz, and cross spectral averages were then formed for a two minute time period.

The inversion was performed for a range independent waveguide of depth 2660 m using the geoacoustic model in figure 1. Based on the results of the simulations, the range was included as a search parameter, and four other model parameters were estimated. The insensitive parameters were held fixed at values of ρ_1 = 1.55 g/cm^3, c_1 = 1550 m/s, and α_1 = 0.2 dB/λ for the sediment, and α_2 = 0.1 dB/λ for both attenuations in the upper crust. There were approximately 35 modes for this environment. The estimated values after 10000 iterations are listed in Table 3, along with the sensitivity scale factors. The accepted values for each parameter are plotted versus iteration number in Fig. 2; for this case, the search converged after about 6000 iterations. The VLA range is in excellent agreement with the GPS measurement of 5.45 km. The sound speeds are consistent with recent data for young crust of age 1-2 million years [7], and the density is also within expected limits [8]. The relatively low value for the final MF correlation,

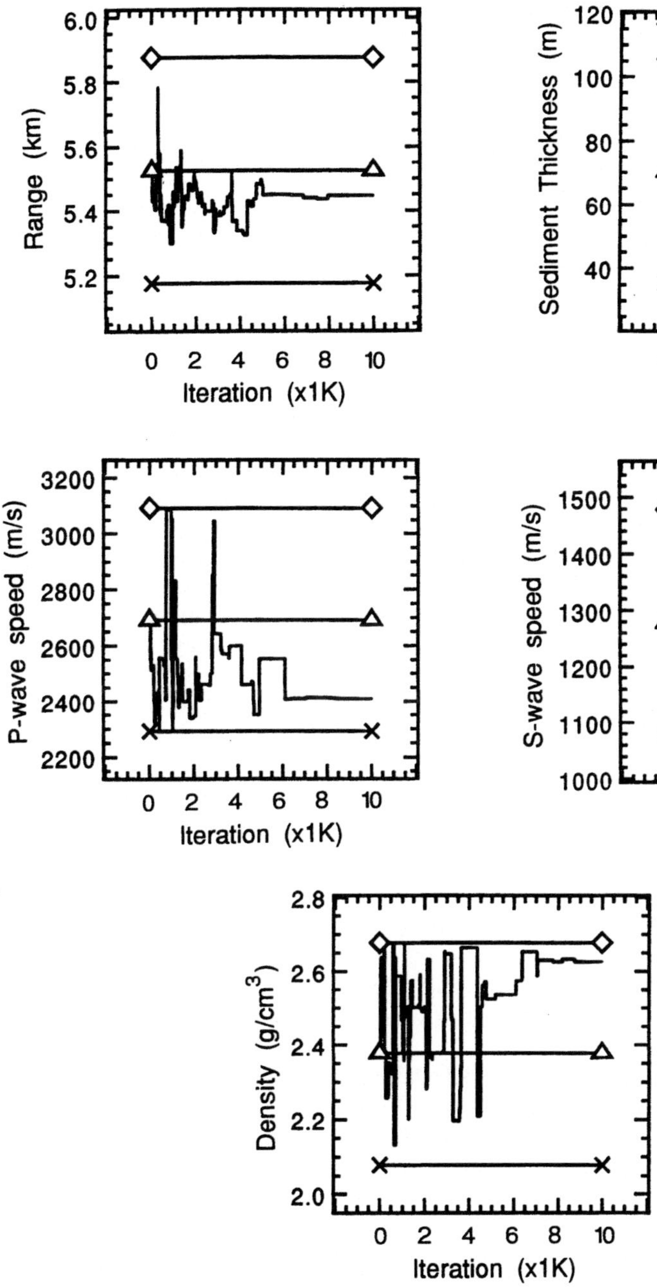

Fig. 2. Accepted values for the range, sediment thickness, and the upper crust p-wave speed, s-wave speed and density. Search limits are shown by the horizontal bars.

0.58, indicates that there is some degree of mismatch. This could be due to effects such as array tilt and range dependence in the waveguide.

Table 3. MF inversion for data from young crustal site.

m_i	Bounds	s_i	Estimate
Range (km)	(5.2 - 5.9)	0.4	5.44
h (m)	(35 - 110)	1.0	79.0
c_{2p} (m/s)	(2300 - 3100)	1.1	2410
c_{2s} (m/s)	(1050 - 1500)	3.0	1128
ρ_2 (g/cm^3)	(2.1 - 2.7)	48.6	2.6

Summary

The effect of uncertainties in experimental geometry has been investigated for matched field inversion for geoacoustic properties of the ocean bottom. Simulations that were carried out for a VLA indicated that the inversion performance was highly sensitive to range errors. However, the uncertainty could be tolerated in the inversion by including the range as a parameter in the simulated annealing search process. The inversion, including the concept of focalization, was applied to estimate a geoacoustic profile for upper oceanic crust at a site near the Juan de Fuca Ridge where the age of the basalt is very young. The estimate for the p-wave speed is consistent with the low values that have been observed for young crust at other sites. In addition, the method provides a means of estimating the shear wave speed.

References

1. Tolstoy, A. (1993) *Matched Field Processing for Underwater Acoustics*, World Scientific Monograph, New York.
2. Collins, M.D., Kuperman, W.A., and Schmidt, H. (1992) Nonlinear inversion for ocean-bottom properties, *J. Acoust. Soc. Am.*, **92**, 2770-2783.
3. Dosso, S.E., Yeremy, M.L., Ozard, J.M. and Chapman, N.R. (1993) Estimation of ocean bottom properties by matched field inversion of acoustic field data, *IEEE J. Oceanic Eng.*, **18**, 232-239.
4. Collins, M.D. and Kuperman, W.A. (1991) Focalization: Environmental focussing and source localization, *J. Acoust. Soc. Am.*, **90**, 1410-1422.
5. Cudrak, C.F. and Clowes, R.M. (1993) Crustal structure of Endeavour Ridge segment, Juan de Fuca Ridge, from a detailed seismic refraction survey, *J. Geophys. Res.*, **98**, 6329-6349.
6. Porter, M.A. and Reiss, E.L. (1985) A numerical model for bottom interacting ocean acoustic normal modes, *J. Acoust. Soc. Am.*, **77**, 1760-1767.
7. Jacobson, R.S. (1992) The impact of crustal evolution on changes of the seismic properties of the uppermost ocean crust, *Reviews of Geophysics*, **30**, 23-42.
8. Hamilton, E.L. (1980) Geoacoustic modelling of the sea floor, *J. Acoust. Soc. Am.*, **68**, 1313-1340.

DETERMINATION OF THE GEOACOUSTICAL PARAMETERS OF A SEDIMENTARY LAYER FROM SURFACE ACOUSTIC WAVES: A NEURAL NETWORK APPROACH

J. GUILBOT and F. MAGAND
Laboratoire d'Acoustique, Systèmes, Signaux et SOnar
CPE-ICPI Lyon
25, rue du plat - 69288 Lyon
France

ABSTRACT. Acoustical characterization of surface acoustic waves which propagate along the water-sediment interface is of great interest for numerous underwater applications. One of the main motivations of such studies is the possible use of interface waves (often called Scholte waves) for the non-invasive characterization of sedimentary bottoms [1]. The aim of this paper is to illustrate the possible estimation of the acoustical and geometrical (thickness) parameters of a sedimentary layer from the analysis of the velocity dispersion of interface waves. The strategy developped to solve the inverse problem is based on a neural network approach.

1. Sea Bottom Configuration

The sea bottom configuration we consider is made up of a sedimentary layer overlaying a semi-infinite hard substratum (figure 1). Both the layer and the substratum are homogeneous and isotropic media. The water layer is an homogeneous semi-infinite layer where the sound velocity and the density are respectively equal to c_0=1500 m/s and ρ_0=1000 kg/m^3. Table 1 gives the limits of the geoacoustical parameters of the sea bottom.

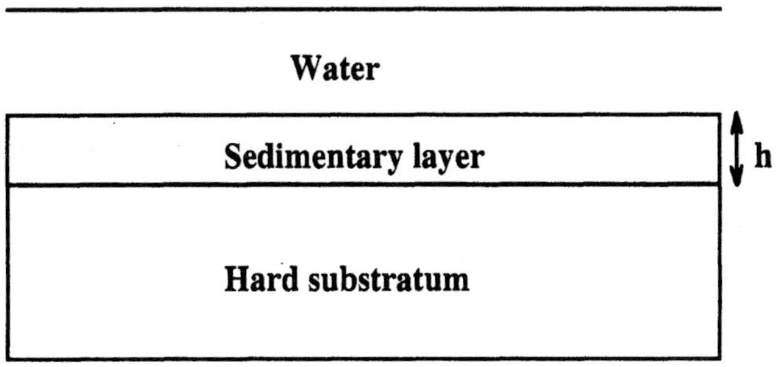

FIGURE 1. Sea bottom configuration

O. Diaschok et al. (eds.), Full Field Inversion Methods in Ocean and Seismo-Acoustics, 171–176.

TABLE 1. Geoacoustical parameters of the sea botttom

Material	thickness (m)	shear velocity (m/s)	compressional velocity (m/s)	density (kg/m^3)
Sediment	$3 \leq h \leq 19$	$140 \leq c_t \leq 460$	$1700 \leq c_l \leq 2900$	$1400 \leq \rho \leq 2200$
Hard substratum	semi-infinite	2935	6300	2700

2. Phase and Group Velocity Dispersion of Interface Waves

Along the plane interface between a perfect fluid and an elastic solid, a surface wave of the Stoneley-Scholte type can propagate. For semi-infinite and homogeneous media, this wave is non-dispersive. Because of different acoustical properties that may exist within the fluid or the solid (velocity and density gradients), the interface wave becomes dispersive. This property offers a relevant approach to recover the acoustical properties of the layer via specific inversion schemes [2]. Velocity dispersion also arises in the case of an homogeneous solid layer with finite thickness overlaying a semi-infinite solid substratum.

In this latter configuration, the formulation of the continuity of stresses and displacements at each boundary (water-sediment and sediment-substratum boundaries) leads to a 7x7 system that has nontrivial solution only when its determinant is equal to zero. The analytic computation of this determinant leads to the phase velocity dispersion equation (1) of the interface wave:

$$a_o + (a_1 b_1 - a_2 b_2)\, ch(q+s)h + (a_3 b_3 - a_4 b_4)\, ch(q-s)h = 0 \tag{1}$$

where $q = \sqrt{k_s^2 - k_l^2}$ and $s = \sqrt{k_s^2 - k_t^2}$,

$k_l = \omega/c_l$, c_l compressional wave velocity of the sedimentary layer,
$k_t = \omega/c_t$, c_t shear wave velocity of the sedimentary layer,
$k_s = \omega/c_\varphi$, c_φ phase velocity of the interface wave and
ω is the angular frequency ($\omega = 2\pi\nu$, ν frequency).

The coefficients a_o, a_1, a_2, a_3, a_4, b_1, b_2, b_3 and b_4 depend on the acoustical properties of the water, the layer and the substratum. Only the coefficients a_i ($1 \leq i \leq 4$) are function of h. Figure 2 illustrates the phase and the group velocity dispersion curves for an example of geoacoustical data extracted from table 1. The group velocity c_g is computed from the phase velocity c_φ thanks to expression (2). Only phase velocity data will be considered through this paper.

$$c_g(\nu) = \frac{c_\varphi(\nu)}{1 - \dfrac{\nu}{c_\varphi(\nu)} \dfrac{\partial c_\varphi(\nu)}{\partial \nu}} \tag{2}$$

When we solve equation (1) a cut-off frequency appears for which the solution of (1) tends towards the shear wave velocity of the sedimentary layer. This low frequency limit results from the condition of real solution for equation (1).

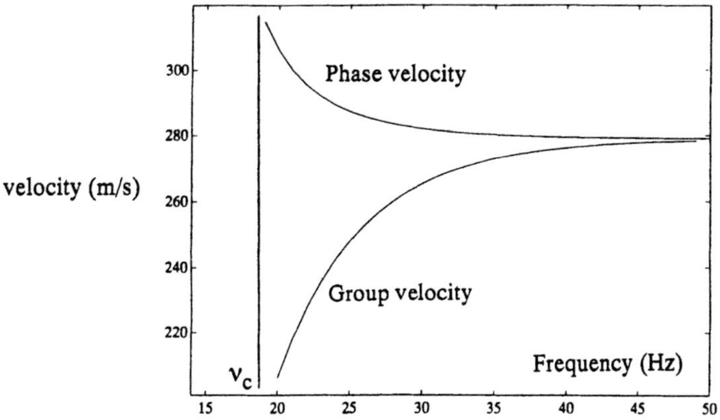

FIGURE 2. Phase and group velocity dispersion of Scholte waves
($h = 8.5$ m, $c_t = 320$ m/s, $c_l = 1900$ m/s, $\rho = 1450$ kg/m^3, cut-off frequency $\nu_c \approx 19$ Hz)

3. A Neural Network Approach to solve the Inverse Problem

Several methods and algorithms have already been investigated to solve the inverse problem [3]. Most of them require numerous *a priori* informations difficult to justify in usual *in situ* applications (one or several parameters are supposed to be known during inversion).
The main motivations that have led to a neural network approach are linked to the basic principles of neural networks. Neural networks are adaptive (take data and learn from them), highly parallel (high speed computation) and nonlinear. The latter point is directly connected to the present work where the geoacoustical parameters nonlinearly depend on the velocity dispersion data.
We used through this work the Gradient Back Propagation algorithm which is a supervised learning algorithm [4]. The aim of this algorithm is to minimize a cost function through an adaptive optimization technique. The whole process is split into two steps:
(i) during the *learning phase*, the network tries to adjust its internal parameters (synaptic weights) using both known phase velocity data (input vector) and the corresponding geoacoustical parameters (output vector),
(ii) during the *testing phase*, the network predicts the geoacoustical parameters using only the given input vectors.

Such a network is made up of several neuron layers, interconnected from the input to the output, so that the state information can propagate forward only from one layer to the next. There are as many neurons in the input layer as the dimension of the input vector and as many neurons in the output as the number of geoacoustical parameters to estimate (figure 3).
The choice of both the number of intermediate layers (hidden layers) and the number of neurons per layer, results from a compromise between two extreme situations:
(i) if those numbers are too high, the network will need a long training duration and will require numerous training examples leading to a poor generalization,
(ii) if the numbers are too small, the network will not be able to learn the training examples.

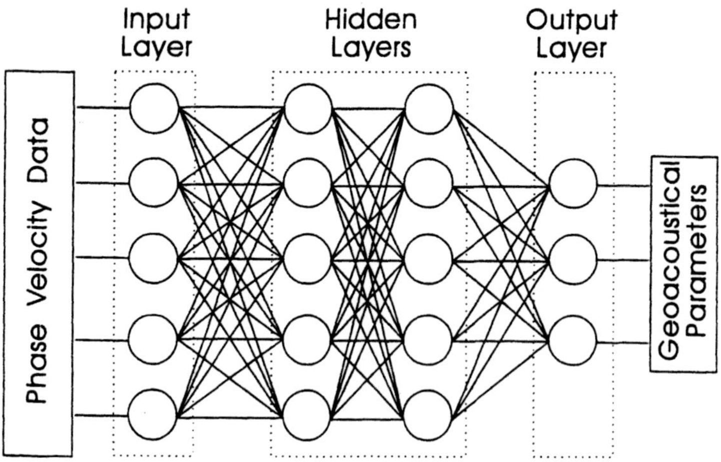

FIGURE 3. Multi-layer network structure

The learning data base has also to be carefully chosen, so as to be representative of the problem. If some parts of the data space are not correctly represented in the training set, one should expect problems during the generalization. During the training phase, synaptic weight are adjusted using the corresponding mean square error in the back propagation algorithm. Performance of estimation are then determined by running the network through the whole data base.

4. Results

In order to make the input data base used by the network during the learning and testing phases, each acoustical parameter and the thickness of the sedimentary layer have been incremented with a constant step to cover the range of possible variations of each parameter. Therefore, 625 input vectors have been obtained. The explicit formulation of the phase velocity dispersion equation (1) is of great interest to build up the data base. It avoids the computation of numerous 7x7 determinants that would be necessary to compute the value of the phase velocity at a given frequency.

As input vector we picked up 7 phase velocity data on the dispersion curve so that their position with respect to the cut-off frequency is similar for each vector. These positions allow a thorough description of the low frequency range (the most dispersive one) and provide enough information about the high frequency asymptotic limit. Hence, before the learning phase, the process has to compute the cut-off frequency for each set of geoacoustical parameters.

The network is a 4-layers network (2 hidden layers) with 7 neurons in the input layer (7 velocity data) and 4 neurons in the output layer (4 parameters to estimate). Among the 625 input vectors, 310 have been devoted (random choice) to the learning phase, the others to the testing phase.

Figures 4, 5 and 6 respectively give the final estimations of the thickness, the density and the shear velocity as a function of the input vector index. In order to quantify the quality of these results, the estimation error is computed for each parameter and then averaged over the whole testing data base. Very good estimates of the density (mean error less than 3 %) and of the shear velocity (mean error less than 5 %) are obtained together with a good estimation of the thickness (mean error less than 10 %).

Nevertheless, a poor estimation of the compressional velocity is obtained (mean error of the order of 15 %). The process gives the mean compressional velocity (table 1) as an estimation of this velocity. This result could be predicted by the analysis of the forward problem where it appeared that the influence of the compressional velocity on the dispersion law was very small.

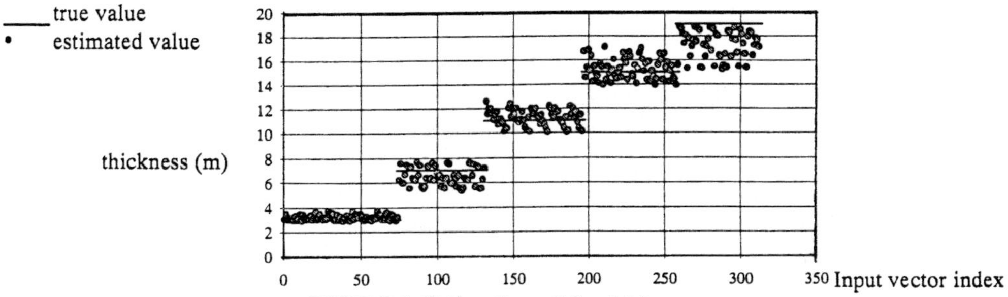

FIGURE 4. Estimation of the thickness

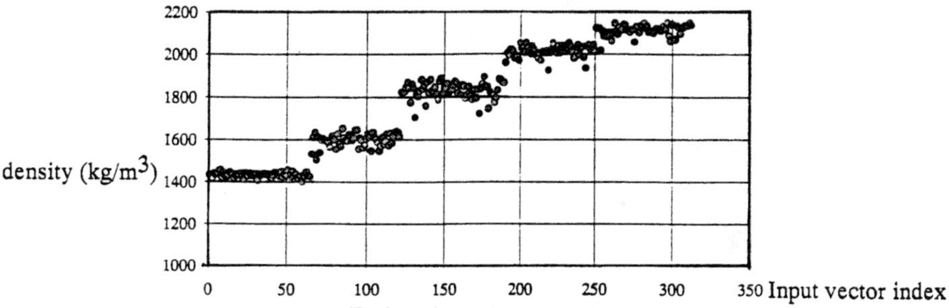

FIGURE 5. Estimation of the density

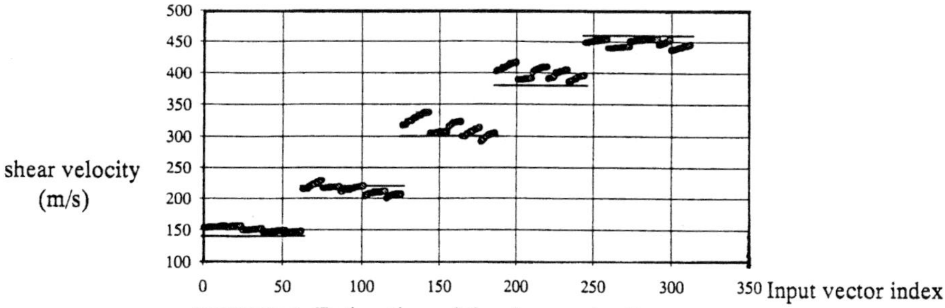

FIGURE 6. Estimation of the shear velocity

In order to illustrate the performance of the inversion process, table 2 gives the estimates of the geoacoustical parameters when these ones do not belong to the learning and testing data base. The basic assumption we make is that the cut-off frequency is known. Nevertheless, an estimation of this frequency can be obtained from the high asymptotic limit of the velocity dispersion curve [5].

TABLE 2. Estimation of the geoacoustical parameters of the layer

	thickness (m)	shear velocity (m/s)	compressional velocity (m/s)	density (kg/m^3)
true value	8.5	320	1900	1450
estimated value	8.45	325	1990	1325

5. Conclusion

We have shown the feasibility of recovering the geoacoustical parameters of a sedimentary layer from a neural network approach. This inversion procedure does not require initial guess about the geoacoustical parameters to estimate. The network only needs representative geoacoustical data to adjust its synaptic weights during the learning phase. When these weights are adjusted, geoacoustical parameters can be easily estimated on line for each set of phase velocity input data. This process can also be seen as an initialization step to others inversion schemes that could be used to recover the compressional velocities.

Very good estimations of the shear velocity, the density and the thickness of the layer have been obtained. The case of a sediment with shear and compressional waves velocity profiles will be studied and further investigations will be carried out to study the influence on the final estimations of disturbed input data (experimental configuration).

References

1. Rauch, D. (1986) On the role of bottom interface waves in ocean seismo-acoustics: a review, in T. Akal and J.M. Berkson, *Ocean Seismo-Acoustics*, Plenum Press, New York, 623-641.
2. Caiti, A., Akal, T. and Stoll, R.D. (1991) Determination of shear velocity profiles by inversion of interface wave data, in J.M. Hovem, *Shear Waves in Marine Sediments*, Kluwer Academic Publishers, Netherlands, 557-565.
3. Rajan, S.D. and Howitt, C.S. (1991) Determination of the sediment shear speed profiles from phase and group velocity dispersion data of SH wave, in J.M. Hovem, *Shear Waves in Marine Sediments*, Kluwer Academic Publishers, Netherlands, 529-536.
4. Davalo, E., Naïm, P. (1990) *Des réseaux de neurones*, Eyrolles, Paris.
5. Guilbot, J. (1992) Dispersion de l'onde de Stoneley-Scholte à l'interface plane séparant un fluide parfait d'un solide à variations continues de célérités. Interprétation des limites asymptotiques, *J. Acoustique*, 5, 573-588.

SEAFLOOR PARAMETER ESTIMATION: APPROXIMATING THE INVERSE MAP THROUGH RBF NETWORKS

A. CAITI (§, ¶), T. PARISINI (¶), AND R. ZOPPOLI (¶)
(§) SACLANT Undersea Research Centre, La Spezia, Italy
(¶) DIST - Univ. of Genova, Genova, Italy

ABSTRACT A method for estimating the seafloor geoacoustic parameters from the measured acoustic field in the water is proposed. In particular, we impose to the inverse function the structure of a Radial Basis Function (RBF) expansion, and identify its coefficients using a set of computer generated input-output pairs. Among the advantages of the method are the robustness to noise in the data and the ability to compute the solution in real time. Succesfully estimated results are presented, with an accuracy that depends on their influence on the acoustic field.

1. Introduction

In order to rapidly estimate seafloor geoacoustic parameters over large areas, several methods have been recently proposed, all based on the measurement of the acoustic field in the water column. The geoacoustic parameters are then determined by inverting the forward map (i.e., the wave equation) that link them to the pressure field. There are two major aspects of this approach: one is experimental, and involves the suitable design of sources, sensors and experimental techniques to accurately measure the field structure; the other is computational, and consists in the algorithmic determination of a unique and stable inverse map able to recover the environmental parameters given the measured field. This work deals with the computational part of the problem. Most of the recent research on this topic has been focused on global optimization methods, and in particular on the minimization of highly nonlinear non-convex functions. In this work we propose a different approach in which the inverse map is given *a priori* the structure of a Radial Basis Function (RBF) series expansion. The coefficients of the expansion are identified from a set of known field-parameter correspondences determined with simulations using a suitable forward model code. The identification procedure can be done off-line even before the measurement is performed, while the inversion results may be computed in real time. It is important to remark that the structure of an RBF expansion is such that, while it is able to approximate highly nonlinear maps, it is ***linear in the parameters***. This means that, once a sufficiently large set of input-output correspondences is given, the optimal choice of the RBF coefficients is determined as the solution of a linear algebraic system. A discussion on the use of RBF in the context of inverse problems can be found in [1].

2. RBF networks and the approximation of inverse maps

In general terms, the inverse problem of interest here can be formalized in the following way: given a known forward relation $\mathcal{F}(\mathbf{x}) = \mathbf{p}$ that links the seafloor geoacoustic parameters

O. Diaschok et al. (eds.), Full Field Inversion Methods in Ocean and Seismo-Acoustics, 177–182.

$\mathbf{x}$ to the pressure field in the water $\mathbf{p}$, and a measurement $\mathbf{p}_o$, determine $\mathbf{x}_o$ such that the cost function $E(\mathcal{F}(\mathbf{x}), \mathbf{p}_o)$ is minimized. In the above, E may be an arbitrary cost function, such as a weighted norm, or, in the case of a matched-field approach, a Bartlett, Capon, etc. mismatch function. The minimization of E is *not* sufficient to uniquely determine $\mathbf{x}_o$, and some constraints on the solution have to be imposed. It is interesting to note that in most of the recent works the constraints do not explicitly appear in the cost function, but are present through the use of boundaries on admissible domain for $\mathbf{x}$, or the use of shaping functions (like Empirical Orthogonal Functions).

A dual approach to the estimation problem may be stated as follows: determine an inverse function $\mathcal{G}(\mathbf{p}) = \mathbf{x}$ such that a cost functional $E'(\mathcal{G}(\mathcal{F}(\mathbf{x})), \mathbf{x})$ is minimized. Also in this case, the minimization of the functional E' is not sufficient to determine $\mathcal{G}$ uniquely, and additional constraints on the structure of $\mathcal{G}$ have to be imposed. Common requirements in inverse problem theory are the boundedness and continuity of $\mathcal{G}$, in order to have also some robustness with respect to noise in the data. So the general functional to be minimized takes on the form: $J'(\mathcal{G}) = E'(\mathcal{G}(\mathcal{F}(\mathbf{x})), \mathbf{x}) + \lambda H(\mathcal{G})$, where H is the constraining operator and λ a Lagrange multiplier.

Let us now suppose that we have a set of N input-output pairs $\{\mathbf{p}_i, \mathbf{x}_i\}_{i=1}^{N}$, with $\mathcal{F}(\mathbf{x}_i) = \mathbf{p}_i$ for every i. Then it is possible to state our problem as: find an *approximation* of the map $\mathcal{G}$ that minimizes the following cost functional:

$$J'(\mathcal{G}) = \sum_{i=1}^{N} \|\mathcal{G}(\mathbf{p}_i) - \mathbf{x}_i\|^2 + \lambda H(\mathcal{G}) \tag{1}$$

By variational arguments [5], it is possible to show that, if $H(\mathcal{G}) = \|R\mathcal{G}\|^2$, where R is a linear operator, then the approximation takes on the following form:

$$\mathcal{G}(\mathbf{p}) = C\Phi(\mathbf{p}_1, \cdots, \mathbf{p}_N; \mathbf{p}) + \mathbf{k}_R(\mathbf{p}) \tag{2}$$

where: Φ is a N-dimension vector, whose i-th component is given by the function $\phi(\mathbf{p}_i; \mathbf{p})$; C is an $m \times N$ coefficient matrix, where m is the dimension of the vector $\mathbf{x}$; $\mathbf{k}_R$ is a m-dimension vector whose components span the null space of R.

The function ϕ is determined as the solution of the differential equation:

$$R^* R\phi(\mathbf{u}, \mathbf{v}) = \delta(\mathbf{u} - \mathbf{v}) \tag{3}$$

where R^* is the adjoint operator of R. In other words, ϕ is the Green function of the self-adjoint operator R^*R. If R is rotationally and translationally invariant, then ϕ is a radial function, i.e. $\phi(\mathbf{u}, \mathbf{v}) = \phi(\|\mathbf{u} - \mathbf{v}\|)$, hence the RBF denomination. Particular instances of RBF often used in interpolation/approximation problems are gaussian functions ($\phi(r) = \exp(-r^2/\sigma^2)$), exponentials, thin-plate and cubic splines, multiquadric.

A most useful property of RBF theory is that, for several constraining operators, the solution of equation (3) is known analytically. Gaussian RBF are derived with $R = \sum_{i=1}^{\infty} \partial^i/\partial \mathbf{p}^i$, exponential RBF with $R = \partial/\partial \mathbf{p}$, whereas splines of different orders can be constructed as intermediate cases. The main point is that, in this way, the choice of a particular set of RBF automatically forces the solution to belong to a specific smoothness class. For the simulations presented in Section 3, we have decided to strenghten the regularity properties of the inverse map, and we have selected gaussian basis functions (in this particular case the term $\mathbf{k}_R$ in equation (2) vanishes). This choice may be debatable: in particular, by smoothing the inverse map, we also obtain a smoothed solution $\mathbf{x}$, i.e. we loose resolution. This may not always be acceptable, and a different set of functions may be required.

The coefficients C can be determined by evaluating equation (2) at the N known input-output pairs, thus leading to the system of algebraic equationss:

$$(M + \lambda I)\mathbf{c}_k^T = \mathbf{b}_k \quad k = 1, \cdots, m \tag{4}$$

where $(M)_{ij} = \phi(\|\mathbf{p}_i - \mathbf{p}_j\|)$, $\mathbf{c}_k$ is the k-th row of matrix C, $\mathbf{b}_k = ((\mathbf{x}_1)_k, \cdots, (\mathbf{x}_N)_k)^T$.

As a last remark, note that any expression like (2), i.e. a weighted sum of kernel functions, can be implemented in hardware in a network fashion, hence the denomination of RBF networks [2], [5].

3. Simulation results

The RBF network described in the previous section has been trained with input-output pairs in which the input $\mathbf{p}$ is the amplitude of the depth-dependent Green function (i.e., the Hankel transform of the pressure field as a function of range, in a range independent environment), and the output $\mathbf{x}$ are the corresponding geoacoustic parameters. Two test environmental models have been selected, made out of horizontally homogeneous layers. For each model the intervals of variation for the parameters of each layer have been taken as in tables 1-2. These intervals are representative of clay, silt-clay, sand and rock or consolidated sediments. It is well known that the different parameters have different influence on the acoustic field. In particular, shear velocities below 400m/s and compressional attenuation below 0.1 dB/λ have very small influence [4], so we do not expect to generate a precise estimate of these parameters. Note that the effect on acoustic propagation depends also on the depth within the bottom and on the acoustic frequency.

In both cases the water depth is 140m, and the sound speed in water is constant and equal to 1500m/s. Propagation at 100 Hz has been considered, with source and receiver at fixed depths in the water, 70m and 100m respectively. Following the suggestion of [3], the Green's function has been severely undersampled, and only 64 equally spaced values have been taken.

In the first stage the RBF approximation has been checked with *known* layer depth, and to estimate a subset of the environmental parameters, in particular compressional velocity and attenuation or compressional and shear velocity. For each case, environmental models have been generated by random selection of the environmental parameters of interest within the pre-specified bounds, and the corresponding depth-dependent Green's function computed with SAFARI [6] and undersampled. 70 input-output pairs have been used to determine the RBF coefficients, and 30 different models ("test set") have been used to check the validity of the RBF approximation. In tables 3-6 the mean errors (absolute and relative) and variances over the test set are reported.

It can be seen that, in CASE 1, the compressional velocity estimate is recovered with a mean error of about 1 %, while the error is about 0.2 % in CASE 2. This may be due to the physical fact that the propagation in CASE 1 is dominated by the hard layer starting at 150m depth, with a coverage of softer sediment that is less than an acoustic wavelenght. As expected, shear velocity of the first two layers in both cases is estimated with large errors (about 20 %), reflecting the negligible influence of shear of these two strata on the acoustic propagation, while the shear velocity of the last layer is well recovered in both cases. Compressional attenuation is acceptably recovered only for the second layer in CASE 1, and

Depth (m)	P vel. (m/s)	S vel. (m/s)	P att. (dB/λ)
0.0	1500	0.0	0.0
140	1490-1600	80-200	0.075-0.15
145	1600-1850	250-500	0.5-1.0
150	2000-3000	800-1000	0.01-0.05

TABLE 1. Environmental parameter space: CASE 1

Depth (m)	P vel. (m/s)	S vel. (m/s)	P att. (dB/λ)
0.0	1500	0.0	0.0
140	1470-1570	30-100	0.07-0.11
145	1500-1650	70-250	0.08-0.12
170	1600-1800	300-500	0.4-0.8

TABLE 2. Environmental parameter space: CASE 2

for all three layers in CASE 2. From these results, we conclude that the estimates obtained with the RBF approach are consistent with the physics of the problem, with increasing accuracy on the parameters most influencing acoustic propagation.

Encouraged by this first result, we have removed the assumption of known layer depth. In particular, for CASE 1 (the "worst case" of our previous simulations), we have done the following: assuming that the depth of the deepest layer is still known (150m), we have discretized in depth the first 10m at steps of 2.5m, and have randomly selected the depth of the second layer. In some cases, this has led to models where layer 1 or layer 2 were absent. The training set has been enlarged to 400 randomly chosen models, while the test set dimension has been kept to 30. The ensemble results over the test set for the case of compressional and shear velocities are reported in table 7.

Depth (m)	P vel.				S vel.			
	Abs. err. (m/s)		Rel. err (%)		Abs. err. (m/s)		Rel. err (%)	
	mean	std	mean	std	mean	std	mean	std
140	18	20	1.2	1.3	38	26	27	19
145	17	14	1.0	0.8	57	71	21	19
150	28	17	1.1	0.6	67	48	7	5

TABLE 3. Mean and standard deviation of the estimation error on the test set for CASE 1, estimation of compressional and shear velocity only, layer depths known

Depth (m)	P vel.				P att.			
	Abs. err. (m/s)		Rel. err (%)		Abs. err. (dB/λ)		Rel. err (%)	
	mean	std	mean	std	mean	std	mean	std
140	13	17	0.9	1.1	0.028	0.027	27	28
145	14	20	0.8	1.0	0.055	0.064	8	10
150	20	21	0.8	0.9	0.017	0.0113	78	96

TABLE 4. Mean and standard deviation of the estimation error on the test set for CASE 1, estimation of compressional velocity and attenuation only, layer depths known

Depth (m)	P vel.				S vel.			
	Abs. err. (m/s)		Rel. err (%)		Abs. err. (m/s)		Rel. err (%)	
	mean	std	mean	std	mean	std	mean	std
140	4	4	0.2	0.3	12	8	23	20
145	3.5	3	0.2	0.2	24	26	18	20
170	8	9	0.5	0.5	14	18	3.7	4.7

TABLE 5. Mean and standard deviation of the estimation error on the test set for CASE 2, estimation of compressional and shear velocity only, layer depths known

It can be noted that the mean error in the compressional velocity slightly increases. This may be due to two concurring effects: one is physical, i.e. the 2.5m steps of the discretization correspond to approximately one sixth of the acoustic wavelenght, and so there is not enough resolution in the data to distinguish more precisely the thin layers; the second is due to the smoothing properties of the method itself. However, the smoothing is a guarantee of robustness when the data are noisy, so that with real data we expect to retain the same performance as with simulated data. Note that the shear velocity of the last layer has a

Depth (m)	P vel.				P att.			
	Abs. err. (m/s)		Rel. err (%)		Abs. err. (dB/λ)		Rel. err (%)	
	mean	std	mean	std	mean	std	mean	std
140	3	3.5	0.2	0.2	0.0077	0.0064	8.5	7.4
145	3	3	0.2	0.2	0.0129	0.0155	13	16
170	2.5	3	0.1	0.2	0.0716	0.0657	14	15

TABLE 6. Mean and standard deviation of the estimation error on the test set for CASE 2, estimation of compressional velocity and attenuation only, layer depths known

Depth (m)	P vel.				S vel.			
	Abs. err. (m/s)		Rel. err (%)		Abs. err. (m/s)		Rel. err (%)	
	mean	std	mean	std	mean	std	mean	std
140	16.5	11	1.0	0.7	27	18	20	14
142.5	31	21	1.9	1.2	36	22	24	18
145	42	30	2.5	1.8	76	57	39	46
147.5	33	28	1.9	1.6	55	41	15	13
150	37	36	1.4	1.3	43	24	4.6	2.5

TABLE 7. Mean and standard deviation of the estimation error on the test set for CASE 1, estimation of compressional and shear velocity only, layer depths *unknown*

smaller error than in the previous case, taking benefit from a larger training set.

4. Conclusions

In this work we have reported the results of a preliminary feasibility study in order to assess if the RBF technique can provide a reliable estimate of the seafloor geoacoustic parameters. It has been shown that, in the cases considered, the compressional velocity can be estimated with an error of at most about 2 % of the true value, while the accuracy on the other geoacoustic parameters decreases depending on their influence on the acoustic field. As expected, the discretization of the environment has to be put in relation with the acoustic wavelength of the propagating field.

Much work still needs to be done. In particular, we have used a "blind" application of RBF, while improvements can be obtained using more sophisticated approaches, like least-squares orthogonalization [2]. This will give an indication on how "large" the training set should be. Also the influence of the particular set of RBF used has still to be investigated.

Acknowledgement. We gratefully thank Å. Kristensen and F. Ingenito for their helpful comments. This work has been supported in part by EEC, under the MAST 2 initiative, contract MAS2-0022.

References

1. Caiti, A., G. Magenes, T. Parisini, R. Simpson, (1994), "Smooth aprroximation by RBFs: three case studies", *J. Appl. Science and Comp.*, vol. 1, n. 1, pp., 88-111.
2. Chen, S., C.F.N. Cowan, P.M. Grant, (1991), "Orthogonal least-squares algorithm for RBF networks", *IEEE Trans. Neural Networks*, vol. 2, n. 2, pp. 302-309.
3. Collins, M.D., W.A. Kuperman, H. Schmidt, (1992), "Non-linear inversion for ocean bottom properties", *J. Acoust. Soc. Am.*, vol. 92, n.5, pp. 2770-2783.
4. Hovem, J.M., Å. Kristensen, (1992), "Reflection loss at a bottom with a fluid sediment layer over a hard solid halfspace", *J. Acoust. Soc. Am.*, vol. 92, n.1, pp.335-340.
5. Poggio, T., F. Girosi, (1990), "Networks for approximation and learning", *Proc. IEEE*, vol. 78, n. 9, pp. 1481-1497.
6. Schmidt, H., (1988), "SAFARI User's guide", *Report SR-113*, SACLANTCEN, La Spezia, Italy.

FEATURE EXTRACTION AND CLASSIFICATION OF UNDERWATER ACOUSTIC SIGNALS

QUYEN HUYNH, WALTER GREENE and JOHN IMPAGLIAZZO
Naval Undersea Warfare Center Division, Newport
Newport, Rhode Island 02841
USA

ABSTRACT. This study explores the ability of wavelets and wavelet packets to extract information from non-stationary signals for the purpose of classification. The signals used were of two types: non-stationary signals from three marine mammals (dolphins, porpoises, and sperm whales), and quasi-stationary signals from an experimental undersea vehicle. This study also focuses on the impact of the compression ratio and the basis selection on the performance of a given classifier.

1. Introduction

High-probability detection and classification of underwater signals by a sensor system is a difficult task. In general, the signals of interest are non-stationary in nature, and traditional Fourier methods are inappropriate for their analysis. Using adaptive wavelet analysis, this study focuses on extraction of features that serve as inputs to a neural network for the purpose of classifying signals. Wavelet, and, more generally, wavelet packet [1] techniques are suitable for transient signal analysis because of their localization properties both in time and frequency. These methods lead to high data compression ratios and are highly adaptive - i.e., the window length varies according to the oscillatory character of the signals. Also, the ability of the wavelet packet algorithm to match the signal to a library of waveforms allows us to represent it using few terms with high fidelity. With a high compression ratio, the neural network classifier can be trained efficiently using few coefficients. Indeed, for all types of transient signals, these coefficients and their variations across scales carry important information such as sharp variation points and singularities.

2. Wavelet Based Classifier

From a classifier point of view, a signal's amplitude and time do not provide adequate information for signal classification. For instance, signal recognition cannot be done easily when the same signal is shifted in time. However, by using the high compression ratios obtainable with wavelets and wavelet packets, the wavelet coefficients coupled with the corresponding time and scale indexes serve as relevant clues for signal recognition.

O. Diaschok et al. (eds.), Full Field Inversion Methods in Ocean and Seismo-Acoustics, 183–188.

Two types of signals were explored for this study: three species of marine mammals (porpoises, dolphins and sperm whales) recorded in open ocean sites represent non-stationary signals, and recordings from an experimental undersea vehicle represent quasi-stationary signals. These signals were classified with a neural network based on the standard layered feed forward network using the back propagation learning algorithm, shown in figure 1. For most of our tests, a data length 8192 was chosen and compressed so that only the 100 largest wavelet, or wavelet packet, coefficients were retained. In the case of the mammals, this length corresponds to approximately one third of a second since the data sampling rate is 25 kHz. By fixing one basis (the standard wavelet basis or best level basis), the coefficients can be labeled with one vector from 1 to 8192. Then, only 100 integers are needed to represent the location indexes. This produces a total of 200 inputs to the network, 100 coefficients and 100 locations. In our computations, we normalized both the coefficients and the location indexes. Since the inputs represent two very different quantities (coefficients and location indexes), the input layer and the first hidden layer are split into two parts, as shown in figure 1. The first hidden layer has 50 nodes for coefficients, 50 nodes for location indexes, and one bias node. The second hidden layer contains 10 nodes and one bias node. For more information on the experimental procedure, see [2].

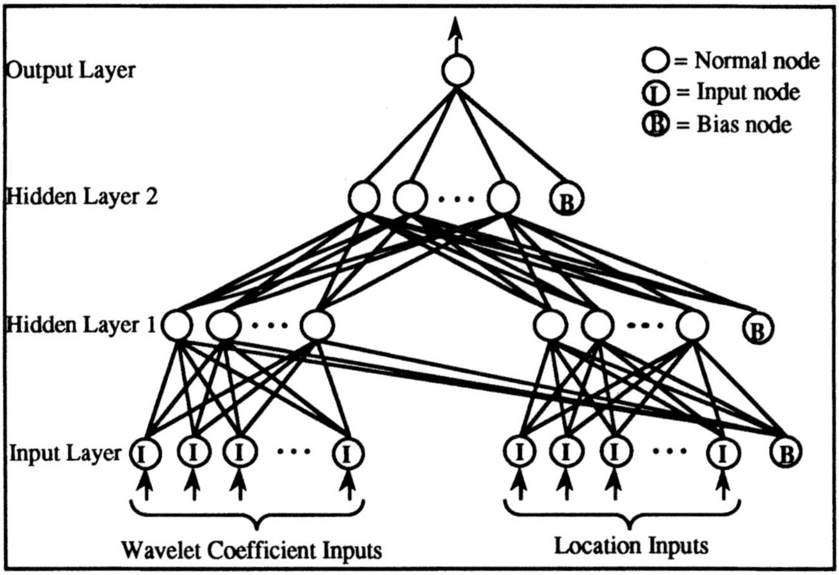

Figure 1. Neural Network

3. Wavelet Packet Transform

Compared to the standard wavelet transform, the wavelet packet transform provides more flexibility in choosing a basis to fit the data. The basis defines how the time-frequency plane is divided. By choosing the proper basis, we obtain a high compression ratio thus reducing the input length to the neural network. From a library of modulated waveforms $W_n(2^l x-k)$ where $l,k \in Z$, $n \in N$, we construct the best basis (collection of waveforms) to fit a specific signal. Each element (waveform) of the library is determined by a scaling parameter l, a localization parameter (in time) k, and an oscillation parameter n. Roughly speaking, the function $W_n(2^l x\text{-}k)$ is centered at $2^{-l}k$, has support of size $\approx 2^{-l}$, and oscillates $\approx n$ times. For fixed l and k, the functions W_n are mutually orthogonal with $n = 0,1,...$ Moreover, each one is orthogonal to all of its integer translates and dyadic rescaled versions. A wavelet packet basis is any orthogonal basis selected among the functions $2^l W_n(2^l x\text{-}k)$. The wavelet packet algorithm is described by using the decomposition tree shown in figure 2. The idea is to apply the quadrature mirror filters described by $W_n(2^l x\text{-}k)$ not only to the "mean" parts but also to the "detail" parts. Note that each level of the tree constitutes a basis and one can choose a best level basis.

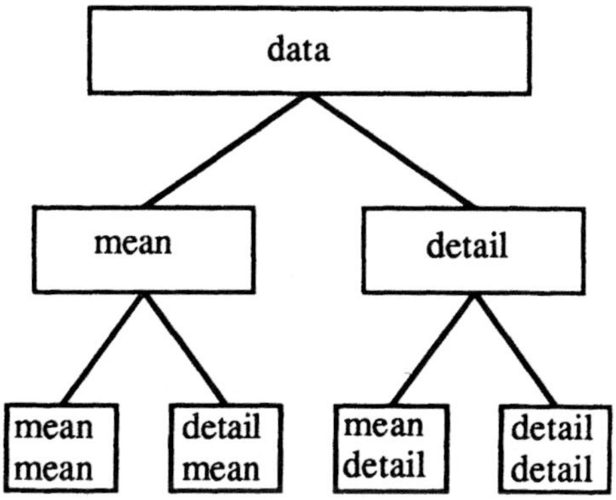

Figure 2. Wavelet Packet Decomposition Tree

Once the decomposition tree has been calculated, a basis can be chosen which spans the time-frequency plane. Each coefficient in the tree is given by an inner product of the original signal with a waveform given by $W_n(2^l x\text{-}k)$, and represents a rectangle in the time-frequency plane whose shape is dependent upon where it is taken from the tree. Higher up (closer to the original data) the coefficients form tall narrow boxes giving good time resolution but poor frequency resolution. Farther down in the tree, the boxes are short and wide giving good frequency resolution but poor time resolution. The area of each box is the same, a constant determined by the Heisenberg uncertainty principle. A basis consists of the same number of boxes as the number of discrete data points in the input vector. Any set of coefficients which completely fill the time-frequency plane without overlapping forms a basis. Notice that the standard wavelet transform produces only one of the many possible basis. In particular, each level of the tree forms a complete basis spanning the time-frequency plane. Since, for a long input data vector there are, for all practical purposes, an infinite number of possible basis, we will restrict our present study to the level basis.

4. Computational Results

The marine mammal sounds (porpoise, dolphins, and sperm whales) were recorded at various locations such as the Gulf of Maine, the Mediterranean Sea, and the Caribbean Sea. Avoiding times when only background noise was present, 400 examples (each 8192 points long) of each type of mammal sound were selected randomly for training the neural network. Background noise alone was also extracted and used for training. For testing the network, and to ensure fairness, the same procedure was repeated with different recordings to create 400 new examples of each type of mammal sound and noise.

In contrast to the mammal data, which is transient in nature, many underwater sources of noise exhibit quasi-stationary characteristics. For example, undersea vehicles with propellers and machinery do not produce purely stationary signals, but do tend to be somewhat harmonic in nature. The compressional properties of the wavelet transform still make it useful for extracting information from quasi-stationary sounds. Recorded passive sounds from an experimental underwater vehicle test run were collected and also background noise from before and after the run. The same procedure as for the mammals was performed to generate training and testing examples for the experimental undersea vehicle versus noise.

	porpoise	dolphin
porpoise	.990	.010
dolphin	.080	.920

Table 1. Porpoise vs. Dolphin using the Wavelet Transform

4.1. BINARY CLASSIFICATION

	porpoise	dolphin
porpoise	.755	.245
dolphin	.252	.748

Table 2. Porpoise vs. Dolphin using the Fourier Transform

Pairs of different mammal types were compressed using the wavelet transform and presented to the network. The output layer of the network consists of one node, which is designed to be a zero for the first type of input data and a one for the second type. The output is considered to be zero if it is less than 0.5 and one if it is greater than 0.5. The training was stopped when the network obtained at least 99% correct. As an example of the results, the confusion matrix for the comparison between porpoise and dolphin is shown in table 1. The values are the percentage of classifications out of 400. Values along the main diagonal are correct classifications while others are false classifications. As a comparison, the same process was followed with the Fourier transform used in place of the wavelet transform. Only the 100 largest Fourier coefficients and their respective location indexes were chosen as inputs. We noted that the wavelet transform performs much better than the Fast Fourier transform.

4.2. FOUR WAY CLASSIFICATION

In this experiment, all four types of data (noise and three marine mammal sounds) formed the training and testing sets. The three output nodes were all zeros for noise. Each mammal sound was represented by a one at one of the three output nodes. Results of this experiment are given in table 3. The 'other' column represents examples where more than one output node has a value of one. The successful rate of recognition is impressive for the wavelet transform. Again, the same test was repeated using the Fourier Transform and these results are shown in table 4. It is obvious that the Fourier coefficients do not carry the necessary instantaneous information for the classifier. Moreover, the deletion of even one Fourier coefficient affects the entire waveform due to the cancellation property. One cannot rely on data compression in Fourier analysis for classification purposes. To better capture transient pulses, we tried the Windowed Fourier Transform, dividing the length of 8192 into 128 segments.

	noise	porpoise	whale	dolphin	other
noise	.995	0	.005	0	0
porpoise	0	.933	.013	.043	.011
whale	.003	.005	.985	.005	.002
dolphin	.080	.045	.045	.820	.001

Table 3. Four-way Mammal Classification using the Wavelet Transform

	noise	porpoise	whale	dolphin	other
noise	.653	.130	.065	.153	0
porpoise	.250	.535	.103	.108	.005
whale	.238	.110	.575	.078	0
dolphin	.393	.065	.063	.480	0

Table 4. Four-way Mammal Classification using the Fourier Transform

Each segment was represented by a dominant frequency amplitude (largest magnitude) with its corresponding time-frequency index. The results are given in table 5. They are an improvement over the Fourier Transform, but not as good as the Wavelet Transform.

	noise	porpoise	whale	dolphin	other
noise	.838	.132	.025	.005	0
porpoise	.132	.770	.010	.078	.010
whale	.113	.110	.663	.078	.038
dolphin	.048	.263	.018	.668	.005

Table 5. Four-way Mammal Classification using the Windowed Fourier Transform

4.3. QUASI-STATIONARY SIGNALS

	noise	vehicle
noise	.990	.010
vehicle	.080	.920

Table 6. Noise vs. an Undersea Vehicle

To explore quasi-stationary signals, we classified an experimental underwater vehicle versus noise (noise is considered to be anything which is not a vehicle). We used the same wavelet transform and neural network as used for the mammals. The output was designed to be one for an underwater vehicle and zero for noise. The Neural Network was trained in the same manner as for the mammal data. The results given in table 6 are percentages out of 400 examples.

4.4. WAVELET PACKET ANALYSIS

The goal here was to determine what is the best basis to use when classifying data types (the previous tests all used the standard wavelet basis). The basis selected should represent the data accurately with the fewest number of coefficients, i.e. the highest compression ratio. Wavelet packets gives us this flexibility. The real objective is to distinguish between two types of signals, and the question is, what basis best discriminates between the two? To answer this, we focus on classification between a dolphin and a porpoise. Examples of each type were transformed using the wavelet packet transform giving us two decomposition trees. The trees were subtracted from each other point by point producing a difference tree. The basis which best fits the difference tree (the largest compression ratio), is the basis which best separates the two types and will therefore best discriminate between them. Due to the large number of possible basis, we limited our search to finding the best level basis. The entire process for a pair of examples, as shown in figure 3, was repeated many times while keeping count of the number of times each level was chosen. From the results shown in figure 4 it is apparent that level 6 is the best level to discriminate between the two data types. The classification results using the level 6 basis are given in table 7.

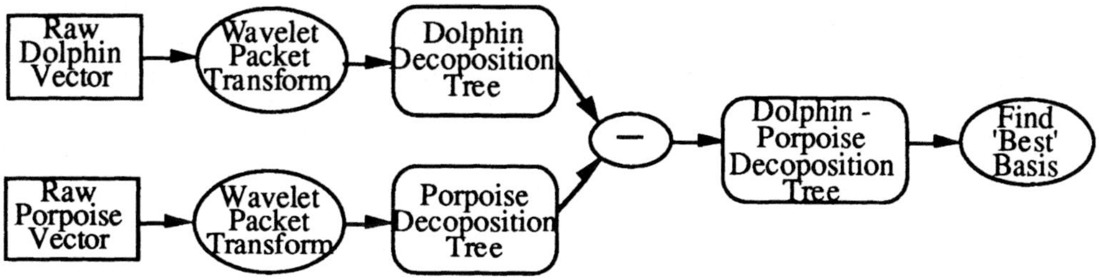

Figure 3. Finding Best Discriminating Basis for One Pair

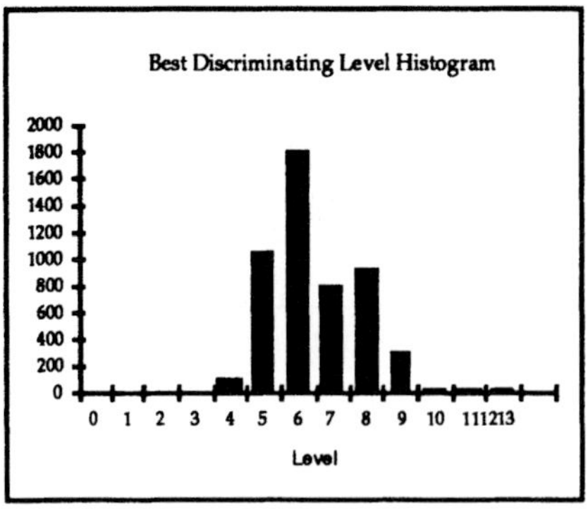

Figure 4. Best Discriminating Level Histogram for Porpoise vs. Dolphin

	dolphin	porpoise
dolphin	.8667	.1333
porpoise	.0283	.9717

Table 7. Wavelet Packets: Dolphin vs. Porpoise using Best Discriminating Basis (Level 6)

5. Conclusions

Extensive numerical experiments indicate that a wavelet based neural network is an effective automatic classifier for transient and quasi-stationary underwater acoustic signals. In this paper, we presented two procedures for feature and parameter extraction. The first procedure is based on the standard wavelet analysis, where the local maxima at each scale and their variations across the scales completely characterize the singularities (transient spikes) of the signals. The relevant clues are represented by these local maxima in terms of magnitude together with their location indexes (time and frequency). The second procedure follows the wavelet packet analysis consisting of a versatile collection of smooth waveforms for the analysis and manipulation of signals of interest. Its *adaptive* property permits the automatic selection of the *best discriminating* basis to separate two given classes of signals. The extracted features of one class of signals are distinctly segregated from the other class of signals. This approach is generally more robust than the standard wavelet analysis where an arbitrary basis is predetermined. Finally, comparing the performance of the Fourier based classifiers and the wavelet based classifiers, it is reasonable to conclude that the wavelet transform more efficiently extracts important clues for classification.

References

[1] R. R. Coifman, Y. Meyer, and M. V. Wickerhauser, Wavelet Analysis and Signal Processing, in Ruskai et al., eds., *Wavelets and their Applications*, Jones and Bartlett, Boston 1992, pp. 153-178.

[2] W. Greene and Q. Huynh, Feature Extraction and Classification of Marine Biological Species, *Proceedings of the International Conference on Theoretical and Computational Acoustics*, D. Lee, J. Ffowcs-Williams, and A. D. Pierce (eds.), World Scientific Press, Singapore (to be published 1994).

MODEL-BASED MATCHED FILTER PROCESSING: A BROADBAND APPROACH TO SHALLOW-WATER INVERSION

J-P. HERMAND

Saclant Undersea Research Centre
Viale San Bartolomeo 400, 19138
La Spezia, Italy

ABSTRACT In this paper, the concept of model-based matched filter (MBMF) processing is extended to the shallow-water inversion problem. Wave propagation effects over a wide range of frequencies are fully integrated into a MBMF receiver by means of modeled range- and depth-dependent impulse responses of the intervening medium (Green's function). Numerical results are presented to demonstrate the performance of a multi-mode MBMF receiver for a summertime shallow-water waveguide southeast of the island of Elba. In contrast to conventional matched-field processing, the characteristics of the received broadband modal structure in both time and space are exploited to infer environmental parameters.

1. Introduction

Deep-ocean tomography techniques cannot be easily applied to the inversion of the shallow-water channel due to the high transmission loss and complex multipath structure incurred by boundary interaction. The classical matched field processing work is being expanded with the goal of remotely sensing the ocean environment [1]. The method relies on exploiting the spatial acoustic field across a vertical receiving aperture and applying inversion techniques to determine the best match between measured and modeled data. In its original formulation matched field processing does not take advantage of frequency-dependent propagation effects. As a consequence, when applied to the inverse problem the method can lack sensitivy to characterize dispersive media such as the shallow-water waveguide and the deep-ocean sound duct.

To date, most of the related shallow-water inversion studies have concentrated on individual frequency components of propagating sound. The dispersion characteristics of the shallow-water channel over a broad *continuum* of temporal frequencies are potentially useful for inferring range- and depth-dependent ocean and bottom parameters [2]–[5]. Recent experimental work has demonstrated that the distortion of time-dispersed wideband signals can be compensated effectively by using a model-based matched filter (MBMF) receiver that incorporates the modeled Green's function of the medium [6]. In this paper the possibility of applying multi-mode model-based matched filter processing to the shallow-water inversion problem is investigated. Numerical examples are presented to assess the efficiency of the method for a multi-mode shallow-water waveguide southeast of the island of Elba.

O. Diaschok et al. (eds.), Full Field Inversion Methods in Ocean and Seismo-Acoustics, 189–194.

2. Intermodal and Intramodal Dispersion: A Key to Inversion

Propagation in a shallow-water channel is frequency-dependent and sensitive to the boundary conditions. Water depth and in-water sound speed structure determine how the propagated energy interacts with the ocean boundaries. The time and frequency dispersion characteristics of the guided wave packets are highly dependent on the waveguide geometry, bottom type, season, and sea state.

In particular, the time dispersion characteristics are intrinsically related to the frequency-dependent geoacoustic properties of the ocean bottom. The trapped modes propagate with different phase speeds (intermodal dispersion) and for a given mode the phase speed depends on frequency (intramodal dispersion). The acoustic signals become increasingly distorted as they propagate through the medium. Waveform distortion increases with range as a result of intermodal dispersion. Interaction with a lossy bottom causes the high order modes to fade out with increasing distance. For sufficiently wide frequency bands intramodal dispersion also contributes to the waveform distortion. The relative contribution of both forms of dispersion is mostly determined by the propagation distance. For further range increase the channel exhibits an optimum frequency of propagation as a result of competing propagation and attenuation mechanisms at low and high frequencies [7]. The optimum frequency depends mostly on the water depth.

As shown in the next sections, the frequency-dependence of the mode shape functions (phase and amplitude) can be exploited to infer ocean and/or bottom parameters of the shallow-water channel, a dispersive waveguide.

3. Multi-Mode MBMF Inversion

In a noise background an indirect measurement of the dispersion characteristics of a shallow-water channel over a broad band of frequencies can be obtained by transmitting large time-bandwidth-product signals and processing the received signals with a model-based matched filter. The frequency band of transmission is determined by medium constraints (specifically the bandwidth about the optimum frequency of propagation). Here follows a description on how MBMF processing is applied to the inverse problem.

Broadband modeling: To obtain the best fit to measured acoustic propagation data between a source and a set of receivers of known spatial positions, environmental data about the intervening medium are specified and fed into a propagation model to predict the frequency-dependent sound field. The broadband transfer function of the medium (Green's function) is obtained by Fourier synthesis of CW solutions calculated at a number of discrete frequencies spaced over the transmission band.

Matched filtering: The resulting impulse responses are convolved with the transmitted signal to develop model-based reference signals (replicas). Hence, the replicas exhibit the predicted propagation-induced distortion as a function of anticipated environmental conditions. The received acoustic data are correlated against the model-based replicas (MBMF processing), and also against a simple replica of the transmitted signal (MF processing). The latter assumes an ideal propagating medium (free-space Green's function) and is taken as reference for the MBMF processing.

Modal extraction: By using normal-mode modeling the transfer function can be decomposed into its modal contributions to develop modal replicas. The MBMF receiver is then implemented mode by mode (multi-mode MBMF processing). Coherent addition of all the modal outputs is equivalent to plain MBMF processing. Incoherent addition accounts only for intramodal dispersion. The low order modes with high temporal and spatial coherence

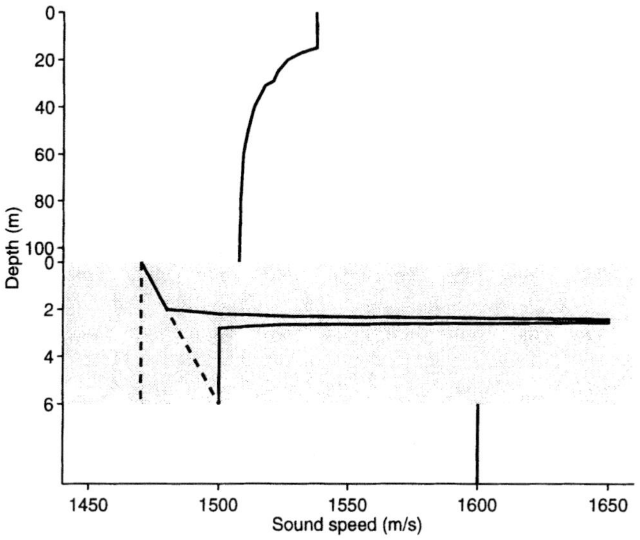

FIGURE 1. Profiles of water sound speed in the summer and bottom compressional speed southeast of Elba island.

can be selected for the inversion. Modal filtering and extraction can be carried out for those modes that strongly differ in their frequency dependence (amplitude and phase).

Parameter inversion: The inversion proceeds by searching for the ocean and/or bottom parameters that maximize the processing gain (or signal-to-noise ratio) at the multi-mode MBMF outputs (from which the cost function of a global optimization method is derived). The fact that the modal replicas carry independent information about different scales or depths of ocean/bottom features can be exploited to search more efficiently the parameter space by considering the gain obtained with the individual modes. In the case where MBMF processing is applied to vertical array data, the cost function can be derived from the sum of the highest correlation peaks (energies) obtained from each hydrophone-channel MBMF receiver.

4. Numerical Examples

To assess the potential of multi-mode MBMF inversion we consider a range-independent shallow-water waveguide in the basin southeast of the island of Elba.

4.1. THE SOUTHEAST ELBA WAVEGUIDE

Figure 1 shows the range-average properties of the water column and bottom for a constant-depth track (106 m water depth). In the summer the water column is characterized by a downward refracting profile (solid line) which causes strong bottom interaction. The bottom southeast of Elba is characterized by a low-speed sediment layer of average 6-meter thickness with no shear. A high-speed ($c = 1650$ m/s) sand layer of average 1-meter thickness is mixed into the sediment layer. The depth-averaged density and compressional attenuation in the bottom ($\rho = 1.5$, $\beta = 0.06$ dB/Λ) and in the subbottom ($\rho = 1.8$, $\beta = 0.15$ dB/Λ) were determined from geological/geoacoustic data and broadband acoustic transmission loss data/model comparison [8],[9].

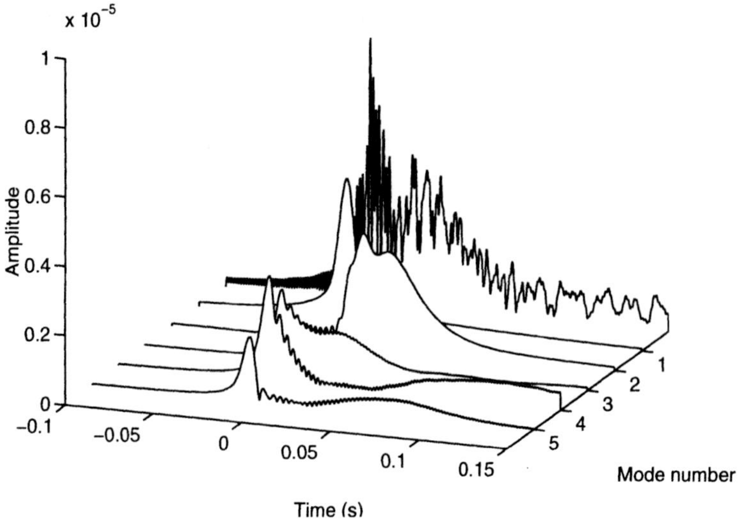

FIGURE 2. Broadband modal responses of the southeast Elba waveguide.

4.2. BROADBAND MODAL RESPONSES

Channel impulse responses were calculated for a source and receiver depths of 53 m and 57 m, respectively, a distance of 30 km, and a frequency band of 100–800 Hz. For these conditions the shallow-water channel acts as an acoustic waveguide that supports modal propagation. The band-limited complex frequency responses were Fourier synthesized from CW solutions computed with the SNAP normal-mode propagation model [10].

In Fig. 2 the broadband impulse response is decomposed into its modal contributions. The envelopes of the first five trapped modes with significant energy are shown. The individual modal responses differ in energy, travel time and time dispersion as a result of the frequency dependence of the modal functions. Most of the energy is carried by modes 1 and 2. Mode 2 exhibits severe time dispersion. The impulse response results from coherent addition of the individual modal responses. The total energy time spread is about 100 ms.

Figure 3 shows the modal impulse responses for the hypothetical case where there is no sand layer. A constant gradient was taken for the bottom compressional speed (oblique dotted line in Fig. 1). Mode-by-mode comparison with Fig. 2 shows that the broadband modal structure was highly affected by the thin sand layer lying in the soft bottom. Mode interference results in a much different broadband impulse response. The frequency-dependent modal functions were such that the response was mostly determined by modes 3 and 4 instead of modes 1 and 2 for the sand layer case. The total time dispersion is smaller than in the sand layer case. The travel time difference between the "fastest" mode 3 and the "slowest" mode 4 is about 10 ms. It is noted that the higher order modes are the most affected by the sand layer.

Figure 4 shows the phase of the frequency response for three different cases of bottom: speed gradient with high-speed sand layer, speed gradient only, and isospeed (vertical dotted line in Fig. 1). The nonlinear part after removal of the linear trend in the unwrapped phase is shown. It can be seen that the phase response is highly dependent on the compressional speed structure in the bottom. When compared with the first two cases an isospeed bottom produces an even more dramatic change in phase response. This demonstrates that the group delay and second-order derivative of the phase with respect to frequency (group de-

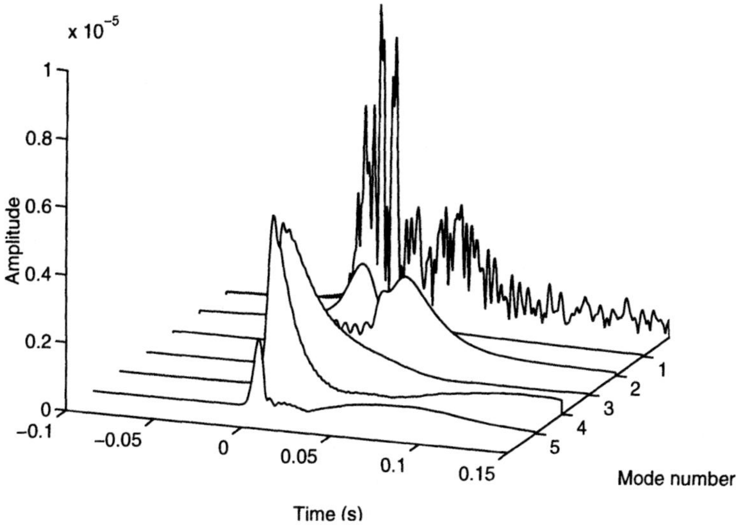

FIGURE 3. Broadband modal responses for no sand layer.

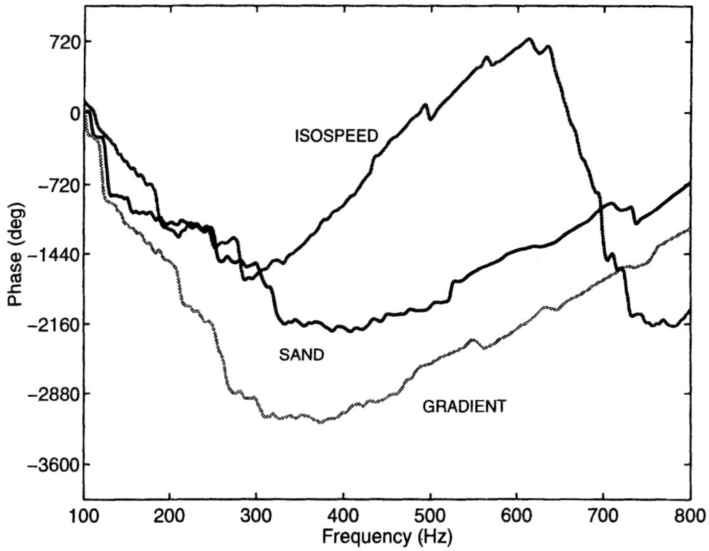

FIGURE 4. Phase nonlinearities for three bottom models.

lay distortion) can be exploited to infer the compressional speed profile in the top sediment layer. Comparison of the magnitude responses (not shown) for the three cases considered above indicated differences exceeding 20 dB across the whole frequency band. The larger differences were above 300 Hz where the spectrum is dominated by modes 3 to 4. Hence information about bottom (and ocean) parameters is embedded in both phase and amplitude of the broadband modal responses.

4.3. MBMF PERFORMANCE

The MBMF performance for the sand layer case was calculated from the formula (46) in Ref. [6]. In that formula, the coefficient γ corresponds to the processing gain (with respect to the MF receiver) of an ideal MBMF receiver that is perfectly matched to the received waveform and totally recaptures the dispersed energy.

Including (coherently) all propagating modes in the model-based replica, the obtained MBMF gain was $\gamma = +9$ dB. The gain was proportional to the energy time spreading measured at the MF output, which is related to the receiver depth. Near the surface (9 m depth) and near the bottom (97 m depth) the gains were respectively +18 dB and +8 dB. When the two bottom models with no sand layer were applied to sand-layer simulated acoustic data the MBMF receiver showed a loss with respect to the MF receiver (formula (47) in Ref. [6]).

5. Conclusion

The proposed inversion method fully exploits the frequency-dependent attenuation and dispersion characteristics of the shallow-water channel. Compared to the space-domain single-frequency matched field processing, it is a time-domain broadband approach based on the distortion of propagated large time-bandwidth-product waveforms. In this concept, ocean and/or bottom parameters are inferred on the basis of the peak energies obtained at the outputs of a multi-mode model-based matched filter receiver that incorporates modeled broadband modal responses of the channel. The method was demonstrated for a waveguide characterized by a soft bottom with a sand layer at a shallow-water site southeast of the island of Elba. In a forthcoming experiment *Yellow Shark '94* acoustic and environmental data will be obtained at the site described in this paper for validation of the multi-mode MBMF inversion method.

References

1. Baggeroer, A.B., Kuperman, W.A., and Mikhalevsky, P.N. (1993) An Overview of Matched Field Processing in Ocean Acoustics, *IEEE Journal of Oceanic Engineering*, **18**, 401–424.
2. Lynch, J.F., Rajan, S.D., and Frisk, J.V. (1991) A Comparison of Broadband and Narrow-Band Modal Inversions for Bottom Properties at a Site near Corpus Christi, Texas, *Journal of the Acoustical Society of America*, **89**, 648–665.
3. Gindler, I.V. and Kozel'skii, A.R. (1992) Feasibility of Reconstructing the Geoacoustical Parameters of the Ocean Bottom on the Basis of Experimental Information about the Green's Function for an Underwater Acoustic Waveguide in a Broad Frequency Band, *Soviet Physics Acoustics*, **38**, 13–15.
4. Hermand, J-P., and Roderick, W.I. (1994) Application of Model-Based Matched Filter Processing to Underwater Acoustic Inverse Problems, in *Proceedings of the Second European Conference on Underwater Acoustics*, Bjørnø, L., Ed. Luxembourg:European Commission, 381–386.
5. Badiey, M., and Jaya, I. (1994) Shallow-Water Acoustic/Geoacoustic Experiments at the New Jersey Atlantic Generating Station Site, *Journal of the Acoustical Society of America*, **96**, 3593–3604.
6. Hermand, J-P., and Roderick, W.I. (1993) Acoustic Model-Based Matched Filter Processing for Fading Time-Dispersive Ocean Channels: Theory and Experiment, *IEEE Journal of Oceanic Engineering*, **18**, 447–465.
7. Jensen, F.B., and Kuperman, W.A. (1983) Optimum Frequency of Propagation in Shallow Water Environments, *Journal of the Acoustical Society of America*, **73**, 813–819.
8. Michelozzi, E. (1994) Quaternary Sediments in the Elba-Formiche Islands Basin, SACLANTCEN Internal Note IN-678, SACLANT Undersea Research Centre, La Spezia, Italy.
9. Jensen, F.B. (1974) Comparison of Transmission Loss Data for Different Shallow-Water Areas with Theoretical Results provided by a Three-Fluid Normal-Mode Propagation Model, in *Sound Propagation in Shallow Water, Vol. II*, Hastrup, O.F., and Olesen, O.V., Eds., SACLANTCEN CP-14, SACLANT Undersea Research Centre, La Spezia, Italy, 79–92.
10. Jensen, F.B., and Ferla, M.C. (1979) SNAP: the SACLANTCEN Normal-Mode Acoustic Propagation Model, SACLANTCEN Rep. SM-121, SACLANT Undersea Research Centre, La Spezia, Italy.

MODEL-BASED INVERSION FOR A SHALLOW OCEAN APPLICATION

J.V. CANDY
Lawrence Livermore National Laboratory
P.O. Box 808, L-495
Livermore, CA 94550

and

E.J. SULLIVAN
Naval Undersea Warfare Center
Code 82101, Bldg 679
Newport, RI 02841

ABSTRACT A model-based approach to invert or estimate the sound speed profile (SSP) from noisy pressure-field measurements is discussed. The resulting model-based processor (MBP) is based on the state-space representation of the normal-mode propagation model. Using data obtained from the well-known Hudson Canyon experiment, a noisy shallow water ocean environment, the processor is designed and the results compared to those predicted using various propagation models and data. It is shown that the MBP not only predicts the sound speed quite well, but also is able to simultaneously provide enhanced estimates of both modal and pressure-field measurements which are useful for localization and rapid ocean environmental characterization.

1. Introduction

The rapid assessment of environmental conditions is extremely important in a tactical as well as experimental situation. The incorporation of ocean acoustic propagation models into signal processing schemes has long been the hope to improve overall processor performance. Current applications are concerned with the localization/detection of an acoustic source as well as inversion and the proposed methods are based on bringing the predictions of a propagation model in concert with pressure measurements from an array. This is called the matched-field problem and much work has been accomplished [1-4]. Model-based techniques offer high expectations of performance, since a processor based on the predicted physical phenomenology that inherently has generated the measured signal must produce a better (minimum variance) estimate than one that does not [5]. However, if the model embedded in this processor is inaccurate or for that matter incorrect, than the model-based processor (MBP) can actually perform worse. Therefore, it is necessary, as part of the MBP design procedure, to estimate the model parameters either through separate experiments or jointly (adaptively) while performing the required processing [6,7]. Note that the introduction of the recursive, on-line MBP can offer a dramatic detection improvement in a tactical passive or active sonar-type system especially when a rapid environmental assessment is required.

In ocean acoustics we are usually concerned with an environmental model of the ocean and how it effects the propagation of sound through this noisy, hostile environment. The problem of estimating the environmental parameters characterizing the ocean medium is called the *ocean tomography* or equivalently the *environmental inversion* problem and has

O. Diaschok et al. (eds.), Full Field Inversion Methods in Ocean and Seismo-Acoustics, 195–201.

long been a concern because of its detrimental effect on various detection/localization schemes [8,9]. Much of the work accomplished on this problem has lacked quantitative measures of the mismatch of the model with its environment. In a related work [10] it was shown how to quantify "modeling errors" both with a known ocean environmental model available and without. In the first case, it was shown how to calculate standard errors for modal/pressure-field estimates as well as an overall measure of fit based on the innovations or residuals between the measured and predicted pressure-field. In the second case, only the residuals are used. These results quantify the mismatch between the embedded models and the actual measurements both on simulated as well as experimental data.

In this paper we concentrate on the design of a MBP to solve the environmental inversion or oceanographic tomography problem while jointly estimating the underlying signals — which we term *model-based inversion*. That is, the MBP is designed to estimate the sound speed profile (SSP) as well as enhance the corresponding modal/pressure-field signals with its accompanying performance statistics quantified using both an established model and the corresponding residuals. More specifically, we are concerned with whether or not it is possible to estimate the SVP from noisy hydrophone measurements in a real, hostile ocean acoustic experimental environment. Theoretical work on the design of MBP for this problem has been accomplished recently indicating that a solution exists, however, most of the results presented were theoretical and supported by a simple shallow water simulation [7]. Here, in contrast, we actually *apply* these techniques for inversion to measured data from the Hudson Canyon experiment — a well-known and substantiated area for such an application. Note that there is presently a growing literature on oceanographic tomography, since it can deal with the estimation of many different parameters relative to ocean acoustics. For more information on these issues, see [8,9] and references therein.

We briefly review the state-space propagator of the normal mode model for the Hudson Canyon experiment which will be used in this paper in sec. 2. In the next section we design a model-based processor to estimate the SVP from noisy experimental hydrophone measurements for this shallow water, ocean acoustic problem.

2. State Space Propagators for Hudson Canyon

In this section we discuss the development of the propagators for the Hudson Canyon experiment performed in 1988 in the Atlantic. Excellent agreement was determined between the model and measured data indicating a well-known, well-documented shallow water experiment with bottom interaction and yielding ideal data sets for investigating the applicability of a MBP to measured ocean acoustic data [11]. A fixed vertical hydrophone array consisting of 24 phones spaced 2.5m apart extending from the seafloor up to a depth of about 14m below the surface was used. SSP measurements were made at regular intervals and the data were collected under carefully controlled conditions in the ocean environment. The normalized horizontal wave number spectrum for a 50Hz temporal frequency is dominated by 5 modes occurring at wave numbers between 0.14 to 0.21 m^{-1} with relative amplitudes increasing with increased wave number. A SNAP [12] simulation was performed and the results agree quite closely, indicating a well-understood ocean environment.

In order to construct the state-space propagator, we require the set of parameters which were obtained from the experimental measurements and processing (wave number spectra). The horizontal wave number spectra were estimated using synthethic aperture processing [13].

For this study we assume a horizontally-stratified ocean of depth h with a *known* source position (x, y) and assume that the acoustic energy from a point source can be modeled as a trapped wave governed by the wave equation. We obtain the solution to this equation following the approach of Pekeris [14] leading to the well-known normal-mode equations.

A solution of the range dependent relation is given by a Hankel function and the *depth* relation is an eigenvalue equation in z with

$$\frac{d^2}{dz^2}\phi_m(z) + \kappa_z(m)\phi_m(z) = 0, \;\; m = 1, \cdots, M \tag{1}$$

whose eigensolutions $\{\phi_m(z)\}$ are the so called *modal functions* and κ_z is the wavenumber in the z-direction. These solutions depend on the sound speed profile, $c(z)$, and the boundary conditions at the surface and bottom as well as the corresponding *dispersion* relation.

The normal-mode solutions can easily be placed in state-space form and we refer the interested reader to Refs. [6,7] for the detailed theory. Our approach for the Hudson Canyon experiment is to place the model into a Gauss-Markov representation which includes the second order statistics. The measurement noise can represent the near-field acoustic noise field, flow noise on the hydrophone and electronic noise. The modal noise can represent SSP errors, distant shipping noise, errors in the boundary conditions, sea state effects and ocean inhomogeneities. By assuming that the horizontal range of the source r_s is known *a-priori*, we can use the Hankel function $H_0(\kappa_r r_s)$ which is the source range solution; therefore, we reduce the state-space model to that of "depth only" and the Gauss-Markov model for this constrained problem is given by

$$\frac{d}{dz}\mathbf{x}(z) = A(z)\mathbf{x}(z) + B(z)\mathbf{u}(z) + \mathbf{w}(z) \tag{2}$$

with the pressure field measurement model given by

$$p(z) = \mathbf{C}^T(r_s, z)\mathbf{x}(z) + \mathbf{v}(z) \tag{3}$$

where

$$\mathbf{C}^T(r_s, z) = [\beta_1(r_s, z) \quad 0|\cdots|\beta_M(r_s, z) \quad 0] \tag{4}$$

with $\beta_i(r_s, z) = q\phi_i(z_s)\int_0^h \phi_i^2(z)dz \;\; H_0(k_r(i)r_s)$. The random noise vectors $\underline{w}$ and $\underline{v}$ are assumed gaussian, zero-mean with respective covariance matrices, R_{ww} and R_{vv}.

Since our array spatially samples the pressure-field, we choose to discretize the differential state equations using a finite (first) difference approach, that is, for the mth-mode we use

$$\frac{d}{dz}\underline{x}_m(z_\ell) \approx \frac{\underline{x}_m(z_{\ell+1}) - \underline{x}_m(z_\ell)}{\Delta z_\ell} \tag{5}$$

where $\Delta z_\ell = z_{\ell+1} - z_\ell$. Therefore, substituting into Eqn. 2, we obtain

$$\underline{x}_m(z_{\ell+1}) = [I + \Delta z_\ell A_m(z_\ell)] \;\; \underline{x}_m(z_\ell) + \Delta z_\ell \underline{B}(z_\ell)u(z_\ell) + \underline{w}(z_\ell) \tag{6}$$

Assuming we have a vertical line sensor array to measure the pressure-field, the measurement model for the mth mode becomes

$$p_m(r_s, z_\ell) = \underline{C}_m^T(r_s, z_\ell)\underline{x}_m(z_\ell) + v_m(z_\ell) \tag{7}$$

The vertical wave numbers are functions of the sound speed profile through the dispersion relationship and can be further analyzed through knowledge of the SSP. Since our processor will be sequential, it is recursing over depth. We would like our processor to improve the estimation of the SSP "in-between" sensors. With a state-space processor we can employ two spatial increments in z simultaneously: one for the measurement system Δz_ℓ and one for the modal state-space $\Delta z_j = \frac{\Delta z_\ell}{N_\Delta}$ where N_Δ is an integer. Therefore in order to propagate the states (modes) at Δz_j and measurements at Δz_ℓ we must have values of the SSP at each Δz_j (subinterval) as well. Suppose we expand $c(z)$ in a Taylor series about a nominal depth, say z_0, then we obtain

$$c(z) \approx \theta_0(z_0) + \theta_1(z_0)(z - z_0) + \cdots + \theta_N(z_0)\frac{(z - z_0)^N}{N!} \tag{8}$$

where $\theta_i(z_0) := \left.\frac{\partial^i c(z)}{\partial z^i}\right|_{z=z_0}$ $i = 0, \cdots, N$. In this formulation we therefore have a "model" of the SSP of the form $\hat{c}_N(z_\ell) = \underline{\Delta}_N^T(z_\ell)\underline{\theta}(z_\ell)$ where the set of $\{\theta_i(z_\ell)\}$ are only known at a sparse number of depths, $\ell = j, j+1, \cdots, j + N_\theta$. More simply, we have a set of measurements of the SSP measured *a-priori* or *on-line* at specific depths — not necessarily corresponding to all sensor locations $\{z_\ell\}$, therefore, we use these values as initial values to the MBP enabling it to sequentially update the set of parameters $\{\theta_i(z_j)\}$ over the layer $z_{\ell-1} \leq z_j < z_\ell$ until a new value of $\theta_i(z_j)$ becomes available, then we re-initialize the parameter estimator with this value and continue our SSP estimation until we have recursed through each sensor location. In this way we can utilize our measured SSP information in the form of a parameter update and improve the estimates using the processor. Thus, we can characterize this SSP representation in an approximate Gauss-Markov model which is nonlinear when we constrain the SSP parameters to the set $\{\theta(z_\ell)\}$, $z_\ell = z_j, \cdots, z_j + N_\theta$, that is,

$$\underline{\theta}_N(z_{\ell+1}) = \begin{cases} \underline{\theta}_N(z_\ell) + \Delta z_\ell \underline{w}_\theta(z_\ell) & z_\ell < z_j < z_{\ell+1} \\ \\ \underline{\theta}_N(z_\ell)\delta(z_\ell - z_j) & z_\ell = z_j \end{cases} \tag{9}$$

where we have $w_\theta \sim N(0, R_{w_\theta w_\theta})$, $\underline{\theta}_N(z_0) \sim N(\underline{\theta}_N(0), P_\theta(0))$, and $\underline{\Delta}_N^T(z_\ell) = [1 \;\; \Delta z_{\ell-1} \cdots \frac{z_{\ell-1}^N}{N!}]$.

It is this model that we use in our MBP to estimate the sound speed and solve the environmental inversion problem. We can now "augment" this SSP representation into our normal-mode/pressure-field propagation model to obtain an overall system model. The augmented *Gauss-Markov (approximate) model* for M-modal functions in the interval $z_\ell < z_j < z_{\ell+1}$ is given by

$$\begin{bmatrix} \underline{x}(z_{\ell+1}) \\ --- \\ \underline{\theta}_N(z_{\ell+1}) \end{bmatrix} = \begin{bmatrix} A(z_\ell, \theta) & | & 0 \\ - & - & - \\ 0 & | & I_{N+1} \end{bmatrix} \begin{bmatrix} \underline{x}(z_\ell) \\ --- \\ \underline{\theta}_N(z_\ell) \end{bmatrix} + \begin{bmatrix} B(z_\ell) \\ --- \\ \underline{0} \end{bmatrix} \underline{u}(z) + \begin{bmatrix} \underline{w}(z_\ell) \\ --- \\ \underline{w}_\theta(z_\ell) \end{bmatrix}$$

with corresponding measurement model

$$p(r_s, z_\ell) = \begin{bmatrix} \underline{C}^T(r_s, z_\ell) & | & \underline{0} \end{bmatrix} \begin{bmatrix} \underline{x}(z_\ell) \\ ---- \\ \underline{\theta}(z_\ell) \end{bmatrix} + v_p(z_\ell) \tag{10}$$

This completes the development of the state-space propagator for the experiment; next we discuss the design of the MBP for the Hudson Canyon data.

3. Model-Based Inversion: Sound Speed Estimator

In this section we develop a solution to the environmental inversion problem by designing a MBP to estimate the sound speed, on-line, from noisy pressure-field measurements. The processor is based on the state-space representation of the normal-mode model and a vertical line array of sensors along with the augmented model discussed in the previous section. We briefly discuss the approach, then the algorithm and apply it to the Hudson Canyon experimental data for a $500m$ range at a $50Hz$ temporal frequency. The *environment inversion problem* can be defined in terms of our previous models as: **GIVEN** a set of noisy acoustic (pressure-field) measurements $\{p(r_0, z_\ell)\}$ and a set of sound speed parameters $\{\underline{\theta}(z_\ell)\}$, **FIND** the "best" (minimum variance) estimate of the SSP, $\hat{c}(z_\ell)$. For this problem we have a *sparse* set of SSP measurements available at $N_\theta = 9$ depths with complete set of pressure-field measurements. The solution to the inversion problem can be obtained using the extended Kalman filter (EKF) algorithm or its variants [5] as a joint state/parameter estimator. Here we choose discrete EKF available in *SSPACK_PC* [15].

The experimental measurements consists of sound speed in the form of discrete data pairs $\{c(z_j), z_j\}$ which can be utilized in the estimator for correction as it processes the acoustic data. We use the first two terms (N=2) of the Taylor series expansion of the SSP (piecewise linear) for our model where both θ_0 and θ_1 are space-varying gaussian random functions with specified means and variances, and therefore, through linearity, so is $c(z_\ell)$. Thus, our Gauss-Markov model for this problem is given by Eq. 10. We will use a spatial sampling interval of $\triangle z_j = \frac{\triangle z_\ell}{10}$ in the state propagation equations as discussed previously. The corresponding EKF estimator evolves from the algorithm (see [7]) with all of the appropriate functions and jacobians.

We observe the performance of the model-based SSP processor. Here we use only the acoustic measurements and the 9 sound speed data values $\{c(z_j)\}$, $j = 1, \cdots, 9$ to set hard constraints on the parameter estimator and force it to meet these values only when the appropriate depth is achieved. The results of the runs are shown in Figure 1. Here we see

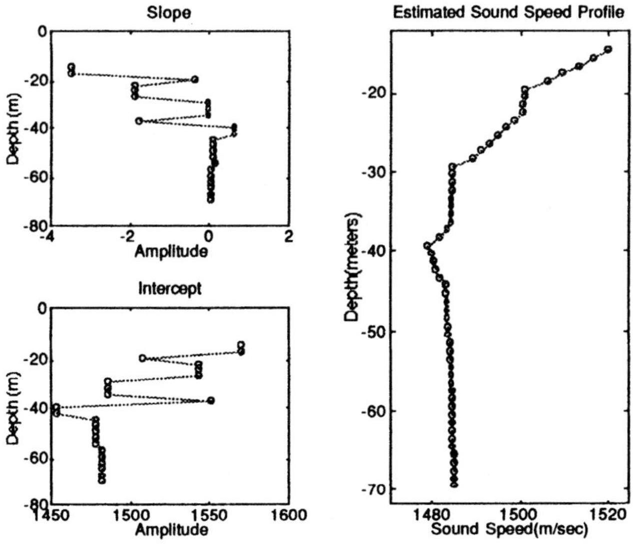

Figure 1.

Model-Based Inversion:

(a) Slope Estimation.

(b) Intercept Estimation

(c) Sound Speed Estimation.

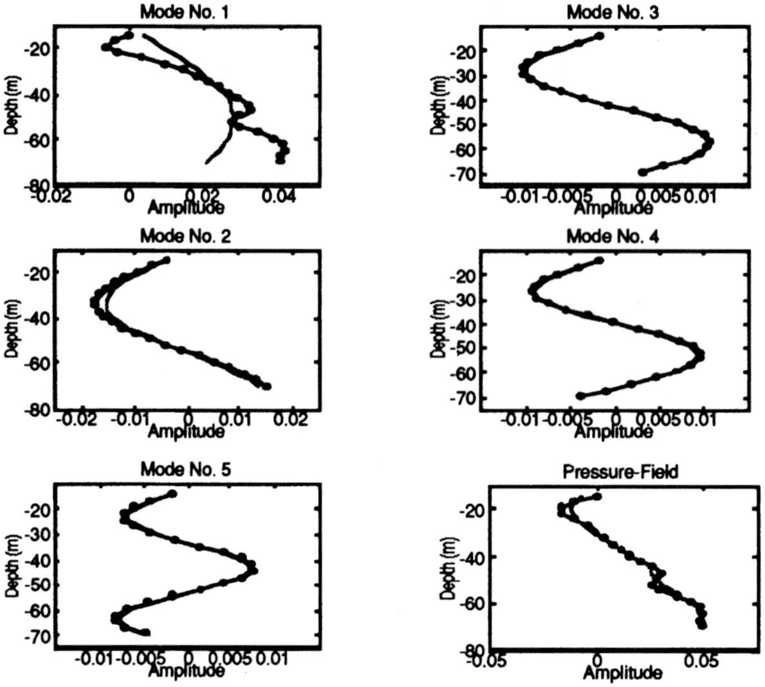

Figure 2. *Model-Based Enhancement: (a) Est. Modes 1 & 3. (b) Est. Modes 2 & 4. (c) Est. Mode5 & Pressure-Field*

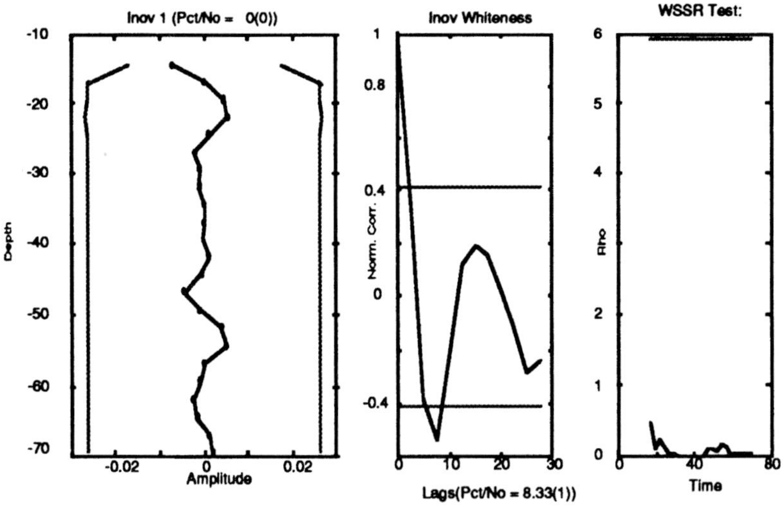

Figure 3. *Model-Based Residuals: (a) Innovation/Residual. (b) Whiteness Test (8.3% Out). (c) WSSR Test (Passed).*

the estimated SSP parameters and reconstructed SSP. The estimator appears to track the SSP parameters as well as the profile. The standard rms modeling errors (see [10] for details) for the SSP parameters and profile are respectively: 1.6×10^{-4}; 2.7×10^{-5}; 1.0×10^{-2}. Since we are using a joint estimator, the enhanced estimates of both modal functions and the pressure field are shown in Figure 2 along with the corresponding mean predicted using the SNAP model parameters. Here we see excellent agreement with all of the modes (except mode 1). The standard rms modeling errors for each mode, respectively, are: 1.0×10^{-2}; 1.3×10^{-3}; 2.3×10^{-4}; 2.2×10^{-4}; 3.5×10^{-4}. The innovations or residuals are shown in Figure 3 where we see that they are zero mean and reasonably white (8.3% out of bounds). The rms standard error for the residuals is given by 2.9×10^{-3}. So we see that the processor is clearly capable of jointly estimating the SSP and enhancing the modal/pressure-field.

4. Discussion

In this paper we have developed an on-line, adaptive, model-based solution to the evironmental inversion problem, that is, a sound speed profile estimation scheme based on coupling the normal-mode propagation model to a functional model of the SSP evolving from Taylor series expansion about the most current sound speed measurement available. The algorithm employed was the extended Kalman filter which evolved as the solution to the minimum variance estimation problem when the models were placed in state-space form. With the inavailability of complete set of sound speed measurements, we showed that the processor performs reasonably well even for very sparse sound speed information.

References

1. Hinich, M.J. (1973) Maximum likelihood signal processing for a vertical array, *J. Acoust. Soc. Am.*, **54**, 499-503.
2. Bucker, H.P. (1976) Use of calculated sound fields and matched-field detection to locate sound in shallow water, *J. Acoust. Soc. Am.*, **59**, 329-337 .
3. Sullivan, E.J. and Middleton, D. (1993) Estimation and detection issues in matched-field processing, *IEEE Trans. Ocean Eng.*, **18**, (3), 156-167.
4. Baggeroer, A.B., Kuperman, W.A., and Schmidt, H. (1988) Matched-field processing: source localization in correlated noise as an optimum parameter estimation problem, *J. Acoust. Soc. Am*, **83**, (2), 571-587.
5. Candy, J. V. (1986) *Signal Processing: The Model-Based Approach*. New York:McGraw-Hill.
6. Candy, J.V. and Sullivan, E.J. (1992) Ocean acoustic signal processing: a model-based approach, *J. Acoust. Soc. Am.*, **92**, 3185-3201.
7. Candy, J.V. and Sullivan, E.J. (1993) Sound velocity profile estimation: a system theoretic approach, *IEEE Trans. Ocean Eng.*, **18**, (3), 240-252.
8. Tolstoy, A. (1992) *Review of matched field processing for environmental inverse problems* , *Int. J. Modern Physics*, **3**, (4), 691-708.
9. Tolstoy, A. (1993) *Matched Field Processing for Ocean Acoustics*. New Jersey:World Scientific Publishing.
10. Candy, J.V. and Sullivan, E.J. (1994) Model-based processor design for a shallow water ocean acoustic experiment, *J. Acoust. Soc. Am.*, **95**, (4) 1-14.
11. Rogers, A.R., Yamamoto, Y. and Carey, W. (1993) Experimental investigation of sediment effect on acoustic wave propagation in shallow water, *J. Acoust. Soc. Am.*, **93**,1747-1761.
12. Jensen, F.B. and Ferla, M.C. (1982) SNAP: the SACLANTCEN normal-mode acoustic propagation model," *SACLANTCEN Report*, **SM-121**, SACLANT Undersea Research Centre, La Spezia, Italy.
13. Sullivan, E.J., Carey, W. and Stergiopoulos, S., Eds. (1993) Special Issue on Acoustic Synthetic Aperture Processing, *IEEE Trans. Ocean Eng.*, **17**.
14. Clay, C. S. and Medwin, H. (1977) *Acoustical Oceanography*. New York:Wiley.
15. Candy, J.V. and Candy, P.M. (1993) SSPACK_PC: A model-based signal processing package on personal computers, *DSP Applic.*, **2**,(3), 33-42.

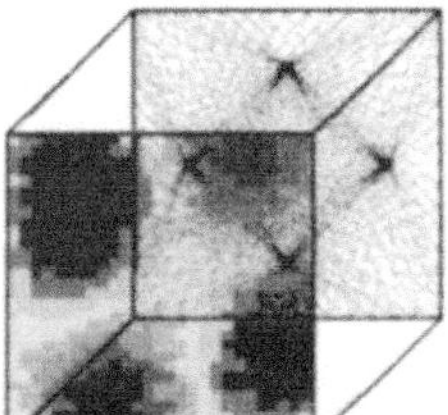

section 3

Sonar and Source Technology

A LARGE DYNAMIC RANGE VERTICAL ARRAY OF ACOUSTIC SENSORS

W.S. Hodgkiss, J.C. Nickles, G.L. Edmonds, R.A. Harriss, and G.L. D'Spain
Marine Physical Laboratory
Scripps Institution of Oceanography
San Diego, CA 92152-6400

Abstract. A large dynamic range vertical line array of acoustic sensors has been designed and four of these arrays have been fabricated and deployed at sea. Each of the four arrays consists of 64 hydrophone elements with 1.875 m interelement spacing ($\lambda/2$ at 400 Hz). The time series from the array elements is sampled at f_s = 1.5 kHz using 24 bit A/D converters and multiplexed onto a single digital data stream for transmission to the data recording hardware. In addition, there are 16 navigation receivers embedded in each array for localizing the array elements. These navigation receivers detect transponder arrivals in the 7.5-14.5 kHz band.

1. Introduction

Active acoustic experiments place particularly demanding requirements on data acquisition systems. In preparation for a major bottom reverberation experiment in summer 1993, the Marine Physical Laboratory designed a large dynamic range vertical line array of acoustic sensors and four of these arrays were fabricated for deployment during the experiment. The focus of this paper is on the engineering aspects of the array design and performance characteristics (clipping level and noise floor) of the large dynamic range preamp-A/D's. Interesting features of the design include a dedicated preamp-A/D-telemetry module for each hydrophone, acoustic array element localization (AEL), lightweight cable design, and low power electronics suitable for battery powered operation and autonomous recording.

2. Design Considerations

A number of requirements needed to be satisfied by the array design. These included acoustic characteristics, deployment options, array element localization, power constraints, and modularity.

The acoustic specifications dictated a bandwidth of approximately 10-600 Hz for the low-frequency (LF) sensors. Thus, a sampling rate f_s=1.5 kHz was selected. Given that the primary frequencies of interest were in the vicinity of 200-400 Hz, the interelement separation was set equal to 1.875 m ($\lambda/2$ at 400 Hz). Furthermore, it was desired to achieve a beamwidth in the vicinity of 2° at 400 Hz. This could be achieved with a 64-element aperture. A particularly significant acous-

O. Diaschok et al. (eds.), Full Field Inversion Methods in Ocean and Seismo-Acoustics, 205–210.

tic requirement was to maintain a large dynamic range. This led to the use of 24 bit A/D technology and the dedication of a single preamp-A/D-telemetry module at each hydrophone.

Several deployment scenarios were envisioned for the arrays. For the summer 1993 experiment, each array was to be buoyed above a bottom-moored, autonomous recording package. In other experiments, an array would be suspended in the vertical from the R/P FLIP. Finally, one or more arrays could be deployed in a horizontal configuration on the sea floor with the data being recorded autonomously or cabled back to FLIP or a support ship for recording and real-time analysis.

Array element localization (AEL) needed to be performed regularly (at least every few minutes). An acoustic transponder net approach was selected with the transponder replies being detected by 16 separate AEL modules embedded in the array.

Power considerations were a significant issue due to the desire for the arrays to operate in an autonomous mode with batteries. Although by no means small, the power budget target for each array was set at 0.5 watt/sensor (averaged across both LF and AEL sensors) or 40 watts total. The measured power consumption of each array is 48 watts.

Modularity was desirable in order to facilitate the fabrication and check-out of the arrays as well as flexibility during deployment. Thus, each array was split into four identical subarrays (16 LF and 4 AEL channels per subarray).

3. Hardware Description

3.1. OVERVIEW

Each of the arrays and recording packages deployed during the bottom reverberation experiment in July 1993 were configured as shown in Figure 1. The autonomous digital recording package (similar in concept to the recording capsule described in [1]), was buoyed 70 m off the bottom. The head module of the array was separated from the recording package by a 30 m umbilical cable. The array was kept vertical by a support float attached just above the tail module. (The terminology "head" and "tail" are carry-overs from towed array design.) Both the recording package and support float were equipped with flashers and radio beacons. For recovery, the array and support float were released from the recording package prior to the recording package releasing the anchor.

The arrays are composed of four identical subarrays, separated by intermediate modules, along with head and tail modules (fewer than four subarrays can be configured if a smaller aperture is desired). Timing for the array is generated in the tail module. Each intermediate module regenerates the system timing and inserts data from the array section above it into the global data stream. Similarly, the head module inserts data from the last subarray into the global data stream. In addition, the head module formats the global data stream for the umbilical cable. The umbilical cable consists of two twisted pairs - one pair for power and one pair for the digital data stream.

The configuration of each subarray is shown in Figure 2. The subarrays are 30 m in length. Each consists of 16 LF modules (spaced 1.875 m apart) and 4 AEL modules (spaced 7.5 m apart). The array cable consists of 7 twisted pairs - global power, global data, local +8 v power, local -8 v power, local clock, local sync, and local data.

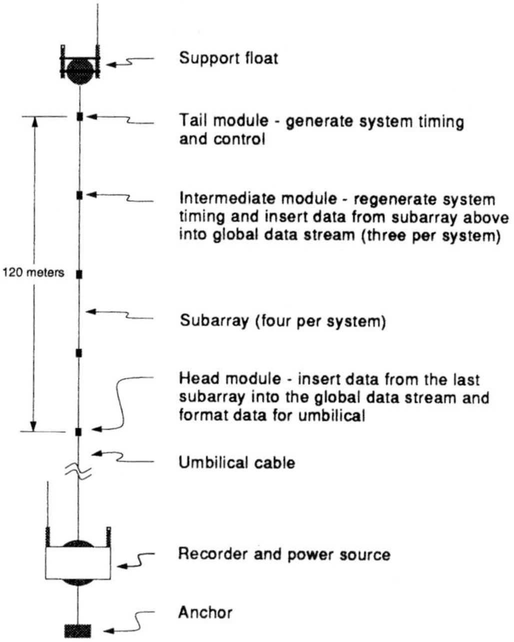

Figure 1. Vertical array and autonomous recording package

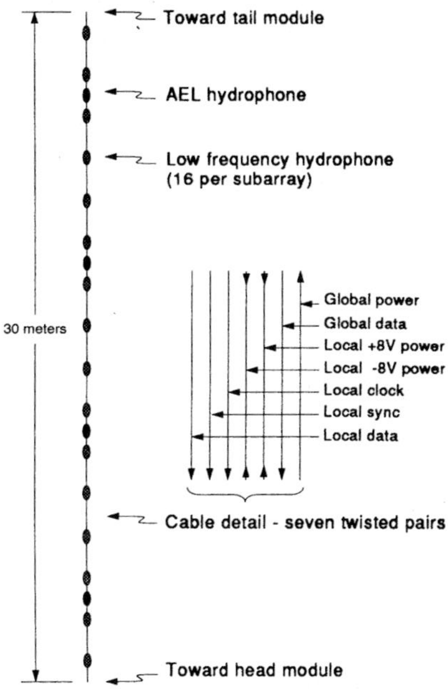

Figure 2. Subarray configuration

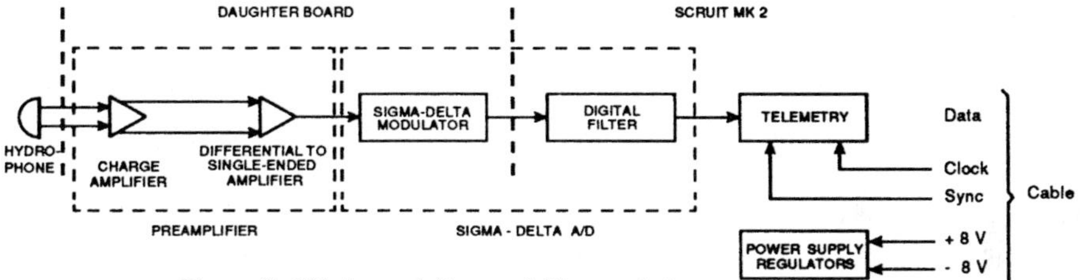

Figure 3. LF sigma-delta acquisition module

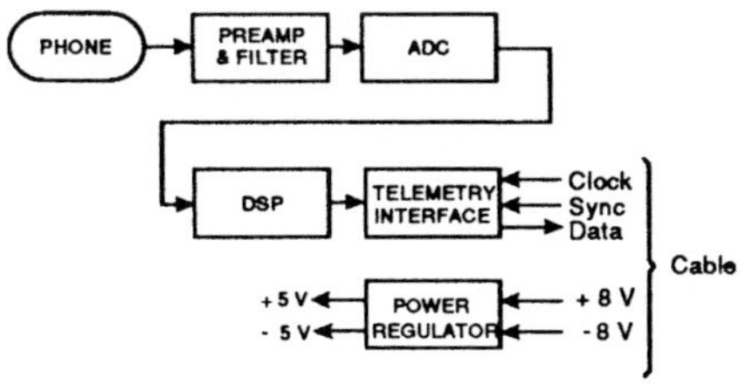

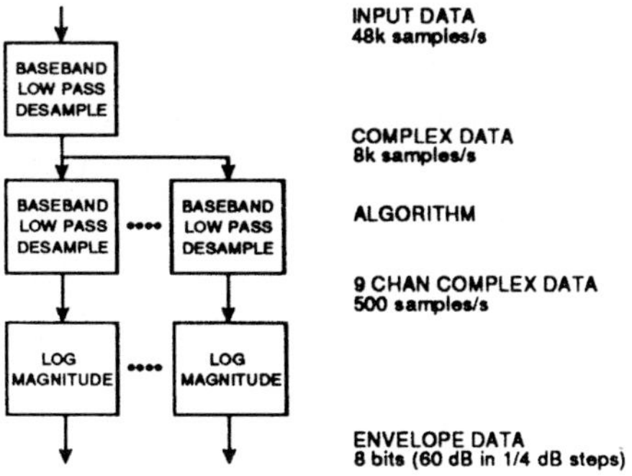

Figure 4. AEL processor

3.2. LOW FREQUENCY HYDROPHONE MODULES

A description of the low frequency data acquisition modules is shown in Figure 3. Each hydrophone has a dedicated preamp, A/D, and telemetry unit. The hydrophones were designed and fabricated by MPL.

The data acquisition units were supplied by GEC-Marconi Avionics (U.K.). They consist of a preamplifier (differential charge amplifier and differential to single-ended amplifier), 24-bit sigma-delta A/D, and telemetry unit (see [2] for a discussion of sigma-delta A/D's). The preamplifier and sigma-delta modulator sit on a daughter board above the board containing the SCRUIT MK 2 ASIC device which performs both the digital filtering function for the A/D as well as data telemetry. There are 5 twisted pairs (data, clock, sync, +8/0 v, and -8/0 v) which go to each GEC data acquisition module.

The hydrophones have a depth-dependent sensitivity (calibrated by NRL USRD in Orlando) of -194.0 dB re 1 V/μPa at 70 m and -196.7 dB re 1 V/μPa at 3500 m. In addition to the depth-dependent sensitivity of the hydrophones, the charge amplifiers themselves also have a depth-dependent gain due to the depth dependency of the hydrophone capacitance. This depth-dependent gain is 31.4 dB at 70 m and 28.1 dB at 3500 m.

The sigma-delta A/D yields a 24 bit output. While the entire 24 bit dynamic range is not realized in practice, a total measured data acquisition module dynamic range of approximately 105 dB is obtained. The noise floor is a combination of the noise contributed from both the preamplifier and the sigma-delta modulator.

3.3. AEL MODULES

A block diagram of the AEL processor hardware and DSP algorithm is shown in Figure 4. Each AEL hydrophone output first is amplified, filtered, and then sampled at f_s=48 kHz by a conventional 12 bit A/D converter. The resulting time series then is processed by a Motorola 56001 DSP chip. The DSP carries out several stages of complex bandshifting, low pass filtering, and desampling.

In the first stage, the time series is complex basebanded around f_c=10.875 kHz with an output sampling rate of f_s=8 kHz. In the second stage, the complex time series is complex basebanded (in parallel) to 9 selectable center frequencies within the band 7.5-14.5 kHz with an output sampling rate of f_s=500 Hz. In the current implementation, 0.5 kHz steps between 9.0 kHz and 13.0 kHz have been selected. Changing the center frequencies involves recompiling the DSP program which is stored in a PROM in the tail module. The magnitude-squared of each complex time series then is computed followed by a log operation. The envelope data for each of the 9 receive frequencies is represented by 8 bits corresponding to a 60 dB total dynamic range in steps a little under 0.25 dB/step.

3.4. TELEMETRY

Array telemetry is divided between local data within each subarray and the global data stream. Each of the receiving sensors (both LF and AEL modules) generates a 32 bit word containing 24 bits of data at a 1.5 kHz rate. The AEL data is submultiplexed such that 3 receive frequencies at a time (8 bits/each) are reported at this data rate (thus, reporting all 9 frequencies in 3 LF sensor sampling intervals). These 32 bit words are in a bit serial format and are time-division multiplexed into the subarray data stream. There are 22 time slots available (1 header information, 16 LF sensors, 4 AEL sensors, and 1 unused) yielding a local data bit rate of 1.056 Mb/s.

The global data stream begins at the tail module. Here, clock and sync for the array are generated as well as header information for each frame of data (where a frame of data corresponds to a single sample of data from each LF and AEL sensor at the 1.5 kHz rate). The intermediate modules and head module (see Figure 1) are responsible for inserting the local data from the subarray just above it (i.e. towards the tail module) into the global data stream. At each of the intermediate modules and head module, the frame header information is passed through, the data from the previous intermediate module is delayed, and the data from the new subarray is inserted just before it. The global bit rate is 4.224 Mb/s or four times the bit rate within each subarray. The head module also puts the global data stream into a PCM format for transmission on a twisted pair over the umbilical cable.

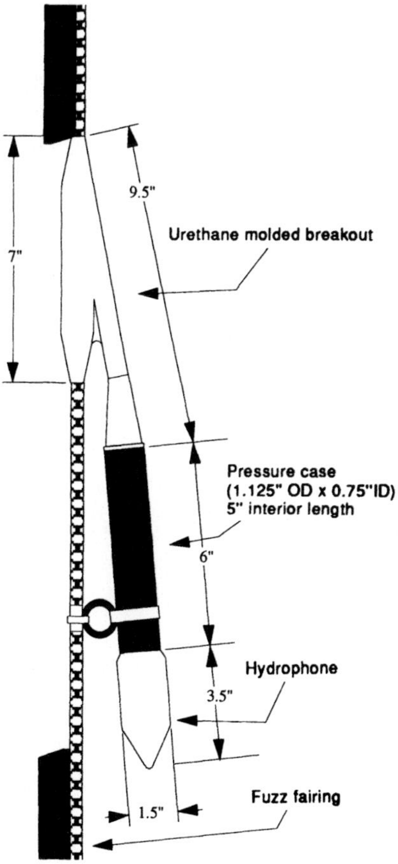

Figure 5. LF hydrophone.

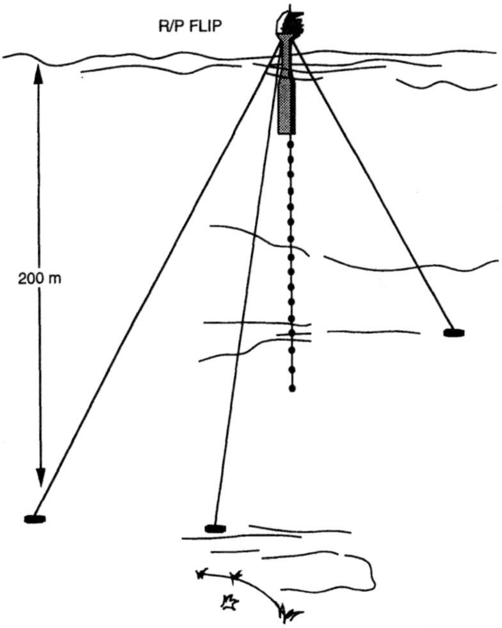

Figure 6. Array deployment from the R/P FLIP.

3.5. ARRAY CONSTRUCTION

The array was made lightweight for ease of deployment and recovery. The backbone cable consists of 7 twisted pairs, a small Kevlar strength member, and fuzz fairing which are held together with a loose-woven overbraid.

The dimensions of the LF hydrophone breakout, pressure case, and hydrophone are shown in Figure 5. At each sensor, 5 twisted pairs need to be broken out (data, clock, sync, and +/- 8 v power). The breakouts are covered with polyurethane with the endcap of each pressure case being part of the breakout. The end cap at the other end of the pressure case is part of the hydrophone.

4. Summary

A large dynamic range vertical line array of acoustic sensors has been described. The array consists of 64 hydrophone elements with 1.875 m interelement spacing ($\lambda/2$ at 400 Hz). The time series from the array elements is sampled at $f_s = 1.5$ kHz using 24 bit A/D converters and multiplexed onto a single digital data stream for transmission to the data recording hardware. In addition, there are 16 navigation receivers embedded in each array for localizing the array elements. These navigation receivers detect transponder arrivals in the 7.5-14.5 kHz band. Interesting features of the design include a dedicated preamp-A/D-telemetry module for each hydrophone, acoustic array element localization (AEL), lightweight cable design, and low power electronics suitable for battery powered operation and autonomous recording.

Four of these arrays, along with their companion autonomous recording packages, were deployed during a bottom reverberation experiment in July 1993 near the Mid-Atlantic Ridge. Subsequently, one of the arrays was deployed from the R/P FLIP during shallow water ambient noise and signal propagation experiments in August and September 1993 off the coast of southern California (see Figure 6). Data recording in the August and September experiments was carried out similar to the approach discussed in [3].

Acknowledgments

This work was supported by the Office of Naval Research, Code 1125OA, under grants N00014-90-J-4042 and N00014-93-1-0062.

References

[1] R. Currier, R. Harriss, C. Nickles, and W. Hodgkiss, "An Autonomous Seafloor Recording Capsule," Proceedings OCEANS'91: 1681-1686 (1991).

[2] S. Park, "Principles of Sigma-Delta Modulation for Analog-to-Digital Converters," Motorola Application Note APR8/D (1990).

[3] W.S. Hodgkiss and J.C. Nickles, "Real-Time Data Management in a UNIX Network Environment," Proceedings OCEANS'90: 294-297 (1990).

LOW-FREQUENCY TRANSDUCER TECHNOLOGY IN UKRAINE

V. T. GRINCHENKO, I. V. VOVK
Institute of Hydromechanics of National Academy of Sciences of Ukraine
Zhelyabov Str., 8/4, Kiev 252057, Ukraine

ABSTRACT

There are many problems associated with the design of low-frequency projectors. Factors such as miniaturization, operation at great depth, small wave dimensions, electro-elastic-fluid interaction, etc., are of particular importance. Some approaches to designing harmonic sound radiators for frequency bandwidths from tens to a thousand Hz are investigated. This frequency band is too large to use a single kind of mechanical vibrating structure. In the upper part of this frequency band one can use a traditional type of radiator, for example a Tonpilz or free-flooding cylindrical transducer. The possibility of reducing the size and weight characteristics of these transducers is considered. To overcome the design problems of low-frequency radiators, the use of flexural modes is effective. In the band of 50–300 Hz, the possibilities of flextensional flat-surfaced transducers are considered. These transducers offer the possibility of effective control of the working frequency without changing the mechanical vibrating system of the transducer.

1. Introduction

The quest for reliable low-frequency underwater sound radiators has been stimulated by many ocean research fields. The creation of effective low-frequency transducers for these applications often demands the satisfaction of conflicting requirements.

The analysis of the basic problems of this area has been given in reference [1], which includes a discussion of new types of radiators. Within the range of radiators applicable to low-frequencies, the flextensional and flexsional ones hold a central position. These types of radiators provide a fundamental way of producing mechanical vibrating systems with resonances in the vicinity of 100 Hz.

Many different drivers can be used to produce acoustic radiation in water. A general review of corresponding transducer structures, and the physical limits important for the design process are given in reference [2]. This work summarizes most of the extensive research in the field of low-frequency transducer design in the former Soviet Union.

This article addresses only radiators that use piezoceramics as the active element of electro-mechanical transformation. The analysis of practical achievements in the field of low-frequency radiator design has shown [8] that transducers with other types of transformation are not competitive with them. This deduction appeared to be fairly convincing in the middle 80s. However, the use of new materials [7], the rare earth-based magnitostrictors, opens up new possibilities in this field. All of the methods of acoustic characteristics control, discussed here, will be of use to the new types of transducers. The problems involved in designing low-frequency radiators can be divided into two groups. Some of them are related to the peculiarities of sound radiation by sources which are small with respect to wavelength. The problems have the character of fundamental physical restrictions. The

O. Diaschok et al. (eds.), Full Field Inversion Methods in Ocean and Seismo-Acoustics, 211–216.

other type of difficulties are related to the attempts of direct extension of high-frequency design experience to the low-frequency area. It is often possible to find effective approaches to overcome this second type of difficulty. A short outline of such approaches will be given here. New methods have been developed [3] to obtain numerical characteristics for such kind of problems.

The paper is organized as follows: Section 1: Introduction. Section 2 discusses the possibility of controlling resonant frequencies of Tonpilz and cylindrical transducers. The main idea is to show the possibility of reducing the resonant frequency on retention of mass of the radiators. The approach of using cylindrical piezoceramic transducers to produce direct low-frequency radiators is justified in Section 3. Section 4 discusses the properties of free-flooding cylindrical radiators. The study of array properties provides the basis for illustrating the importance of interaction effects of low-frequency radiators. Some approaches for controlling the characteristics of the flextensional transducers are considered in Section 5.

2. Control of properties of Tonpilz transducers

When the general problems of low-frequency radiators were discussed by R.S. Woollett [1], the Tonpilz transducer (Fig.1) was taken as an example to illustrate specific characteristics. Such transducers are common for the frequency band from 2 to 80 kHz [8]. The important element of the construction is the tail mass M. The requirement of a high ratio of the tail to head mass is essential for effective working of the transducer. If this is the case, the tail end of the piezoceramic stack is practically motionless at all frequencies. But in practice this condition is only necessary in a frequency band near the resonance frequency. In this situation it is possible to replace the passive mass M with a dynamic system with an inlet impedance of mass kind. The impedance has to be at a maximum value at the resonance frequency of the transducer. The simplest constructional solution is to replace the mass M with an elastic plate with a correspondingly selected eigenfrequency (Fig.1b).

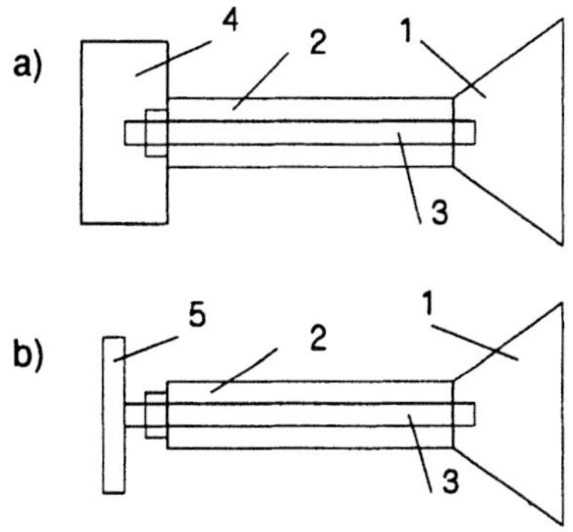

Fig. 1.

1.Head mass 2.Piezoceramic stack 3.Bolt for bies 4.Tail mass 5.Plate

To illustrate the effectiveness of the proposed method a comparison between the experimentally determined characteristics of the two methods of transducer design, viz. tail plate and traditional tail mass, are given in Table 1. As can be seen from this data comparison there is practically a complete preservation of the main electroacoustical parameters of the Tonpilz transducer when using a plate, and it is substantially (about 1.5 times) lighter than a transducer with a passive tail mass. One can see that the proposed method provides the possibility of significantly reducing (about 2.3 times) the resonant frequency of a transducer and simultaneously realizing the gain associated with the mass. Note that the observed reduction in the efficiency η and sensitivity α $[Pa/V]$ of a transducer is caused by the decrease in its dimension. Since in the experiments only one physical transducer was used (with only the tail mass replaced by the corresponding plates), the wave dimension of the head mass decreased with reduction of its resonant frequency. Naturally, this inevitably

TABLE 1. Comparison of the transducers characteristics

Tonpilz transducer	Mass, kg	$f_{res.}$	α	d/λ	η
With tail mass	22.0	3.3	21.8	0.5	0.28
Plate ($f_{res.}$=3.4 kHz)	14.5	3.25	20.5	0.5	0.28
Plate ($f_{res.}$=1.5 kHz)	14.6	1.35	12.0	0.22	0.19

led to the reduction of radiation impedance and efficiency.

3. Two-mode piezoceramic cylindrical radiator

A radiating acoustical system consisting of a radially polarized piezoelastic cylindrical shell filled with liquid, with an asymmetric internal insertion is investigated next (see Fig. 2). Vibrations were excited by a harmonic voltage applied to the electrodes. Although the surfaces of the shell were considered to be completely coated by electrodes, the asymmetric reaction of the medium in the internal cavity of the shell provided the possibility for simultaneous effective excitation of both the pulse and oscillating modes of the shell.

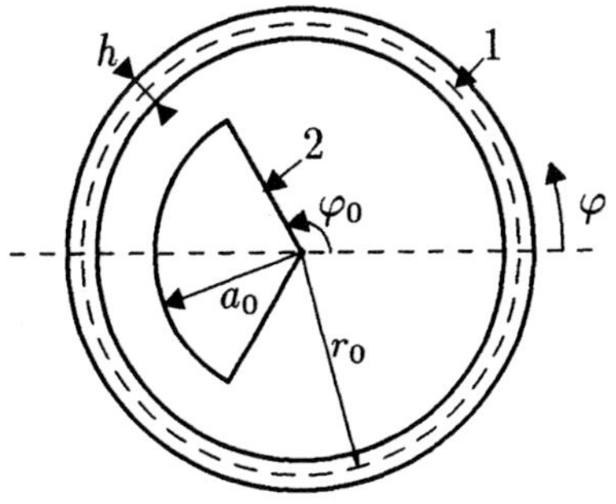

Fig. 2. Piezoceramic shell (1) with insertion (2).

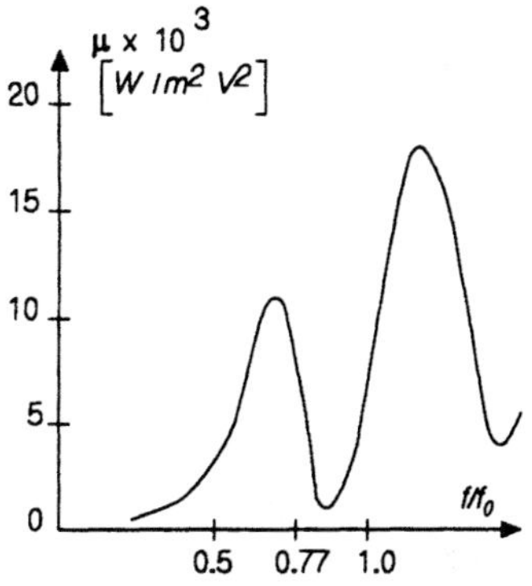

Fig. 3. Energy efficiency of the system.

The analysis of the acoustical properties of the system under consideration was conducted on the basis of exact solutions of the equations of the corresponding mathematical model. The influence of geometrical, acoustical and coupling effects on the energy efficiency of the system, the characteristics of the shell vibrations, and the directivity properties of the radiator were examined. In particular, variants of couplings, such as hard sector and hard cylinder positioned excentrically with respect to the radiating shell were considered from practical points of view.

It was established that for these types of couplings the frequency dependence of acoustic power radiated by the system has a new maximum situated significantly below the resonance frequency of the pulsating shell (viz. at $f/f_0 \approx 0.77$, as illustrated in Fig.3).

Thus, a vibrational system has been formulated, in which the shell plays the role of a mass connected with the elasticity of the liquid filling the internal cavity. In this way the appearance of a new specific resonance of the system, absent in the eigenfrequency spectrum of the shell vibrations in vacuum, can be explained.

As a result of these investigations the possibility of creating a low-frequency directed hydroacoustic radiator, having high energy efficiency and small wave dimensions with the use of traditional piezoceramic rings, has been demonstrated.

4. Free-flooding cylindrical radiator

The two distinctive features of the free-flooding cylindrical radiator, that are especially interesting in the context of this article, are the depth capability and the potential of using the internal fluid column to control acoustic characteristics. We will consider a finite array of free-flooding coaxial piezoceramic cylindrical shells. The details of the method [9] will be omitted, and only some quantitative data will be discussed.

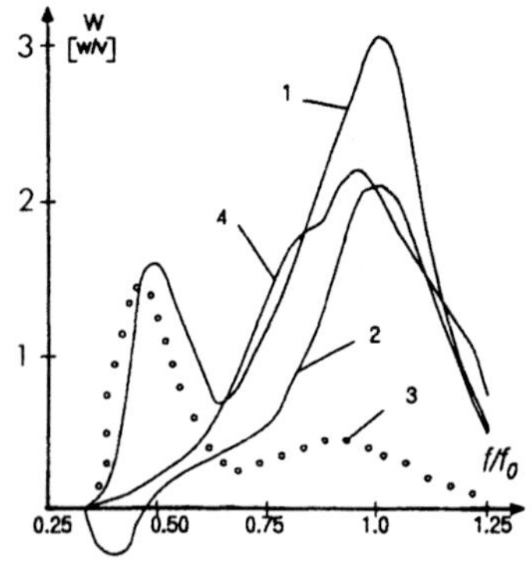

Fig. 4. Frequency dependencies of radiated acoustic power. Total power (1). Radiated by outer and inner walls (2,3). Radiated by closed grating (4).

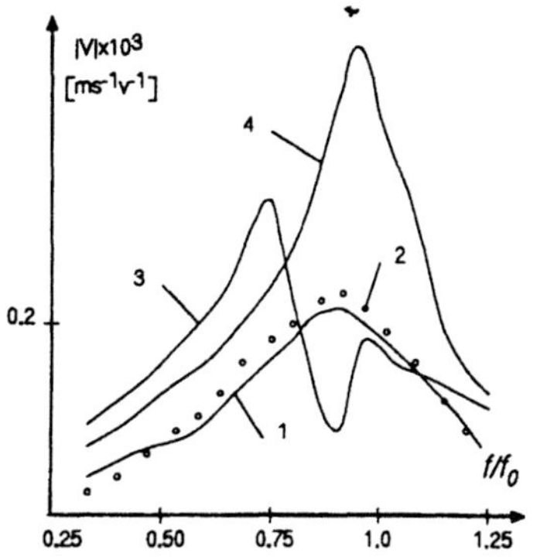

Fig. 5. Frequency dependencies of shell oscillation velocities. Free-flooding grating (1,2). Closed grating (3,4).

The frequency dependence of the normalized (with respect to electrical voltage V) acoustic power radiated by a grating made of three shells is shown in Fig. 5. Curves 1 and 4 characterize the total radiated power of the free-flooding and closed shell. In the latter case rigid caps are placed at the ends of the array. The grating of free-flooding shells has two sharp radiation maxima. The high-frequency maximum is caused by the shell's resonance, and the low-frequency maximum is due to resonance of the liquid column in the grating.

The factors associated with the liquid column resonance have been well studied. This topic is very important because the level of radiated power in the vicinity of the resonance is very high. As a result, a grating of free-flooding shells has a higher bandwidth and is energetically more efficient than one of closed shells.

Comparison of curves 2 and 3, shown in Fig. 4, reveals that there is a sort of 'division of labour' between the inner and outer walls. In the vicinity of the high-frequency maximum, the outer walls provide the main contribution. In the vicinity of the low-frequency maximum, the main part of the energy is radiated by the inner walls and the contribution of the outer walls decreases to zero level at $f/f_0 = 0.43$. The special feature is that at $f/f_0 \leq 0.43$ only the inner walls emit sound and the outer walls consume energy.

Figure 5 demonstrates the frequency dependence of normalized (to the voltage V) amplitudes and phases of the shell oscillation velocities in the three-component grating for the cases when the shells are free-flooding and closed. Amplitudes and phases of the free-flooding shells depend comparatively weakly on their positions in the grating. It is different in the case of closed shells, especially in the region of their resonant frequencies, where the velocity amplitudes of outermost and central shells can differ by 3.5 times and the phase difference can attain $2/3\ \pi$. It is important to note that maximum values of velocity amplitudes of free-flooding shells are significantly lower than for closed ones. This fact can be very important in practice. We can see that the internal fluid substantially attenuates the injurious effects of interactions in the array.

5. Flextensional flat-surfaced transducer

The flextensional electroacoustical transducer has been extensively studied as a candidate for the construction of low-frequency sonars [5], [6]. Much of this research has been conducted in the former Soviet Union.

Here we shall consider only one aspect related to the use of such transducers. One of the approaches to construct flextensional transducers was proposed in reference [4]. The use of a 'flexbar' as an element of the transducer enabled the creation of a purely monopole source. Our laboratory has extensive experience in designing radiators at frequencies of about 100 Hz. The flat-surfaced flextensional transducer offers both advantages and disadvantages over transducers with oval shells. Our interest here is the possibility of using these radiators in an array, as illustrated in Fig. 6. We have here a set of two 'flexducers' [4] Under this arrangement of the transducers, the effect of a hydraulic transformer can manifest itself. The directions of active surface motion and fluid particle motion on the radiating surface are different. As the distance d becomes relatively small, the added mass grows. So by changing only the distance d, we can vary the frequency of the radiator with no change of the mechanical elements of the array.

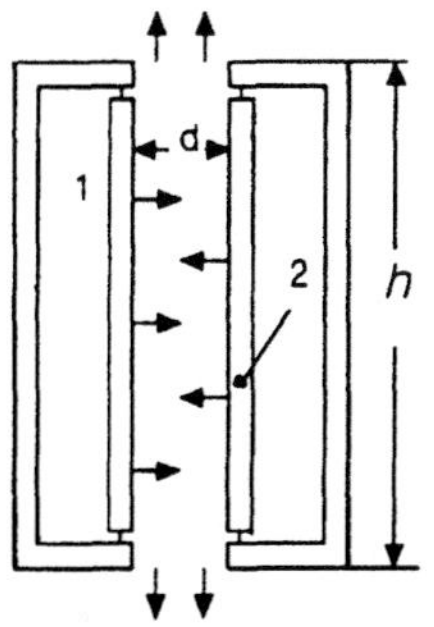

Fig. 6. Element of radiator. Air filled case (1). Radiating surface (2).

This effect was exploited in experimental equipment. We could create arrays with a working frequency smaller than the resonance frequencies of the component elements by a factor three. In Fig. 7 the mechanical part of such an experimental array is shown. The elements of the array have an eigenfrequency of 250 Hz.

Two examples of practical applications of this type of array follow. By changing the inter-element spacing, it is possible to compensate for: a) the change in the eigenfrequency of an air compensated radiator if the depth changes; b) the physical/technological differences between the elements of the array.

Fig.7. The mechanical system of an array of flexural transducers.

6. Conclusion

Research and development in the area of low-frequency transducer design has been concerned with the establishment of basic physical limits. The solution of the associated theoretical and applied problems can yield large dividends in transducer design. This paper has provided a short description of the approaches to solving such problems in Ukraine. The limited size of the article permits only a general description of the results.

References

1. Woollett, R.S. (1980). Basic problems caused by depth size constraints in low-frequency underwater transducers, *J.Acoust.Soc.Am.*, **68**, 1031–1037.
2. Rimsky-Korsakov, A.V., Yamshichikov, V.S., Zhulin, V.I., Rechman, V.I. (1984). Low-frequency Hydroacoustic Transducers. Sudostroenie, St.Petersburg (in Russian).
3. Grinchenko, V.T., Vovk, I.V. (1986). Wave problems of sound scattering by elastic shells. Naukova Dumka, Kiev (in Russian).
4. Massa, F. (1989) Sonar transducers, *Sea Technology*, **30**, (11), 39–48.
5. Brigham, G.A., Glass, B. (1980). Present status in flextensional transducers technology, *J.Acoust.Soc.Am.*, **68**, 1046–1052.
6. Rolt, K.D. History of the flextensional electroacoustic transducer, *J.Acoust.Soc.Am.*, **87**, 1340–1349.
7. Proceedings NATO Advanced Study Institute, (1981), **66**, 225–243.
8. Gribakina, N.M. (1984) Foreign low-frequency hydroacoustic radiators, *Foreign Shipbuilding*, **8**, 18–26 (in Russian).
9. Vovk, I.V. (1992) Sound radiation by a finite grating formed by free-flooding cylindrical piezoceramic shells, *Sov.Phys.Acoust.*, **38**, (3), 234–238.

LOW-FREQUENCY SOURCES FOR OCEAN ACOUSTIC TOMOGRAPHY

M. M. SLAVINSKY, B. N. BOGOLUBOV
Department of Ocean Acoustics Institute of Applied Physics
46 Uljanov St., Nizhny Novgorod, 603600 Russia

and

J. L. SPIESBERGER
Department of Meteorology, Penn State University
512 Walker Building, University Park PA 16802, USA

ABSTRACT The comparison of various types of low-frequency acoustic sources with small wave parameters shows that monopole resonant sources have the highest quality. It is shown how to provide contradictory requirements: broad bandwidth, high efficiency, small dimensions and weight, and how to overcome the main engineering problems. This technology was certified in the creation of a number of low-frequency sources. The results of source tests show that sources are suitable for the acoustic tomography.

The efficiency of low-frequency sources is essentially determined by their wave parameter kL (k is the wave number, L is the characteristic dimension of the source). At $kL \gg 1$ the efficiency of the source is large and depends on the oscillating surface area. The water reaction is practically of the active character, therefore the source has a wide frequency band ΔF and a high efficiency η.

For low frequencies the condition $kL \gg 1$ is often due to inadmissable increase of dimensions and mass. A compromise solution can be found by creating sources with small wave parameters $kL < 1$. In this case the water reaction is of the inertial character and the radiation energy constitutes only part of the energy stored by the source. The ratio between the energies radiated and stored by the source during a quarter of the oscillation period is equal to the limiting coefficient of the radiation efficiency G. This coefficient enables one to compare various types of sources in their pure forms, since only the energy stored in the water is taken into account, while the construction is assumed to be ideal. The coefficient G characterizes the maximum admissable ΔF at maximum η of the given source type. Further extension of ΔF can only be due to losses in a real construction with corresponding decrease of η.

Let us give expressions for the coefficient G of various source types:

- monopole $G_{\mathrm{m}} = [1 + (\pi kL)^{-1}]^{-1}$
- dipole $G_{\mathrm{d}} = [1 + (\pi kL)^{-1} + 4\pi^{-1}(kL)^{-2} + 4\pi^{-1}(ka)^{-3}]^{-1}$
- Helmholtz resonator $G_{\mathrm{H}} = [1 + 3(ka)^{-2}]^{-1}$
- resonant 1/4 wavelength tube $G_t = [1 + 0,5(kL)^{-1} + \pi(kL)^{-2}]^{-1}$.

At small values of L the monopole efficiency multiply exceeds the dipole effect, always remaining better than that of the Helmholtz resonator and of the resonant tube, since the reactive impedance of the latter is much larger due to the energy additionally stored in the internal volume of these resonators.

O. Diaschok et al. (eds.), Full Field Inversion Methods in Ocean and Seismo-Acoustics, 217–222.

The difficulties caused by a large reactive impedance of the monopole can be eliminated by electric matching in a bounded frequency band which is still sufficient for practical application. Thus for low frequencies the monopole resonant source is potentially more effective.

The main engineering problems are caused by the necessity of creating an efficient and reliable single-mode oscillating system having the largest shift between the frequencies of the fundamental (monopole) and parasitic (multipole) modes. Parasitic modes are practically not excited here and maximum value of the oscillating column at the operating frequency is provided. At the same time this oscillating system should have small losses and withstand multiple cyclic alternating loads.

It is convenient to express the monopole source parameters through the radius of the equivalent sphere Re

$$\mathrm{Re} = \left(\frac{n}{\pi}\int\int_S g(x,y)dS\right)^{\frac{1}{2}},$$

where $g(x, y)$ is the normalized profile of the radiating surface RS, S and n are the area and number of RS. The acoustic impedance is expressed through the active and reactive components

$$Z = X + iY = r_\mathrm{a} + i\omega m_\mathrm{a},$$

where $r_\mathrm{a} \sim (k\mathrm{Re})^2\mathrm{Re}^2 \sim \omega^2\mathrm{Re}^4$ is the radiation resistance, and m_a is the attached mass of water.

The reactive radiation band is determined as

$$\frac{\Delta F}{F} \sim \frac{r_\mathrm{a}}{\omega m_\mathrm{a}} \sim k\mathrm{Re}.$$

The acoustic radiation power equals

$$W_\mathrm{a} = \frac{2\pi\rho c^5}{\omega^2}(k\mathrm{Re})^6\delta^2 = \frac{2\pi\rho}{c}\omega^4\mathrm{Re}^6\delta^2,$$

where $\delta = \frac{\Delta}{\mathrm{Re}}$ is the relative deformation of RS, and Δ is the displacement amplitude of RS.

The set of equations governing the harmonic oscillations of a real source has the form

$$\sum_{f=1}^{n}\left[-\omega^2(m_\mathrm{c}\delta_{df} + m_{df}) - i\omega(r_\mathrm{c}\delta_{df} + r_{df}) + K_{df}\right]x_f = 0,$$

$$\delta_{df} = \begin{cases} 0 & d \neq f \\ 1 & d = f \end{cases} \qquad d = 1, 2, \ldots n.$$

Here the coefficients K_{df} are responsible for the natural elasticity of RS and for their internal elastic bonds. In their absence $K_{df} = 0$ at $d \neq f$; r_c and $r_{df} = r_\mathrm{a}$ are the losses by the internal friction and by the radiation, respectively,

$$m_\mathrm{c} = \rho_\mathrm{c}\int\int_S h(x,y)g^2(x,y)dS$$

is the reduced RS mass; ρ_c and $h(x, y)$ are the density and thickness of RS,

$$m_{df} = \frac{1}{\omega^2}\int\int_S p_f g(x,y)dS$$

is the reactive component of the induced pressure.

For equal RS($K_{df} = K, m_{df} = m_a$) and negligibly small losses the set of equations has two real solutions $\omega_1^2 = \frac{m_c+m_a}{m_c+nm_a}\omega^2$ and $\omega_2^2 = \frac{m_c+m_a}{m_c}\omega^2$, the solution ω_2 being a multiple of $(n-1)$. The solution ω_1 corresponds to pulsating oscillations and is the operating frequency of the source. The solution ω_2 corresponds to multipole radiation. To provide a single-mode regime one should maximize the ratio of these frequencies:

$$\left(\frac{\omega_2}{\omega_1}\right)^2 = 1 + n\frac{m_a}{m_c},$$

which is fulfilled under the condition of minimized oscillating mass of the source $m_c \ll m_a$. The maximum frequency band at maximum efficiency $\frac{\Delta F}{F} = \frac{r_c+r_a}{m_c+m_a}$ can be ensured under the same condition $m_c \ll m_a$ and the natural requirement of minimizing the internal losses $r_c \ll r_a$. The acoustic parameters of the source, the construction of which satisfies these conditions are very close to optimum ones.

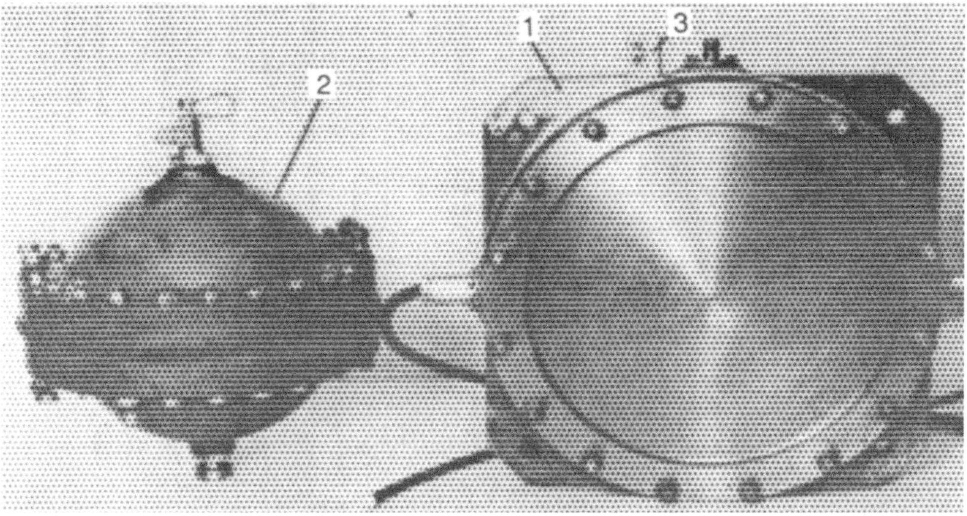

FIGURE 1. Source with pressure compensation system, source 1, pressure compensation system 2, match box 3. Source: weight 123 kg, diameter 540 mm; PCS: weight 10 kg, diameter 450 mm.

The service life of the source is mainly determined by the capability of the elastic element to withstand multiple cyclic alternating loads. Two periods of increasing service life dependent on the fatigue strength have been used:

- one, is to decrease the mechanical loads by optimization of the geometry of the mechanical oscillating system;
- the second, is to use the best materials and optimize the strengthening technology.

The variational problem of optimizing the shape of the elastic element of the oscillating system which maximizes the oscillation profile, i.e., Re, and minimizes the stress in the construction, has been solved. A set of computer programs for optimization of the elastic elements has been developed. These programs are based on the use of equal loads on the

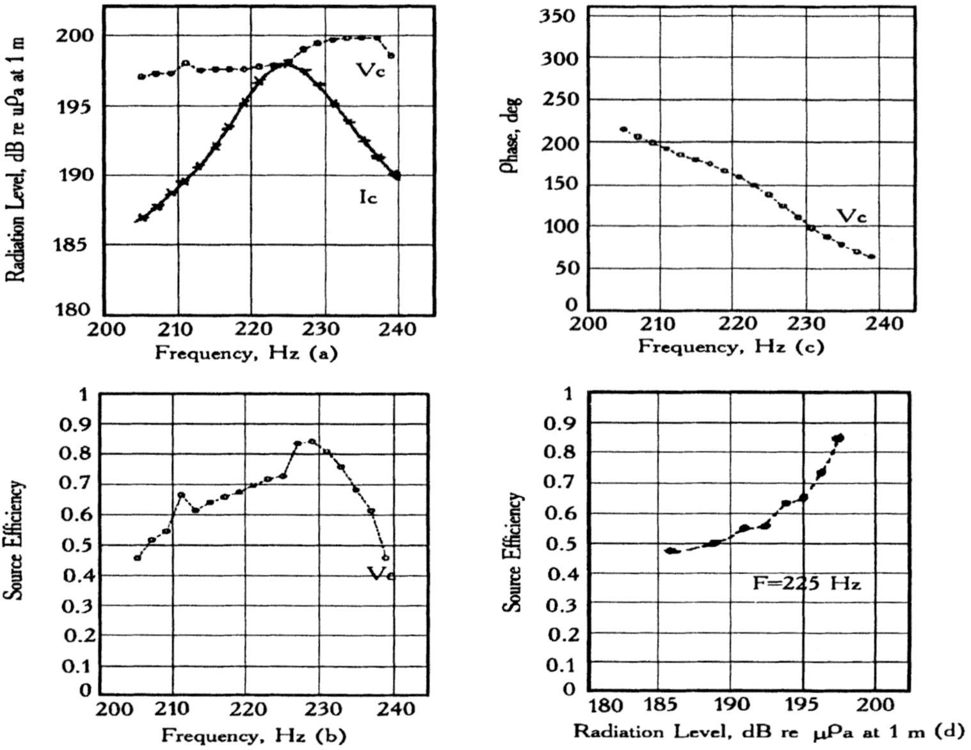

FIGURE 2. Frequency dependences of the source output level (a), efficiency (b), emitted signal phase (c) at the constant current I_c and voltage V_c, and efficiency as a function of the output level at the resonant frequency (d).

dangerous sections. The optimization is performed for elements of variable thickness and small bending.

An original technique has been elaborated and the equipment has been created to test the fatigue strength. The testing base is 2×10^8 oscillation cycles. The dimensions of the test samples in thickness are close to those of the parts of an actual construction. At present we have the results of tests for a number of steel qualities, titanium alloys and duralumin samples using various methods of hardening and strengthening. The fatigue curve has been found for each material which permits us to predict the behavior of the material at various stresses and large number of oscillation cycles. Therefore, the proper choice of the material and shape of the elastic element provides an essentially lower stress level in a real construction than the maximum admissable one for this material for a given number of cycles of the source oscillations.

The design of similar sources operated at different frequencies are based on a perfected and many times approved method of calculation and optimization of a single-mode oscillating sytem. The main advantage of this design is that it maintains the reciprocating motion of the oscillating system (operating mode), and the frequencies of the remaining (parasitic) modes are much greater than the operating frequency. Consequently, the repumping of energy from the operating mode into parasitic modes is almost absent and the latter are

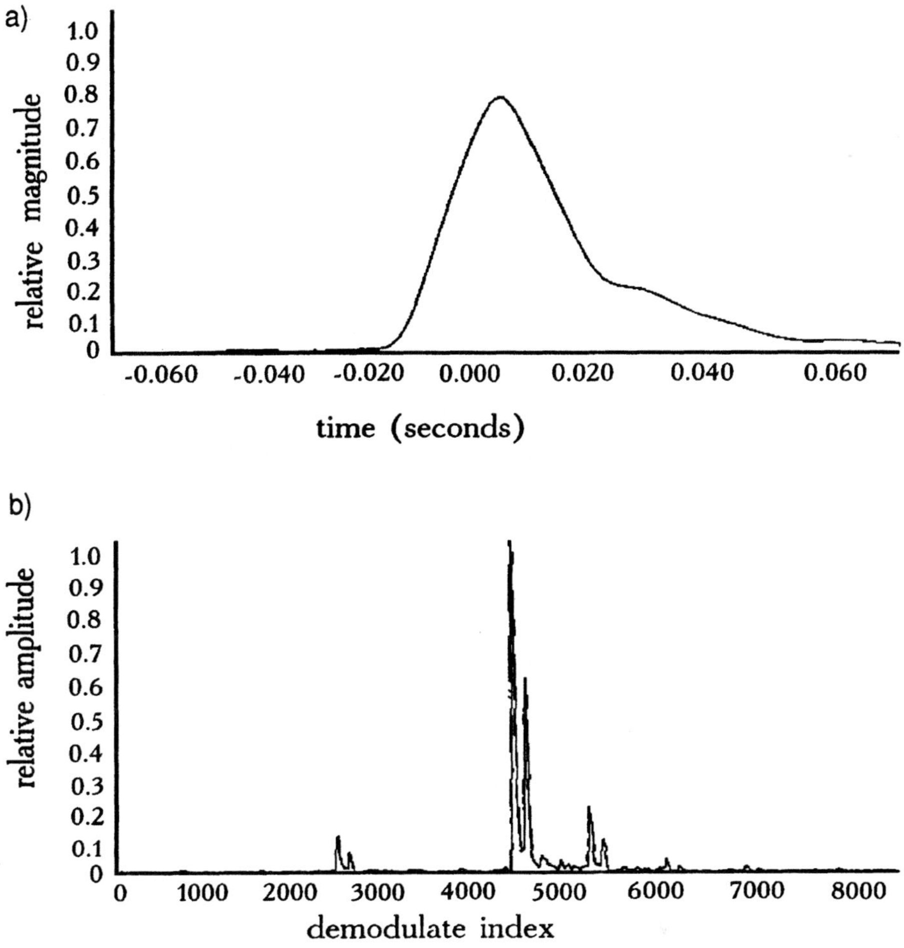

FIGURE 3. Correlation coefficient for $Q = 4$. Submergence of the source at (a) 23 m and (b) 100 m. The sampling frquency of the demodulates is 900 Hz. The surface arrival is predicted to arrive $\sim$ 0.14 s or $\sim$ 126 demodulates after the direct arrival.

therefore not excited.

This method and these technologies were certified in the creation of a number of low-frequency sources, including the source with a resonance frequency 225 Hz, which was tested together with American experts at Lake Seneca in 1992 (Fig. 1).

The tests have shown that resonance monopole sources emit radiation of acoustic power of the order of 200 dB with an efficiency of the order of 75% at the resonance frequency, 225 Hz (Fig. 2). The transmission system synthesizes a wide class of coded signals using amplitude, phase, and frequency modulation. The system radiates coded signals with a pulse resolution of 22.8 ms (Q-factor equals 4) and an efficiency of 50% (Fig. 3), which are admissable for acoustic tomography [1, 2].

References

1. Slavinsky, M.M., Bogolubov, B.N. and Spiesberger, J.L. (1992). Low-frequency, high efficiency sources for acoustic monitoring of climatic temperature changes in ocean basins, *Journal of the Acoustical Society of America*, **92**(4), 2349–2350.
2. Slavinsky, M.M. and Spiesberger, J.L. (1993). Evaluation of electromagnetic source for ocean climate acoustic thermometry at Lake Seneca, Technical Report WHOI-93-09. Woods Hole, MA, Woods Hole Oceanographic Institution.

TRANSDUCERS FOR LOW-FREQUENCY COMMUNICATIONS

M. LETICHE[1], S. MARSELLI[2], P. ERIKSEN[3], R. COATES[4], Y. QINGSHAN[5]
[1] *Thomson Sintra ASM Valbonne (F),* [2] *USEA S.p.A., La Spezia (I),*
[3] *Reson System A/S Slangerup (DK),* [4] *University of Birmingham (UK),*
[5] *Technical University of Denmark, Lyngby (DK)*

ABSTRACT Large scale underwater sound transmission (e.g., ocean acoustic tomography) requires transducers operating in the very low-frequency range (< 500 Hz) to cope with the unavoidable medium losses. The LF-projectors should be highly efficient in converting the driving electrical energy into powerful structural vibrations, and then in transmitting these vibrations to the outgoing acoustic field. Thus, those devices should have a very high source-level, and, moreover, they should operate in a broad frequency band, and at great depth if required. To facilitate their handling and/or installation on fixed or moving platforms they should also be not too bulky and as light-weight as possible. This paper reviews the present most interesting techniques for the realization of these LF-transducers, as well as the primary motor technologies as the secondary matching methods for maximizing transducer performance.

1. Introduction

Until about 15 years ago the use of very low-frequency (VLF) underwater sounds has been restricted almost exclusively to passive systems like e.g. flank or towed arrays and moored or drifting sensors-configurations (the frequency-bands of interest are here defined as follows: ELF: $15 \div 150$ Hz; VLF: $150 \div 1500$ Hz; LF: $1.5 \div 15$ kHz). To cope with the steadily improved silencing of military vessels, and to realize

(i) Long-range underwater communication (some hundred to some thousand km),
(ii) Acoustic telemetry to and from industrial off-shore installations or marine eco-system monitors,
(iii) Ocean acoustic tomography,

there was a strong need for active VLF/LF-systems, and thus for the development of suitable projectors.

By nature efficient VLF/LF-transducers tend to be large, heavy, and quite expensive, and, moreover, these trends are worse if mid- to deep-water operations are required. For that reason a suite of problems had to be faced and investigated step by step to find more acceptable solutions from the operational, and, above all, economical point of view. Unfortunately some of these requirements (small, light-weight, low-cost, etc.) are in conflict with each other. Thus the only way out of this dilemma consists in a series of suitable compromises.

At (very) low frequencies the dimensions of reasonably-sized transducers will always be quite small compared to the wavelength in water and thus afflicted by an intrinsic handicap we have to live with. In spite of that, to realize the desired high radiation resistance with respect to the water load $\mathrm{Re}(\mathbf{Z}_r) = \mathrm{Re}\mathbf{F}/\mathbf{v}$ ($\mathbf{Z}_r$ radiation impedance, $\mathbf{F}$ = force exercised by the vibrating surface; $\mathbf{v}$ = its normal velocity) and thus a small radiation-Q

O. Diaschok et al. (eds.), Full Field Inversion Methods in Ocean and Seismo-Acoustics, 223–228.

($Q_r = |\operatorname{Im}(\mathbf{Z}_r)/\operatorname{Re}(\mathbf{Z}_r)|$) the vectors $\mathbf{F}$ and $\mathbf{v}$ should be in phase over the active surface, which should be as large as possible. Beside structural fitting of the projectors' vibratory action, there is also the need for high acoustic output power and thus large amplitude deflections which require an adequate selection and lay-out of the driving mechanisms.

Together with the final frequency-tuning, these matching and motor technologies also have to counterbalance all collateral effects of those mechanical modifications which are imposed by the depth-requirements (e.g., higher structural stiffness and/or bigger vibrating masses). To operate deeper than about 500 m most projector-types have to be fitted with properly designed pressure-compensation (air-bottles, compliant tubes, liquid filling). Aiming mainly at tomographic applications the following specifications

- Frequency range: $250 \div 500$ Hz
- Source level: > 190 dB//μPa
- Operating depth: > 500 m
- Q-factor : < 4
- Efficiency : $> 50\%$
- Total mass: < 250 kg

have been chosen as guidelines for an international research programme which is presently carried out for the European Community Marine Science and Technology (MAST) Directorate under the LOFTOMO-Contract MAS2-CT91-0010-C.

2. Motor technologies

The most-used 'active materials' for the conversion of electrical into mechanical energy are:

(a) Piezoelectric ceramics.
(b) Electrostrictive materials.
(c) Magnetostrictive alloys.

The materials of groups a) and b) generate relative small primary deflections and big forces. As depicted in Fig. 1a, those of group (a) (e.g., lead titanate zirconate) produce expansions and contractions according to the applied electrical polarity while those of group (b) (see Fig. 1b) always give rise to expansions and require a biasing field (to operate on the linear flanks of their response-curves).

To realize 'motors' with sufficiently large displacements a set of those elements (disks, rings, etc.) has to be assembled as a stack (mechanically switched in series), and driven via suitable transformers. Besides this bias-field and its various technical implications it is also because of the stronger temperature-dependence of group (b) that preference is usually given to piezo-ceramics (a).

The response of the relatively soft or compliant alloys (sound velocity $c \approx 1700$ m/s) made from materials of group (c), to a bias and an oscillating magnetic field, is quite similar to electrostrictive ceramics, group (b), but much bigger under pre-stress conditions as reproduced in Fig. 1c. Probably the best material of group (c) is 'TERFENOL' (ternary alloy of terbium-dysprosium-iron). Due to their low-voltage operation, magnetostrictive motors can be directly driven by solid state amplifiers. These latter motors are the most efficient strain-generators and have the highest power-handling capacity but are not only quite sensitive to biasing and pre-stressing, but are also affected by unavoidable losses (eddy currents). Furthermore, their costs are still high.

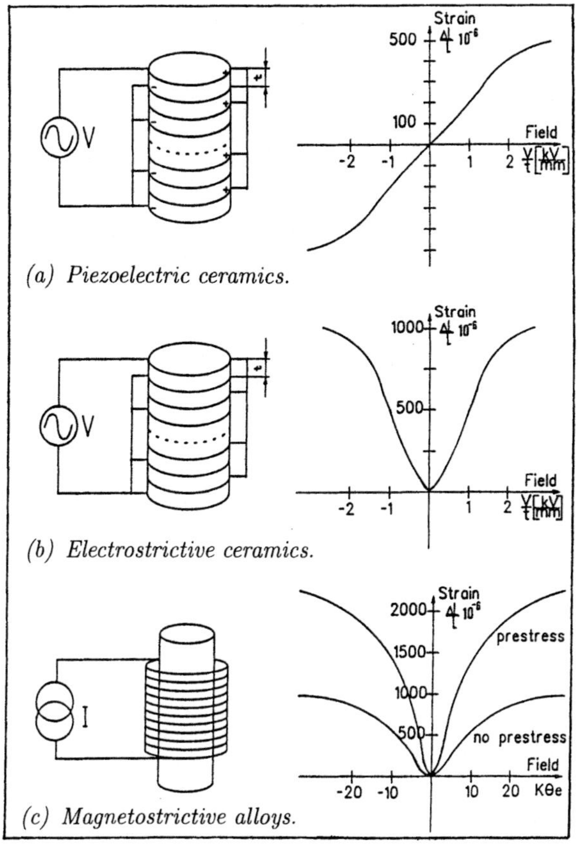

Figure 1 *Active materials for the three motor technologies.*

3. Matching technologies

Suitable structural links have not only to transmit the motor-deflection as efficiently as possible on the active surfaces of the projector, but simultaneously also to act as mechanical transformers to match the motor's impedance to the shape- and size-dependent radiation impedance. By keeping the global mechanical Qm-factor as small as possible and selecting properly the electrical characteristics of the motor, a reasonably large bandwidth of the projector can be realized. Figure 2 exhibits the most important matching-configurations which have been investigated by the five working-groups of the LOFTOMO-project. Based on classical Transmission-Line Models (TLM) and Finite Element codes (FEM) modelling studies have been carried out to predict the electro-acoustic performance and the static and the dynamic behaviour at various operational depths. The most promising projector-types are the following: the Janus-Helmholtz transducer (JH), *developed by ERAMER, Toulon, France*, the Class III Flextensional transducer (FT) and the Capped Ring transducer (CR).

Type JH is a double-sided tonpilz (see Fig. 3a) having a vibrating mass at both ends and the centre of its axis (tail-mass) fixed to the support-structure. This type is usually a liquid-filled cylinder acting simultaneously as a Helmholtz-resonator whose natural resonance-frequency is strongly decreased by inserting air-filled compliant tubes. Under a pressure-load (down to about 500 m without additional compensations) those devices can be operated in a quite broad frequency-band, which is determined at its lower end by the cavity-resonance and at its upper end by the drive's structure. Here we are restricting our interest to some design-features of a very compact tonpilz without a resonant cavity. Subdividing the driving stack of ceramics into a system of two concentric columns joined by two properly designed mechanical bushes (e.g., made from special aluminium alloys) the overall length of a 400 Hz-projector can be reduced to about 50 cm as illustrated by Fig. 3b. The complex admittance and the efficiency of this transducer have been calculated with help of FEM codes. Figure 3c shows the mesh chosen for a static analysis in a particularly critical zone. A magnetostrictive version was investigated as well (with 6 Terfenol bars), but is dismissed as it would require pressure compensation at small operational depths. To realize projectors with larger bandwidth (and higher source-level) a certain number of two-sided tonpilz can be mounted into large and thick baffles or other suitable support-structures (Janus-arrays).

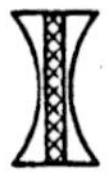	**Class III Flextensional**	Ideal for spar-buoy configuration. Depth capability requires further research. Otherwise meets all requirements.
	**Class IV Flextensional**	Mechanically more difficult than class III to fit in spar-buoy configuration. Difficult to meet frequency, depth, weight requirements. Rejected.
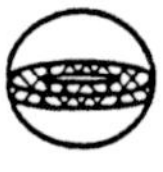	**Class V Flextensional**	Acceptable design developed. Potential problems with depth capability have to be analysed.
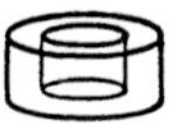	**Flooded Ring**	Resonant frequency depends upon loaded surface area. Thus unacceptably massive. Rejected.
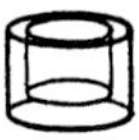	**Helmholtz**	Size, cost conflict at design frequency. Rejected.
	**Driven Pipe**	Standard "Webb" source. High Q-factor. Rejected.
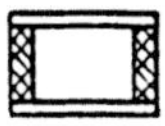	**Capped ring**	Better option than flooded ring. Still massive at design frequency. Depth requirement asking further research.
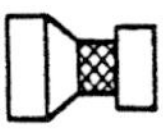	**Tonpilz**	Inappropriate choice in comparison with Janus, for operation at design frequency. Q-factor relatively high. Rejected.
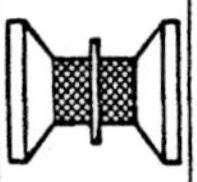	**Janus**	Acceptable design developed. Capable of operating to full depth requirement.

Figure 2 *Matching technologies.*

The FT-projector consists of a concave or convex cylindrical shell having the shape of a 'basque' or a 'barrel' as depicted in Figs. 4a and 4b. Placing the motor on its axis both (relative stiff) lids are slightly displaced relative to each other, which results in a much bigger lateral deflection of the hull, which reaches maximum amplitudes (especially along the meridian) if driven at frequencies near resonance. To achieve resonance-frequencies below 500 Hz with very short cylinders (40-80 cm) the hull needs to be made more compliant by a suitable number of equidistant longitudinal cuts or by assembling it from flat staves without cross-curvature.

As sketched in Fig. 2 the last type, the CR-transducer, consists of a (rigidly) capped short cylinder (diameter and height approx. 50 cm) which is only radially vibrating or pulsating with an almost uniform amplitude-distribution all over its outer mantle (reaching Q_r-values as low as 2.5 at 450 Hz). Its relatively thick cylindrical hull is made from 8 longitudinal sectors or staves (see Fig. 4c) which are rhythmically cleaved asunder and drawn together (push-pull action) by 16 or 32 circumferentially inserted piezoelectric drives (motors). This projector-type, which can be used up to a depth of 500 m without any compensation, was investigated in detail via FEM-modelling. Figure 4d shows a typical FEM-mesh for a sector.

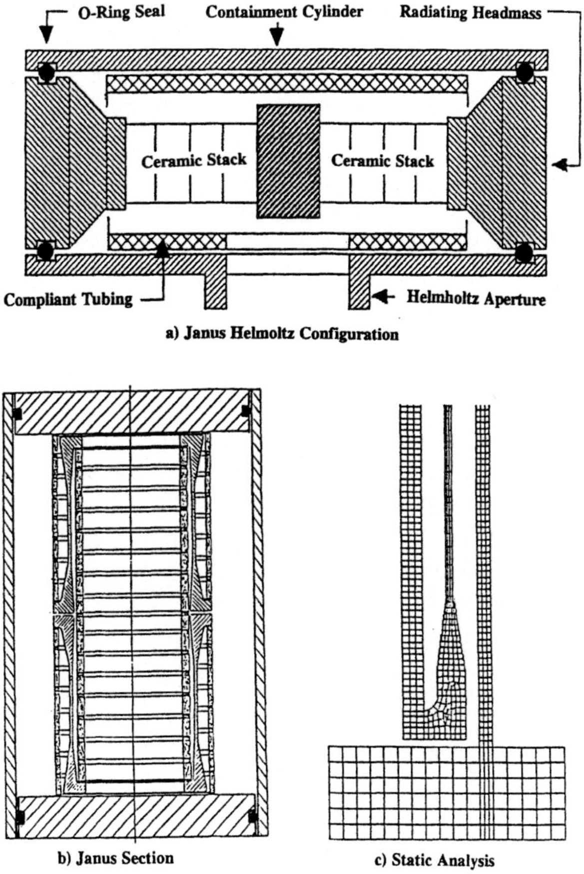

Figure 3 *Janus transducers.*

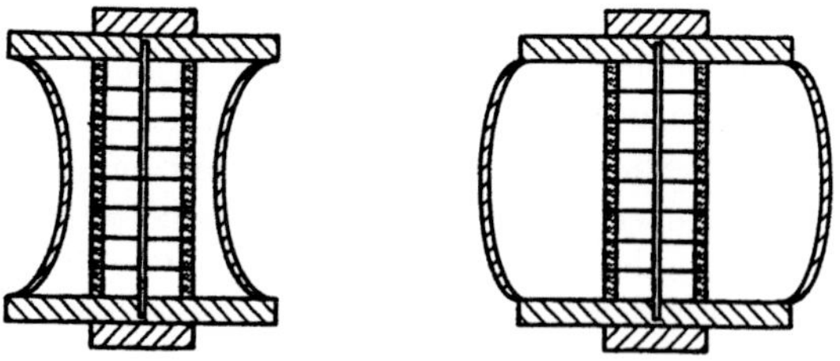

(a) 'Basque' flextensional transducer. *(b) 'Barrel' flextensional transducer.*

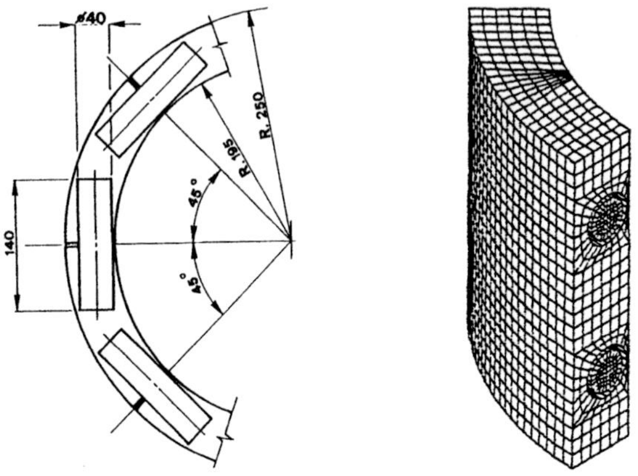

(c) Capped ring section. *(d) FEM mesh for a sector.*

FIGURE 4. Flextensional and capped ring transducers

4. Conclusions

The above described feasibility-study, started two years ago, achieved the following milestones:

(a) Initial definition of the acoustic load for a chosen projector-shape and size.
(b) Choice of the proper mechanical transformer to match this load to the driving unit.
(c) Selection of the most suitable motor-technology.

Our studies have confirmed that the above-described types JH, FT, and CR are the best compromises. The on-going research activities will be terminated by the realization and careful testing of at least one of these three projectors.

5. Acknowledgement

The authors gratefully acknowledge the extremely valuable contributions of T. Francis, J.R. Dunn, J. Ahmed, C. Baylis, M. Peselli and F. Silvano.

References

1. LOFTOMO: Annual Report, First Year, (1993) European Comunity Marine Science and Technology (MAST) Directorate, Brussels (Belgium).
2. LOFTOMO: Progress Reports, Second Year, (1994).

A LOW DIFFRACTION BEAM TECHNIQUE FOR SCANNING IN THE SHALLOW OCEAN

D. Kent LEWIS
Mail Stop L - 495
Lawrence Livermore National Laboratory
Livermore, CA 94550, U.S.A.

ABSTRACT The Localized Wave pulse project has created the concept of highly localized, pulsed wave energy. We have confirmed, and developed the technique in a series of experimental and numerical tests. This technique generates radiation, mechanical or electromagnetic, from an array. LW generation sends a broad frequency bandwidth pulse of energy along the array axis. The pulse has no sidelobes and maintains the frequency bandwidth and pulse amplitude better than any other method proposed to date. Confirmation of the theory by acoustic experiments show that the effect is real, and that the advantages of the technique can be realized in a laboratory environment. Numerical studies have shown that this effect is extendible to the microwave (vector field) and elastodynamic (tensor field) regimes as well as the stratified ocean environment.

The main features of the LW beam are 1) the pulses are broad bandwidth and maintain that bandwidth over their entire working range; 2) the beam generates no side lobes; 3) the LW effect is accomplished in the linear regime, so that small errors in beam generation result in a less effective beam not the loss of the effect; 4) the beam can be generated using present technology in the acoustic realm.

The LW beam has a wide range of applications. As an active source, it offers a unique method of variable beam forming from spatially wide to the narrowest possible beam. Its uses include oceanographic sound velocity profile estimation and attenuation measurements. Since the pulse characteristics are changeable through the source signal parameters, it also offers the possibility of adaptive beam forming in real time.

Passively, the LW array can use a wide spatial width listening pattern until something of interest occurs. Instantaneously, the listening pattern could be changed to a very narrow, steerable beam with a broad bandwidth. These uses also apply in the electromagnetic realm.

Specific ocean applications include imaging in the shallow ocean, characterization of the ocean bottom and inclusions embedded in it, secure communications and broad bandwidth imaging over long ranges. The lack of side lobes and the wide bandwidth lend the beam to characterization of scattering targets as SONAR cannot do.

1. Introduction

The Localized Wave (LW) is a pulsed, broad bandwidth beam which maintains its beam width and bandwidth better than any other beams currently available. It is generated without side lobes and depends on individually addressable array technology to put the correct energy at the correct place at the correct time. It uses simple solutions to the wave equation which are superposed to create the desired pulse characteristics. The LW concept is applicable to any linear wave regime, acoustic, electromagnetic or elastodynamic.

The main features of the LW beam are 1) the pulses are broad bandwidth and maintain that bandwidth over their entire range; 2) the beam generates no side lobes; 3) the LW effect is accomplished in the linear regime, so that small errors in beam generation result in a less effective beam not the loss of the effect; 4) the beam can be generated using present technology in the acoustic realm.

O. Diaschok et al. (eds.), Full Field Inversion Methods in Ocean and Seismo-Acoustics, 229–234.

The beam is the result of converging interests. Jim Brittingham at LLNL searched for years to find a transverse solution to Maxwell's equations. When he found it, he termed it the Focus Wave Mode (FWM)(1). The FWM is a perfect waveform since it travels forever without changing its shape or losing its energy. It is also physically impossible in exactly the same way that infinite plane waves are, since it requires an infinite amount of energy and an infinitely large aperture to create.

Rick Ziolkowski, also at LLNL, reasoned that while it is true that both the FWM and plane waves were impossible to create, we use plane waves to simulate wave fields constantly, so if he could find a method of combining the FWM family of basis functions into a physically realizable pulse, he could create a highly localized beam. He searched for a spectrum which would let him do that, and he finally found one that works. He called this spectrum the Modified Pulse Spectrum (MPS)(2).

The resulting source signals are rather simple as illustrated in figure 1. These source functions are for a 1 cm diameter circular array consisting of a central spot and 5 surrounding rings. The effective frequency of these signals was such that the effective wavelength was approximately equal to the radius of the array.

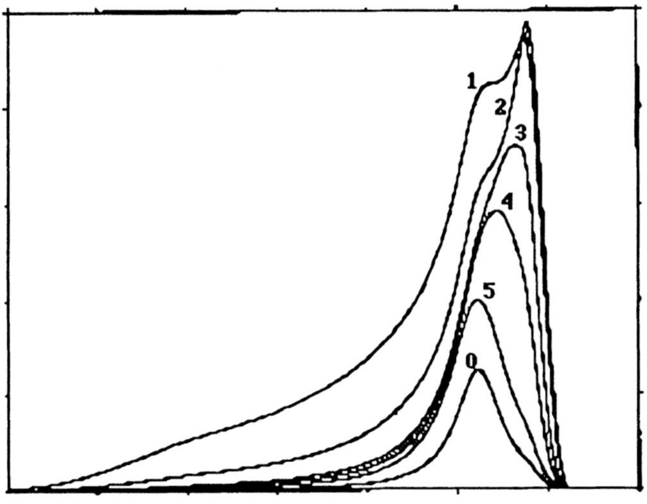

Figure 1. LWT source signals for a 25 element square array. Relative linear amplitudes for central element (0), nearest neighbors (1) etc. Time axis is 7 microseconds.

At the same time I was searching for a broad bandwidth acoustic beam to improve the imaging capabilities of a scanning system. Hearing of the LW beam from a consultant, Bill Cook of the University of Houston, I talked with Ziolkowski about trying to create the pulses in the acoustic domain. This meant rescaling the frequencies into the MHz range and launching the pulse in water. This was possible since we had a programmable digital to analog converter that worked in this frequency range. This began a period of tightly coupled simulations and experiments which were to confirm the feasibility of the LW pulse (3, 4, 5, 6).

2. Signal Preprocessing

The pulse was subjected to both numerical and experimental testing. Numerically, we examined the predictions of simulations in both time and frequency based models. In the time domain we used both point to point phasing and ray tracing techniques. Frequency simulations launched the different frequencies of the source array and recombined them to create the pulse. All of these methods produced the same results, and these results agreed with the theoretical predictions.

One aspect of the LW project was the use of array folding and time gating. Array folding assumes that the aperture is larger than the actual aperture. By using a single folding point on the axis as a reference, each signal from the virtual part of the array is mapped into a part of the existing array. Using this technique, there is only one point, the folding point, at which the actual field is that which would have been created by the larger array. But near this point the field is approximately the same.

A measure of the accuracy of the field can be represented by what we can imagine as a similarity parameter. Obviously at the array surface, the field created is infinitely different than that of the virtual array. Similarly as the field approaches infinity the field will be different since the signals launched are different than the virtual signals. But over the range of the beam, the agreement is sufficiently good to increase the localization of the pulse.

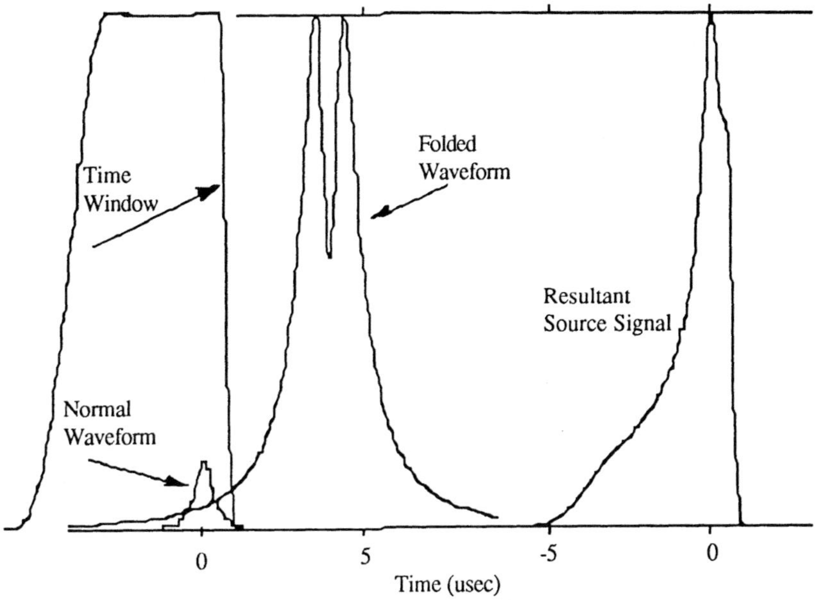

Figure 2. Formation of the folded signal for an array element. Normal Waveform is the signal prescribed by theory fo rhte element, Folded Waveform is the signal from a virtual element time shifted to arrive at the folding point at the proper time and magnified by the ratio of virtual element area to actual area. The signals are summed and windowed as shown to arrive at the Resultant Waveform.

Time windowing is a treatment of the combined real element and virtual element signals which increases the efficiency of the beam. When we used the folding technique, we enhanced the localization of the pulse dramatically, but we noticed that there was now more energy in the off axis space of the pulse. This energy was from that part of the folded signals which were not needed for the selected range of the particular designed beam. In order to increase the efficiency without decreasing the localization, we experimented with different time windowing schemes.

As shown in figure 2, the small, original signal and the larger, folded signal, are modified by another signal which throws away most of the folded signal. The resulting signal, different from both the original signals, is shown also with a normalized amplitude. These signals, like those of the first illustration, are still very simple signals, and very easy to generate.

3. Experimental Tests

The experimental methods included acousto-optical field measurements, two transducer simulation measurements, and sound field measurements from a 5% active area array and an 80% active area array. The first two experiments were simulated array measurements in that the source signals were generated separately and combined in a computer. The last two experiments were of active arrays actually launching the LW pulse.

During the development of this beam, we have developed the technologies of individually addressable acoustic arrays and of source signal pre-processing. Our multi-channel electronics are applicable to existing acoustic array equipment and lend them selves to massively parallel computer interfaces.

The pre-processing technique uses sampling of the source signals and inverse filter processing to tailor the source signals for the optimum element response. This allows us to launch as many individual signals as necessary to our array elements and to overcome element to element feedback interactions. Preliminary discussions with vendors encourages us to believe that present ocean arrays may be of sufficient bandwidth to be used in further development.

The experimental tests of the LW beam have involved several different techniques. In the first set of tests, to determine if the source signals could be accurately launched, an acousto-optic technique was used. In this technique a laser beam is used to probe the sound field so that a measurement can be made without changing the field. Predicted and measured fields were extraordinarily similar.

Since the acousto-optic technique is actually measurement of the change to a light beam over a distance caused by the sound field, the model of this experiment was of an array of line sources radiating to a point receiver. The confirmation of the theory in these tests, 1 dimensional array, led us to try the next logical step of a circular array.

The simulated 2 dimensional array results were also positive, so preparations were made to design and test a real array. The first array tested, with all the source signals launched at the same time, was a 25 element array consisting of 1.2 mm current feeds at 2.5 mm center to center distances. The elements were fed signals to simulate a central spot and 5 surrounding rings. The active piezoelectric was a PVDF sheet with a ground plane on the outer face. The active area was only 5% of the total array size.

In this series of tests, the beam achieved better than 10 times the near field distance compared to a conventional beam of the same effective frequency. We then constructed an array of a central spot and 10 surrounding rings. This array had an active area which was 80% of the total array surface. This beam did not perform as well as the previous one, and achieved only a factor of slightly greater than 3 times the near field expected. We believe that much of the energy went into deforming the PVDF co-polymer between the very close rings.

The results of the latest test were clearly better than expected as seen in the final figure. Figure 3 shows the fields from the 2 cm array at a distance of 100 cm downrange for the LW pulse at an effective frequency of 205 kHz, a 250 kHz, 500 kHz, 1 MHz and 2 Mhz tone bursts. It should be noted that the LW beam achieved nearly the localization of the 2 MHz beam without any side lobes, with a broad bandwidth, and with smooth behavior close to the array in contrast to the complicated behavior of a single frequency toneburst.

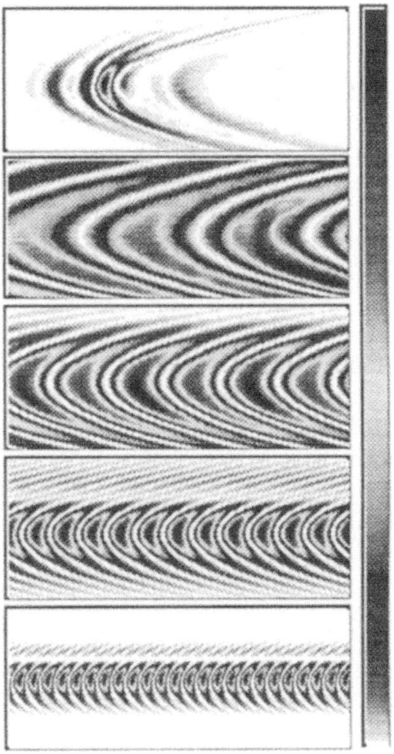

Figure 3. Linear intensity fields of LW pulse (top), 250 kHz, 500 kHz, 1 MHz and 2 MHz tonebursts 100 cm downrange from a 2 cm diameter array. The field widths are 6 cm and the effective frequency of the LW pulse is 205 kHz. Intensity color map is at right.

4. Applications

The LW beam is immediately applicable for shallow water conditions. It is useful for identifying and locating floating, submerged and buried mines, scanning for and identification of submarines and characterization of ocean parameters such as the SVP, attenuation caused by reverberation, sediment and ocean floor wave speeds. Since it is a narrow pulse with no side lobes, it is possible to avoid many of the interactions that plague SONAR arrays.

In Biot solids, such as saturated sand or sediment, the beam offers a tight profile which overcomes the limitations of single frequency beams. A relatively high frequency beam suffers scattering loss since the particles are large compared to the wavelength. A tight focus is maintained, but insufficient power returns for effective scanning. A low frequency beam overcomes the scattering limits but loses power through beam spreading since the wavelength is large relative to the array size. The LW beam is a tight beam pattern caused by superposition of low frequency waves, so if the beam behaves as one would expect, it can return more power in a tighter beam.

For passive detection using an array, the techniques of the LW pulse, specific frequency filtering for each element, can be used for improved beam forming. The same lack of side lobes and higher localization over the effective range can be brought to passive listening techniques. Combined with the beam generator, there is the possibility of better than inverse fourth power scanning.

The LW beam is also useful for nondestructive evaluation applications such as inspection of composite structures. Carbon composites or metal matrix composites offer the same problems as a Biot solid and the high frequency behavior of the low frequency LW pulse offers better penetration, broader bandwidth and higher return field for these difficult to inspect materials. The passive technique is also possible with present day technology.

5. References

1. "J. B. Brittingham, J. Appl. Phys. **54**, 1179 (1983)
2. "R.W.Ziolkowski, J. Math. Phys. **26**, 861 (1985)
3. "R.W.Ziolkowski, D. K. Lewis and B.D.Cook, Phys. Rev. Let. **62**, 147 (1989)
4. "J. V. Candy, D. K. Lewis and R. W. Ziolkowski, J. Acous. Soc. of Am. **88**, 2235 (1990)
5. "J. V. Candy, D. K. Lewis and R. W. Ziolkowski, J. Acous. Soc. of Am. **88**, 2248 (1990)
6. R. W. Ziolkowski and D. K. Lewis, J. Appl. Phys. **68**, 6083 (1990)

6. Acknowledgement

This work was performed in part by the Lawrence Livermore National Laboratory fo rthe U. S. Department of Energy under Contract No. W-7405-ENG-48.

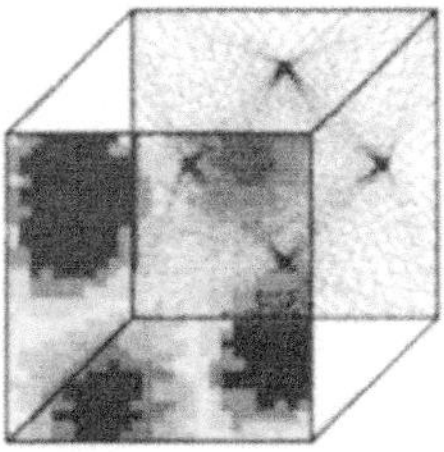

section 4

Ocean Parameter Estimation

FORWARD MODELING OF THE BARENTS SEA TOMOGRAPHY VERTICAL LINE ARRAY DATA AND INVERSION HIGHLIGHTS

CHING-SANG CHIU and JAMES H. MILLER
Naval Postgraduate School
Monterey, CA 93943

WARREN W. DENNER
EOS Research Associates
Monterey, CA 93940

and

JAMES F. LYNCH
Woods Hole Oceanographic Institution
Woods Hole, MA 02543

ABSTRACT A coupled normal-mode approach for the prediction of pulse propagation in a complex coastal environment was used to model the modal arrival structure in the 1992 Barents Sea Coastal Acoustic Tomography Test. Model results as well as the comparison to the data measured by a vertical hydrophone array are discussed. The predicted arrival structure was used for beamforming, identification and tracking of normal modes. The modal travel time series has been inverted in conjunction with the beamformed ray travel time data to produce high-resolution sound speed maps of the Barents Sea Polar Front at five-minute intervals over a tidal period. The inversion results are highlighted.

1. Introduction

The Chiu-Miller-Lynch Coastal Acoustic Tomography (CAT) scheme [1, 4] involves the use of real-time telemetered vertical line hydrophone arrays, space-time signal processing techniques and a hybrid ray/mode inverse method. In shallow water, the low modes/small-angle rays often arrive at the receiver from a distant tomography sound source first. These early arrivals are often loud but are closely spaced in time. On the other hand, the higher modes/higher-angle rays, which arrive later, are well separated in time but with much weaker signals. Vertical line arrays and space-time beamforming, capable of separating the early overlapping arrivals and enhancing the signal of the later arrivals, are therefore required to resolve and identify individual rays and modes. In CAT, the inversion of the travel time perturbations of the beamformed individual arrivals is accomplished using a stochastic method that minimizes the mean square errors of the sound speed maps. By augmenting multiple rays and modes, the tomographic maps can attain a high spatial resolution. A high temporal resolution is achievable through frequent signal transmissions.

Accurate forward propagation modeling of ray and modal arrival structures in a complex shallow-water environment with spatially varying sound speeds, bottom bathymetry and sediment properties is central to the CAT scheme. The forward model runs with a "background" ocean field provide the key acoustic quantities, such as raypath geometries, mode functions, mode coupling coefficients, travel times, etc., for the subsequent signal pro-

O. Diaschok et al. (eds.), Full Field Inversion Methods in Ocean and Seismo-Acoustics, 237–242.

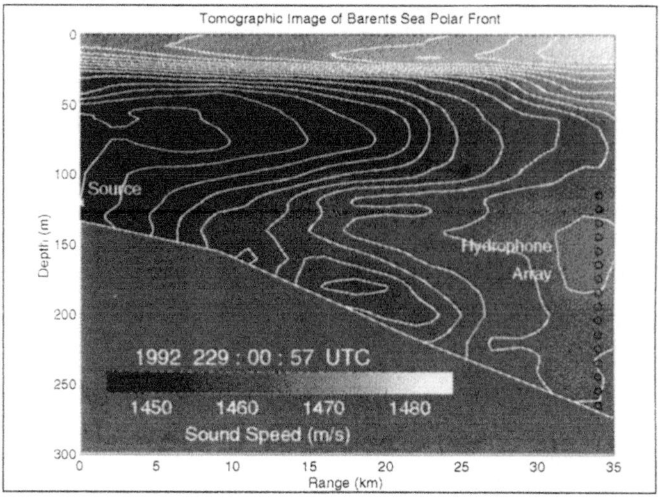

FIGURE 1. A tomography snapshot of the Barents Sea Polar Front.

cessing and inversion of the data for the "perturbed" ocean. The hybrid ray/mode inverse method, the beamforming techniques, and the ray arrival structure modeling approach for CAT have been presented in previous papers by Chiu et al. [1], Miller et al. [4] and Mykyta [5], respectively. The thrust of the present paper is on the forward modeling of modal arrival structure for CAT. Specifically, the quality of the broadband, coupled normal-mode approach presented by the authors in a companion paper [3] is tested against the vertical line array data from the 1992 Barents Sea Coastal Acoustic Tomography Test.

2. Forward Modeling Results and Comparison with Measurements

A MATLAB computer code has been generated based on our coupled-mode approach outlined in a companion paper [3]. This code was used to model the propagation of normal modes through the Barents Sea Polar Front in the 1992 Barents Sea Test. Figure 1 shows the complexity of the sound speed structure (at time 229:00:57 UTC) mapped tomographically. It also shows where the tomography sound source and the vertical hydrophone array were deployed relative to the location of the front and the bottom bathymetry. The bathymetry was charted by a ship-mounted Acoustic Doppler Current Profiler. Based on geoacoustics information from the US Naval Oceanographic Office (NAVOCEANO), the seabed of the experimental area consists of a very thin sediment layer (at most a few meters) overlying a rigid basement which has a sound speed of about 3200 m/s and density of about 2600 kg/m^3.

Neglecting the very thin sediment layer and using the hydrographically mapped water-column sound speeds, acoustically charted bathymetry and bottom information from NAVOCEANO, an environmental model was build for input to the acoustic model. The sound speed field mapped by hydrographic stations in the experiment is similar (except for a coarser horizontal resolution and is temporally aliased) to the tomographic snapshot shown in Figure 1. Furthermore, modal attenuation coefficients were empirically estimated based on the measured omni-directional sound pressure time series and the predicted travel times

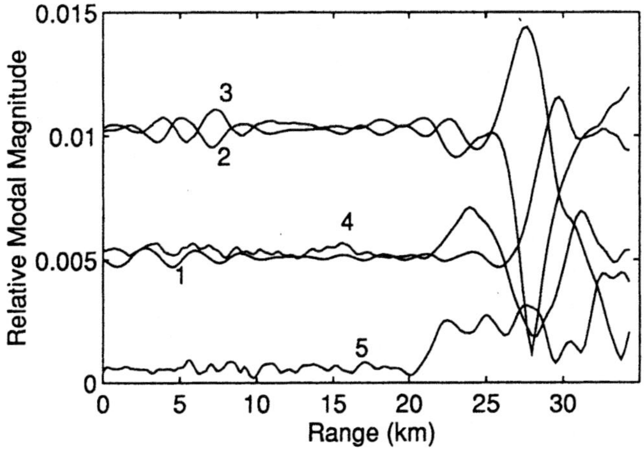

FIGURE 2. Modal magnitude of the first five modes versus range at a frequency of 224 Hz.

of the modes. Since the transmission was almost directly cross front and downslope, little horizontal refraction and transverse mode coupling could occur.

The sound source was located at a depth of 122 m transmitting phase-encoded signal with a carrier frequency of 224 Hz and a bandwidth of approximately 32 Hz. The autocorrelation of the transmitted signal corresponds to a 62.5-ms pulse. Since our interest was to compare the forward modeling results to the matched filter output, it was appropriate to model the source signal as a 62.5-ms Blackman window modulating a 224-Hz sinusoid. The vertical array, 35 km away from the source, consisted of 16 hydrophones (Channel 0 to 15) with a 10-m even spacing. The top phone (Channel 0) was located at a depth of 124 m.

A total of 30 modes were used in the propagation model. This number corresponds to the maximum number of modes trapped by the water column at the source location. The magnitudes of the first five modes versus range at the center frequency are shown in Figure 2. Energy transfer between modes is evident. For example, mode 2 and mode 3 are seen to be tightly coupled. As the energy in one mode increases, the energy in the other mode decreases. At the initial ranges away from the frontal zone, the coupling of these low modes were mild and were caused mainly by the bathymetric variations. However, the coupling for these low modes became extremely significant in the frontal zone where the horizontal sound speed gradients were large. On the contrary, the higher modes were affected more by the bathymetry and less by the oceanic front.

In Figure 3, we display both the modeled and measured omni-directional arrival structures at one of the channels (Channel 2), showing remarkable agreement between the two. Similar quality of agreement between model and measurements is found for all 16 channels. Both the model results and measurements consistently show that the received signal at a single phone was dominated by a big initial arrival of energy followed by many smaller arrivals. As we shall see in the modal beamformed results that the dominant initial arrival was constructed by the low modes that arrived closely in time whereas the smaller arrivals were nonoverlapping individual modal arrivals.

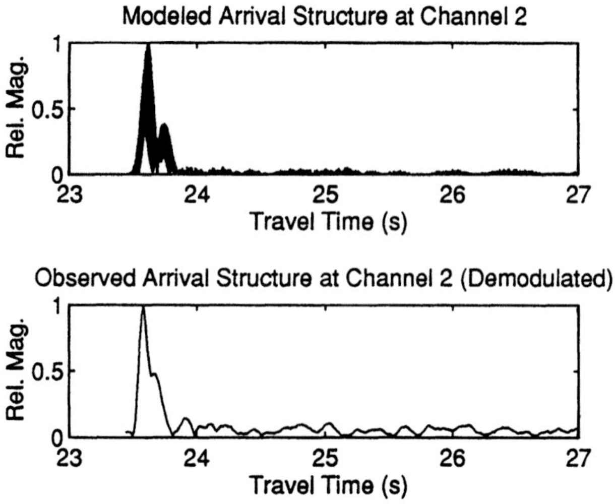

FIGURE 3. Modeled (upper panel) and measured (lower panel) omni-directional arrival structures at Channel 2.

Figure 4 is a gray-scale plot of the modeled and measured omni-directional arrival structures at each of the 16 channels. For each channel in the display, 28 consecutive sequences of the pulse responses of the sound channel are stacked together. It is interesting to point out that both the predicted and measured vertical distribution of energy clearly show the existence of three peaks and two nulls with the strongest peak near the bottom. This indicates that the vertical structure of the acoustic wavefield near the receiver was dominated by mode 1 and mode 3. This dominance is also apparent in the predicted range variations/coupling of the modal magnitudes (Figure 2).

Both the predicted and measured modal beamformed arrival structures are displayed in Figure 5. In the figure, modal magnitude is plotted against mode number and travel time in a linear gray-scale map with light shade corresponds to high values and dark shade corresponds to low values. At each mode, the beamformed data is seen through a window of 250 ms centered at the expected arrival time calculated by the model. Moreover, the data in each window were normalized so that the higher modes, which have significantly less energy, can be visualized. Both the model and measured data show that the first 10 modes arrive between 23.6 s and 23.8 s and then followed by the higher modes which were more separated in time. Clearly, the first 10 modes could not be resolved using a single phone indicating the importance of spatial processing.

3. Highlights of Inversion Results

Although forward propagation modeling is the focus of this paper, we would like to briefly mention the 3-D (range, depth and time) inversion results of the 1992 Barents Sea Coastal Tomography Test [2]. A time series of cross-frontal images of sound speed was produced by applying the hybrid ray/mode inverse technique [1] to the observed travel time perturbations of rays and modes. These travel time perturbation data were derived from the differences between the beamformed vertical array data and the propagation model predic-

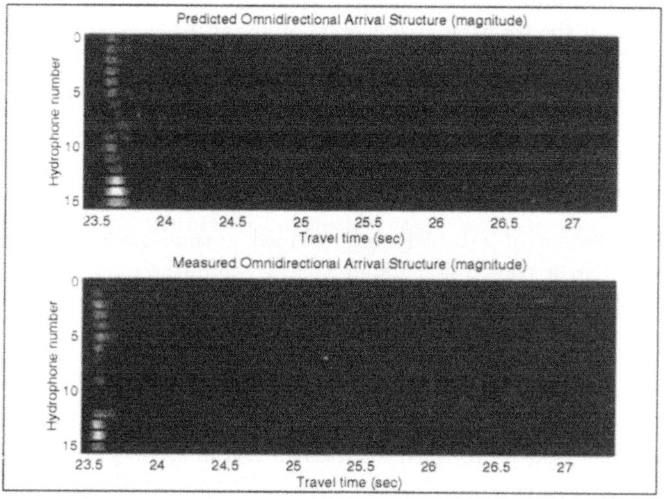

FIGURE 4. Modeled (upper panel) and measured (lower panel) omni-directional arrival structures at all 16 phones. Light shade corresponds to high values and dark shade corresponds to low values.

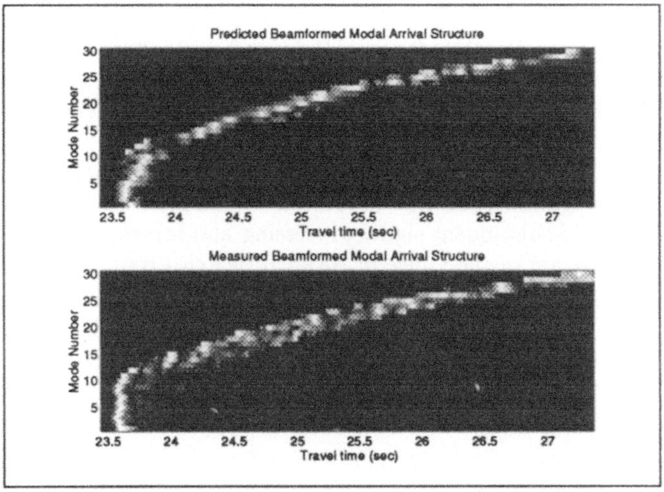

FIGURE 5. Modeled (upper panel) and measured (lower panel) modal beamformed arrival structures. Light shade corresponds to high values and dark shade corresponds to low values.

tions in the background ocean. The background modal arrival structure was modeled using our 3-D coupled mode model and the results are presented in the previous section. The background ray arrival structure was simulated using a 3-D raytracing method and the ray results have been presented by Mykyta [5].

The tomographically derived sound speed distributions, covering the ocean from the surface to the bottom over a range of 35 km, were produced every 5 min over a tidal period. The distributions show thermal structures consistent with vertical turbulent mixing in the lower portion of the front on a time scale of about 2 hours, wave-like motions in the thermocline, and frontal oscillations consistent with local barotropic and baroclinic tidal forcing. One of the tomography snapshots is shown in Figure 1. Strong interleaving in the intermediate front is observed.

4. Concluding Remarks

Based on our formulation of a broadband, coupled normal-mode propagation modeling approach (presented in a companion paper [3]), a computer model has been written in MATLAB. This computer model is thus machine independent and easy to use. The model has been used to predict the modal arrival structure in the 1992 Barents Sea Coastal Acoustic Tomography Test. Model predictions were compared to the data measured by a vertical hydrophone array showing remarkable agreement. Subsequently, the predicted arrival structure was used for beamforming, identification and tracking of normal modes. The modal travel time series has been inverted in conjunction with the beamformed ray travel time data to produce high-resolution sound speed maps of the Barents Sea Polar Front at five-minute intervals over a tidal period. These acoustic images show that a coastal ocean can have significant variability in short time and space scales. Hence, there is a need for a tomographic monitoring system to adequately measure those scales.

Acknowledgments

We would like to acknowledge Bob Bourke, Robin Muench, Keith von der Heydt, John Kemp, Steve Liberatore, John Bouthillette, Arthur Newhall, Marla Stone, Rich Pawlowicz and Nick Witzell who participated in the 1992 Barents Sea Coastal Tomography Test. Special thanks go to Chris Miller, Phil McLaughlin, Rost Parsons, John Mykyta and Stefan Hudson for their assistance in data processing, modeling and graphics. The development of the MATLAB propagation code was sponsored by EOS Research Associates. The experimental work and the subsequent signal processing and inverse analysis are support by the High Latitude Program of the Office of Naval Research (Tom Curtin) under Research Project RR032-05-01. The first author is grateful to the inversion Conference Organizing Committee (particularly to the Chairman, Orest Diachok) for the invitation to present this paper.

References

1. Chiu, C-S., J.H. Miller, and J.F. Lynch, "Inverse Techniques for Coastal Acoustic Tomography," in *Environmental Acoustics*, (D. Lee and M. Schultz editors), World Scientific, 1994 (in press).
2. Chiu, C-S., J.H. Miller, R.H. Bourke, J.F. Lynch, and R.D. Muench, "Acoustically-Derived Images of the Barents Sea Polar Front," *EOS Transactions, AGU*, Vol. 75, NO. 3, supplement, p118, 1994.
3. Chiu, C-S., J.H. Miller, W.W. Denner, and J.F. Lynch, "A Three-Dimensional, Broadband, Coupled Normal-Mode Sound Propagation Modeling Approach," same issue.
4. Miller, J.H., C-S. Chiu, and J.F. Lynch, "Signal Processing for Coastal Acoustic Tomography," in *Environmental Acoustics*, (D. Lee and M. Schultz editors), World Scientific, 1994 (in press).
5. Mykyta, John L., "Prediction of planewave beamformed arrival structure for the acoustic transmission in the 1992 Barents Sea Polar Front Experiment ," Naval Postgraduate School Master's Thesis, September 1993.

LONG VERTICAL ARRAYS IN ACOUSTICAL SOUNDING OF THE OCEAN

A. VORONOVICH
NOAA/ERL/ Environmental Technology Laboratory
325 Broadway, Boulder, CO 80303

ABSTRACT The use of mode-resolving long vertical arrays has potential for solving inverse problems in ocean acoustics. In particular, new possibilities are offered by coherent processing of signals received by a system of such arrays. Theoretical aspects of applications of vertical arrays are discussed. A new scheme of acoustical tomography of the ocean, based on measuring horizontal refraction of modes by two acoustic interferometers, is presented.

1. Introduction

Sound is the only known usable type of radiation able to propagate through the ocean over large distances. For this reason ocean acoustics is nearly monopolist in remote sensing of the ocean inner structure. There are three grounds which make large scale acoustic monitoring of the ocean feasible. The *first* is the existence of the SOFAR sound channel which traps the sound in the interior of the ocean; the *second* is the very low attenuation of sound waves at frequencies below a few hundred Hz; the *third* is the high sensitivity of the sound field to ocean inhomogeneities.

The presence of the ocean wave guide leads to the fact that the sound field, far from its source, can be represented as a finite sum of propagating normal modes. The vertical structure of the sound field at any given point can be completely described by N complex numbers—the mode amplitudes A_m for a range independent environment. An alternative description is the ray approximation, where the field is a sum of contributions from rays. The ray approach is approximate, and the accuracy degrades with decreasing frequency. Also, the number of rays propagating to a point increases with the distance between the source and the receiver. Thus the ray approach is sufficient for small distances and high frequencies. For low frequencies and large distances the modal description of the field is more convenient.

To determine A_m, it is enough in principal to measure the field at N different points. Two main possibilities exist: to use horizontal arrays with hydrophones on the sea floor or to use vertical arrays. Horizontal arrays are widely applied but for modal decomposition they have some drawbacks: 1) the sound field is affected by the bottom; 2) because sound in the ocean propagates mainly in horizontal direction, much longer horizontal arrays are required to receive the same information provided by shorter vertical arrays; 3) to receive the sound propagating in the wave guide, horizontal arrays should be deployed at the depth of the sound channel.

O. Diaschok et al. (eds.), Full Field Inversion Methods in Ocean and Seismo-Acoustics, 243–248.

2. Vertical arrays

Vertical arrays are free from the above disadvantages. Suppose that amplitudes and phases of the sound field are measured at N points in the ocean. It is clear that for reliable modal analysis, the array should span a significant part of the ocean wave guide. This means practically that the length of the array should be of the order of kilometer(s). We shall refer to such arrays as *long vertical arrays* (LVA). Normally the LVA cannot be deployed strictly vertically in the ocean, and the following linear set to determine mode amplitudes arises

$$p_\alpha = \sum_{m=1}^{N} A_m \exp\left[i\xi_m(\omega)r_\alpha\right] \cdot u_m(z_\alpha, \omega) \ .$$

Here r_α is the horizontal coordinate of the αth hydrophone in the plane of sound propagation, $\xi_m(\omega)$ is the wave number of the mth mode, and u_m is the mode profile. Thus to determine entries of the corresponding matrix, r_α should be measured to within a fraction of a wavelength. Practically this accuracy should be of the order of 1 m. Taking into account the total length of the array, achievement of this accuracy offers some technical difficulty. This problem can be dealt with using a special subsidiary acoustic system which makes use of short high-frequency pulses to measure travel times between HF transponders and the hydrophones of the array.

A few publications concerning experiments on modal decomposition of the sound field in the deep ocean are available. Probably one of the first measurements was made in the Arctic in 1982 [1]. Another example was the 1990 experiment in the Norwegian sea where a vertical array was deployed from a drifting vessel [2]. Modal decomposition was controlled in this experiment by 'emitting' the sound numerically from the array in the backward direction. The inverted field focused at a point situated not far from the real position of the source. Thus modal decomposition can be performed practically for situations of interest.

Consider the possibilities which such modal analysis offers. Let the source and the array be situated in a layered waveguide which includes some inhomogeneities between them. The source, deployed at depth z_s, produces a field at the array consisting of modes with known amplitudes. As a result of interactions with inhomogeneities, the spectrum of modes changes and the modified spectrum is measured at the array. Setting the source at N different depths, and each time measuring the spectrum of modes, one can calculate S-matrix. The entry S_{mk} of this matrix represents the scattering amplitude of the kth mode into the mth one. For homogeneous wave guides the S-matrix degenerates into the unit matrix. When the S-matrix is known, the resulting field beyond the inhomogeneities can be calculated for any source position:

$$\Psi(r, z; r_s, z_s) = \frac{e^{i\pi/4}}{\sqrt{8\pi}} \frac{1}{\sqrt{r - r_s}} \sum_{m,k}^{N} \frac{1}{\xi_m^{1/2}} e^{i\xi_m \cdot r} u_m(z) \cdot S_{mk} \cdot u_k(z_s) e^{-i\xi_k \cdot r_s} \ .$$

3. Inverse problems

The inverse problem has been well studied in quantum mechanics, but it has not been extensively studied for wave guides. For the 1D Schrödinger equation, the potential can be reconstructed if the reflection coefficient is known as a function of frequency. For this the Gelfand-Levitan-Marchenko (GLM) linear integral equation should be solved. The physical meaning of the inverse problem is as follows. Assume the potential is zero for $x < 0$. Scattering data allows reconstruction of the reverberation signal at $x = 0$ from an incident δ-impulse: $\Psi^{(in)} = \delta(x - t)$. The scattered signal at time t depends on the potential in the interval $[0, t/2]$ as the wave travels both forwards and backwards. The scattered field at time $t + dt$ depends on the potential at the adjacent layer $[t/2, t/2 + dt/2]$ along with multiple scattering within $[0, t/2]$. If the potential in $[0, t/2]$ has been reconstructed the multiple scattering in $[0, t/2]$ can be calculated and the reflections from the layer $[t/2, t/2 + dt/2]$ separated. Proceeding this way one can retrieve the whole potential. This procedure is the classical *layer-peeling algorithm.* It is nonlinear, but can be reduced to the linear GLM equation. The reflection coefficient is significant for reconstruction of the media in those cases; the transmission coefficient allows us to retrieve only the average potential.

The layer-peeling procedure can be applied to wave guides as well. Because horizontal scales of the inhomogeneities significantly exceed available acoustic wavelengths, it is clear that backscattering, due to water mass inhomogeneities, is too weak to be detected. However, the forward scattering amplitudes contain much more information. For weak inhomogeneities we obtain:

$$S_{mk} = \delta_{mk} + \frac{1}{2i(\xi_m \xi_k)^{1/2}} \int u_m(z) \Delta \left(\frac{\omega^2}{c^2(r,z)} \right) u_k(z) dz \cdot e^{-i(\xi_m - \xi_k) r} dr$$

If the set of wave number differences $(\xi_m - \xi_k)$ spans a large range the inhomogeneities $\Delta\left(\omega^2/c^2(r,z)\right)$ can be reconstructed. Unfortunately, the Born approximation is often inadequate in our case. In any case, one can conclude that the multi-dimensional case differs significantly from 1D one. In particular, for multi mode wave guides the layer-peeling procedure can be generalized to reconstruct inhomogeneities using forward scattering only.

Practical formulation of the inverse problem should take into account that the S-matrix can rarely be measured completely and is only available at discrete frequencies. On the other hand, the structure of the inhomogeneities usually is not completely arbitrary and can be described by a small number of parameters x_s (e.g., related to empirical orthogonal functions). The inverse problem is a solution of the following equations with respect to x_s:

$$\begin{aligned} F_{ij,k}(x_1, x_2, \ldots, x_s) &= |S_{ij}(\omega_k)| \\ \varphi_{ij,k}(x_1, x_2, \ldots, x_s) + 2\pi N_{ij,k} &= \arg S_{ij}(\omega_k) \end{aligned}$$

Here F and φ are respectively the amplitudes and phases of the S-matrix, calculated by a model, and the RHS represents measured values. Using absolute values of phases results in the unknown integers $N_{ij,k}$. To eliminate them one can subtract adjacent phase equations corresponding to adjacent frequencies. In this case considering the phases of diagonal entries only leads to the modal type of Munk-Wunsch acoustic tomography:

$$t_{ii}(\omega) = \frac{d\varphi_{ii}(\vec{x})}{d\omega} = \frac{1}{2\pi} \cdot \frac{\Delta\varphi_{ii}(\vec{x})}{\Delta f}$$

If, for instance, $\Delta f = 3\,Hz$ and the accuracy of the modal phase is $10°$ then the travel time accuracy is $\delta t \sim 10°/360° \cdot 1/3 \sim 10^{-2}\,s$ This is sufficient for reasonable reconstruction of the medium.

Unfortunately, the functions F, and φ (or $\nabla\varphi$) are nonlinear in many situations. A general analytical procedure, related to the invariant embedding approach, is suggested for solution of this problem. Let us introduce the dimensionless parameter γ:

$$F_{ij,k}(\vec{x}(\gamma)) = \left|S_{ij}^{(\gamma)}(\omega_k)\right|$$
$$\tfrac{d}{d\omega}\varphi_{ij,k}(\vec{x}(\gamma)) = \tfrac{d}{d\omega} \arg S_{ij}^{(\gamma)}(\omega_k, \gamma)$$

The point $\gamma = 0$ corresponds to a reference layered waveguide. For this case $\widehat{S}^{(\gamma=0)} = 1$. The experimentally determined matrix differs from the unit matrix; it corresponds to the point $\gamma = 1$. Let us connect the points $\widehat{S}^{(\gamma=0)}$ and $\widehat{S}^{(\gamma=1)}$ in the multidimensional matrix space with some continuous segment corresponding to values $0 \leq \gamma \leq 1$. Now we can solve our equations making small steps along this curve and considering at each step a linear problem for small deviations of the media related to small deviations in scattering amplitudes:

$$\frac{d\widehat{F}}{d\vec{x}} \cdot d\vec{x} = d_\gamma \widehat{S}$$

At the terminal point, $\gamma = 1$, the initial unit matrix transforms into the experimentally measured one and the initial layered medium into an inhomogeneous one. Adiabatic mode approximation depends to a greater degree on space averaged characteristics of the waveguide rather than on details of its inner structure. This results in difficulties for unambiguous tomographic reconstruction of the media when this theory is used [3, 2]. Thus a model that can calculate accurately both the S-matrix and its derivatives with respect to describing the media parameters x_s is essential for this approach. A non-parabolic algorithm which does not use any significant approximations was developed recently [4].

The approach to the inverse problem described above differs from the Munk-Wunsch (MW) ray-type tomography scheme in the following respect. Interference of different rays in MW ray tomography is an effect which hinders the resolution of arrivals from different rays and necessitates the application of rather broad-band signals. In the above scheme, it is in fact just this interference which is measured (because interference is responsible for formation of mode structure). The scheme of acoustical tomography of the ocean, based on cw-signals and modal description of the field, was suggested by E.C. Shang in 1989 [5]. An attempt of experimental reconstruction of the ocean inhomogeneities which used the effects discussed was performed in the previously mentioned experiment in the Norwegian sea [2].

4. Horizontal antennas out of vertical arrays

New possibilities are offered by coherent processing of signals received by a system of mode-resolving LVA's. It provides, in particular, a new scheme for acoustical monitoring of the ocean: horizontal refraction mode tomography [6, 7]. Horizontal gradients of sound speed transverse to the acoustic path deflect horizontal rays in the (x, y) plane associated with appropriate acoustic modes. The corresponding horizontal refraction angle (HRA) α_n of the nth mode can be calculated as follows

$$\alpha_n = \int_0^R \frac{x}{R}\frac{\partial}{\partial y}\frac{[\Delta c]_n}{c_0} dx \ ,$$

where x, y is the distance along the path and transverse to it, respectively, and $[\Delta c]_n$ is a deviation of the sound speed from a reference profile $c^{(0)}$ averaged over the depth spanned by the nth mode:

$$[\Delta c]_n = \int (c - c^{(0)}) u_n^2 dz \ .$$

For typical mesoscale eddies we assume $\alpha \sim 5\cdot 10^{-3}$. This small value can be easily measured with the help of an acoustic interferometer consisting of two LVA's deployed at a distance $L \sim 10$ km apart:

$$\alpha_n = -\frac{\Delta\varphi_n}{\xi_n L \sin\beta},$$

where β is the direction to the source and $\Delta\varphi_n$ is the phase difference between the complex amplitudes of the nth mode at the two arrays due to horizontal refraction:

$$\exp(i\Delta\varphi_n) = A_n^{(1)} \left(A_n^{(2)}\right)^* \left|A_n^{(1)} A_n^{(2)}\right|^{-1} \exp\left(-i\xi_n L \cos\beta\right)$$

Estimates show that using a weak tonal source (185 db re 1 μPa at 1 m) and common satellite navigation it is possible to achieve accuracies of the order of 10^{-4} for α_n at distances $R \sim 10^3$ km. This is smaller than values of interest. Estimates also reveal that, in contrast to the relatively high-frequency ray case distortions caused by internal waves (IW) for low frequency low order modes are small enough. Interaction with relatively short IW results in negligible attenuation of the coherent component of the sound field. On the other hand, short IW contribute mainly to the diffusion of corresponding rays with respect to angles. Thus IW do not significantly smear the angular distribution of the acoustic field (in the horizontal plane). Additional improvement can be achieved by using signals consisting of discrete frequencies. Such signals can be obtained even from resonant sources by modulation of the power supply. Obviously, using multi-frequency signals also offers a number of other possibilities for tomographic reconstruction of the ocean.

An experiment could be performed as follows. A vessel tows a tonal source at a distance of 1000 km from *two* acoustic interferometers (two pairs of mode-resolving LVAs) which are separated from each other by about 1000 km. The above integral of the cross-gradient of sound speed is known along straight lines connecting the source with the interferometers. Due to the moving source, the set of lines will cover the area under investigation. Using known procedures of 2D tomography, one can retrieve 2D profiles of depth averaged sound speed deviation from the reference stratification $[\Delta c]_n$. Comparing 2D profiles for different modes, one can obtain a 3D reconstruction of the media. Duration of the experiment for an area of the order of 1000 km $\times$ 1000 km for a tow speed of 2.5 m/s is about 1000 km/2.5 m/s $\sim$ 5 days.

We summarize the advantages of using horizontal refraction angles as follows: 1) Only simple tonal signals of relatively low intensity are involved. 2) There is no need for exact positioning of the source. The source can be towed and the accuracy provided by common satellite navigation is sufficient. 3) Interpretation of data and the tomographic reconstruction of the media is simplified significantly. The measured value (i.e., HRA) is zero for layered media (cf. with time-of-flight tomography in which a complicated sequence of arrivals appears both for layered and inhomogeneous oceans). Qualitative 2D profiles of inhomogeneities can be obtained without refined post processing of the data. 4) The duration of the experiment is rather short. The experiment would require no new instrumentation.

5. Acknowledgments

This work was supported by Environmental Technology Laboratory, NOAA, Boulder, Colorado and by the Associateship Program of the National Research Council.

References

1. Yang T.C. A Method of Range and Depth Estimation by Modal Decomposition, *J. Acoust. Soc. Am.*, **82**, 1987: 1736–1745.
2. Goncharov V.V. and Voronovich A.G. An Experiment on Matched Field Acoustic Tomography with Continuous Wave Signals in the Norway Sea, *J. Acoust. Soc. Am.*, **93**, 1993: 1873–1881.
3. Tolstoy A. Linearization of the Matched Field Processing Approach to Acoustic Tomography, *J. Acoust. Soc. Am.*, **91**, 1992: 781–789.
4. Voronovich A. Non-Parabolic Marching Algorithm for Calculating *S*-Matrix of the Ocean Waveguide, *Journal of the Acoustical Society of America*, submitted.
5. Shang E.C. Ocean Acoustic Tomography Based on Adiabatic Mode Theory, *J. Acoust. Soc. Am.*, **85**, 1989: 1531–1537.
6. Voronovich A.G. and Shang E.C. A New Scheme of the Acoustic Tomography of the Ocean, *J. Acoust. Soc. Am.*, **95**, 1994. (Suppl.).
7. Voronovich A.G. and Shang E.C. A note on the Horizontal Refraction Modal Tomography, *J. Acoust. Soc. Am.*, submitted.

STUDY OF THE SIGNIFICANCE OF AN IMPROVEMENT IN TOMOGRAPHY METHODOLOGY THROUGH A JOINT ASSESSMENT OF ARRIVAL ANGLES AND TIMES.

C. NOEL [1)] - F. EVENNOU [2)] - Y. STEPHAN [2)] - M.C. PELISSIER [3)]

[1)] SEMANTIC TS Cy - 8 Av du 11 Novembre - 83150 Bandol - France
[2)] Military Center for Oceanography. E.P.S.H.O.M. - BP 426 - 29275 Brest Cedex - France
[3)] University of Toulon - MS/ETMA - BP132 - 83957 La Garde Cedex - France

Abstract:
The purpose of this study is to theoretically evaluate the performance of a vertical receiving array in a fluctuating medium by means of arrival time-angle diagrams.
First, the suitability of using the eikonal approximation to monitor the sound speed structure in a fluctuating ocean and especially to probe meso-structure in the North-East Atlantic Ocean are discussed. Then, a study of the evolution of tomographic parameters: travel time, temporal spreading, temporal resolution and angular density, based on ray theory is conducted through a synthesis of recent theoretical and experimental studies. Its aim is to quantify and precisely define the undesirable effects induced by the fluctuating medium. These results will be compared with those of a statistical study, and then the effectiveness of an array processing scheme dealing with a joint assessment of arrival angles and times will be investigated. Finally, the feasibility of realizing predicted improvements are considered, particularly for the case of the North-East Atlantic Ocean.

Although numerical inversion techniques are becoming more and more efficient, their results depend on the accuracy of the measured input data. These data, known as the invertible parameters of tomography (travel time fluctuations of an acoustic pulse, arrival angle...), are deduced from observables identified as a line integral of the desired field.

Fluctuations of the medium noticeably modify the invertible parameters. In a modelization of acoustic propagation with geometric rays, this generates general problems of identification and separability of acoustic rays. Mathematical inversion methods may be ill-conditionned and it becomes difficult to yield information on ocean structure. This is the case of measurements obtained during the acoustic tomography experiment carried out in 1990 by the SHOM (Hydrographic and Oceanographic Center,

O. Diaschok et al. (eds.), Full Field Inversion Methods in Ocean and Seismo-Acoustics, 249–254.

French Navy) in the Bay of Biscay. Because of a very specific sound velocity profile, illustrated in Figure (1), received rays are not separable and identifiable, and the stability of geometric rays is questionable.
In order to improve tomographic techniques in the North-East Atlantic Ocean, the mechanics of the influence of propagation conditions on tomography results must be well known, summed up and compiled. Furthermore, it seems that non resolvable paths might be separated with a vertical reception array.

Deterministic aspect

First, we checked the validity of ray equations in a double-channel characteristic of NEA (North-East Atlantic) by considering the order of magnitude of the spatial celerity gradients. They are in fact insufficient to explain unstable motion of ray paths.
On the other hand, it has been previously shown that acoustic ray trajectories are expected to exhibit chaotic motion, i.e. extreme sensitivity to initial and environmental conditions, in range-dependent ocean models. J.Yan (1993) defines , as follows, an instability criterion of Hamilton's ray equations:

$$\frac{C^2_0}{C(z_0)^4}\left\{C(z_0)\left[\frac{\partial^2 C(z_0)}{\partial z^2}\right]-3\left[\frac{\partial C(z_0)}{\partial z}\right]^2\right\}+\frac{\partial^2 g(z_0,r_0)}{\partial z^2}<0$$

where Co is the reference sound speed, C is the sound speed, g is the range-dependent perturbation in potential function, z is the depth, zo and ro are the reference depth and range. By analyzing instability he shows that this criterion is a necessary condition for ray chaos, and so that chaos may be induced even if there is no fluctuation (g = 0). The Figure (2) presents, as a function of depth, the evolution of this criterion in a case of a deterministic sound speed profile characteristic of the NEA zone. Even in this deterministic case there are two areas where necessary conditions for chaos are satisfied (z ≈ 1200m (channel axis) and z ≈ 70m).

Figure (1) : Mean sound speed

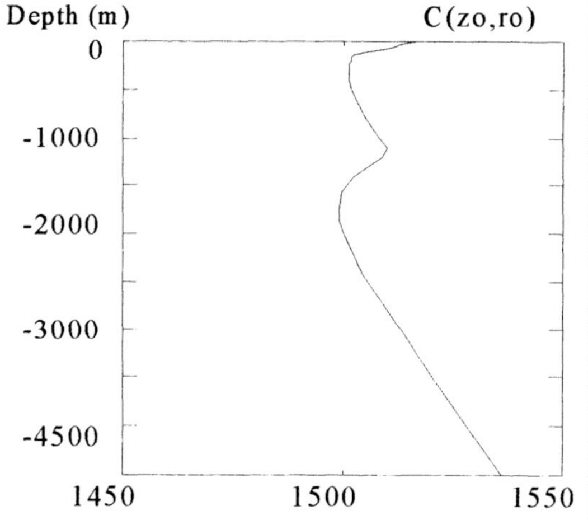

Stochastic aspect

When some fluctuations are introduced in the sound speed profile the criterion is more likely to be satisfied if the derivative $\partial^2 g/\partial z^2$ is negative and admits a large absolute value. This is the case with NEA fluctuations, see Figure (2), near the channel axis.

This is not a sufficient condition, but the conclusion is that NEA fluctuating profiles require the existence of chaotic motions. To investigate and characterize more precisely ray chaos, we should evaluate, as in [K. Smith, M. Brown, F. Tappert 1992], the Lyapunov exponents in order to quantify the degree of chaos in this area.

Figure (2): Instability criterion [1]

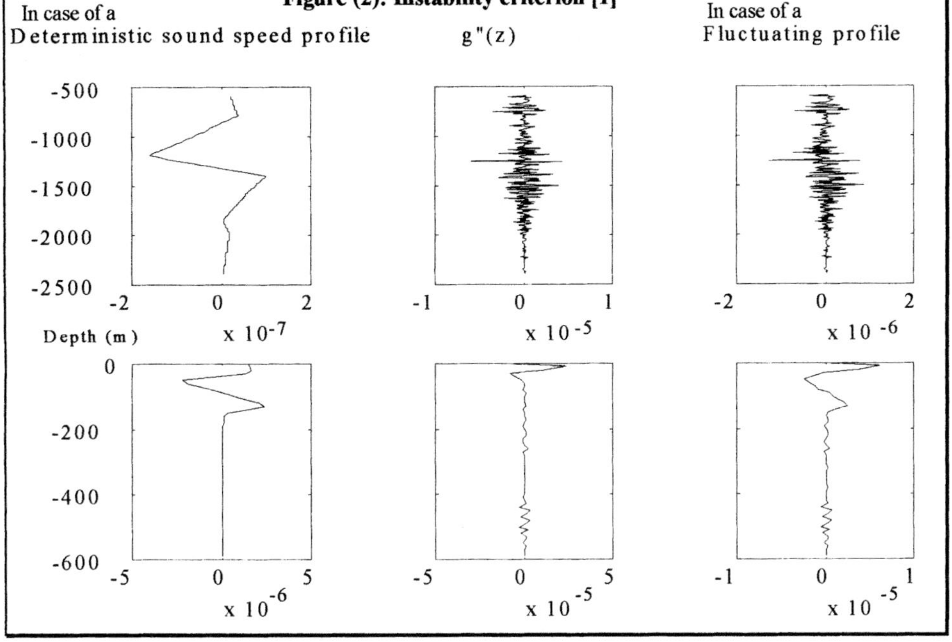

In order to characterize sound propagation features with a ray model in a fluctuating ocean, we use theories developed by Flatté et al, based on path - integral methods [S. Flatté, R. Dashen, W. Munk, K. Watson, F. Zachariasen (1979)]. Their treatment of ocean medium includes the effects of anisotropy and the background sound channel as well as statistical inhomogeneity and internal-wave spectra. They define two parameters representing the strength and size (spatial extent) of the inhomogeneity that control the character of the fluctuations in a wave field crossing these perturbations. These parameters may be evaluated only from in-situ measurements of sound speed fluctuations by the mean of their variance and correlation lengths for some propagation configurations (source-receiver pair).

SOUND SPEED FLUCTUATIONS IN NEA

Fluctuations known from in-situ measurements in the NEA zone are compared in their order of magnitude and vertical correlation lengths to those considered in Flatté et al's theories. In fact NEA fluctuations are weaker in the first channel. Because of the order of magnitude and the closeness of continental shelves, the assumption that fluctuations are essentially due to internal waves seems to be consistent. In addition, we suppose, but with no possible confirmation (because of too sparse in-situ measurements), that the horizontal correlation length was approximatively equal to 10km as it is usually taken for that kind of perturbation.

CALCULATION OF THE STRENGTH AND DIFFRACTION PARAMETERS ϕ AND Λ.

The parameters defined as follow: $\phi^2_{ray} = q_0^{\ 2} \int_0^R dx \left\langle \mu^2(z_{ray}) \right\rangle L_P(\theta, z_{ray})$ and

$\Lambda_{ray} = \phi^{-2}_{ray}\, q_0^{\ 2} \int_0^R dx \left\langle \mu^2(z_{ray}) \right\rangle L_P(\theta, z_{ray}) \left| q_0\, A\, L_V^{\ 2} \right|^{-1}$ where μ is the relative sound speed fluctuation, Lp the correlation length and A the phase curvature function, are numerically computed as in [R. Esswein, S. Flatté 1980][R. Leung, H. DeFerrari 1980] along some unperturbed eigenrays localised in the upper channel. We use a Runge-Kutta method to carry on the calculus of the phase curvature function. Some convergence difficulties appear in computing these parameters: their equations are shown to have instability criterion dependent on $\partial^2 g/\partial z^2$ too.

The results obtained are summarized in **Figure** (3) for a propagation range of 180km and compared with results in Cobb, Bermuda, Azores in [Flatté et al] p238 and with Atlantic in [Esswein-Flatté] p1530. They seem to be in good agreement and consistent with those obtained in the nearest previously studied area: in the Azores.

Figure (3) **Regions in Λ–ϕ space**

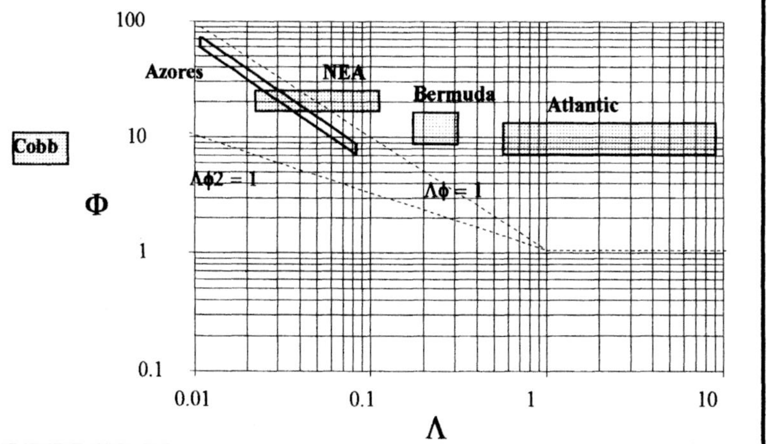

So, the whole NEA zone seems to be at its limits of tractability with the basic equation of geometrical optics.

TIME-ANGLE PARAMETRIC DIAGRAM

These parameters give some informations about the number of multipaths and their spatial spreading. They are also used to evaluate temporal and angular spread and wander. Temporal characteristics are presented in Figure (4) where eigenrays are referenced by their emission angle (r-0.3° corresponds to an emission angle of -0.3°, r3.6 to 3.6°) and to the first temporal arrival (eigenray r-0.3°). The parametric values obtained are from 10ms to 50ms and correspond to those observed with in-situ measurements.

Figure (4): Temporal spread and wander

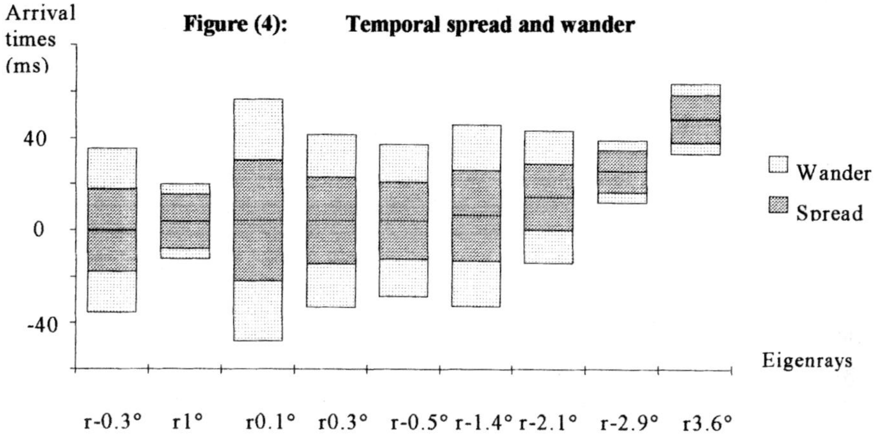

Fluctuations in arrival angle, Figure (5), seem to be very important in our area, especially for axial rays but they remain consistent with the order of magnitude given in Stoughton-Flatté [1988].

Figure (5): Arrival angles

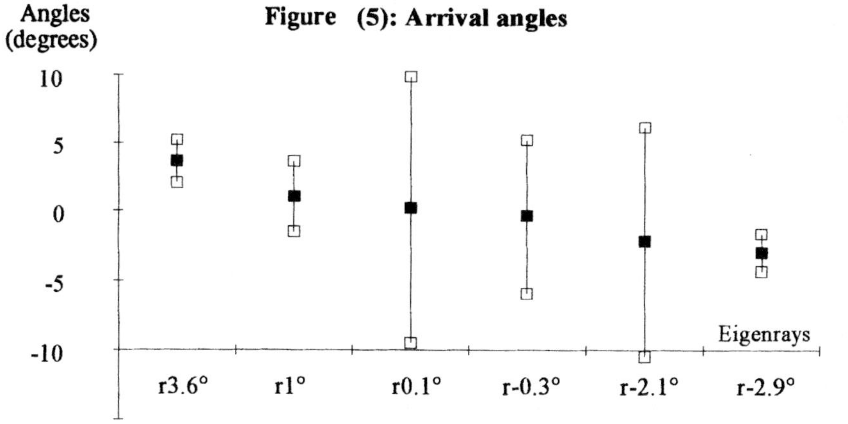

For all of these calculations we take into account the variance of celerity fluctuations over the entire NEA area, which is probably the worst case. In order to compare the parametric results with some measured travel times and arrival angles over 180km range propagation, we should have considered only the variance of the fluctuations of the crossed area instead of the one of the whole zone.

CONCLUSIONS

This study defined the propagation characteristics in the worst fluctuating case. A study of sound speed fluctuations in the NEA area concludes that the order of magnitude is of the same order as those usually observed in other oceans where rays are stable and identifiable. So, sound speed perturbations can't alone explain the complexity of temporal arrivals detected in-situ during the tomographic experiment. Considering chaotic instability, we have shown that the deterministic double-channel satisfies the necessary condition for chaos and this aspect is reinforced in case of fluctuating celerity profile.

In conclusion of the theoretical study, problems due to chaos seem to be the more important because chaotic rays may arrive at any time and anywhere and perturb all the arrival scheme without any logic. Interpretation may then become impossible. An issue might be to use eigenrays that do not cross chaotic areas by making judicious choice of source-receiver depth or by limiting the source to small emission angles. More information than the parametric one obtained here on arrival angles, is needed to predict the potential of an array to separate eigenrays in this tomographic experiment.

This study will be completed with a statistical approach. This will consist of constructing from in-situ celerity measurements, a great number of realistic realizations of the random oceanic process, which are in fact maps in 2D of the sound speed profile. For each of these maps, the corresponding trajectories and arrival times of rays are then computed by solving a deterministic propagation equation with a range dependent ray model. This method makes no assumption on the fluctuations because it only treats deterministic problems and will yield more realistic results in specific subareas concerning some tomographic paths instead of giving general results for the whole NEA as those of the parametric study.

REFERENCES

R. Esswein, S. Flatté May (1980) "Calculation of the strength and diffraction parameters in oceanic sound transmission". JASA 67 (5),

S. Flatté, R. Dashen, W. Munk, K. Watson, F. Zachariasen (1979) "Sound transmission through a fluctuating ocean.", Cambridge University Press.

R. Leung, H. DeFerrari (Jan 1980) "ϕ and Λ computations for real and canonical oceans." JASA 67 (1).

K. Smith, M. Brown, F. Tappert (April 1992) "Ray chaos in underwater acoustics." JASA 91 (4). "Acoustic ray chaos induced by mesoscale ocean structure." JASA 91 (4).

J. Yan (Nov 1993) "Rays chaos in underwater acoustics in view of local instability." JASA 94 (5), p2739.

R. Stoughton, S. Flatté (Oct 1988) "Prediction of internal-wave effects on ocean acoustic coherence, travel time variance, and intensity moments for very long-range propagation". JASA 84 (4).

ACOUSTIC MODELING AND MEASUREMENTS IN THE BAY OF BISCAY IN A DOUBLE SOFAR CHANNEL

Y. STEPHAN, F. EVENNOU and F.R. MARTIN-LAUZER
Military Center for Oceanography, EPSHOM
13 rue du Chatellier, 29275 BREST Cedex
France

ABSTRACT. This paper gives a general description of the underwater acoustic propagation in the Bay of Biscay. The propagation is characterized by a deep double SOFAR channel due to Mediterranean Waters intrusions. Comparisons are drawn between data collected during the GASTOM 90 tomography experiment and simulated signals computed from various propagation models. Our goal is to better «understand» the propagation in those peculiar conditions in order to define the best strategy for applying matched-field techniques.

Introduction

The French tomography experiment GASTOM 90 was carried out in the Bay of Biscay by the French *Service Hydrographique et Océanographique de la Marine* (SHOM) in cooperation with IFREMER [1]. The main objective of the cruise was to check the feasibility of Ocean Acoustic Tomography (OAT) in an oceanic area with a tough acoustic environment. The Bay of Biscay is a high variability area. Both mesoscale features (as fronts, eddies or jets) and high frequency phenomena (internal tides and waves, fine structures) strongly influence the sound propagation. In addition, Mediterranean Water intrusions induce a deep maximum in the Sound-Speed Profiles (SSP) resulting to a double SOFAR channel.

This paper is motivated by the following remarks. Firstly, the GASTOM experiment has shown the complexity of applying usual tomographic methods as originally proposed by Munk and Wunsch [2] because of the non-separability of acoustic arrivals. In that sense, the ray approach for tomography may not be the optimal way to tackle the problem. Secondly, the methods based on similarity estimations between measured and simulated signals (temporal or amplitude matching methods) need to have a reliable propagation model as well as a rather precise knowledge of the environment. A preliminary step is then to «understand» the propagation, which means to be able to find reliable models and strategies for simulations in a given area.

This paper describes the acoustic conditions of propagation in the Bay of Biscay from two points of view. The first section of the paper deals with the acoustic environment and recalls some generalities on the propagation in a deep double SOFAR channel. In the second part, we draw comparisons between acoustic measurements collected during the experiment and the results of simulations with various propagation models.

1. Acoustical environment in the Bay of Biscay

The Bay of Biscay is an oceanic area of high interest. From an oceanographic point of view, many phenomena are involved covering the whole spectrum of scales. From an acoustic point of view, it is characterized by a deep double SOFAR channel leading to peculiar propagation conditions.

O. Diaschok et al. (eds.), Full Field Inversion Methods in Ocean and Seismo-Acoustics, 255–260.

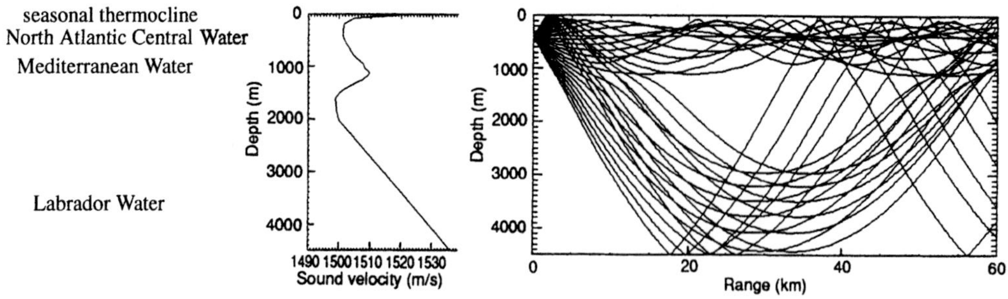

Fig 1. Typical sound-speed profile in the Bay of Biscay in July and associated ray tracing

1.1 GENERAL DESCRIPTION AND OCEAN VARIABILITY

One major characteristic of the north-east Atlantic Ocean stems from the various origins of water masses (modal Atlantic Waters, Mediterranean Waters, Labrador Waters). In particular, the oceanic circulation is characterized by Mediterranean Waters (MW) intrusions. They are expanding westward from the Straits of Gibraltar and northward toward the Bay of Biscay. MW are stabilized around 1000 meters deep. Since they have a higher salinity than surrounding waters, a deep maximum in the sound velocity profile is induced (figure 1) splitting the usual SOFAR channel in two parts. The upper channel is up-limited by the seasonal thermocline ST or by the ocean surface when ST is not present (in winter for instance). It is down-limited by the upward refracting gradient due to MW intrusions (between 600 and 1000 meters depth). The main channel is defined by the ST or the surface in its upper part and by the deep layer (hydrostatic gradient in Labrador Waters) between 2500 meters and the bottom.
The variability in the Bay of Biscay is rather important. Two main sources of sound-speed fluctuations can be quoted. The first is the mesoscale variability, often linked to the MW intrusions which induce streams and eddies. They particularly affect the value of the deep celerity maximum. The second is related to internal tides which strongly affect the propagation by leading to semi-diurnal periodic displacement of the density gradients [3].

1.2 ACOUSTIC PROPAGATION

We consider an acoustic source in the upper channel (source at 400 meters depth). For the profile shown on figure 1, we can define four types of rays. The first type is composed by purely refracted rays in the upper channel (type denoted RR^+). Their turning depths are located in the seasonal thermocline and in the low North-Atlantic Central Water. They are rays with small grazing angles and are travelling faster. The second type is composed by purely refracted rays in the main channel between the seasonal thermocline and the deep layer. They correspond to the convergence zones in a deep channel and are denoted RR^-. RR^- arrive later than RR^+ and have greater grazing angles. The third type, very similar to RR^-, is composed by surface reflected rays (SR). Their low turning depth is in the deep layer. SR arrive later but very close to the RR^-. Lastly, totally reflected rays on surface and bottom are denoted SB rays. They arrive significantly later than the others. The various types are presented on figure 1 and an example of temporal pattern is given in figure 2. It is noticeable that the arrival order of shallow and deep rays is inverted compared with deep channels (Western Atlantic, Pacific Ocean).
For ray tomography, a disagreeable effect of the double SOFAR channel is the focusing of the arrivals even on long ranges: it is often impossible to separately estimate the arrivals. In order to perform temporal tomographic inversions, a solution has been proposed in [4] to estimate signal

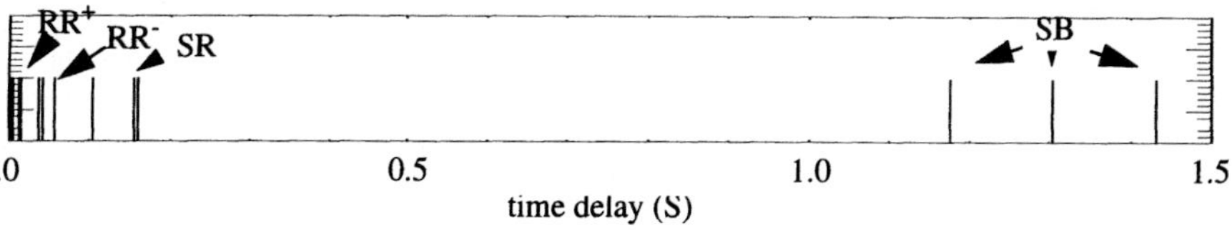

Fig 2: temporal pattern computed from the geometric approximation for a 149 km propagation (amplitudes have been arbitrarily removed)

characteristics (front and maximum in the received energy) and to relate them to particular rays (for instance the ray with the minimum time-delay). An alternative approach is to perform inversions on the shape of the signal. The problem stems from the fact that not only temporal but reliable energy estimations are required. Therefore, it is important to have a good modeling of the propagation in the double channel environment. This encompasses the availability of reliable propagation models and a precise knowledge of the real environment in order to validate the adiabatic approximation or the range-independence approximation.

2. Comparisons

In this section, we will consider two cases of propagation from the GASTOM experiment. A «short» range propagation experiment is given assuming the independence of the SSP between the source and receiver. Another example is presented with a range-independent and a range-dependent medium.

2.1 REAL DATA

The tomographic network was composed of six transceivers moored 400 meters deep. The distances of propagation extend from 134 kilometers to 307 kilometers. The sources periodically emitted a phase modulated broadband signal centered on 400 Hz. The received signal was averaged, demodulated and intercorrelated with the transmitted code. The transceivers had a 100 Hz bandwidth inducing a theoretical temporal resolution of 10 ms. Transmission was repeated from June to September every 2, 3 or 6 hours depending on the tidal amplitude. Both signal-to-noise ratio and time delays were drastically fluctuating, especially at the semi-diurnal frequency because of the internal tides effects. The comparisons will therefore be drawn on signals with high signal-to-noise ratio and on lined-up averaged sequences over several days to filter tidal effects.

2.2 RANGE-INDEPENDENT MEDIUM

The first case we consider is described in annex A (134 km). A mean sound-speed profile is built from the CTD measurements in the area in mid-July. We notice a general good matching between simulations and data, especially with the normal-mode model. The geometric ray theory which is known to be irrelevant near caustics and shadow areas does not put forward the double maximum in the energy ratio. This double maximum appears in real data and the normal mode prediction. In the averaged sequences, it is smoothed by temporal spreading. We notice that RR^- are fluctuating but are significantly strong when averaged.

2.3 RANGE-DEPENDENT MEDIUM

The comparisons in the case of propagation from 182 km are given in annex B. Simulations have been performed with a local mean SSP with ray theory and normal mode. An additional prediction has been made with gaussian beam theory [5] which is based on a ray description with validity in and near the caustics. The range-dependent medium was built from CTD measurement between the source and the receiver. All predictions are rather similar but don't match well the data. The problem stems from arrivals between the RR^+ packet and RR^-. These arrivals can be seen on data and are particularly evident in averaged sequences. In addition, they seem to have a structured form (three rays) over a four days averaging. We notice that this structure is present with a lower importance in the first case (annex A).

3 Discussion

We have performed comparisons on all transceivers pairs (15). The major part of the received signal (RR+ packet and RR-) is well predicted. Nevertheless, there are some exceptions for some pairs especially at long ranges (over 180 kilometers). The difference between simulations and measurements mainly consists in non-predicted energy between RR^+ packet and RR^-. The difference is more significant when reception sequences are lined-up and averaged. As can be seen in example B, some acoustical energy is present between RR^+ and RR^-. We have not managed to predict these arrivals with a gaussian beam model and a range-dependent ray model (not shown). This may be explained by energy exchanges between the upper channel and the main channel. In the modal representation, three kinds of modes are present. Some are in totality trapped in the upper channel. Some are trapped in the lower channel and are not excited by a source at 400 meters. The remaining modes are trapped in the main channel. The modal pattern in very sensitive to fluctuations of the deep maximum. We think that range variations in the deep maximum (which appears rather obvious at mesoscale ranges) leads to energy exchanges between trapped modes in the upper channel and trapped modes in the main channel. An intuitive approach is to consider that the former modes are less stable than the latter ones. With perturbation effects in the ray channel some acoustical energy may be transmitted to more stable modes. These modes have lower group velocities so that the energy arrives later between purely up-trapped energy and RR- (RR- rays correspond to a system of stable modes trapped in the main channel). Unfortunately, a computational check of this explanation in the case of the GASTOM signals is a hard task. Indeed, it would be necessary to run a coupled-mode model to long ranges, with high frequencies (350-450 Hz). The computation time would be prohibitive at present.

Conclusion

We have presented in this paper some comparisons between measurements of tomographic signals and corresponding predictions. We have put forward that a part of the signals remains non-predicted with codes that do not take into account mode coupling. Nevertheless, major features as the first arrivals (purely refracted rays in the upper channel) and late arrivals (purely refracted rays in the main channel) are well predicted. The continuation of this work will be twofold. On one hand, we will continue to work on the modal description in the double SOFAR channel. On the other hand, we will used the predicted part of GASTOM signal to develop a temporal matched-field method for inverting GASTOM data. In addition, we will consider the possibility of applying matched-field tomography [6] in the Bay of Biscay.

References

1. Martin-Lauzer F.R. (1991), «GASTOM 90: an oceanographic and ocean acoustic cruise in the Bay of Biscay», *Proceedings of OSATES 91*, A08.

2. Munk W.H., et Wunsch C. (1979). «Ocean acoustic tomography: a scheme for large scale monitoring», *Deep-sea research*, 26A,123-161.

3. Martin-Lauzer F.R., Stéphan Y., and Evennou F. (1994) «Analysis of tomographic signals to retrieve tidal parameters», *Proceedings of the 2nd European Conf. on Underwater Acoustics*, Vol. II, pp 1051-1056.

4. Stéphan Y. and Martin-Lauzer F.R. (1994) «GASTOM 90 tomography experiment. Data inversion», *Proceedings of the Int. Conf. on Acoustics, Signal and Speech Processing*, Vol 2, pp 301-304.

5. George T. (1987) Une approximation à haute fréquence de l'équation des ondes : la sommation de faisceaux gaussiens (a high frequency approximation for the wave equation: the gaussian beam summation), Thèse de l'université de Paris 6.

6. Tolstoï A., Diachok O. and Frazer L.N. (1991) «Acoustic tomography via matched field processing», *J. Acoust. Soc. Am.* 89(3).

ANNEX A

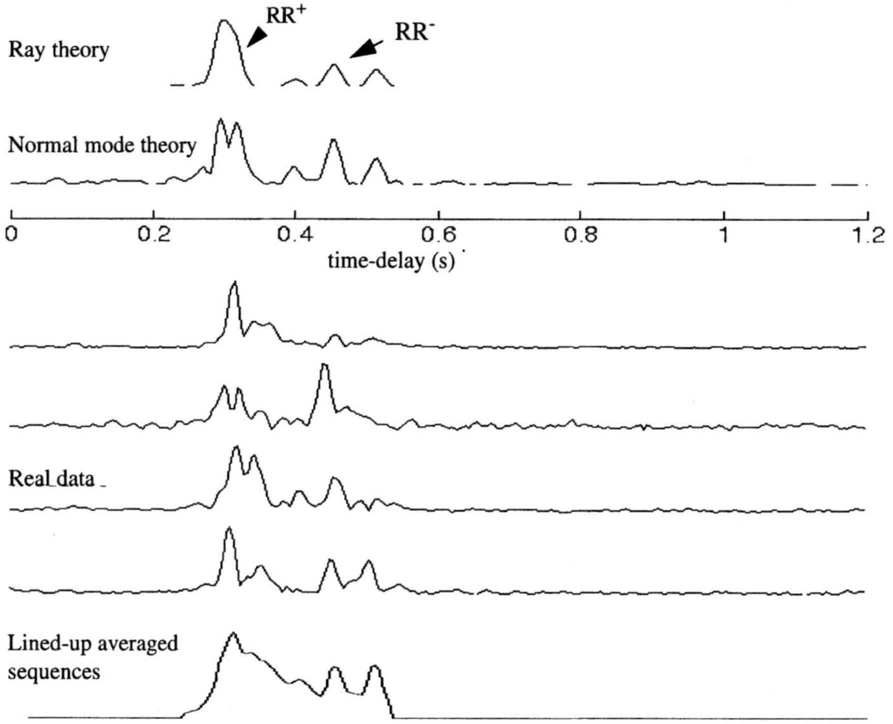

Comparisons between data and simulations (134 kilometres)
Top: predictions with the geometric approximation and normal mode theory in a mean sound velocity profile in July. Middle: set of four receptions in July. Bottom: lined-up averaged signals over five days. All Y-axis represent the normalized energy.

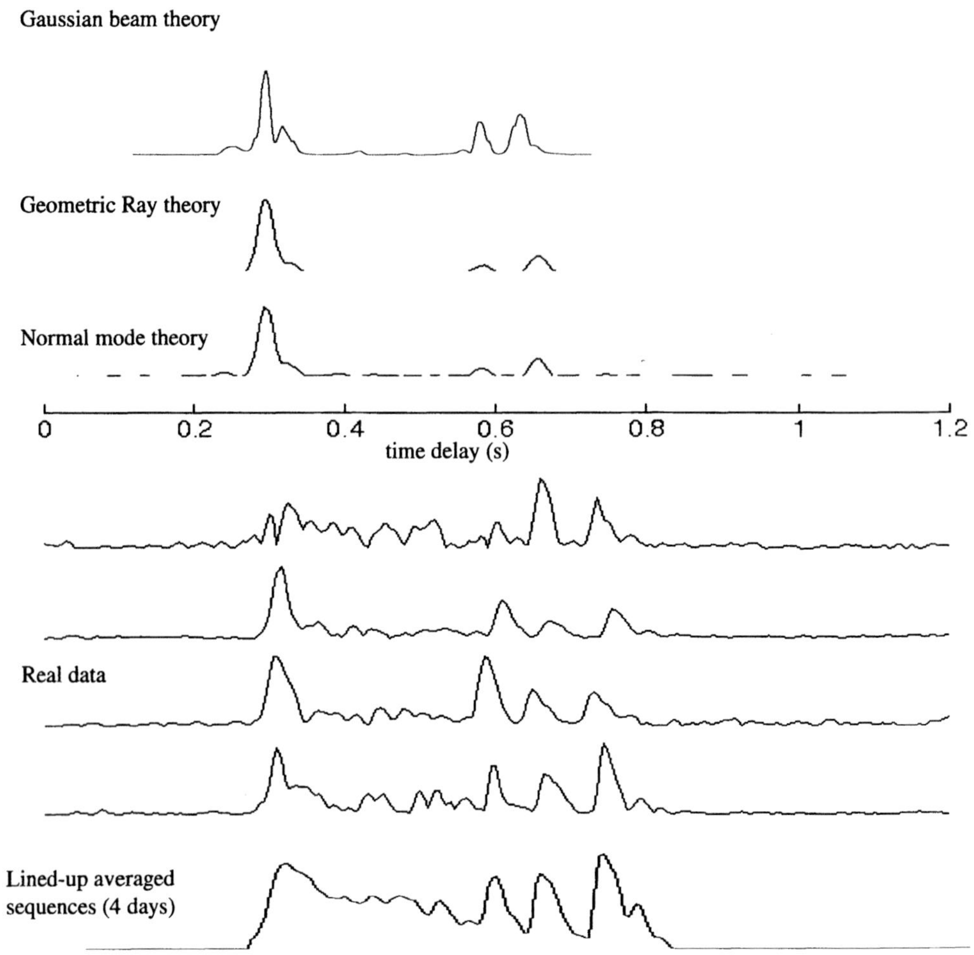

Comparisons between data and simulations (182 kilometres)
Top: predictions with the gaussian beam theory , with the geometric approximation and normal mode theory in a mean sound-speed profile in July. Middle: set of four receptions in July. Bottom: lined-up average received signals over four consecutive days in July.

A FULL FIELD INVERSION METHOD FOR ACOUSTIC TOMOGRAPHY OF OCEAN CURRENTS

O.A.GODIN, D.YU.MIKHIN and A.V.MOKHOV
Acoustic Wave Propagation Lab.
P.P. Shirshov Oceanology Institute
23 Krasikova str., Moscow 117851, Russia

ABSTRACT An extension of the matched field inversion technique to include oceanic current tomography is considered. Matching of the non-reciprocity of acoustic fields propagated in opposite directions (instead of matching sound fields themselves) makes it possible to separate relatively subtle acoustic effects of the currents from those due to sound speed variations, bottom topography etc. A set of numerical simulations is carried out under environmental conditions typical of the Gulf Stream to evaluate the stability of the inversions with respect to mismatches in transceiver positions and in sound speed for various choices of physical quantities to be matched. Matching of non-reciprocity in phase is shown to result in sufficient sensitivity to current velocity and high stability with respect to mismatches in other parameters.

1. Introduction

Ocean currents are capable of influencing underwater acoustic fields considerably. However, to separate effects of currents and sound speed variations on sound waves in the ocean one should use reciprocal transmissions. Usually differences in time-of-flight in opposite directions along identified pairs of rays serve as input data for inversion schemes. Applicability of this approach, which we will further refer to as travel time tomography (TTT), relies on availability of a sufficient quantity of identified ray pairs. This condition is violated at least in two cases of practical importance: (i) sound propagation through powerful flows such as the Gulf Stream, where differences in propagation conditions up and down current can change the geometry of rays with shallow turning points significantly; (ii) sound propagation in shallow seas where rays are not resolvable. To overcome difficulties with ray resolution and identification it is natural to use a full field approach.

It is well-known that long-range sound propagation in the ocean with sound speed c and current velocity $\mathbf{u}$ is very similar to propagation in a motionless medium with an effective sound speed $c_e = c + \tilde{u}$, where $\tilde{u}$ is the projection of $\mathbf{u}$ on a vertical plane passing through the sound source and receiver [1]. This is because of the smallness of the Mach number, $M < 2 \times 10^{-3}$, of oceanic currents and representative grazing angles of acoustic waves. Hence, one can find c_e but not c, and/or any projection of $\mathbf{u}$ based on data in one-way sound propagation. In addition, absolute errors that are quite tolerable in c inversion are absolutely inadmissible in current tomography. That is why usual matched field processing (MFP) is hardly useful for acoustic monitoring of oceanic currents. However, using data on sound propagation in opposite directions for full-field inversion and matching the non-reciprocity of sound propagation, one can separate acoustic effects of currents from those

O. Diaschok et al. (eds.), Full Field Inversion Methods in Ocean and Seismo-Acoustics, 261–266.

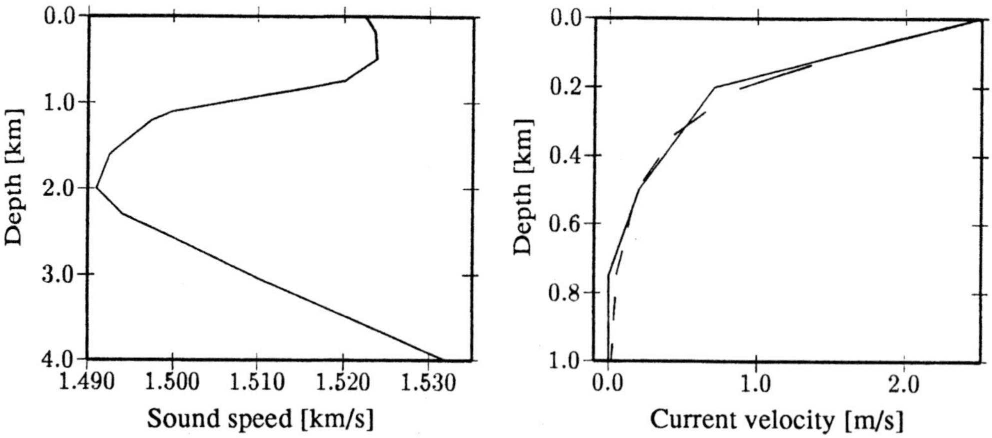

FIGURE 1. Environmental model used in numerical simulations.

due to sound speed variations etc. quite similar to the TTT case. Note that cw or narrow-band signals could be used for such matched non-reciprocity tomography (MNT).

The aim of the present study is to demonstrate by numerical simulations the feasibility of current velocity inversion following the MNT approach under realistic environmental conditions, when 'experimental data' are contaminated by some errors.

2. Three Versions of MNT

Our numerical experiments correspond to the following scenario: There are two identical transceivers (i.e. transmitter-receiver pairs). One transceiver is fixed whereas the position of the other varies. In this way a synthetized aperture array (below assumed vertical for simplicity) may be created for acoustic signals propagating up- and downstream.

Mathematically, the inverse problem of current velocity reconstruction resulting from MNT consists of minimizing a functional F that describes the discrepancy between measured and calculated non-reciprocities of an acoustic quantity.

Both the possibility and the quality of the inversions depend on choice of the functional F. A successful functional should meet, at least, the following conditions:

- F should be written in terms of acoustic quantities that can be reliably measured in field experiments, and which desirably possess clear physical meaning.

- Inherent errors of acoustic measurements, particularly errors in transceiver positioning, should not give rise to inversion instability.

- Inversion should be possible in spite of underdeterminancy of environmental parameters, such as sound speed and bottom topography, within reasonable limits.

In what follows we will consider 3 versions of the functional F:

$$F_a = 2\frac{\langle |p_e^{(+)}| - |p_e^{(-)}| - |p_t^{(+)}| + |p_t^{(-)}|\rangle}{\langle |p_e^{(+)}| - |p_e^{(-)}|\rangle + \langle |p_t^{(+)}| - |p_t^{(-)}|\rangle} \tag{1}$$

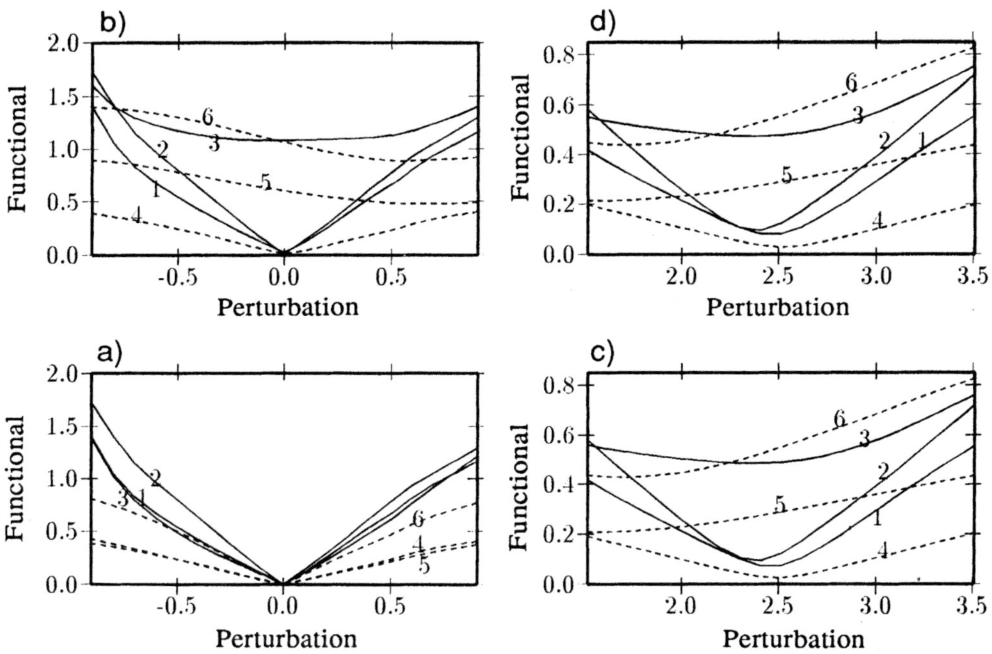

FIGURE 2. Functionals F_a (1), F_p (2), F_c (3) (solid lines), Φ_a (4), Φ_p (5) and Φ_c (6) (dashed lines) vs trial current velocity parameters α (left) and a (right). Propagation range mismatch equals 0 (bottom) and 25 m (top) (the same ordering of curves is used throughout the paper)

$$F_p = 2\frac{\left\langle \arg\left(\frac{p_e^{(+)}}{p_e^{(-)}}, \frac{p_t^{(-)}}{p_t^{(+)}}\right) |p_e^{(+)} p_e^{(-)} p_t^{(+)} p_t^{(-)}|^{1/4} \right\rangle}{\left\langle \arg\left(\frac{p_e^{(+)}}{p_e^{(-)}}\right) |p_e^{(+)} p_e^{(-)}|^{1/2} \right\rangle + \left\langle \arg\left(\frac{p_t^{(+)}}{p_t^{(-)}}\right) |p_t^{(+)} p_t^{(-)}|^{1/2} \right\rangle} \tag{2}$$

$$F_c = 2\frac{\langle |p_e^{(+)} - p_e^{(-)} - p_t^{(+)} + p_t^{(-)}| \rangle}{\langle |p_e^{(+)} - p_e^{(-)}| \rangle + \langle |p_t^{(+)} - p_t^{(-)}| \rangle} \tag{3}$$

corresponding to minimizing the discrepancy between experimentally measured and theoretically predicted non-reciprocity in (1) sound intensity, (2) wave phase, and (3) complex acoustic field. Here $p^{(\pm)}$ is the complex acoustic pressure for propagation down/up the flow, subscripts t and e refer to theoretically calculated and experimentally measured quantities, $\arg(w)$ is an argument of a complex number w, $-\pi < \arg(w) \leq \pi$. Angular brackets stand for the rms value of the corresponding quantity over N positions of the mobile transceiver.

To compare results of MNT and MFP, in addition to F_c, F_a and F_p three more functionals will be calculated that correspond to matching acoustic intensity, phase and complex field, respectively, on one-way propagation (to be specific, downstream propagation is assumed). Respective functionals Φ_a and Φ_c can be obtained formally from (1) and (3) by substituting $p_t^{(-)} = p_e^{(-)} = 0$, and Φ_p from equation (2) by substituting $p_t^{(-)} = |p_t^{(+)}|$, $p_e^{(-)} = |p_e^{(+)}|$.

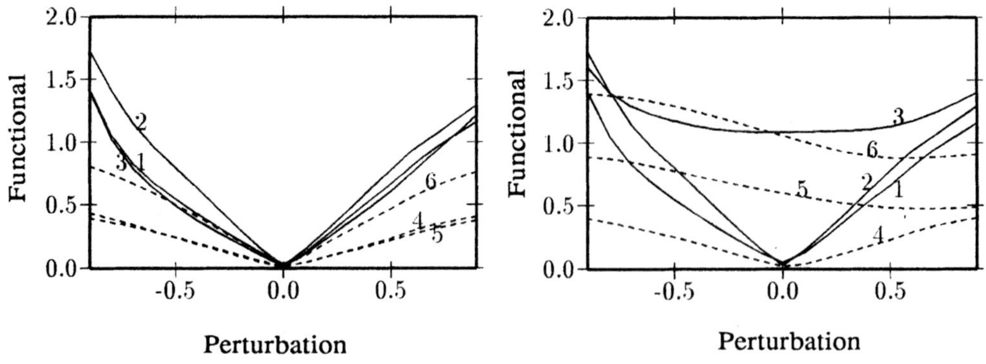

FIGURE 3. MNT and MFP functionals vs perturbation amplitude α for SSP with random error, $\Delta = 0.05$ m/s; propagation range mismatch is 0 m (on the left) and 25 m (on the right).

3. Numerical simulations.

All numerical simulations were made for the range-independent enviromental model taken from [2], that represents conditions of the North-West part of the Gulf Stream. The ocean depth was taken to be $H = 4$ km. Profiles of c and $\tilde{u}$ are shown in Fig. 1. The cw acoustic field was calculated within the narrow-angle parabolic approximation for moving media using a computer code described in [3]. The frequency was taken to be 50 Hz.

In the first numerical simulation presented below the goal was to study the impact of mismatch in propagation range on currents inversion. The current velocity profile was represented in parametric form as $\mathbf{u} = \mathbf{u}^{(0)}(1+\alpha)$, where $\mathbf{u}^{(0)}$ is the real dependence of the current velocity vs depth shown in Fig. 1 and constant α is the perturbation amplitude. The required parameters of sound fields propagating up and downstream were calculated at a distance of 60 km from the source for α varying sequentially from -1 to $+1$ with a step of 0.1. These dependencies of complex acoustic pressure fields vs depth were substituted in (1–3) and the corresponding equations for MFP functionals as 'theoretical' ones. The same calculations carried out under the environmental conditions of Fig. 1 gave 'experimental' vertical sections of sound fields for source-receiver separations of 60.0 km and 60.025 km to simulate propagation range mismatch of 0 and 25 m respectively.

Calculated functionals are represented in Figs. 2a and 2b as functions of α. Of course, in the no-mismatch case of Fig. 2a all six functionals turn into zero at $\alpha = 0$. For a propagation range mismatch of 25 m (Fig. 2b), however, the behavior of functionals differs dramatically. First, MFP functionals Φ_c and Φ_p that use data on sound field phase in one way transmission show no signs of minima in the vicinity of $\alpha = 0$. That is what one should expect because the range mismatch, 25 m, is comparable to the wave length, 30 m, and results in significant phase changes. Degrading of the naive functional F_c that seemingly takes full advantages of MNT is for the same reason. Indeed, the difference between $p^{(+)}$ and $p^{(-)}$ as well as these functions themselves is locally periodic in range with the period of the order of the acoustic wave length. That is why to match non-reciprocity of complex fields one should know the transceiver separation with the same accuracy as for the matching of $p^{(+)}$. Inversion based on the functionals (1),(2) has not been significantly affected even when a 200 m mismatch in the propagation range occurs.

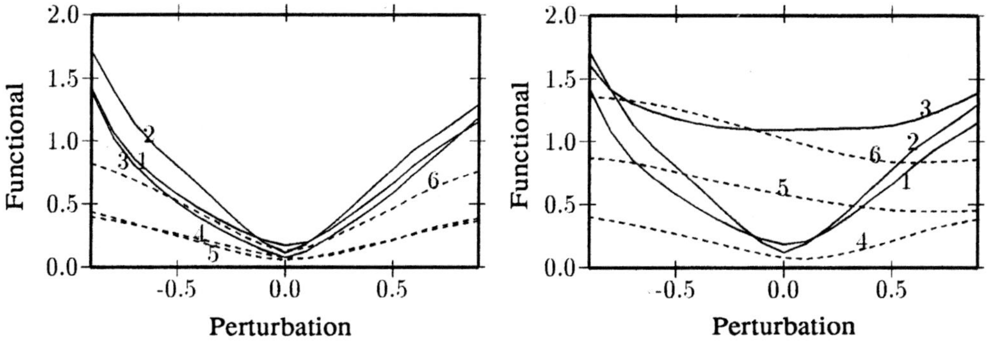

FIGURE 4. The same as in Fig. 3 for $\Delta = 0.25$ m/s.

So far we have assumed that trial and real current velocity profiles were proportional. To override this limitation the following computer experiment was conducted: The trial current velocity profile was represented in exponential form as $u = a\exp(-bz)$ with $b = 5.192\,\mathrm{km}^{-1}$, all other conditions being the same. Pre-exponential factor a (i.e., current velocity at the surface) varied from 1.51 m/s to 3.51 m/s with steps of 0.1 m/s. In the real environment the flow speed at the ocean surface was 2.51 m/s. The dependencies of functionals considered vs a are shown in Figs. 2c and 2d for propagation range errors 0 and 25 m, respectively. It turned out that MFP functionals Φ_c and Φ_p gave very poor results, while F_a and F_p worked well. Some displacements of F_a and F_p minima toward slower surface flow are quite easy to explain. Previous computations [2] showed that, for the problem in hand, MNT is most sensitive to the barotropic component of u. The trial u profile for $a = 2.51$ m/s is shown in Fig. 1 with the dashed line. The trial current velocity is usually higher than the real one, so, the best match should be reached at smaller a.

The last group of numerical simulations is devoted to studying the impact of unavoidable uncertainties in our knowledge of environmental conditions on the results of current velocity inversion. In the examples presented below sound speed profiles (SSP) were chosen to be $c = c^{(0)} + \Delta c$, where $c^{(0)}$ is the real profile shown in Fig. 1 and Δc – some perturbation. The original SSP was given in 13 points z_i with linear interpolation in between. In the first stage, in order to simulate random errors in c, sound speed values at z_i (for depths z_i less than 1.2 km) were substituted with $c^{(0)}(z_i) + \Delta(-1)^i$, the deviation Δ being 0.05 m/s, 0.25 m/s and 1.25 m/s. Results of u inversion by different methods are presented in Fig. 3–5. Trial current velocity profiles were taken to be $u = u^{(0)}(1+\alpha)$. Calculations were made for propagation range mismatch of 0 m and 25 m. Presented plots show that functionals (1) and (2) are relatively stable with respect to the given error while MFP solutions and the results of (3) degrade dramatically for $\Delta = 1.25$ m/s and cannot tolerate a 25 m distance mismatch.

In the second example the perturbation Δc varied linearly with depth from 1.2 m/s at the ocean surface to 0 at 1.2 km depth, other conditions being equal. The dependencies of functionals considered vs current velocity perturbation amplitude α are given in Fig. 6 for range mismatches 0 and 25 m. Only functional F_p managed to handle this task and reconstructed the current fairly well. It is noteworthy that the systematic error considered in the sound speed turned out to be much more significant for current inversion than the random one.

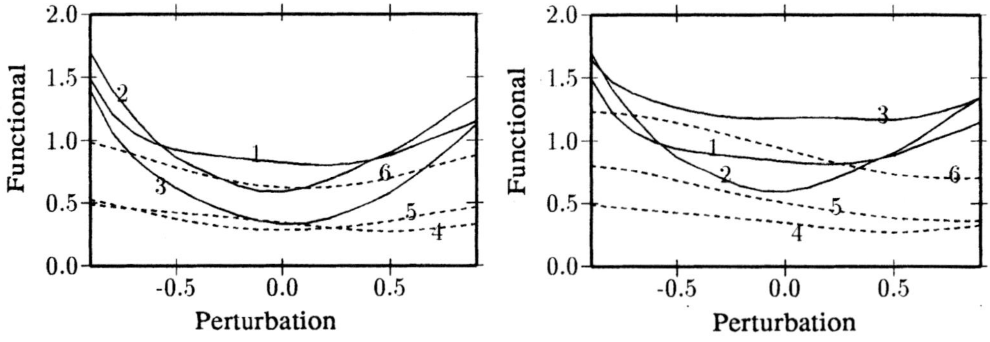

FIGURE 5. The same as in Fig. 3 for $\Delta = 1.25$ m/s.

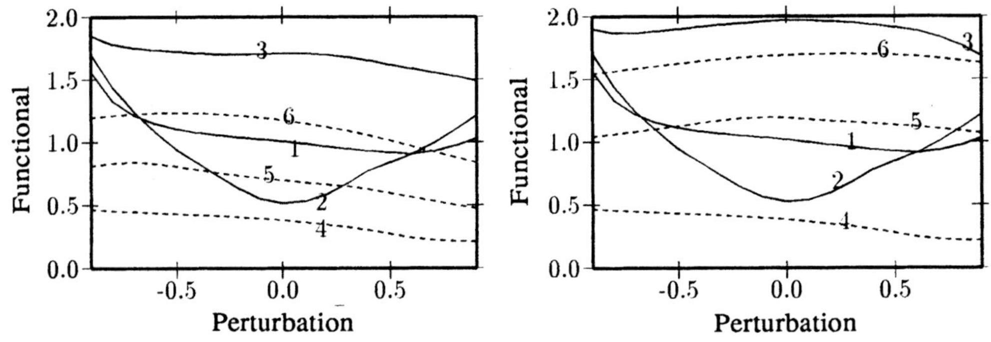

FIGURE 6. The same as in Fig. 3 for SSP with linear perturbation.

4. Conclusion

Results of numerical simulations of matched non-reciprocity tomography are presented. It was demonstrated that an appropriate choice of acoustic non-reciprocity parameters to be matched permits efficient reconstruction of current velocity fields in spite of significant 'experimental' errors. Although further studies are necessary to evaluate the proposed approach in more detail, especially for shallow water, matched non-reciprocity tomography seems to be a promising practical tool for acoustic monitoring of oceanic currents.

References

1. Godin, O.A., Mikhin, D.Yu. and Molchanov, S.Ya (1993) On effective sound speed approximation in acoustics of moving media *Izv. Acad. Sci. Atm. and Ocean Phys.* **29**, N 2.
2. Godin, O.A. and Mokhov, A.V. (1992) Parabolic equation modelling of ocean currents influence on acoustic field, in M. Weydert (ed.), *European Conference on Underwater Acoustics, Elsevier Applied Scince*, pp. 280-283.
3. Godin, O.A. and Mokhov, A.V. (1992) Narrow-angle parabolic approximation for sound in moving medium, *Sov. Phys. Acoust.* **38**, N 3

DEPTH VARIATION OF ACOUSTIC HORIZONTAL REFRACTION THROUGH A COLD-CORE OCEAN EDDY

M. V. HALL
Maritime Operations Division
Defence Science and Technology Organisation
17 Jones Bay Road, Pyrmont, NSW, 2009
Australia

ABSTRACT. Adiabatic mode theory is used to calculate the path in the horizontal plane of vertical acoustic normal-modes (at frequencies of 25 and 50 Hz) that have propagated through an idealized cold-core eddy of radius 125 km, inside which the sound-speed along the SOFAR axis varies by 15 m/s. The eddy centre is at a fixed distance (625 km) from the source, and receiver positions beyond the eddy on a line at 5° to the source-centre axis are considered. The arrival directions of the normal-modes, in conjunction with their relative amplitudes as functions of depth, are used to obtain the apparent transverse displacement of the source for various transducer depths. External to the eddy, the depth of the SOFAR axis is 1250 m. The transverse displacement is found to be a slowly varying function of range beyond the eddy and appears to approach a finite value as the range becomes infinite. For source and receivers at the SOFAR axis, for example, the dominant mode (at any frequency) is Mode 1, and at both 25 and 50 Hz the transverse displacement of Mode 1 is 9 km at ranges of 1000 km and beyond. For the eddy model considered, the degree of horizontal refraction of normal-modes at a given frequency decreases as the mode number is increased. The transverse displacement therefore decreases as the depth of the transducers is moved away from the SOFAR axis. As transducer depth decreases to 300 m for example, Mode 16 becomes dominant at 25 Hz, and the transverse displacement at 1000 km range is only 4.3 km, while at 50 Hz the dominant mode is 27 and the corresponding displacement is 4.7 km. Thus the transverse displacement at a given depth is also a slowly varying function of frequency.

1. Introduction

In order to localise a distant sound source with passive sonar, it is necessary that the bearing of the source be accurately known. Mesoscale ocean features such as fronts and eddies can cause sound to be refracted in the horizontal plane, and hence cause the source's measured bearing to differ from its true bearing. The major cause of horizontal refraction in deep water is horizontal variation of the sound speed.

Several previous studies have reported the effect of eddies on horizontal refraction. A parabolic equation calculation has been reported by Baer (1981) for propagation at a frequency of 25 Hz between transducers both 600 m deep. Baer used one example of the idealized cold-core eddies described in the work of Henrick, Siegmann, and Jacobson (1977). In a deep ocean eddy, the sound speed in the vicinity of the SOFAR axis is the main determinant of the eigenvalues of the vertical

O. Diaschok et al. (eds.), Full Field Inversion Methods in Ocean and Seismo-Acoustics, 267–272.

modes (unlike shallow water where the sea-floor depth is the dominant factor). At a range of 750 km (just beyond the eddy), Baer's calculations yielded a bearing deviation of around 0.5°. This would give rise to a 6.5-km apparent transverse displacement of the source. Hall and Irving (1989) applied adiabatic mode theory to the same eddy as studied by Baer, and obtained a bearing deviation about 8% less than Baer's.
The purpose of the present paper is to describe some further results for horizontal refraction induced by the above eddy, using adiabatic mode theory. An additional frequency of 50 Hz is considered, and the transducer depths extend from 300 m to 1500m.

2. Basic Theory

The basic theory has been described by Hall and Irving (1989). The main assumption is that the ocean is an "almost stratified" medium (Pierce, 1965), in which sound speed (c) varies slowly with horizontal co-ordinates (x, y) but arbitrarily with depth z. A source gives rise to an acoustic pressure that may be represented by a series of mode terms labelled by the mode number n. Each mode term is a product of a function $H_n(x, y)$ of position in the horizontal plane and a depth-dependent function $Z_n(z \mid x, y)$ that varies slowly with horizontal position. Each H_n can be shown to satisfy a coupled differential equation, the right-hand side of which contains inter-mode coupling coefficients. The $Z_n(z \mid x, y)$ are the usual normalised eigenfunctions of z (for given values of x and y) which satisfy the Helmholtz equation

$$d^2 Z_n / dz^2 + [(\omega / c)^2 - \xi_n^2] Z_n = 0$$

where $\xi_n(x, y, \omega)$ are the horizontal wave-number eigenvalues of the modes.
The adiabatic approximation is that the modes adapt to the gradual changes in the properties of the medium without inter-mode coupling. This is equivalent to assuming that the coupling coefficients are zero, and leads to the horizontal functions H_n satisfying a scalar wave equation in horizontal co-ordinates. In the geometric acoustics approximation (which will be applied to the horizontal plane only), the functions $H_n(x,y)$ may be represented by paths in the horizontal plane. The vertical modes Z_n travel along these horizontal paths. If the coupling coefficients are neglected, the effective horizontal propagation speed of a mode (V_n) is related to the eigenvalue by $V_n = \omega / \xi_n$.

2.1 GEOMETRIC ACOUSTIC THEORY IN THE HORIZONTAL PLANE

Since the phase-speed of each mode varies differently across the eddy, the modes will follow different paths. Geometric acoustics should provide a satisfactory model of the effect of mesoscale features on the horizontal dependence since: (i) the length scale over which the mode phase-speeds change is long relative to the wavelengths involved; and (ii) the mode phase-speeds do not increase sufficiently (even in a warm-core eddy) to cause the values of ξ_n to become small. A corollary of the latter is that caustics and shadow-zones should be absent, since the horizontal paths are not trapped within any region.
Considering the special case of propagation within a circular eddy, of radius r_1, the problem can be expressed in polar co-ordinates (r, ϕ) with the origin situated at the centre of the eddy. It is assumed that the phase-speeds $V_n(r)$ are functions of radius only. Munk (1980) showed that, in the case of circular symmetry, the equation for the horizontal paths is

$$\mu_n(r)\, r \sin \beta(r) = \Gamma \qquad (1)$$

where $\beta(r)$ is the (anti-clockwise) angle from the radius vector to the path direction, Γ is a constant for each path, and $\mu_n(r) = V_n(r_1)/V_n(r)$. For cold-core eddies, paths are refracted towards the centre. If the distance of the point of closest approach (PCA) for mode n is denoted by ρ_n, then $\beta(\rho_n) = \pi/2$ and $\Gamma = \mu_n(\rho_n)\,\rho_n$.

2.2 TRANSVERSE DISPLACEMENT AND RANGE OF INDIVIDUAL MODES

Consider a circular eddy centred at the origin O. Polar angle, bearing and direction are all measured anti-clockwise relative to the axis OX. The sound-source is at position S, at a fixed distance of r_S from the eddy centre and a polar angle of π. The locus of receiver positions is a straight line at a direction of b_R. Consider a straight path SA taken by a particular mode that has a direction of b_A ($\approx b_R$) and is incident on the rim of the eddy at point A.

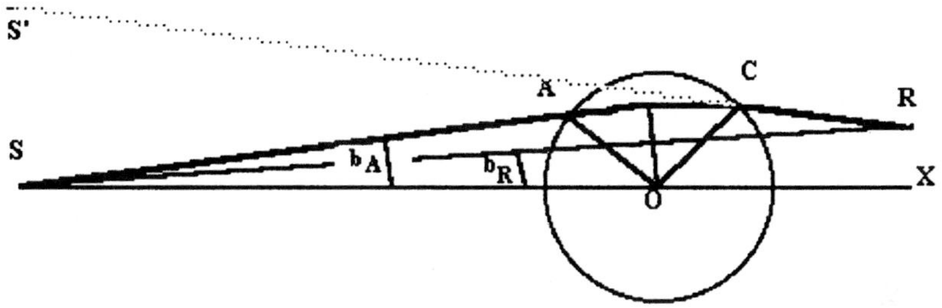

From the simple geometry of triangles, the angle β_A (the value of β at the incident point A) is given by $\sin\beta_A = r_S \sin b_A / r_1$. The value of the constant Γ is therefore:
$\Gamma = \mu_n(r_1)\, r_1 \sin\beta_A = r_S \sin b_A$, which is the perpendicular distance from O to the line SA. For a given incident path, Γ is therefore independent of both the eddy's characteristics and mode number. If the refractive index is a known function of radius r, then for a specified minimum distance from the centre to the curved path through the eddy, the angle b_A is obtained exactly by solving:
$\mu_n(\rho_n)\,\rho_n = r_S \sin b_A$.
The angle between the radii to the PCA and to the incident point A is given by (Munk, 1980):

$$\phi_0 = -\int_{r_0}^{\rho_n} \tan\beta(r)\, r^{-1} dr \text{, where } \beta(r) \text{ is given by Eq. (1).}$$

Since the polar angle of the incident point A is $\phi_A = \pi - \beta_A + b_A$, the polar angle of the PCA is $\phi_B = \pi - \beta_A + b_A - \phi_0$.
If the model is applied only to positions external to the eddy, then the path will leave the eddy (at point C) before it returns to the receiver locus. In this case the path is symmetrical about the radius to the PCA, and the polar angle of the exit point is $\phi_C = \phi_A - 2\phi_0$. Again making use of the symmetry of the path about the radius to the PCA, it is straightforward to show that the direction of the straight line that emerges from the eddy is $b_C = \phi_C - \beta_A = \pi - 2\beta_A + b_A - 2\phi_0$ (b_C will be negative for receiver points not far from the eddy rim). The emerging path intersects the receiver locus at an anti-clockwise angle of $e_R = b_C - b_R$. This difference between bearings is the eddy-induced bearing-deviation of the mode. The apparent transverse displacement (SS' in the diagram) is given by $D_T = L \tan e_R$, where L is the length of the straight line between source and receiver (SR).
In order to determine D_T, the remaining step is to derive an expression for L. This was obtained by Hall and Irving (1989) from the geometry of the triangles formed by the line from source to

receiver (SR), the line from the source to the eddy centre (SO), and the radius from the eddy centre to the path exit point (OC).

2.3 ACOUSTIC FIELD BEARING

In order to detect a bearing, it is necessary to use a sensor that divides the horizontal plane into beams. At a given range, each mode arrives at a distinct bearing and therefore is contained within a particular beam. If there is only one mode in a beam, then that mode will not interfere significantly with other modes. If there is more than one mode in a beam however, then the (complex) amplitudes of the contained modes should be summed coherently, after being modulated by the beam-response. The acoustic field bearing would be the angle of the beam that has the largest amplitude. For this paper, an idealized beam-response will be assumed, such that only one mode is contained within any beam.

3. Application of Theory

The scenario to be examined has been described by Baer (1981) and Hall and Irving (1989). A circular cold-core eddy has a radius of 125 km. An omni-directional CW source is at range r_S = 625 km from the eddy centre, on a bearing of π. The locus of the receiving array positions is a straight line on a direction of $b_R = \pi/36$ (5°) from the source. At its PCA to the eddy centre, the distance between the receiver locus and the eddy centre is therefore 625 sin $\pi/36$ = 54 km (44% of the eddy radius). This is the region of maximum magnitude of the sound-speed horizontal gradient. The eddy's rim crosses the receiver locus at distances of 510 km (near) and 735 km (far) from the source.

Sound-speed profiles (SSP) at radii of 50 to 125 km in steps of 12.5 km have been calculated from analytic expressions presented by Henrick, Siegmann, and Jacobson (1977). As radius increases from 50 to 125 km, the depth of the SOFAR axis increases from 880 m to 1250 m, and the SOFAR axis sound-speed increases from 1485 m/s to 1494 m/s. The SSP outside the eddy was taken to be that of the rim.

3.1 MODE EIGENVALUES AND EIGENFUNCTIONS

The Pedersen-Gordon N-LAYER normal mode program (Gordon, 1979) was used to calculate the eigenvalues V_n and depth-eigenfunctions Z_n for the first 31 modes at radii from 50 to 125 km (in steps of 12.5 km). Inspection of the depth-eigenfunctions and damping rates of the high-order modes indicated that the modes beyond Number 31 would be negligible at depths within several hundred metres of the SOFAR axis and ranges of the order of 1000 km. For the SSP at the rim, the modes that have the greatest amplitude at each of several depths, at frequencies of 25 and 50 Hz, are listed in Table 1. An entry such as 20.5 for example, means that the amplitudes of modes 20 and 21 are approximately equal.

For each mode, the phase-speed increases monotonically with radius. Horizontal mode paths therefore refract towards the centre. The increments in phase-speed of modes 1 to 31 in steps of 5, from the 50-km radius to the rim of the eddy, are listed in Table 2. For Mode 1, the phase-speed closely follows the 9.0-m/s increase in SOFAR axis sound-speed (V_1 increases by 8.7 and 8.8 m/s at 25 Hz and 50 Hz respectively). For the higher modes however, it may be seen that V_n does not follow the axial sound-speed so closely. The fractional change in phase-speed across the eddy is therefore largest for mode 1 and becomes progressively smaller as the mode number increases.

TABLE 1. The modes that have the greatest amplitude at a given depth (exterior to the eddy) at frequencies of 25 and 50 Hz. The SOFAR axis is at 1250 m.

Depth (m)	25 Hz	50 Hz	Depth (m)	25 Hz	50 Hz
200	20.5	-	900	2.5	4
300	16	27	1000	2	2.5
400	13	21	1100	1.5	2
500	10	15	1200	1	1
600	7	12	1300	1	1
700	5	9	1400	1	2
800	4	6	1500	2	2

TABLE 2. The increments in phase-speed of modes 1 to 31 in steps of 5, from the 50-km radius to the rim of the eddy. Frequencies of 25 and 50 Hz.

Mode	25 Hz	50 Hz	Mode	25 Hz	50 Hz
1	8.7	8.8	21	3.9	5.6
6	6.7	7.7	26	3.4	5.1
11	5.5	6.8	31	3.0	4.7
16	4.6	6.1			

3.2 MODE AMPLITUDES

For determining the relative amplitudes of the modes, a range of 1000 km was selected as being representative of the ranges of interest. The predominant factor which determines the amplitudes is the magnitude of the product of the depth eigenfunctions at the source and receiver depths (Hall and Irving, 1989). Mode damping was found to negligible except for Mode 31, the relative amplitude of which was damped to about half of its value at zero range (higher-order modes would be even further damped). The focussing correction factor for each mode was calculated by determining the end-point of a path slightly displaced in horizontal source angle from the exact path. For the majority of ranges the correction factor differed by less than 1% from unity.

4. Results for Apparent Transverse Displacement

The values of D_T for modes 1 to 31 have been calculated at various ranges beyond the eddy. The procedure was similar to that described in Hall and Irving (1989), except that a polynomial fit was obtained for the variation of each mode's phase-speed as a function of radius (a third-order polynomial was found to be satisfactory). As expected from Table 2, D_T at any range decreases monotonically with mode number. An interesting feature of the results, which is easily described in the normal-mode formalism, is the strong depth dependence of the displacement. The stronger the lowest modes are excited (in relation to the higher modes), the larger will be the predicted average D_T. The lower order modes have a greater concentration in the SOFAR duct, and much smaller magnitudes near the surface, than do high order modes. From Table 1, the corresponding values of D_T would be as listed in Table 3

TABLE 3. Apparent Transverse Displacement (km) at a range of 1000 km from the source, as a function of depth of the transducers. Frequencies of 25 and 50 Hz.

Depth (m)	25 Hz	50 Hz	Depth (m)	25 Hz	50 Hz
200	3.62 ± 0.06	-	900	8.25 ± 0.27	8.4
300	4.3	4.7	1000	8.5	8.79 ± 0.14
400	4.8	5.3	1100	8.75 ± 0.25	8.9
500	5.5	6.1	1200	9.0	9.2
600	6.3	6.6	1300	9.0	9.2
700	7.1	7.2	1400	9.0	8.9
800	7.5	7.9	1500	8.5	8.9

Conclusions

The largest values of apparent Transverse Displacement occur for sources and receivers near the SOFAR axis, since at this depth Mode 1 dominates. They are around 9 km for the eddy examined. Conversely, sources and receivers at shallow depths will excite predominantly higher modes, which will not be deviated to the same extent. The apparent displacements at a frequency of 50 Hz differ only slightly from those at 25 Hz.

References

Baer R N (1981) Propagation through a three-dimensional eddy including effects on an array *Journal of the Acoustical Society of America* 69, 70 - 75

Gordon D F (1979) *Underwater sound propagation-loss program,* Tech Report 393, Naval Ocean Systems Center, San Diego, California

Hall M V and Irving M A (1989) Application of adiabatic mode theory to the calculation of horizontal refraction through a mesoscale ocean eddy. *Journal of the Acoustical Society of America,* 86, 1465-1477.

Henrick R F, Siegmann W L, and Jacobson M J (1977) General analysis of ocean effects for sound transmission applications *Journal of the Acoustical Society of America* 62, 860 - 870

Munk W H (1980), Horizontal deflection of acoustic paths by mesoscale eddies, *Journal of Physical Oceanography* 10, 596-604.

Pierce A D (1965) Extension of method of normal modes to sound propagation in an almost stratified medium, *Journal of the Acoustical Society of America* 37, 19 - 27

STOCHASTIC BATHYMETRIC EFFECTS ON MATCHED FIELD PROCESSING

J. F. SMITH, III , O. DIACHOK, R. HEITMEYER AND E. LIVINGSTON
U.S. Naval Research Laboratory
Acoustics Division
Washington, D.C. 20375-5000

ABSTRACT This paper presents the results of a theoretical analysis of bathymetric effects on the Matched-Field-Processing (MFP) signal-gain-degradation (SGD). The bathymetry along the source track is characterized by small scale roughness, several small seamounts and one seamount that protrudes several hundred meters above the other seamounts. The analysis was obtained using a parabolic equation model to describe the "observed" acoustic field, the measured bathymetry along the source track, bottom characteristics representative of the region, and an experimentally obtained range dependent sound speed field for the water. The SGD results are computed over the full source track at low frequency. The simulations suggest that the SGD is primarily due to the interaction of energy with the nearby seamount. This degradation appears as isolated peaks at convergence zone separations measured relative to the nearby seamount.

1. INTRODUCTION

The ocean bottom in the Northeast-Pacific can be characterized as groups of abyssal hills with a thinly sedimented basaltic crust . To rigorously describe the MFP mismatch degradation due to bottom interactions in such an area requires precise bathymetry measurements and a highly accurate acoustic model. In particular, the bathymetry must be accurately known in two dimensions. Furthermore, for each point along the source track the acoustic model must be capable of correctly describing three- dimensional reflection, refraction and diffraction. Neither the precise two-dimensional bathymetric measurements nor the exact three-dimensional acoustic model are currently available. Consequently, it is important to assess the mismatch degradation that results from using an approximate bathymetry and an approximate acoustic model to compute the replica field.

In this article, we present such an assessment for a Northeast-Pacific sound-speed source track. The simulation results are obtained using the bathymetry measurements conducted along this track together with a wide-angle parabolic-equation model (FEPE) [1, 2] to compute the true field. The replica field is computed using an adiabatic-normal-mode model (ANM) [3] with a flat bottom. The bottom depth (5400 m) was chosen to be large enough so that if there were no bathymetry variations, the number of modes used to describe the replica field would exceed the number required for the true bathymetry. The mismatch degradation is then computed as a function of the number of modes used in the replica field with the minimum number of modes determined by the shallowest feature in the bathymetry. The number of modes that yields the minimum mismatch degradation is referred to here as the optimum mode-number and the corresponding degradation is referred to as the optimum degradation. The rationale underlying this computation is based on the

O. Diaschok et al. (eds.), Full Field Inversion Methods in Ocean and Seismo-Acoustics, 273–278.

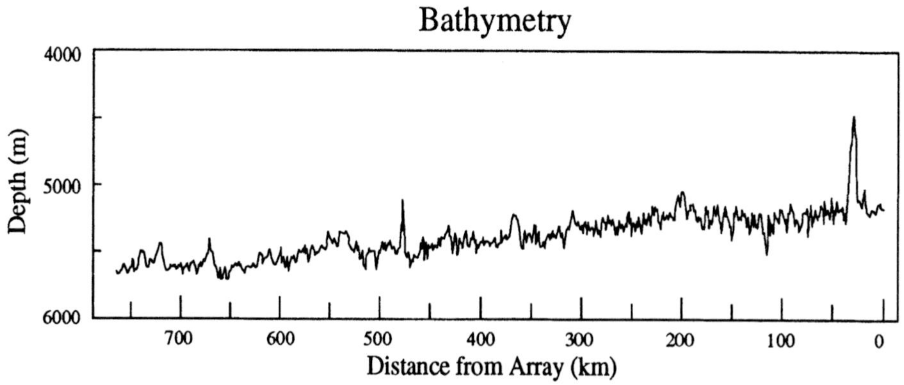

FIGURE 1. Northeastern-Pacific bathymetry plotted as a function of source range over a 4400 to 5400m depth interval

proposition that the mismatch degradation resulting from the mode-coupling introduced by the bathymetric features can be reduced by eliminating some of those high-order modes in the replica field.

This paper has the following organization. In Sec. II the environment is described. Sec. III examines the properties of the signal gain degradation as a function of range and bathymetry. In Sec. IV conclusions are given.

2. THE ENVIRONMENT

To simulate the true field and calculate replica fields a sound speed field must be provided for the water column, and bottom conditions specified. The sound speed field was described by sound speed profiles every 10 km obtained from measurement [4]. The bottom is characterized by a rugged thinly sedimented bathymetry also obtained from measurement [5].

The bathymetry measured along the track is plotted as a function of source range over a 4400 to 5400 m depth interval in Fig. 1 . As seen in the figure, the bathymetry is characterized by one large seamount and many intermediate to small sized hills. The large seamount is approximately 1 km in relief and located 30 km from the array site. The intermediate sized hills are from 1/2 to 1/4 km in relief. The small hills are distributed fairly uniformly along the track.

The bottom beneath the thin (less than 50 m) sediment layer consist of basalt with a shear critical angle [6, 7] of about 5 degrees. This environment was simulated with a fluid bottom with a compressional critical angle of 5 degrees and no overlying sediment layer. This highly idealized geo-acoustic model was then used in conjunction with the parabolic equation program FEPE to simulate mode conversion effects over sloping bathymetry. The resulting computations are expected to be of heuristic value for wavelengths greater than the sediment thickness, i.e. for frequencies below about 15 Hz.

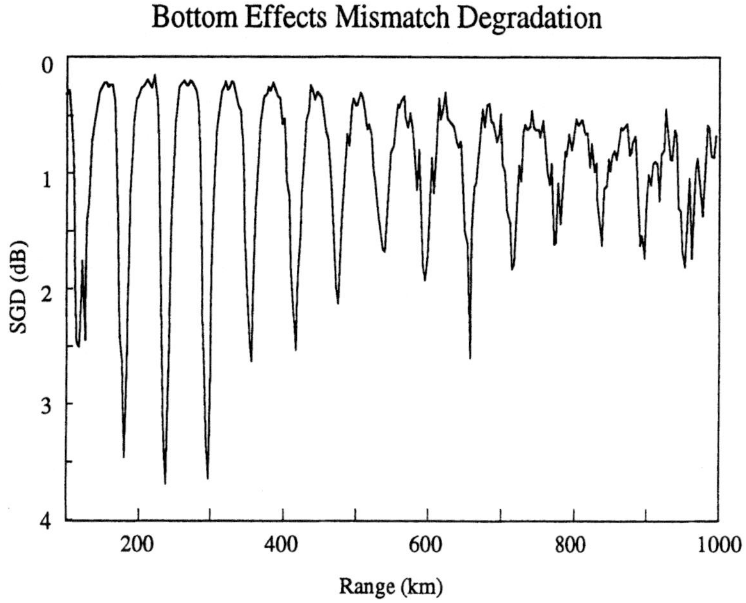

FIGURE 2. The continuous curve is the SGD plotted as a function of source range at 14 Hz. The optimum number of modes was used to do the replica field calculations.

3. FIELD and SIGNAL GAIN DEGRADATION COMPUTATIONS

In this section simulated case studies are presented to describe the properties of the SGD for a source at a depth of 100 m for the environment described above. In particular, the SGD is considered for full and small scale bathymetries as a function of range.

The calculations in this section involve three components: the simulation of the true field "measured" on the hypothetical vertical array, calculation of replica field, and finally the matched field processing step. The true field was simulated by a wide angle parabolic equation program FEPE, the replica fields calculated using adiabatic normal modes (KRAKENC), and a Bartlett processor was used for the MFP step.

In the remainder of this section, we develop properties of the mismatch degradation. Examples are presented that illustrate the low frequency dependence of the degradation on the source range, the small-scale and large scale features in the bathymetry and the number of modes used in the replica field.

The range dependence for the *optimum* mismatch degradation obtained using the full bathymetry along the track is plotted in Fig. 2. Optimization over replica field mode number can reduce SGD at low frequency by 1 dB or more at ranges less than 500 km. As seen in the figure, the SGD exhibits large peaks that occur at convergence-zone intervals and a low-level degradation that increases with source range. The convergence-zone peaks decrease with range from close to 2 dB for ranges less than 200 km to between 1.0 and 1.5 dB for ranges between 900 km and 1000 km. The minima between peaks increase from about 0.1

dB around 100 km to 0.3 dB at 1000 km. These features can be interpreted as follows. The acoustic field interacts with the bottom only at convergence zone spacings so that the number of interactions increases with range as the ratio of the range to the convergence-zone spacing. Furthermore, there are only specific ranges where the energy interacts with the large seamount. At these sea-mount ranges, only one of these interactions will be with the large seamount, the remainder will be with the intermediate and small-scale bathymetric features. For short seamount ranges, the percentage of the total energy that is corrupted by reflection from the sea mount is large, resulting in the comparatively large SGD peaks observed in the figure. Conversely, for short ranges that are not sea-mount ranges, the only degradation is due to the small number of interactions with the small-scale features and the resulting degradation is small. As the source range increases, the number of bottom interactions increases so that the degradation for the non-sea-mount ranges increases. On the other hand, for the sea- mount ranges, the percentage of the total energy that strikes the large seamount will be reduced due to the attenuation of the high-angle energy from the other bottom interactions. Consequently, the degradation at the longer sea-mount ranges is reduced. Thus, as the range becomes very large, the difference in the SGD between sea-mount and non-sea mount ranges is diminished. It is reasonable to speculate that, for extremely long ranges, so much of the bottom interacting energy will have been attenuated by the multiple-bottom interactions that this difference will disappear entirely.

In the final case we illustrate the effect of the large bathymetric features on the optimum mismatch degradation at long range. To this end, we compare the optimum mismatch degradation computed using the full bathymetry with that obtained by omitting the large sea-mount and including only the small-scale features. The results are shown in Fig. 3. The full-bathymetry degradation is shown by the solid curve; the small-scale bathymetry degradation is shown by the dashed curve. The solid curve shows two large peaks with a convergence zone separation at 890 km (1.1 dB) and 950 km (1.3 dB). The dashed curve is roughly constant over the range interval shown with a value of about 0.2 dB. This curve represents the mismatch degradation that would be found along a track with only small scale bathymetry features. Note that the degradation with the full bathymetry is always higher than the degradation without the small-scale bathymetry, the minimum separation between the two curves being about 0.2 dB. Thus, at the sea-mount ranges, the sea-mount adds about 1 dB to the degradation of 0.2 dB obtained for the small scale features only.

Seamounts are to a rough approximation, conical structures. Hence, sound interacting with seamounts is diffracted into three dimensions, with only a portion of the energy reflected into the forward direction. In contrast, the computations exhibited here assume two dimensional seamounts that reflect all of the incident energy into the plane of propagation. Consequently, the "with-seamount" curve shown in Fig. 3 represents an exaggerated estimate of the effect of the seamount.

Abyssal hills, on the other hand are, to a good approximation, two dimensional structures, resembling gratings. The bathymetry shown in Fig. 1 was measured perpendicular to the long axis of the hills. Consequently, the two dimensional "without seamount" computations, shown in Fig. 3 , are probably realistic, assuming that a critical angle of 5 degrees is reasonable.

Since abyssal hills are two dimensional structures, it follows that sound propagating along directions which are not normal to the hill will be deflected into out-of-plane directions and will not contaminate high order waterborne modes. For such propagating directions, the expected degradation due to seamounts is hypothetically small. Consequently the "without-

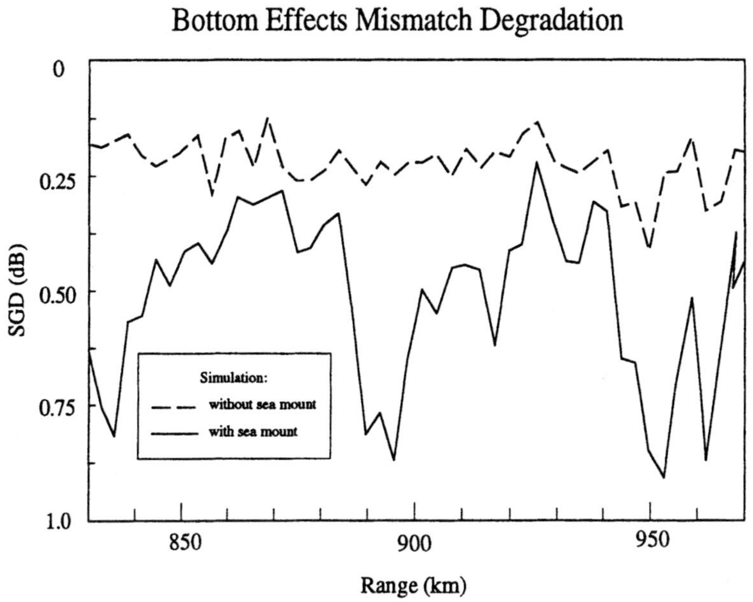

FIGURE 3. SGD vs. range curves: the solid curve is for full bathymetry and the dashed curve small scale bathymetry.

seamount" curve shown in Fig. 3 , should be considered representative only for the "normal to seamount" direction.

A more realistic physical model of the reflection process in two dimensions would include a sediment layer overlying the basalt, a range-dependent shear critical angle and shear wave attenuation in the basalt, and random roughness at both water-sediment and sediment-basalt interfaces. Computations, based on such a model are possible but beyond the scope of this paper.

The implication of these observations on matched field tomography experiments (as described in reference [8]) in the Pacific Ocean follow. Seamounts can probably be treated simply as "mode strippers", whereas abyssal hill may corrupt high order waterborne modes along "normal to hills" directions. Data along such directions may have to be mode filtered to remove contaminated modes prior to application of inversion algorithms. Reasonably realistic geoacoustic models and detailed bathymetry would facilitate estimation of the number of contaminated modes. The magnitude of this effect increases with increasing critical angle and range, and may be significant at ranges of interest to investigators concerned with global change.

4. CONCLUSIONS

This paper presents an analysis of bathymetric effects on the matched-field-processing (MFP) signal-gain-degradation (SGD) for a Northeast-Pacific environment. The cases considered were in the range interval 100 km - 1000 km at 14 Hz.

Low frequency SGD range curves exhibit large peaks (with convergence zone separations) due to interaction with the seamount 30 km from the array. The absolute magnitude of the peaks decrease with source range, whereas the local minima between peaks increases with source range due to increased interaction with smaller scale bathymetric features. Optimization over replica field mode number can reduce SGD at low frequency by 1 dB or more at short range.

To better understand the effect of the large seamount, the SGD has been calculated as a function of range at 14 Hz with the large seamount present (full bathymetry) and in its absence (small scale bathymetry). If the seamount is removed from the bathymetry the SGD is approximately constant with a value of 0.2 dB. Comparison between 14 Hz SGD vs. range curves for small scale and full bathymetry clearly show the large seamount contributes more than 1.0 dB in some cases.

Consideration of three dimensional bathymetric effects, however, suggests that the two dimensional treatment of reflection of sound from "conical" seamounts probably exaggerates this effect, whereas the "without seamount" computations are probably representative only of directions that are normal to the abyssal hills.

5. ACKNOWLEDGMENTS

The authors would like to acknowledge the valuable assistance provided by M. Collins.

References

1. Collins, M.D. (1988) *Fepe User's Guide* , NORDA Technical Note 365.
2. Collins, M. (1989) "Applications and time-domain solution of higher-order parabolic equations in underwater acoustics", *J. Acoust. Soc. Am.* **86**, 1097-1102
3. Porter, M.,B. (1992) *The KRAKEN Normal Mode Program*, Saclant Undersea Research Centre.
4. Cornuelle, B.D. , Worcester, P.F., Hildebrand, J.A., Hodgkiss, W.S., Duda, T.F., Howe, B.M., Mercer, J.A. and Spindel, R.C. , "Vertical slice ocean tomography at 1000-km Range in the North- Pacific Ocean", to be submitted.
5. Private communication with Peter Lonsdale.
6. Diachok, O., Wales, S., Dicus, R. , Feirtag, F. , Shirley, D. and Siegel, J. (1991) "Compressional and Shear Wave Velocities in the Upper Crust", in Shear Waves in Marine Sediments, edited by J. M. Hovem et al., Kluwer Academic Publishers, 387-394.
7. Diachok, O. , "New measurements of sound velocities in the upper crust: implications for crustal evolution," to be submitted.
8. A. Tolstoy, O. Diachok and L.N. Frazer, "Acoustic tomography via matched field processing," J. Acoust. Soc. Am. 89, 1119-1127 (1991).

MATCHED-FIELD INVERSION FOR SOURCE LOCATION AND EQUIVALENT BATHYMETRY

S. E. DOSSO and PIERRE ZAKARAUSKAS

Defence Research Establishment Pacific
FMO Victoria B.C. Canada V0S 1B0

ABSTRACT

A good knowledge of the bathymetry is required for reliable matched-field source localization in shallow-water environments. However, in Arctic environments the bathymetry is often poorly known and difficult to measure. This paper describes two methods of matched-field inversion for source location with an inadequate knowledge of the bathymetry. These methods are based on determining an acoustically-equivalent bathymetry model as part of the localization procedure.

INTRODUCTION

Matched-field processing (MFP) for source localization requires knowledge of the acoustic properties of the ocean environment. In realistic applications of MFP, environmental properties are required throughout the two- or three-dimensional region suspected of containing the source. In practice, however, sufficiently complete and accurate measurements of the environment may not be available; this leads to a condition known as mismatch, and can preclude reliable localization using standard MFP methods. In particular, uncertainty in bathymetry can strongly affect localization performance in shallow-water environments [1, 2]. This is a practical problem in Arctic waters, where the bathymetry is often poorly known and difficult to measure due to ice cover.

This paper describes two recent methods [3, 4] of matched-field inversion for source location with an inadequate knowledge of the bathymetry: bathymetry focalization and bathymetry-optimized ambiguity surfaces. Both methods are based on determining, as part of the localization procedure, a bathymetry model which is acoustically equivalent (within the adiabatic approximation) to the true bathymetry. In this paper, the two methods are described using the common concept of equivalent bathymetry, and the similarities and differences between the two approaches are illustrated using a synthetic example of localization with uncertain bathymetry in an Arctic environment.

O. Diaschok et al. (eds.), Full Field Inversion Methods in Ocean and Seismo-Acoustics, 279–284.

METHOD 1: BATHYMETRY FOCALIZATION

The method of focalization [5] provides a general approach to the problem of environmental mismatch based on generalizing the localization problem to include uncertain environmental parameters as well as source location in the search for an optimal match with the acoustic data. This method was originally applied to determine source range in an environment with an unknown range-dependence of the ocean sound-speed profile [5]. Dosso [3] applied the concepts of focalization to determine source range and depth in an environment with an inadequate knowledge of the (range-dependent) bathymetry. This section briefly describes the method and applies it to an example of source localization with uncertain bathymetry.

The localization problem is formulated as an optimization by defining a discrete model of the unknown parameters $\mathbf{m} = \{z_s, r_s; d_1, \ldots, d_N\}$, where z_s and r_s are the source depth and range and the d_i represent water depth values on a predefined range partition. A bounded range of values is assigned to each parameter representing the available *a priori* knowledge of the source location and bathymetry. The optimization involves finding the set of parameter values which maximize an objective function E representing the match or correlation between the measured acoustic fields p and replica fields $p(\mathbf{m})$ modelled for all realizations of $\mathbf{m}$ within the defined parameter search space. For the case of a vertical array of receivers, we have used an objective function E consisting of the Bartlett or linear correlator, scaled to within $[0, 1]$, with a value of unity indicating a perfect match.

In practice, the parameter space over which E is optimized can be extremely large with many local maxima; therefore, we have applied the method of simulated annealing (SA) [6] which provides an efficient optimization that avoids becoming trapped in sub-optimal solutions. The SA algorithm consists of a series of iterations involving random perturbations of the unknown parameters of $\mathbf{m}$. After each iteration a control parameter T is decreased. Perturbations which increase E are always accepted; perturbations which decrease E are accepted with a probability which decreases with T according to the Boltzmann distribution:

$$P(\Delta E) = \exp(-\Delta E/T). \tag{1}$$

Accepting some perturbations which decrease E allows the algorithm to escape from local maxima in search of a better solution. However, as T is decreased, the probability of accepting decreases in E decreases, and the SA algorithm eventually converges to a solution which should approximate the global maximum. The rate of reducing T, and the number and type of the perturbations define the annealing schedule. SA is a heuristic technique, and there are a variety of methods of defining an effective annealing schedule: the schedule used here is similar to that described in [3].

The SA algorithm requires the replica fields to be computed a large number of times; therefore, it is important that the propagation modelling be performed efficiently. To accomplish this, the adiabatic normal mode solution (valid for moderate range dependence) was used to model the acoustic fields:

$$p(z_h) = \sum_{j=1}^{M} \phi_j(d(r), z_s)\, \phi_j(d(0), z_h)\, \frac{\exp[i \int_0^r k_j(d(u))\, du]}{\sqrt{\int_0^r k_j(d(u))\, du}}, \tag{2}$$

where z_h and z_s are the hydrophone and source depths respectively, and r is the range. The mode functions ϕ_j and wave-numbers k_j depend on the local water depth d, and M represents the minimum number of modes occurring between source and receiver. For rapid repetitive computations of $p(\mathbf{m})$, mode functions and wave-numbers are pre-computed for a grid of values of water depths and source depths, and stored in 'look-up' tables for fast reference. Within the SA algorithm, mode function and wave-number values for a particular realization of $\mathbf{m}$ are calculated by linear interpolation between the stored grid values. This procedure is a great deal more efficient than re-computing the mode values each time $\mathbf{m}$ is perturbed.

According to the adiabatic normal mode model (2), the acoustic field depends on the mode functions at the source and receiver, but only on the range-integrals of the wave-numbers over the intervening region. Therefore, the bathymetry $d(r)$ in this region cannot be determined uniquely from the acoustic fields. However, matched-field inversion can provide an effective model for the bathymetry which is acoustically equivalent to the true bathymetry (within the adiabatic approximation) and allows accurate source localization. This equivalent bathymetry is applicable only to the localization problem in which it was determined.

To illustrate the method of source localization in uncertain bathymetry, consider the environment shown in Fig. 1. Figure 1(a) shows the true bathymetry (solid line) and source location (large cross) at 37-km range and 450-m depth. Figures 1(b) and (c) show the ocean sound-speed profile (typical of the Arctic) and bottom model: these properties were assumed to be known and did not vary with range. Synthetic acoustic data p were computed at 20 Hz for an array of 16 receivers spaced at 25-m intervals from 100- to 500-m depth using the adiabatic normal mode model (2). The parameters to be determined were taken to be the source depth and range and water-depth values each kilometre in range (the water depth at the array was assumed known), i.e., $\mathbf{m} = \{z_s, r_s;\, d_1, \ldots, d_{50}\}$. The bathymetry was known only to lie

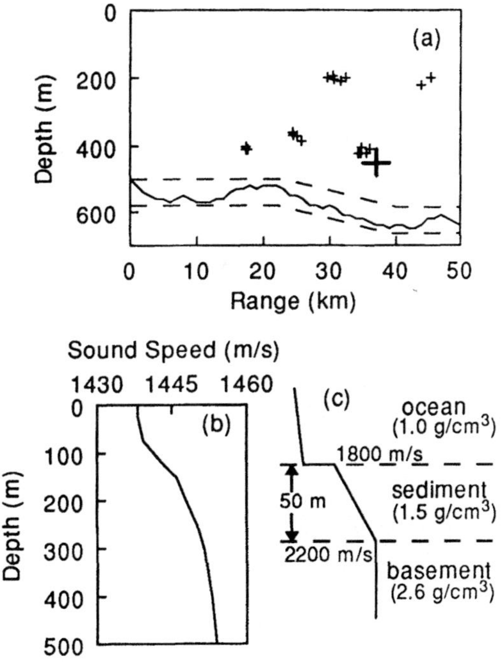

Fig. 1. Localization environment.

between the bounds indicated by the dashed lines in Fig. 1(a); this represented a total uncertainty of ~80 m. The source search region extended from 100- to 500-m depth and from 10- to 50-km range.

The importance of the bathymetry in MFP localization for this example is illustrated in Fig. 1(a). Twenty random bathymetry models were constructed within the dashed-line bounds, subject only to the constraint that the bottom slope nowhere exceeded 3° to avoid violating the assumptions of the adiabatic approximation. In each case, the bathymetry model was then held fixed and the SA algorithm used to determine the source range and depth which maximized the Bartlett objective function E. These locations are indicated by small crosses in Fig. 1(a). There are a number of clusters of solutions, one of which is near the true source location; however, it is clear that the uncertainty in bathymetry precludes reliable localization.

To locate the source given the uncertainty in bathymetry, a joint inversion was carried out for source location and bathymetry. Figure 2 shows the convergence of the SA inversion in terms of the mean match E averaged over all accepted models at a given iteration, and the source depth and range. The final values of $E = 0.999998$, $z_s = 450$ m, and $r_s = 36.9$ km represent an extremely good match and excellent estimates to the true source depth and range. Figure 2(d) shows the effective bathymetry model determined in the SA inversion. According to (2), minimizing the mismatch for the bathymetry between source and receiver requires only that the wave-number integrals approximate those of the true bathymetry. The effective model in Fig. 2(d), which oscillates about the true bathymetry, fulfils this requirement. Note, however, that at the true source range where mode functions are evaluated, the effective bathymetry coincides with the true water depth.

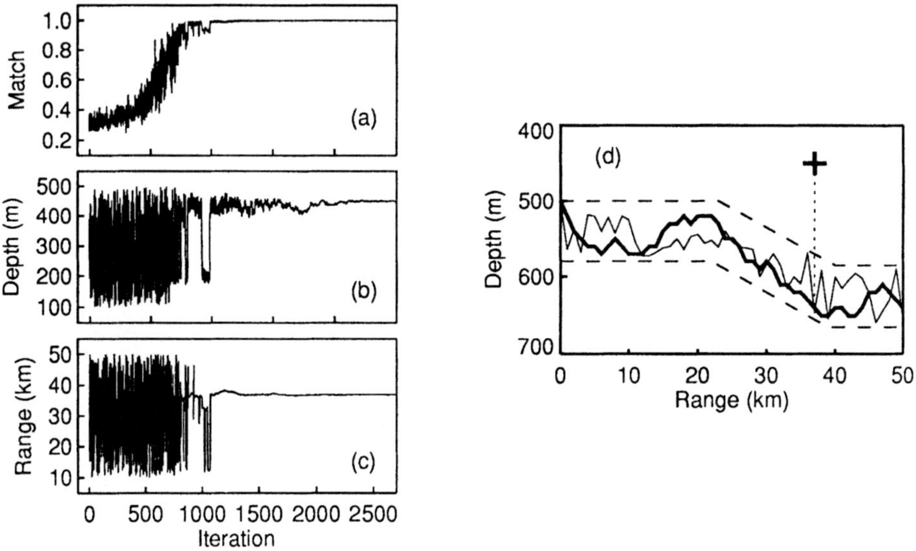

Fig. 2. Convergence of SA inversion in terms of (a) match, (b) source depth, and (c) source range; the equivalent bathymetry model is shown by the light line in (d).

METHOD 2: BATHYMETRY-OPTIMIZED AMBIGUITY SURFACES

The bathymetry focalization method presented in the previous section is applicable to any environment where the adiabatic approximation applies: large variations in water depth and/or long-range propagation can be accommodated by an appropriate choice of the range partition for the bathymetry model. The approach in Method 1 is based on searching for a general equivalent bathymetry model by including a sufficiently large number of discrete water depths in the model. However, other approaches can be formulated for cases where the variation in water depth and the propagation range are such that simple bathymetry models are sufficient. Within the adiabatic approximation (2), the acoustic fields depend only on the water depth at the source and receiver (through the mode functions ϕ), and on the range-integrals of the wave-numbers k, which depend on the bathymetry $d(r)$ between source and receivers. Therefore, assuming that the water depth at the receiver array is known, the simplest equivalent bathymetry model consists of just two values: the water depth at the source, and a water depth value representing the region between source and receivers. Whether or not a single parameter is sufficient to represent the bathymetry between the source and receiver in a particular environment can be determined in a practical manner as follows. Generate a synthetic acoustic data set for a bathymetry model with the maximum variation in water depth allowable within the assumed upper and lower bounds on the bathymetry. Then, holding the source depth fixed at the true value, carry out a one-dimensional search for the equivalent water depth value which produces the best match to the acoustic data. If an equivalent depth is found which satisfactorily reproduces the data, the representation is adequate. Zakarauskas *et al.* [4] provide a more mathematically rigorous statement of this criterion based on representing the wave-number integrals by a single-pole expansion.

In cases where a two-parameter bathymetry model applies, matched-field inversion for equivalent bathymetry is reduced to a two-dimensional search. This search can be carried out very efficiently, and applied to a grid of values for source range and depth to produce a bathymetry-optimized ambiguity surface. This surface essentially represents the envelope of the ambiguity surfaces that could be generated using all possible realizations of the bathymetry within the assumed bounds. Compared to the single, best-match source location of the bathymetry focalization method, the bathymetry-optimized ambiguity surface has the advantage of displaying the matches of the entire search region simultaneously, while obtaining the best matches over all possible bathymetries.

Figure 3 shows a perspective view of the bathymetry-optimized ambiguity surface computed for the Arctic localization example of Fig. 1. To represent the value of the match to the acoustic fields over a number of orders of magnitude, the vertical axis represents the match as $-10\ \log_{10}(1-E)$, where E represents the normalized Bartlett processor. The ambiguity surface has a strong peak at the true source location of 37-km range and 450-m depth, clearly indicating that over all possible bathymetries this location provides the best match with the acoustic data. In addition, it is interesting to note that the positions where the false source localizations cluster in Fig. 1(a) correspond to sidelobes of the bathymetry-optimized ambiguity surface.

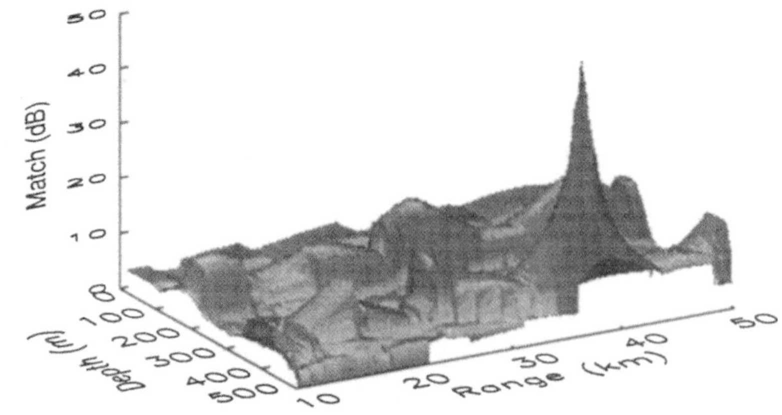

Fig. 3. Method 2: Perspective view of bathymetry-optimized ambiguity surface for the Arctic localization example of Fig. 1.

SUMMARY

This paper describes two methods of matched-field inversion for source localization with an inadequate knowledge of bathymetry: bathymetry focalization and bathymetry-optimized ambiguity surfaces. Both methods were illustrated with an example of localization with uncertain bathymetry in an Arctic environment .

REFERENCES

[1] Hamson, R. M. and Heitmeyer, R. M. (1989) "Environmental and system effects on source localization in shallow water by the matched-field processing of a vertical array," J. Acoust. Soc. Am. **86**, 1950-1959.

[2] Del Balzo, D. R., Feuillade, D., and Rowe, M. M. (1988) "Effects of water-depth mismatch on matched-field localization in shallow water," J. Acoust. Soc. Am. **83**, 2180-2185.

[3] Dosso, S. E. (1993) "Matched-field inversion for source localization with uncertain bathymetry." J. Acoust. Soc. Am. **94**, 1160–1163.

[4] Zakarauskas, P., Dosso, S. E., and Fawcett, J. A. (1994) "Matched-field inversion for source location and optimal equivalent bathymetry." submitted to: J. Acoust. Soc. Am.

[5] Collins, M. D. and Kuperman, W. A. (1991) "Focalization: Environmental focusing and source localization," J. Acoust. Soc. Am. **90**, 1410-1422.

[6] Kirkpatrick, S., Gelatt, C. D, and Vecchi, M. (1983) "Optimization by simulated annealing," Science **220**, 671-680.

THE INVERSION KERNEL IN GROUP TRAVEL TIME TOMOGRAPHY

B.E. MCDONALD
Naval Research Laboratory
Washington, DC 20375 USA

ABSTRACT Perturbative inversion of modal group travel time data has been investigated by a number of researchers for acoustic monitoring of large scale trends in the ocean's thermal structure. In regions where adiabatic mode theory is applicable, the inverse problem for each vertical eigenmode consists of an integral equation whose kernel contains the eigenfunction and its frequency derivative. The frequency derivative of the eigenfunction is obtained from a singular two point boundary value problem. We illustrate kernel structure for the lowest 8 modes of three types of SSP's, and also give the kernel correlation matrix, which connects the mode slowness perturbations with inverted sound slowness perturbation profiles.

1. Introduction

Acoustic tomography of the ocean seeks to use signal travel time data to infer changes in the sound speed structure (and thus the thermal structure) of the ocean. Pulse arrival times are determined by the horizontal group velocities of the ocean waveguide modes excited by the pulse. It was observed in the Heard Island Feasibility Test[1] that roughly eight acoustic eigenmodes survived propagation across a great expanse of the Pacific Ocean. We investigate the inversion of modal group travel time data by perturbation analysis applied to local acoustic eigenmodes. The complications of strong range dependence and multipath will be deferred for later investigation. The approach taken here is similar to that of Shang[2], but we give an explicit expression for the frequency derivative of the eigenmodes appearing in previous formulations [2, 3]. In addition, our numerics result in greater vertical resolution. We show results for three types of sound speed profiles (SSP): a smooth deep water SSP, a mostly upward refracting SSP with strong surface interaction, and a double- ducted SSP.

2. Definitions

In a weakly range dependent ocean environment, the acoustic field is described by adiabatic mode theory with vertical eigenmodes $\psi_n(z;x,y)$ where the vertical coordinate is z, and the dependence of the environment on (x,y) is weak. The eigenmodes satisfy

$$\begin{aligned} L\ \psi_n &= \kappa_n^2(x,y)\ \psi_n, \\ \Big(\psi_n,\ \ \psi_m\Big) &= \delta_{n,m}, \end{aligned} \tag{1}$$

where the linear operator L is

$$L = \rho\partial_z\rho^{-1}\partial_z + k^2(z;x,y), \tag{2}$$

with boundary conditions $\psi_n = 0$ at the pressure release surface $z = 0$, and $\psi_n \to 0$ for $z \to -\infty$, corresponding to mode trapping. Also $k(z;x,y) = \omega/c(z;x,y)$, ρ is density

O. Diaschok et al. (eds.), Full Field Inversion Methods in Ocean and Seismo-Acoustics, 285–290.

relative to $10^3 kg\ m^{-3}$, and the inner product $\left(\cdot\ ,\ \cdot\right)$ in (1) is

$$\left(A,\ B\right) \equiv \int \rho^{-1} A^* \ B\ dz, \tag{3}$$

where * denotes complex conjugate. We will supress the weak (x, y) dependence in what follows, so that functions $f(z)$ really represent $f(z; x, y)$.

Define slowness as the reciprocal of velocity: $s(z) = c(z)^{-1}$. A basic relation which results from taking the inner product of ψ_n with the ω- derivative of eq. (1) is

$$s_n^p s_n^g = \left(\psi_n,\ s(z)^2\psi_n\right) + \left((L^\dagger - L)\psi,\ \partial\psi_n/\partial\omega\right), \tag{4}$$

where phase and group slownesses of mode n are $s_n^p = \kappa_n/\omega$ and $s_n^g = \partial\kappa_n/\partial\omega$ respectively. We assume a basic state in which the operator L is self adjoint, so that the second term in (4) vanishes. We retain this term in case the perturbed slowness profile is complex to allow for attenuation. For the sake of generality, we allow for complex ψ_n.

3. Perturbation of Group Slowness

Upon perturbing the sound slowness profile, $s(z) \to s(z) + \delta s(z)$ one has from (4)

$$\delta s_n^p s_n^g + s_n^p \delta s_n^g = \left(\delta|\psi_n|^2,\ s(z)^2\right) + 2\left(|\psi_n|^2,\ s(z)\delta s(z)\right) + 2i\omega^2\left(\ s(z)\delta s_i(z)\psi(n),\ \frac{\partial\psi_n}{\partial\omega}\right) \tag{5}$$

where δs_i is the imaginary part of the slowness perturbation. Perturbing (1), then taking the inner product of the result with ψ_n leads to an expression for the perturbed phase speed: $s_n^p \delta s_n^p = (\psi_n, s(z)\delta s(z)\psi_n)$. Using this and eq. (5) one finds

$$s_n^p \delta s_n^g = 2\left(\delta\psi_n,\ s(z)^2\psi_n\right) + \left(\psi_n,\ s(z)\delta s(z)\psi_n\right)(2 - \frac{s_n^g}{s_n^p}) + 2i\omega^2\left(\ s(z)\delta s_i(z)\psi(n),\ \frac{\partial\psi_n}{\partial\omega}\right). \tag{6}$$

An alternate form of the group speed perturbation was derived by Rajan *et al*[3]. It was not clear at first that their form was identical to the result (6).

Momentarily adopting the notation of Rajan *et al.* (except for ψ in place of φ for the eigenfunction), their perturbed group speed was of the form

$$\left(\frac{1}{v_m^g} - \frac{1}{v_m^{g(0)}}\right) = \frac{1}{k_m}\int_0^\infty dz \frac{k(z)}{\rho(z)}\frac{\Delta c(z)}{c(z)^2}\left\{\omega_0 \frac{d}{d\omega}|\psi_m(\omega_0, z)|^2 - \left(2 - \frac{\omega_0}{k_m v_m^{g(0)}}\right)|\psi_m(\omega_0, z)|^2\right\}. \tag{7}$$

The response $\delta\psi$ of the eigenfunction to the sound speed perturbation Δc in this expression is not explicit, but a frequency derivative term appears in its place. We recently showed[4] the equivalence of eqs. (6) and (7) by demonstrating that for arbitrary slowness variation $\delta s(z)$, one has

$$(s(z)^2\psi_n,\ \delta\psi_n) = \omega\left(\frac{\partial\psi_n}{\partial\omega},\ \psi_n s(z)\delta s(z)\right). \tag{8}$$

The proof of this assertion, having not been published, will be given here. Combine $s(z)$ and ω into a single variable

$$\begin{aligned} q(z) &\equiv s(z)^2\omega^2 \\ &= k(z)^2 \end{aligned} \tag{9}$$

Then rewrite (1) as

$$\begin{aligned} L\psi_n &= \rho\partial_z\rho^{-1}\partial_z\ \psi_n + q(z)\ \psi_n \\ &= \kappa_n^2\ \psi_n \end{aligned} \tag{10}$$

Variation of (10) and the orthonormality condition in (1) with respect to q (sound slowness profile and/or frequency) yield:

$$(L-\kappa_n^2)\delta\psi_n = \psi_n(\delta(\kappa_n^2) - \delta q(z)) \tag{11a}$$

$$\delta(\kappa_n^2) = (\psi_n,\ \ \delta q(z)\psi_n) \tag{11b}$$

The combination of (11a,b) leads to

$$\delta\psi_n = -(L-\kappa_n^2)^{-1}(\psi_n\delta q)_\perp, \tag{12}$$

where $\perp$ denotes projection onto the space $\mathbf{R}_{\perp n}$ of functions orthogonal to ψ_n. Assuming the eigenvalue κ_n is non degenerate, then the operator $(L-\kappa_n^2)^{-1}$ is nonsingular on $\mathbf{R}_{\perp n}$.

Varying ω gives $\delta q = 2s^2(z)\omega\delta\omega$. This and (12) lead to

$$\frac{\partial\psi_n}{\partial\omega} = -2\omega(L-\kappa_n^2)^{-1}(s^2(z)\psi_n)_\perp \tag{13}$$

Varying $s(z)$ on the other hand gives $\delta q = 2s(z)\delta s(z)\omega^2$, and with (12) we have

$$\delta\psi_n = -2\omega^2(L-\kappa_n^2)^{-1}(s\delta s(z)\psi_n)_\perp. \tag{14}$$

Inner products of $\psi_n s\delta s$ with (13) and $\psi_n s^2$ with (14) then yield

$$(\frac{\partial\psi_n}{\partial\omega},\ \ \psi_n s\delta s) = -2\omega\left(\ (L-\kappa_n^2)^{-1}(s^2\psi_n)_\perp,\ \ (\psi_n s\delta s)_\perp\right) \tag{15}$$

$$(\psi_n s^2,\ \ \delta\psi_n) = -2\omega^2\left((\psi_n s^2)_\perp,\ \ (L-\kappa_n^2)^{-1}(s\delta s\psi_n)_\perp\right). \tag{16}$$

Invoking the self adjointness of L and thus $(L-\kappa_n^2)^{-1}$ establishes the assertion made in (8). From (8), the agreement between (6) and (7) follows.

4. The Group Slowness Kernel $K_n(z)$

Returning to the issue of relating changes in group travel time to the sound slowness changes, one has from (6) and (14)

$$\delta s_n^g = \frac{\omega}{\kappa_n}\left(2-\frac{s_n^g}{s_n^p}\right)(\psi_n,\ \ s(z)\delta s(z)\psi_n) - 4\frac{\omega^3}{\kappa_n}\left(\left(L-\kappa_n^2\right)^{-1}(s^2(z)\psi_n)_\perp,\ s(z)\delta s(z)\psi_n\right) \tag{17}$$

One can rewrite (17) as in integral equation

$$\delta s_n^g = \int dz\ \delta s(z)K_n(z), \quad \text{where} \tag{18}$$

$$K_n(z) = \rho^{-1}\frac{\omega}{\kappa_n}\psi_n^* s(z)\left\{\left(2-\frac{s_n^g}{s_n^p}\right)\psi_n - 4\omega^2\left(L-\kappa_n^2\right)^{-1}(s^2(z)\psi_n)_\perp\right\} \tag{19}$$

The shape of the kernel $K_n(z)$ determines the sensitivity of the modal group slowness to perturbations in slowness profile $\delta s(z)$.

Normalization of $K_n(z)$ is obtained by multiplying (19) by $c(z)$ and integrating. The second term in the curly brackets of (19) makes no contribution to the normalization, since it resides on $\mathbf{R}_{\perp n}$. The result is

$$s_n^p \int K_n(z)c(z)dz = \left(2 - \frac{s_n^g}{s_n^p}\right) \simeq 1, \tag{20}$$

the latter near equality being due to the closeness between phase and group slowness for well- trapped modes. For the results to be presented below, the left hand side of (20) falls in the range 0.9917 to 0.9996.

Project the slowness perturbation $\delta s(z)$ onto the space of functions $K_n(z)$:

$$\delta s(z)\Big|_K = \sum_n A_n K_n(z), \tag{21}$$

where $\mathbf{A}$ is determined by least squares minimization of the difference between $\delta s(z)$ and the right hand side of (21). Invoking (18), the result is

$$\mathbf{A} = \mathbf{K}^{-1}\{\delta \mathbf{s_g}\}, \tag{22}$$

where the kernel correlation matrix $\mathbf{K}$ is

$$\mathbf{K}_{nm} \equiv \int K_n^*(z)K_m(z)dz. \tag{23}$$

Equations (21 - 23) give an approximate inversion of group slowness data to yield the slowness perturbation in the water column, and are consistent with the range- integrated formulation of Shang[2].

5. Numerical Results

Numerical results for $K_n(z)$ at a frequency of 70Hz are shown in Figure 1 for three types of sound channels. The frequency choice 70Hz was made in consideration of proposed trans-ocean acoustic transmission experiments. Additional calculations (not shown here) carried out in the frequency band 60 - 80Hz showed little variation from Figure 1. Finite differences were used on a vertical grid of 201 points with 10m separation for a total water depth of 2km. The numerical inversion of the singular tridiagonal matrix representing $(L-\kappa_n^2)^{-1}$ at first revealed a sensitivity to roundoff error (certain kernels were anomalously larger than others). This sensitivity was removed by temporarily substituting $\kappa_n^2 \rightarrow (\psi_n, L\psi_n)$ in the definition of tridiagonal matrix elements. None of the results shown here contain bottom interaction, so the density ρ was taken to be unity throughout.

6. Discussion

In Figure 1 (a) are shown the first eight eigenmodes for a smoothly varying deep water sound channel typical of January near Ascension Island[5]. The group kernels $K_n(z)$ in Figure 1 (b) have dimensions of inverse meters, with the full width of the abscissa representing 0.5m^{-1}. The $K_n(z)$ occupy approximately the same vertical extent as the eigenfunctions $\psi_n(z)$, but oscillate roughly twice as many times. This is to be expected from the quadratic dependence of k_n upon ψ_n in (17). The negative extrema in the higher modes result from

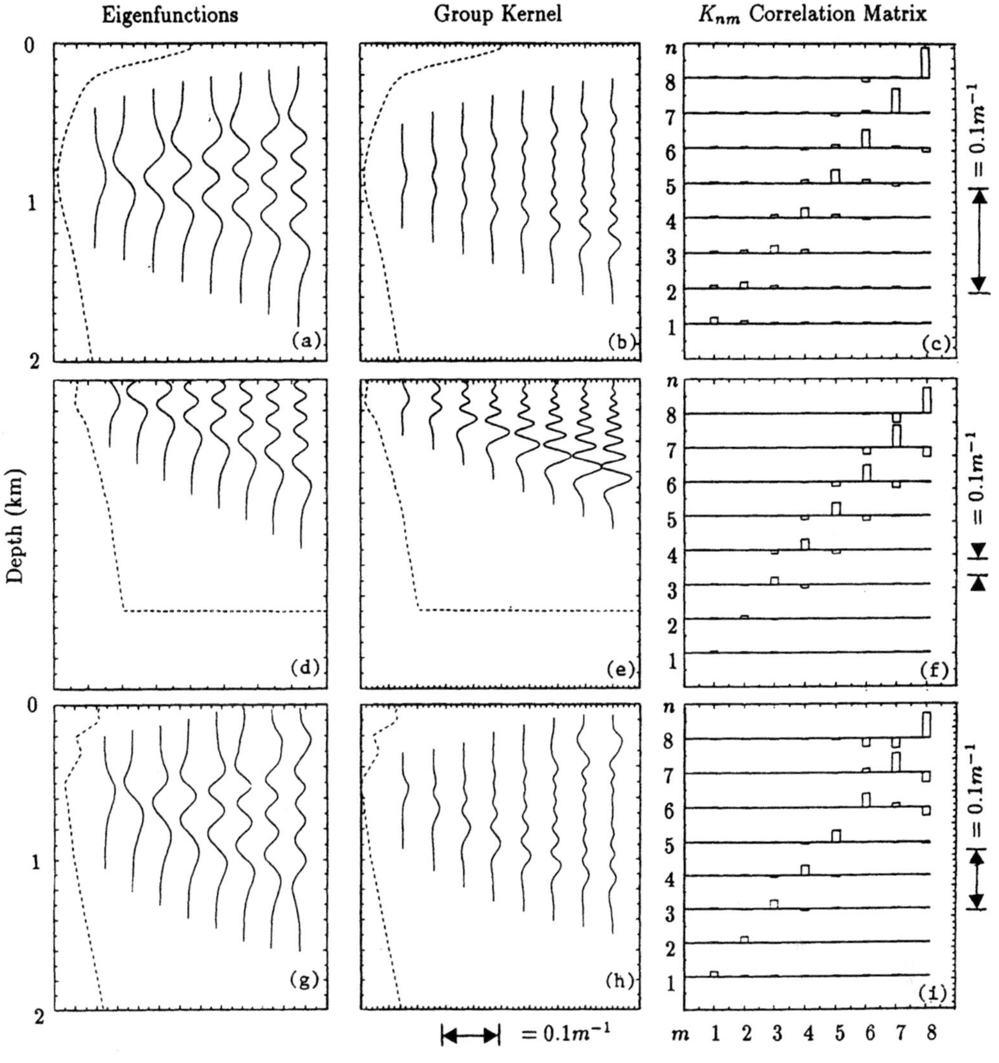

FIGURE 1. Eigenfunctions (left panels), group kernels (center panel; full width of abscissae represents 0.5 m^{-1}), and kernel correlation matrices (right panels; labels on the right translate bar heights to inverse meters) for three types of SSP. Each line plot is terminated where the amplitude of the function falls below 0.001 of its maximum amplitude.

the operator inverse term in (17). In all the examples considered here, the group kernel has its first minimum (coming up from the bottom) just below the lower ray turning point corresponding to the mode under consideration (*i.e.*, just below the lowest inflection point in ψ_n), and then has a large maximum just above the first extremum in ψ_n. Strong surface interaction and double ducting to be considered below complicate the near- surface behavior of K_n. Figure 1(c) illustrates the structure of the correlation matrix **K**. The scale on the right of 1(c) gives the relation between the height of the bars representing matrix elements and their numerical value (in inverse meters). The matrix is dominated by its diagonal elements, and strongly so for the higher modes.

Figures 1(d-f) illustrate a sound speed typical of the near- polar oceans with strongly surface interacting modes. The SSP is from Levitus[6] for a point near Heard Island (53S, 74E), where the ocean is 1500m deep. The bottom is represented by a fluid- fluid interface with bottom sound speed 1600m/s. The modes shown are not bottom interacting, so the representation of the bottom does not affect the results shown. The amplitude of K_n for this surface- limited SSP is larger than either of the two deep ocean SSP's illustrated, indicative of high dispersion.

Figures 1(g-i) show an SSP with a double duct typical of a point (44S, 41E) northwest of Crozet Island in January[7]. Modes 7 and 8 are partially trapped in the duct, and show increased group slowness sensitivity near the secondary duct.

For a smooth deep water SSP such as that of Figure 1a, modal group slowness sensitivity to water column slowness is greatest near outer inflection points (corresponding to ray turning points). The highly dispersive SSP of Figure 1 (d-f) shows about three times the sensitivity as the smooth deep water case. But again, maximum sensitivity occurs just inside the ray turning point locations. Double ducting (Figure 1g-i) complicates the relation between water column perturbations and group slowness response. There is only weak sensitivity in the upper water column for modes 2 - 6 (Figure 1h), but modal resonances with the secondary duct produce large responses of opposite sign for modes 7 and 8.

In summary, the results of Figure 1 support the unsurprising conclusion that group speed perturbations are most sensitive to sound speed changes near turning points in a ray description, but with the added complication that the sign of the inversion kernel near the turning points may be positive or negative in a double ducted SSP.

References

1. Baggeroer, A. B., K. Lashkari, J. Miller, C. S. Chiu, P. N. Mikhalevsky, and K. von der Heydt, (1994) Vertical Array Receptions off Monterey for the Heard Island Feasibility Test Signals, *J. Acoust. Soc. Am.* (in press).
2. Shang, E. C. (1989) Ocean Acoustic Tomography based on Adiabatic Mode Theory *J. Acoust. Soc. Am.* **85**, 1531–1537.
3. Rajan, S. D., Lynch, J. F., and Frisk, G. V. (1987) Perturbative inversion methods for obtaining bottom geoacoustic parameters in shallow water, *J. Acoust. Soc. Am.***82**, 998–1017.
4. McDonald, B. E. and Baggeroer, A. B. (1993) Modal Eigenfunction Perturbations and Group Speed Tomography, *J. Acoust. Soc. Am.* **94**, 1794.
5. Palmer. D. (1993) personal communication.
6. Levitus, S. (1982) Climatological Atlas of the World Ocean, *NOAA prof. paper 13*, U.S. Government Printing Office, Washington DC, 173pp.
7. Sergei Gorshov, ed. (1979) *World Ocean Atlas*, Pergamon Press.

TRAVEL TIME INVERSION OF MARINE SEISMIC DATA FOR OCEAN SOUND SPEED RECONSTRUCTION

G. BOEHM, A. CRISE and A. VESNAVER
OGS, Department of Geophysics of the Lithosphere, P.O. Box 2011, 34016 Trieste, Italy

ABSTRACT Tomographic inversion of marine seismic data is carried out to explore possible hydrocarbons reservoirs in the geological formations below the sea floor. The vertical velocity variations in the water structure are usually neglected, because they are much smaller than those at the targets of the oil and gas industry. In this paper we study the possibility of exploiting data and techniques of seismic exploration for oceanographic purposes. The depth of the sea floor is estimated by the reflection tomography at first, and then the velocity distribution in the sea water is computed by travel time inversion of both transmitted and reflected arrivals. Encouraging preliminary results are presented, obtained from a seismic profile acquired by OGS at the Ross Sea (Antarctica).

1. INTRODUCTION

Exploration and production of hydrocarbons requires an accurate imaging of the earth's interiors, and a large amount of seismic data is acquired world-wide to this goal. In 1991 only, nearly 1,400,000 miles of 2D and 3D marine seismic surveys were carried out, with a total expenditures of over 900 millions of dollars (Riley, 1993). If some information about the physical properties of the sea water could be obtained from this data at no extra cost for acquisition, the knowledge of the large scale oceanographic phenomena could be significantly improved. In fact, the sound speed in water is strictly connected to the density, which is a basic parameter to model the sea currents and features.

The characterisation of the water masses by acoustic tomography has been receiving increasing attention during the last decades (see e.g. et al., 1988). Transmitted waves and ad hoc acquisition geometries were mainly considered till now. In this paper we study the possibility of exploiting both the transmitted and reflected arrivals acquired during a conventional seismic profile. We present some preliminary results obtained applying the dual tomography (Carrion, 1991) and minimum time ray tracing (Vesnaver, 1993) to invert the travel times of reflected and transmitted waves in the sea water.

2. INVERSION STRATEGY AND EXPERIMENTAL RESULTS

The inversion procedure adopted here consists of three steps:

1. travel time picking of seismic signals in the pre-stack domain;
2. estimation of the sea floor depth by reflection tomography;
3. tomographic inversion of sound speed in the sea water.

Firstly, the seismograms are interpreted to get the travel times of the seismic signals to be inverted. Eight shots with 96 receivers each were considered, with a spacing of 100 m

O. Diaschok et al. (eds.), Full Field Inversion Methods in Ocean and Seismo-Acoustics, 291–296.

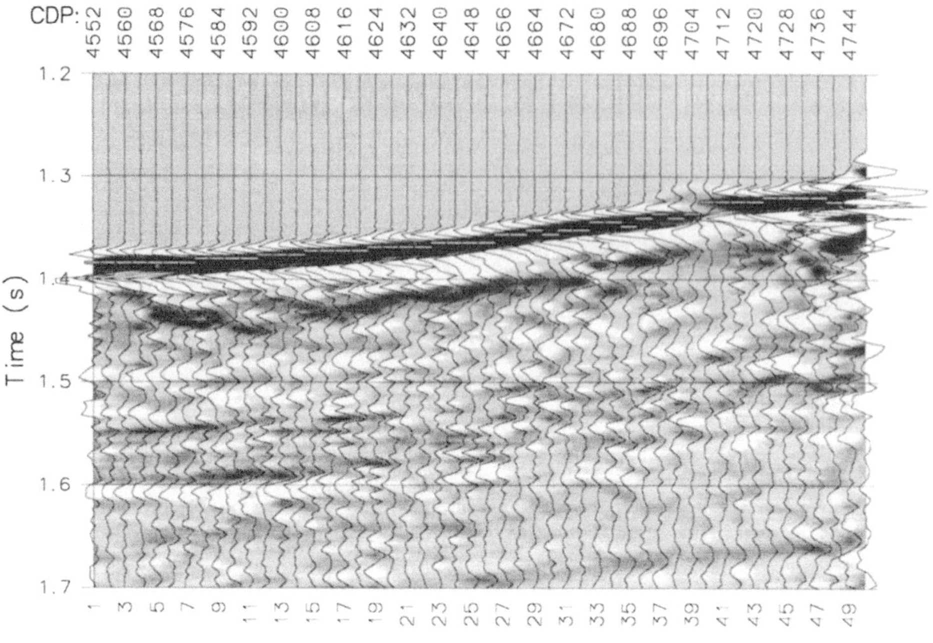

FIGURE 1.

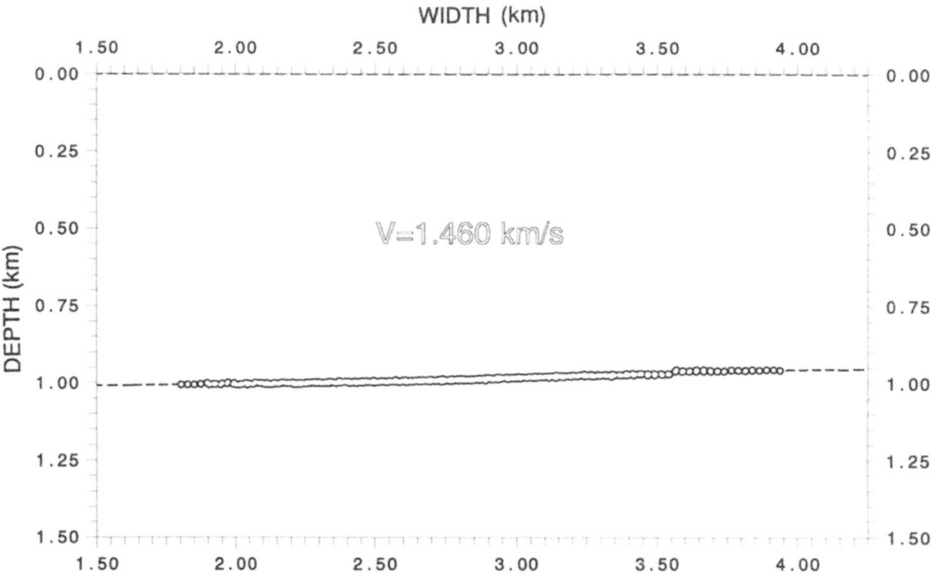

FIGURE 2.

between the sources and 50 m between adjacent receivers. The frequency band of seismic signals is in the range of 5 to 80 Hz, with a sampling rate of 4 milliseconds. Fig.1 shows a common offset section (170 m) of the profile, acquired by OGS at the Ross Sea (Antarctica). The main peaks of the sea floor reflection were picked for all seismograms, indicated by thin white bars. The same was done for the direct arrivals, i.e. the transmitted waves travelling from the source to the receivers close to the sea surface.

The second step was the determination of the sea floor depth directly from seismic data. The joint determination of the velocity distribution and reflectors depth from travel times may be obtained exploiting the multiple coverage of conventional seismic acquisition (Carrion et al., 1993a,b). We start with some initial guess for the velocity distribution and the reflectors depth, which may be both very different from the correct solution. Tracing the rays for each source-receiver couple, we estimate the corresponding reflection points. Since all reflection points must be located along a single reflecting surface (the sea floor), we can update our model iteratively to reduce their dispersion, improving alternately the spatial position of reflectors and the velocity distribution. The final solution is obtained when the dispersion of the estimated reflection points is minimum, and the same velocity-depth model is obtained for all source-receiver offsets.

Fig.2 displays the reflection points estimated assuming a constant velocity in the sea water. The sea floor is slightly dipping, with a small step on the right. An average velocity of 1.460 km/s was obtained, with an error smaller than 4 m/s. In fact, a minimum dispersion is obtained for that value (Fig.3), whilst a sensibly larger one is got with a velocity of 1.456 km/s (Fig.4). In these figures, the different darkness of dots corresponds to the different offsets used to estimate the reflection points. The same reflecting surface (dotted line) is interpolated from all offsets contributions when the optimal velocity is used (Fig.3).

The third step was the estimation of the vertical velocity variations. The basic problem in this case is the choice of the model discretization. A first test was carried out on the Ross Sea data assuming a water stratification discretized in 9 levels only (Fig.5).

We started with an homogeneous velocity field, with the average value of 1460 m/s provided by reflection tomography. The estimated velocity field shows a low velocity zone (1439 m/s) in the uppermost part, extending the mixed layer within the first 200 m, which may sound irrealistic for an Atlantic Basin in the austral summer. We decided therefore to increase to 12 the number of layers, modifying their thickness to fit better the model to a mean profile computed according to real data recently acquired in the Ross Sea. The obtained result is still stable and in good agreement with the known phenomenology in that zone. But increasing to 19 the layers number, a totally unreliable result was obtained. Unlike the previous two tests, where a satisfying convergence was obtained after 9 of iterations only, in the last case quite different results were obtained at each iteration. This means that the acquisition geometry and the low velocity contrasts do not allow a so detailed reconstruction of the velocity gradients.

The velocity values estimated for the three models are reported in Fig.6, where also the corresponding mean velocities V_t can be compared with that V_m used for the water bottom reconstruction. The condition numbers obtained for the different tomographic grids are listed in the Table. We notice that the condition numbers decrease when the layers number is larger, which confirms the decreasing inversion stability observed. We can notice also that the velocity estimated in the uppermost layer is always the same in all three cases. This is due to the fact that this part of the model space is well constrained by the experimental data, i.e., in particular, by the transmitted waves.

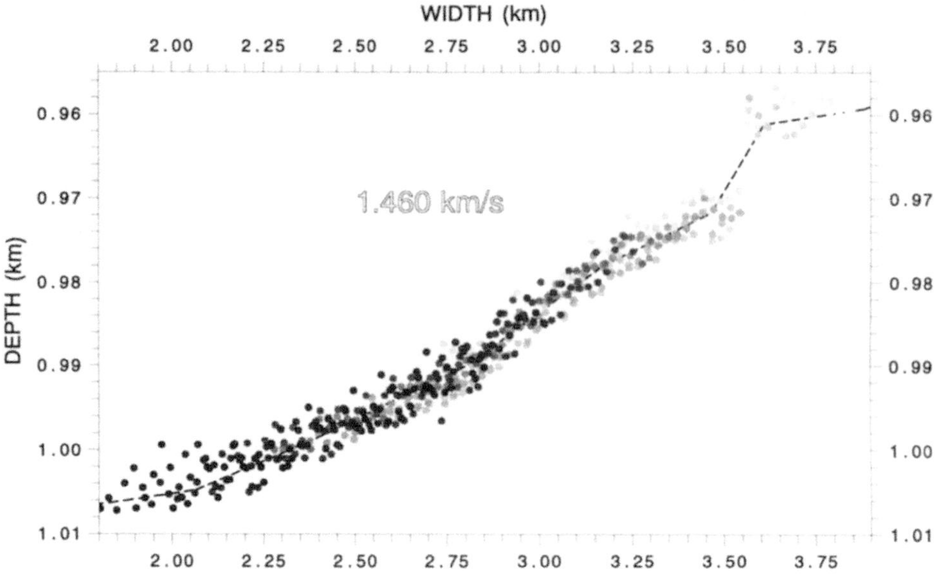

FIGURE 3.

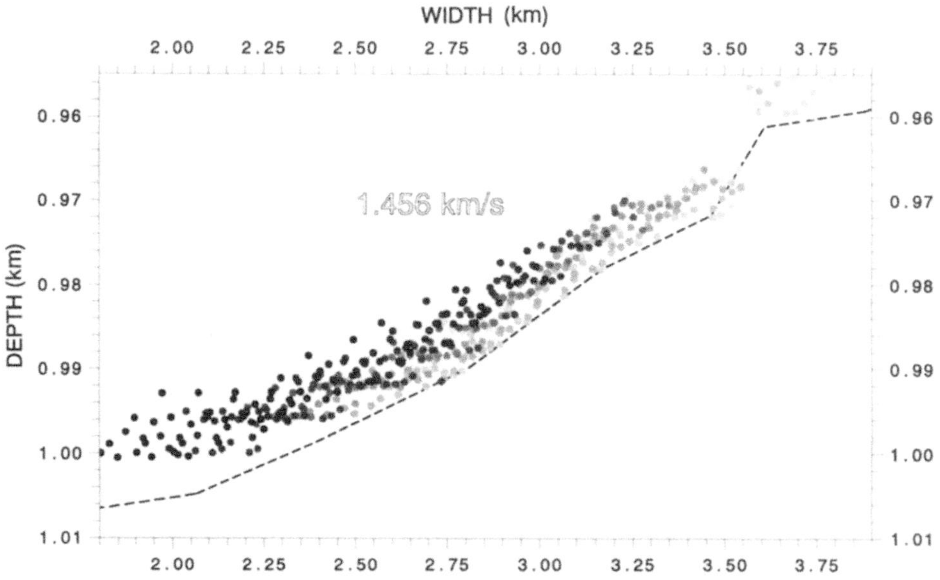

FIGURE 4.

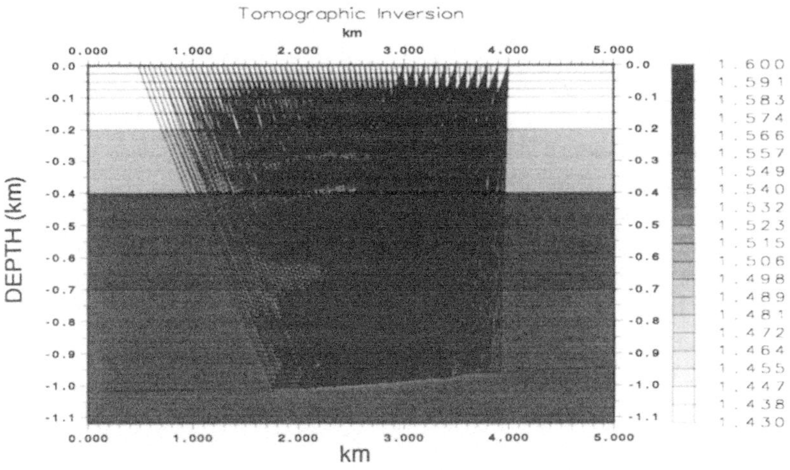

FIGURE 5.

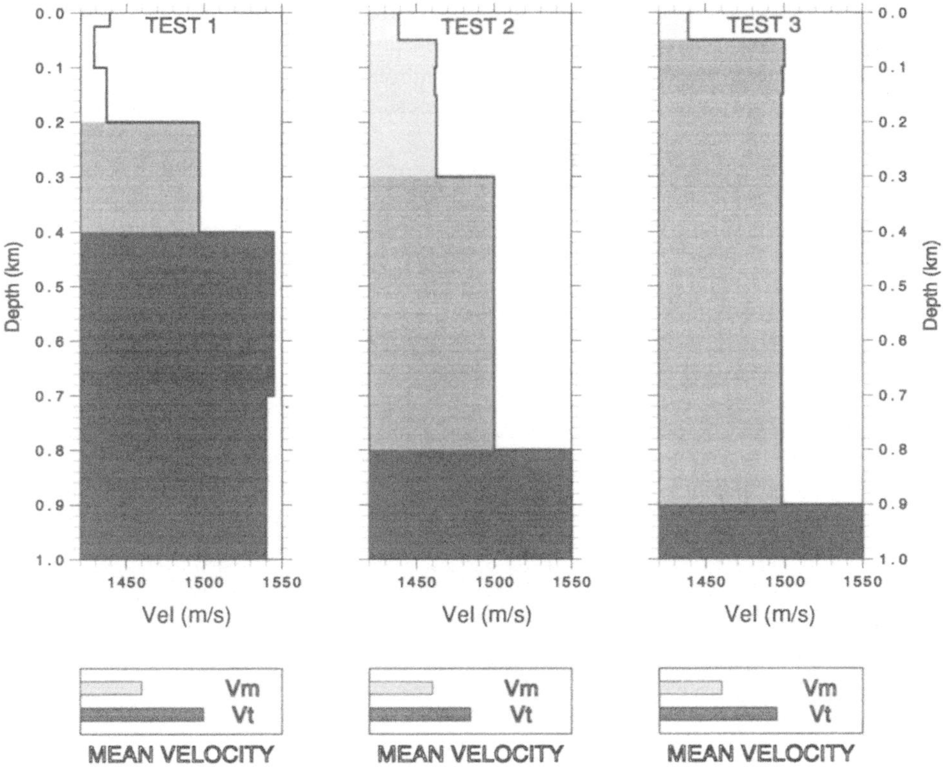

FIGURE 6.

TABLE

Condition number TEST1	Condition number TEST2	Condition number TEST3
1.86E+04	6.39E+04	15.35E+04

Additional tests were carried out recalculating the water depth with the new velocity fields provided by the traveltime inversion, but without getting improvements with respect to the previous results. In fact, the dispersion of the estimated reflection points becomes larger when the average velocity obtained by the tomographic reconstruction diverges from the optimal velocity of 1460 m/s previously found. This fact suggests us the criterion that one should prefer the tomographic solutions with an average velocity consistent with that used for the reflectors reconstruction in depth. According to it, the second model should be preferred in our case.

3. CONCLUSIONS

Marine seismic data can be used to estimate the vertical gradient of the sound speed in the sea water, which is tightly connected to the density. A preliminary application of the proposed method to a real seismic profile provided encouraging results, demonstrating that a typical oceanic water structure may be identified in the Ross Sea.

Since an enormous amount of such data is acquired every year world-wide, mainly for industrial surveys but also for geophysical research, a noticeable additional information source is so available for oceanographic studies.

Since the acquisition techniques of marine seismic profiles is optimised for different targets and, in particular, for much larger velocity contrasts, we may not expect a high resolution. The model discretization for travel time inversion could be guided by a priori knowledge about the expected water layering, or by other independent measurements.

References

1. Carrion, P. (1991) Dual tomography for imaging complex structures, *Geophysics*, **56**, 1395–1404.
2. Carrion, P., Boehm, G. Marchetti, A., Pettenati, F., and Vesnaver, A. (1993a) Reconstruction of lateral gradients from reflection tomography, *Journal of Seismic Exploration*, **2**, 55–67.
3. Carrion, P., Marchetti, A., Boehm, G., Pettenati, F., and Vesnaver, A. (1993b) Tomographic processing of Antarctica's data, *First Break*, **11**, 295–301.
4. Cornuelle, B., Munk, W., and Worcester, P. (1989) Ocean acoustic tomography from ship, *Journal of Geophysical Research*, **94**, 6232–6250.
5. Riley, D.C. (1993) Geophysical activity in 1991, *The Leading Edge*, **12**, 1094–1117.
6. Munk, W., and. Wunsch, C, (1979) Ocean acoustic tomography: A scheme for large scale monitoring, *Deep-Sea Research*, **26A**, 123–161.
7. Vesnaver, A. (1993) Minimum time ray tracing of reflected, refracted and diffracted waves in irregular grids, *Expanded Abstracts of 55th EAEG Meeting*, P059.

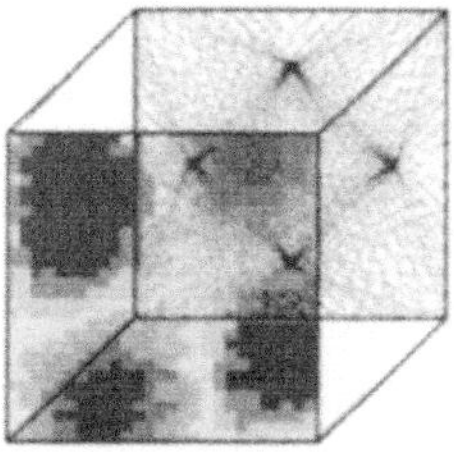

section 5

Geoacoustic Parameter Estimation

Waveform Inversion for Geoacoustic Parameters of the Ocean Bottom in Shallow Water : Matched Mode Method

Subramaniam D. Rajan
Woods Hole Oceanographic Institution
Woods Hole, MA 02543 USA.

ABSTRACT In full field inversion methods, the problem of estimating the geoacoustic parameters of the ocean bottom is generally solved by casting it as a parameter estimation problem in which a cost function $\phi(\boldsymbol{p}) = \sum_n \mid d_n - f_n(\boldsymbol{p}) \mid^2$, where d_n is the measured field at the nth receiver, f_n is the computed field for the assumed bottom model and $\boldsymbol{p}$ is a vector containing the unknown parameters, is minimized. In this paper, we present a modified cost function based on decomposing the field into its contributions from individual modes and then determining the set of bottom parameters that predict this field mode by mode. We investigate the performance of this algorithm using synthetic and field data.

1. Introduction

In shallow water, the geoacoustic parameters of the ocean bottom play an important role in influencing the propagation of sound in the water column. The ocean bottom acoustic parameters are the compressional wave speed (c_p), the compressional wave attenuation (α_p), shear wave speed (c_s), shear wave attenuation (α_s) and density (ρ) which in general are functions of all three spatial coordinates. If we assume horizontal stratification, these parameters are then functions of depth alone. Further by modeling the ocean bottom as a stack of homogeneous layers, the bottom model can be represented by a vector whose elements are the geoacoustic parameters associated with individual layers.

In full field methods, the general approach used to solve the problem of estimating the geoacoustic parameters from field measurements is to cast the problem as a parameter estimation problem. The measured field at a set of receivers is the data which is represented by the vector $\boldsymbol{d}$ where d_n is the pressure field at the nth receiver. This is related to the bottom model through the equation $d_n = f_n(\boldsymbol{p})$, where the vector $\boldsymbol{p}$ is a vector containing the parameters to be estimated. The estimation problem then reduces to the determination of the set of parameters $\boldsymbol{p}$ which minimizes the cost function

$$\phi(\boldsymbol{p}) = \sum_n \mid d_n - f_n(\boldsymbol{p}) \mid^2 . \tag{1}$$

This estimation problem is nonlinear[1] and therefore a global optimization method such as simulated annealing is used to solve the problem[2]. In this paper, we present a modified cost function which we call the modal cost function. Simulations carried out indicate that this cost function is only weakly nonlinear and better behaved than the cost function at Eqn. (1). Calculus based algorithms can therefore be employed to solve the parameter estimation problem.

O. Diaschok et al. (eds.), Full Field Inversion Methods in Ocean and Seismo-Acoustics, 299–304.

In the analysis presented in subsequent sections, we assume that the field has been acquired on a real or synthetic aperture horizontal array using an omnidirectional CW source. This experimental configuration is similar to that used by Lynch et al. [3] for their experiment conducted in the Gulf of Mexico.

2. Modal Cost Function

In shallow water the acoustic pressure in the far field is dominated by contributions from the trapped modes and is given by[4]

$$P(r) \sim \sqrt{\pi}(1+i)\sum_{m=1}^{M} a_m U_m(z_s)U_m(z_r)\frac{\exp(ik_m r)}{\sqrt{k_m r}}, \tag{2}$$

where $P(r)$ is the complex pressure field at range r, U_m is the modal function, a_m is the modal normalization constant, k_m is the modal eigenvalue, and z_s and z_r are the source and receiver depths, respectively.

In a field experiment the field $d_n = P(r_n)$ is measured at discrete ranges r_n which can be decomposed into contributions from individual modes. We then have,

$$d_n = \sum_{m=1}^{M} d_{mn}, \tag{3}$$

where d_{mn} is the field at the nth receiver due to the mth mode. Similarly, the computed field $f_n(\boldsymbol{p})$ at range r_n for a given bottom model $\boldsymbol{p}$ can be written as a sum of contributions from individual modes. Then,

$$f_n(\boldsymbol{p}) = \sum_{m=1}^{M} f_{mn}(\boldsymbol{p}), \tag{4}$$

where $f_{mn}(\boldsymbol{p})$ is the computed field at range r_n due to mode m.

We now define the modal cost function

$$\Psi(\boldsymbol{p}) = \sum_m \sum_n \mid d_{mn} - f_{mn}(\boldsymbol{p}) \mid^2 . \tag{5}$$

Comparing Eqns. (1) and (5), we note that in Eqn. (1) the cost function compares the total measured field to the total field predicted by the model and estimates the bottom parameters that minimizes this difference in a least squared sense. On the other hand, in Eqn. (5) such a comparison is made mode by mode.

To determine whether the modal cost function is better behaved than the cost function at Eqn. (1), we need to consider the linearity of Eqn. (5). We write

$$\begin{aligned} \Psi(\boldsymbol{p}) &= \sum_m \left(\sum_n \mid d_{mn} - f_{mn}(\boldsymbol{p}) \mid^2\right) \\ &= \sum_m \psi_m(\boldsymbol{p}) \end{aligned} \tag{6}$$

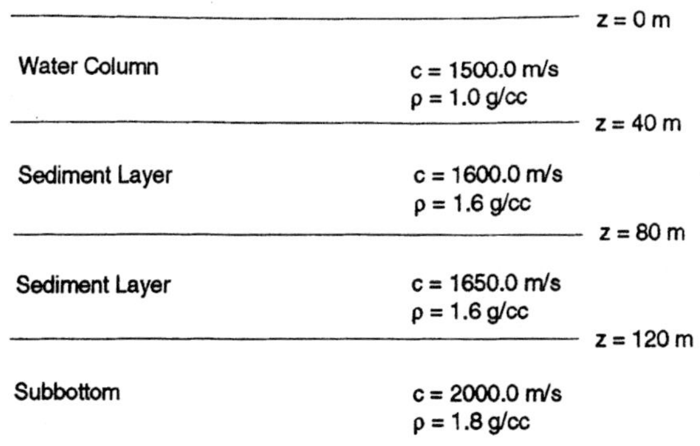

Figure 1: Ocean Model

where $\psi_m(\boldsymbol{p})$ is the cost function for each mode. If each of this cost function is linear in $\boldsymbol{p}$, then the sum of such cost functions will also be linear and the parameter estimation problem can be solved by any calculus based algorithm.

Let k_m^t be the true eigenvalue of the mth mode and $k_m(\boldsymbol{p})$ the corresponding eigenvalue for the bottom model represented by the vector $\boldsymbol{p}$. It can be shown that as one sweeps through the acceptable values of $\boldsymbol{p}$, the cost function has a single minimum (i.e., linear) if the quantity $|k_m^t - k_m(\boldsymbol{p})| \times R$, where R is the range over which the field is summed in Eqn. (5), is less than 2π. However, the resolution attainable with the modal cost function will be less than that attainable if Eqn. (1) is used. In the case of lower order modes, the changes in eigenvalues are small and the above condition can be satisfied. For example consider the ocean model in Figure 1. For source frequency of 50 Hz, the eigenvalue of the lowest mode is 0.2026846. If the layer speed in the top sediment layer is varied from 1510 m/s to 1950 m/s the eigenvalue of the lowest mode changes from 0.2057882 to 0.2002674 giving a maximum value of $|k_m^t - k_m(\boldsymbol{p})|$ = 0.003131036. We see that in this case even when the search is made over a wide range of values for the layer speed, the linearity condition will be satisfied if the range over which the summation in the cost function is carried out is less than 2000m. In the case of higher order modes, the changes in the eigenvalues are larger and the condition for the cost function to be linear can be satisfied by reducing the range R. Reducing the range over which the summation is carried out will, however, reduce the resolution of the estimates.

In simulation studies done with synthetic data, the surficial sediment acoustic properties were obtained by using the lowest order mode in the cost function. The estimates were accurate and, in most cases, were obtained in one step because the problem was almost linear. This, therefore, suggests that starting with the lowest order mode, one can obtain the acoustic properties of deeper layers by adding more and more modes to the cost function.

3. Decomposition of the total field into its modal contributions

The usefulness of the cost function in Eqn. (5) depends on our ability to decompose the field into its modal components. This is accomplished in two steps. In the first step, the eigenvalues

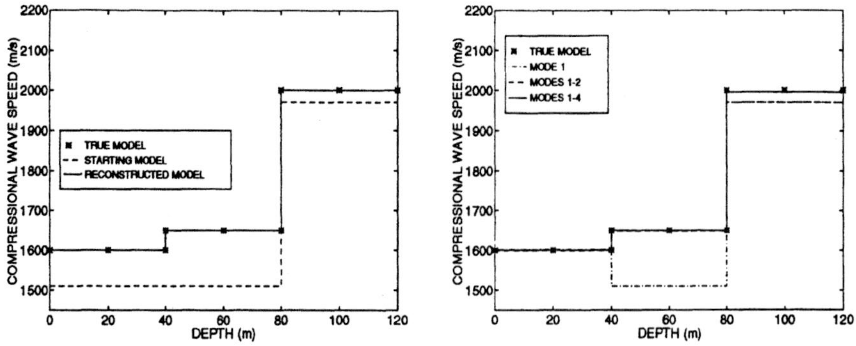

Figure 2: (i)True, starting and reconstructed profiles (ii) Evolution of bottom model

of the propagating modes are obtained by FFT beamforming[5] or by using high resolution frequency estimation methods[6]. The next step is to determine the modal amplitudes. For this, we write Eqn. (2) as

$$y = Ae \tag{7}$$

where $y = [\sqrt{r_1}P(r_1), \cdots, \sqrt{r_N}P(r_N)]^T$, $A_{mn} = \frac{\exp(ik_m r_n)}{\sqrt{k_m}}$, $e = [e_1, \cdots, e_M]^T$, and $e_m = (1 + i)\sqrt{\pi} a_m U_m(z_s) U_m(z_r)$. The modal amplitudes e_m are obtained by matrix inversion of Eqn. (7).

4. Inversion for Bottom Parameters

4.1 SYNTHETIC DATA EXAMPLE

In this section we present an example of inversion using synthetic data. The data were obtained using the ocean model in Figure 1. The other experimental parameters were: frequency = 50 Hz., source depth = 18 m and receiver depth = 22 m. In performing the inversion we assumed that the unknowns were the compressional wave speeds in the two layers and in the terminating half space. The reconstructed profile, true profile, and the starting model for the inversion are shown in Figure 2(i) for the noise free case. We note that the reconstructed model matches the true model. This inversion was performed starting with the lowest order mode. The evolution of the bottom model with the inclusion of more and more modes into the inversion algorithm is shown in Figure 2(ii).

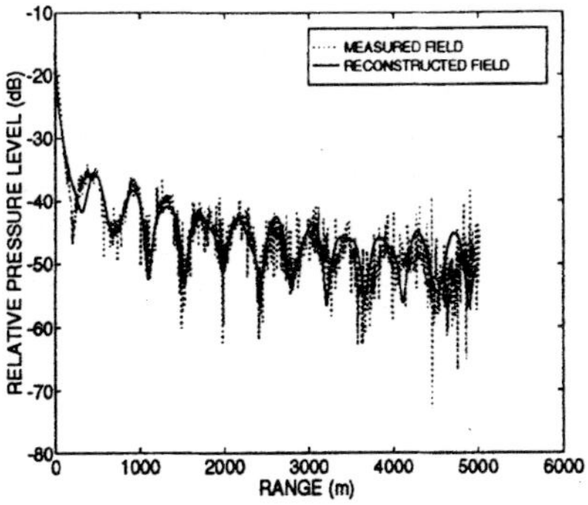

Figure 3: Measured field and reconstructed field

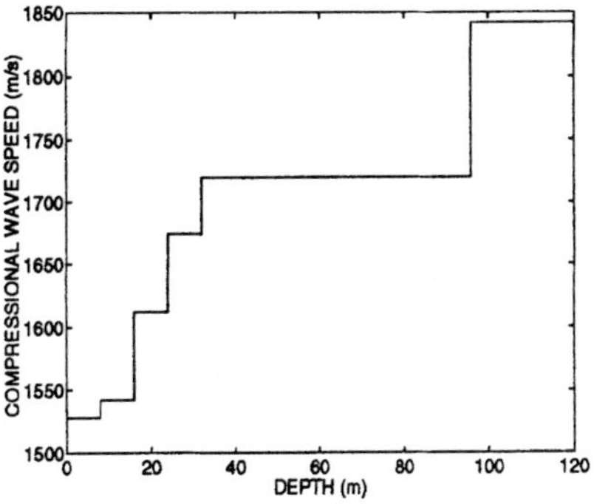

Figure 4: Estimated compressional wave speed profile

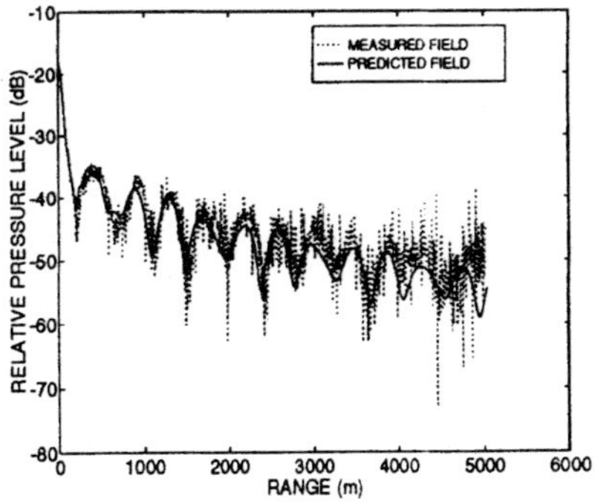

Figure 5: Measured field and field predicted by model

4.2 INVERSION OF FIELD DATA

In this section, we present an inversion using data obtained during an experiment conducted in the Gulf of Mexico[3]. The measured field is shown in Figure 3. The contribution of each mode to the total field was obtained by following the procedure outlined in Section 3. As a check on the validity of the results, the total field was computed by summing the contribution of each mode and then compared with the measured field. The comparison between the measured field and the field reconstructed from its modal contributions is shown in Figure 3.

The data (the contribution of individual mode to the total field) thus obtained were used in an inversion algorithm based on the modal cost function. In performing the inversion, we assumed that the only unknown was the compressional wave speed profile. The bottom was modelled as a fluid with the density and compressional wave attenuation in the sediment layers obtained from archival data. The compressional wave speed profile obtained by inversion is shown in Figure 4. The agreement between the field predicted by this model and the measured field is excellent as seen in Figure 5.

5. Conclusions

We have introduced a modified cost function for estimating the bottom properties from the field acquired on horizontal array. The problem is only weakly nonlinear for lower order modes and therefore calculus based methods can be used to estimate the geoacoustic parameters of the top sediment layers. Having obtained this, one can determine the parameters of deeper layers by including other modes in the cost function. We demonstrated the performance of this method using synthetic and field data.

6. References

1. Rajan, S. D.,(1992), Determination of geoacosutic parameters of the ocean bottom - Data requirements, *J. Acoust. Soc. Am.*, 92,, 2126-2140.

2. Collins, M.D., Kuperman, W.A. and Schmidt, H.(1992) Nonlinear inversion for ocean bottom properties, *J. Acoust. Soc. Am*, 92, 2770-2783.

3. Lynch, J.F., Frisk, G.V. and Rajan, S.D.(1991) A comparison of broad band and narrow-band modal inversions for bottom geoacoustic properties at a site near Corpus Christi, Texas, *J. Acoust. Soc. Am.*, 89, 648-665.

4. Brekhovskhih, L. and Lysanov Yu (1982) *Fundamentals of ocean acoustics*, Springer, Berlin.

5. Frisk, G.V. and Lynch J.F.(1984) Shallow water waveguide characterization using the Hankel transform, *J. Acoust. Soc. Am.*, 76, 205-216.

6. Rajan, S. D. and Bhatta, S.D. (1993) Evaluation of high-resolution frequency estimation methods for determining frequencies of eigenmodes in shallow water acoustic field, *J. Acoust. Soc. Am.*, 93, 378-389.

A REVIEW OF MODAL INVERSION METHODS FOR INFERRING GEOACOUSTIC PROPERTIES IN SHALLOW WATER

GEORGE V. FRISK
Woods Hole Oceanographic Institution
Woods Hole, Massachusetts 02543
U.S.A.

ABSTRACT. A review is presented of modal inversion techniques for determining the acoustic properties of the seabed in shallow water. Attention is focused on methods which utilize measurements of the modal eigenvalues at one or more frequencies as input data. Various inversion schemes for inferring the geoacoustic properties in a horizontally stratified waveguide are discussed. It is shown that these concepts can be extended to laterally varying waveguides as well, based on the notion, originating in adiabatic mode theory, that the local modes adapt to the local environment. Measurements of the modal trajectories in range can then be used to infer the range-dependent variation in geoacoustic properties. The results of the range-dependent inversion of experimental data obtained at 140 Hz in Nantucket Sound are presented.

1. Introduction

It is well known that the normal modes of an acoustic waveguide are fundamental constituents of the sound field and have properties which are intimately related to the characteristics of the waveguide and its boundaries [1]. In particular, the modal eigenvalues are significantly influenced by the acoustic properties of the seabed. It is therefore possible, in horizontally stratified environments, to use measurements of these eigenvalues as input data to inversion schemes for inferring the geoacoustic properties. These concepts can be extended to laterally varying waveguides as well, based on the notion, originating in adiabatic mode theory, that the local modes adapt to the local environment. Here, the modal evolution of the waveguide with range is influenced by lateral variations in the geoacoustic parameters. Measurements of the modal trajectories in range can therefore be used to infer the range-dependent variation in geoacoustic properties. In this paper, we focus on several narrowband inversion schemes developed for both horizontally stratified and laterally varying waveguides. We also describe experimental techniques by which the required eigenvalue data can be obtained. Finally, we present the range-dependent results of inverting experimental data obtained at 140 Hz in Nantucket Sound.

2. The Range-Independent Case

We consider a water column of thickness h, characterized by a constant density ρ and a sound speed $c(z)$, and bounded above $(z=0)$ by a pressure-release surface and below $(z=h)$ by a horizontally stratified bottom. Then, for a point source with cylindrical coordinates $(r=0,\ z=z_0)$

O. Diaschok et al. (eds.), *Full Field Inversion Methods in Ocean and Seismo-Acoustics*, 305–310.

and harmonic time dependence $\exp(-i\omega t)$, the spatial part of the far-field acoustic pressure p is given by [1]

$$p(r,z) \sim \frac{\sqrt{2\pi}e^{i\pi/4}}{\rho} \sum_n u_n^*(z_0) u_n(z) \frac{e^{ik_n r}}{\sqrt{k_n r}} \quad , \tag{1}$$

where the perfectly trapped normal modes, $u_n(z) e^{ik_n r}/\sqrt{k_n r}$, correspond to the natural modes of vibration, or resonances, of the waveguide and are entirely dependent upon its acoustic properties. The vertical eigenfunctions $u_n(z)$ and eigenvalues k_n satisfy the equation,

$$\frac{d^2 u_n(z)}{dz^2} + \left[k^2(z) - k_n^2\right] u_n(z) = 0 \quad , \tag{2}$$

where $k(z) = \omega/c(z)$. These modes must also satisfy appropriate boundary conditions at the upper and lower interfaces, which manifest themselves through the eigenvalue equation. We note that this formalism will accommodate both fluid and elastic solid bottoms, as long as the requirement of horizontal stratification is satisfied.

Suppose that a horizontal receiving array of length L is inserted into the modal field at range R, so that the array extends over the interval $R \le r \le R + L$. For simplicity, we assume that the array is composed of a continuous line of receivers, and we determine the beam response $P(k_r, z)$ by computing the Fourier transform

$$P(k_r, z) = \frac{1}{\sqrt{2\pi}} \int_R^{R+L} \sqrt{r}\, p(r,z) e^{-ik_r r} dr \quad . \tag{3}$$

Substituting Eq. (1) into Eq. (3), we obtain

$$P(k_r, z) \sim \frac{2e^{i\pi/4}}{\rho} \sum_n e^{-i(k_r - k_n)(R+L/2)} \frac{u_n^*(z_0) u_n(z)}{\sqrt{k_n}} \frac{\sin\left[(k_r - k_n) L/2\right]}{(k_r - k_n)} \quad . \tag{4}$$

It is clear from Eq. (4) that there are peaks in the beam response at the horizontal wavenumbers $k_r = k_n$ corresponding to the eigenvalues of the perfectly trapped modes. Furthermore, it is evident that our ability to identify and resolve these peaks increases with the length of the receiving array. Also, if the source and/or the receiver is positioned in a null of a particular mode, then the peak associated with that mode will not appear in the beam-formed output. It is therefore desirable to obtain data at more than one source/receiver depth in order to ensure that all possible waveguide modes are identified. Finally, it has been shown [2] that the beam-forming operation in Eq. (3) is an approximation to the exact Hankel transform relationship between the pressure field and the depth-dependent Green's function for the waveguide [1].

The input data required to perform the beam-forming operation may be obtained using either a synthetic aperture array or a real horizontal towed array [2,3]. Once the data have been transformed into the horizontal wavenumber domain, there are several approaches available to us for obtaining the geoacoustic properties of the seabed. One general class of inversion methods utilizes the beam response function directly as input data to iteration of forward models techniques [4,5]. In these approaches, theoretical beam responses are computed for various values of the bottom parameters until the parameter set is found which provides a best fit to the data. These schemes may involve sophisticated search algorithms through parameter space, such as simulated annealing [5]. Iterating forward models in the horizontal wavenumber domain is clearly advantageous over iterating in the range domain, since calculating the beam response is typically far less computationally intensive than calculating the pressure field. In another approach which utilizes the beam response function, these data are used to obtain velocity estimates for assumed constant-velocity layers in the bottom from the identification of modal critical angles in the variable-offset beam-former response [6].

In a second general class of inversion techniques, only the modal eigenvalues, rather than the entire beam response function, are required as input data. In these approaches, the measured beam response is used only to identify the locations, k_n, of the prominent peaks associated with the dominant modes of propagation. The eigenvalue estimates can then be used as input data to iteration of forward models techniques [7]. They can also be used in perturbative inversion methods in which the variation in bottom parameters about an assumed set of parameters is determined from the variation in the measured eigenvalues about the eigenvalues computed for the background model [8-11]. Finally, a non-perturbative, WKB-based inversion scheme, in which ray-mode equivalence is exploited to infer a geoacoustic smodel, has also been developed [12].

3. The Range-Dependent Case

The methodology described in Sec. 2 can be extended to the case of a laterally varying waveguide, where both the sound velocity in the water column, $c(r,z)$, and the properties of the bounding media are assumed to be slowly varying functions of range. Then the exact mode sum in Eq. (1) is replaced by the approximate, adiabatic mode sum [1]:

$$p(r,z) \sim \frac{\sqrt{2\pi}e^{i\pi/4}}{\rho} \sum_n u_n^*(0,z_0) u_n(r,z) \frac{\exp\left[i\int_0^r k_n(r')\,dr'\right]}{\sqrt{k_n(r) r}} \quad . \tag{5}$$

Here the eigenfunctions $u_n(r,z)$ and eigenvalues $k_n(r)$ satisfy the local eigenvalue equation and local boundary conditions at range r [cf. Eq. (2)]:

$$\frac{d^2 u_n(r,z)}{dz^2} + \left[k^2(r,z) - k_n^2(r)\right] u_n(r,z) = 0 \quad , \tag{6}$$

where $k(r,z) = \omega / c(r,z)$. We see from Eqs. (5) and (6) that adiabatic mode theory allows the vertical and radial eigenfunctions to adapt to the local environment as they propagate along the waveguide.

Suppose that, analogous to the range-independent case, a horizontal receiving array of length L is inserted into the adiabatic modal field at range R, so that the array extends over the interval $R \leq r \leq R+L$. Computing the beam response in Eq. (3), we obtain

$$P(k_r,z) \sim \frac{2e^{i\pi/4}}{\rho} \sum_n e^{i\int_0^R k_n(r')\,dr'} e^{-ik_n(R)R} e^{-i[k_r - k_n(R)](R+L/2)} \frac{u_n^*(0,z_0) u_n(R,z)}{\sqrt{k_n(R)}} \frac{\sin\{[k_r - k_n(R)]L/2\}}{[k_r - k_n(R)]} \quad , \tag{7}$$

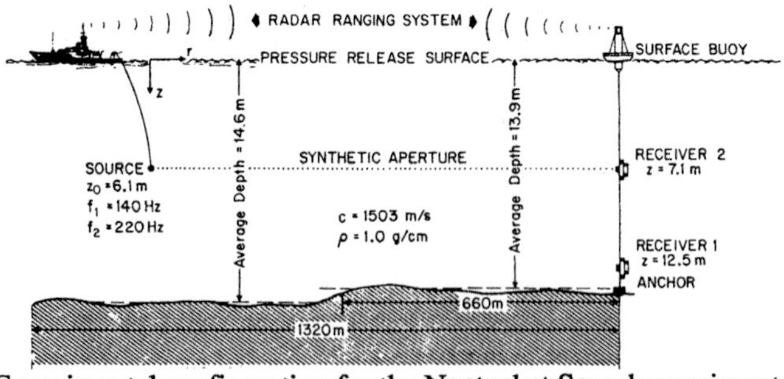

Figure 1. Experimental configuration for the Nantucket Sound experiment.

where we have assumed that the waveguide properties are sufficiently slowly varying that they may be considered range-independent on scales corresponding to the array aperture L. This assumption implies that $u_n(r,z) \approx u_n(R,z)$ and $k_n(r) \approx k_n(R)$ for $R \le r \le R+L$. It is clear from Eq. (7) that the modal peak positions in the beam response are now a function of array range R. As in the range-independent case, our ability to identify and resolve these peaks increases with the length of the receiving array, and certain modes may be nulled out depending upon the source/receiver depths. Equation (7) suggests that the modal evolution of the waveguide with range may be determined by calculating the beam response at successive array ranges. Furthermore, since the waveguide properties are assumed to be range-independent over the array aperture, we can infer the local bottom properties from the beam response at each array range using the range-independent inversion techniques described in Sec. 2. Then the range-dependent geoacoustic parameters are simply synthesized from the resulting sequence of range-independent properties. We note that increasing the array length decreases our ability to resolve range-dependent waveguide characteristics, whereas it increases modal resolution. Therefore, in any given experiment, these trade-offs must be carefully examined in order to arrive at optimal data acquisition and processing schemes.

We now apply these ideas to experimental data obtained in Nantucket Sound, Massachusetts using the experimental configuration shown in Fig. 1 [9]. The experiment consisted of towing a source transmitting pure tones at 140.05 and 220.26 Hz at fixed depth away from two moored receivers over an aperture which extended from zero to 1320 m. Here we will concentrate on the interpretation of the 140 Hz data, which is shown in Fig. 2 for the mid-column receiver. The modulated phase in Fig. 2 corresponds to the measured phase minus the quantity $k_0 r$, where $k_0 = \omega / c_0$ and $c_0 = 1550$ m / s.

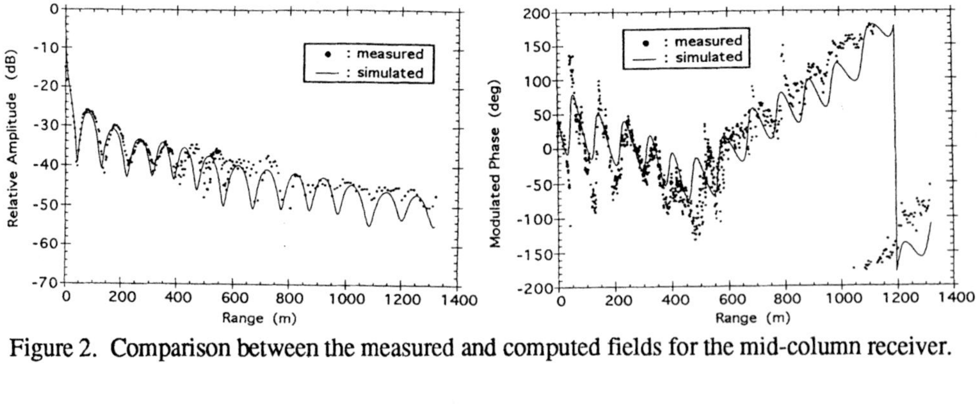

Figure 2. Comparison between the measured and computed fields for the mid-column receiver.

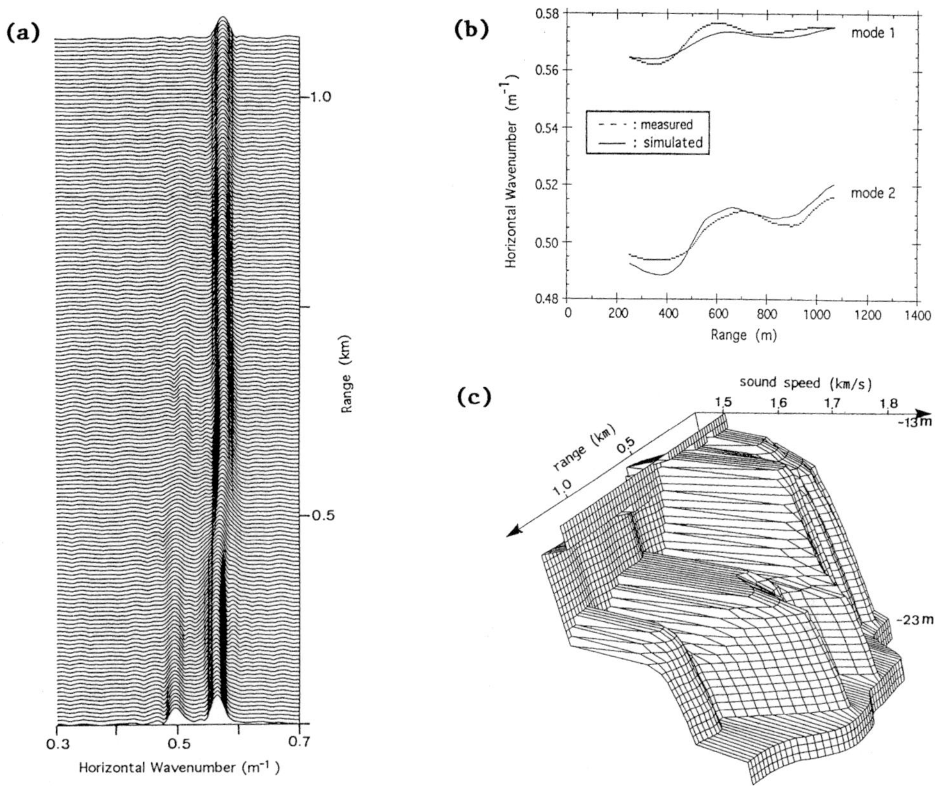

Figure 3. (a) Modal evolution at the mid-column receiver, (b) modal trajectories, (c) and inferred sediment sound speed profiles.

These data were then beam formed over a succession of 500 m sub-apertures, where the position of each sub-aperture was shifted from the previous one by 5 m. In each case, a Hanning window, rather than the rectangular window in Eq. (3), was applied. The results of this processing for the mid-column receiver data are displayed in Fig. 3a, which illustrates the gradual evolution with range of two dominant modes in the waveguide [7]. It is useful to track the modal peak positions as a function of range, and these modal trajectories are shown in Fig. 3b. We note that these trajectories were obtained by incorporating the data from the near-bottom receiver as well, through a process of mode filtering [7]. An adiabatic mode program was then used in an iteration of forward models approach to generate theoretical modal trajectories for various geoacoustic models. After thirty iterations, the level of agreement between theory and experiment shown in Fig. 3b was achieved. The corresponding inferred sound speed profiles in the sediment are shown in Fig. 3c [7]. The density was assumed to have a known value of $\rho = 1.7\ \mathrm{g/cm^3}$ [9]. In order to evaluate the effectiveness of the inversion method and the quality of the inferred result, we computed the pressure field using the adiabatic mode program with the inferred geoacoustic model as input data. The resulting excellent comparisons between the computed and measured fields for the mid-column receiver are shown in Fig. 2.

4. Conclusions

We have seen that modal inversion methods are an effective means for determining the acoustic properties of the seabed in shallow water. They are useful in both range-independent and slowly varying, range-dependent environments and are currently being extended to three-dimensionally varying waveguides [7]. We saw that the input data required to perform these inversion techniques can be obtained using either a synthetic aperture or a real horizontal towed array, and that these methods are sufficiently robust to be successful with experimental data. Finally, we note that, although we concentrated here on a review of narrowband techniques, closely related broadband methods, utilizing modal dispersion curves as input data, have also been developed [8,10].

5. References

1. Frisk, G.V. (1994) *Ocean and Seabed Acoustics: A Theory of Wave Propagation*, Prentice-Hall, Englewood Cliffs, NJ.
2. Frisk, G.V. and Lynch, J.F. (1984) "Shallow water waveguide characterization using the Hankel transform", J. Acoust. Soc. Am. **76**, 205-216.
3. Kuperman, W.A., Werby, M.F., Gilbert, K.E., and Tango, G.J. (1985) "Beam forming on bottom-interacting tow-ship noise", IEEE J. Ocean. Eng. **OE-10**, 290-298.
4. Frisk, G.V., Lynch, J.F., and Doutt, J.A. (1986) "The determination of geoacoustic models in shallow water", in T. Akal and J.M. Berkson (eds.), *Ocean Seismo-Acoustics*, Plenum, New York, pp. 693-702.
5. Collins, M.D., Kuperman, W.A., and Schmidt, H. (1992) "Nonlinear inversion for ocean-bottom properties," J. Acoust. Soc. Am. **92**, 2770-2783.
6. Werby, M.F. and Tango, G.J. (1986) "Characterization of average geoacoustic bottom properties from expected propagation behavior at very low frequencies (VLF) using a towed array simulation", in T. Akal and J.M. Berkson (eds.), *Ocean Seismo-Acoustics*, Plenum, New York, pp. 881-889.
7. Ohta, K. (September 1993) "Analysis of Modal Evolution Caused by a Weakly Range-Dependent Seabed in Shallow Water and its Application to Inversion for Geoacoustic Properties", Ph.D. thesis, MIT/WHOI Joint Program, Cambridge, MA and Woods Hole, MA.
8. Rajan, S.D., Lynch, J.F., and Frisk, G.V. (1987) "Perturbative inversion methods for obtaining bottom geoacoustic parameters in shallow water", J. Acoust. Soc. Am. **82**, 998-1017.
9. Frisk, G.V., Lynch, J.F., and Rajan, S.D. (1989) "Determination of compressional wave speed profiles using modal inverse techniques in a range-dependent environment in Nantucket Sound", J. Acoust. Soc. Am. **86**, 1928-1939.
10. Lynch, J.F., Rajan, S.D., and Frisk, G.V. (1991) "A comparison of broadband and narrow-band modal inversions for bottom geoacoustic properties at a site near Corpus Christi, Texas", J. Acoust. Soc. Am. **89**, 648-665.
11. Rajan, S.D., Frisk, G.V., and Lynch, J.F. (1992) "On the determination of modal attenuation coefficients and compressional wave attenuation profiles in a range-dependent environment in Nantucket Sound", IEEE J. Ocean. Eng. **OE-17**, 118-128.
12. Casey, K.D. (June 1988) "A Modal/WKB Inversion Method for Determining Sound Speed Profiles in the Ocean and Ocean Bottom", S.M. and Ocean Engineer's Degree thesis, MIT/WHOI Joint Program, Cambridge, MA and Woods Hole, MA.

EXPERIMENTAL TECHNIQUES FOR BOTTOM PARAMETER INVERSION IN SHALLOW WATER

R. D. STOLL[1] and T. AKAL[2]
[1]*Lamont-Doherty Earth Observatory of Columbia University*
Palisades, New York, 10964, USA
[2]*SACLANT Undersea Research Center*
19138 San Bartolomeo, La Spezia, Italy

ABSTRACT. In order to calculate transmission loss and other important quantities in a shallow water environment, it is necessary to carefully model the velocity and attenuation profiles of the sediments immediately beneath the seafloor. The most efficient way to study these profiles for the purpose of developing appropriate models is to place both the source and receiver used in geoacoustic experiments directly on the seafloor. In this paper several new experimental techniques designed to excite strong refractions and interface waves in the near-bottom sediments are described. These include a method for generating horizontally polarized shear-waves and Love waves using a source which applies a torsional pulse to a small, circular area of the seafloor and a method of generating vertically polarized shear-waves using an impulsive source powered by 22-caliber blank cartridges of the type used in the construction industry to drive connectors into structural elements.

1. Introduction

One of the principal roles of the marine geoacoustician is to provide an adequate model of velocity and attenuation for the sediments immediately beneath the seafloor. In this model both shear and dilatational waves are important, and it has been shown recently that the high velocity gradients caused by the naturally occuring vertical inhomogeneity tend to cause coupling between dilatational and shearing motion. The result is an overall attenuation somewhat larger than would be expected on the basis of classical Rayleigh reflection coefficients which neglect this coupling. Thus the "water column" acoustician who would like to use a "fluid" bottom as an approximation because of mathematical expediency, may tend to underestimate the bottom loss based on traditional estimates of p- and s-wave attenuation.

Our studies of the physics of wave propagation in water-saturated, porous sediments have advanced to the point where we can now write the coupled equations of motion for a water-saturated medium that will describe the response of a wide variety of sediments ranging from soft clays to compact sands. The velocity profile in these sediments is strongly influenced by the porosity (or voids ratio) and the overburden stress, both of which may change quite rapidly in the near-bottom sediments. Most geotechnical engineers and geoacousticians agree that the dynamic shear modulus, G, can be estimated using a simple equation of the form

$$G = f(e) \cdot p^n \tag{1}$$

O. Diaschok et al. (eds.), Full Field Inversion Methods in Ocean and Seismo-Acoustics, 311–316.

where e is voids ratio, p is vertical effective stress, and n is a pure number in the range of .33 to .5. The form of the function f(e) depends on the type of sediment and to some extent on the preference of the user (e. g., see Bryan and Stoll, 1988). The variation of voids ratio with depth depends partly on the depositional history of the sediment deposit and partly on the amount of consolidation that has occured as a result of the overburden pressure. If this kind of information is available (and it often is not) it is possible to estimate velocity and attenuation of both p- and s-waves using the Biot or Gassmann equations (Stoll, 1990). In many cases the information about the sediment may be limited to a rudimentary, often subjective description, and thus the acoustician is forced to choose a simplified model with very little guidance. Often this choice is made in such a way that the extensive numerical calculations required will be simplified mathematically rather than making the choice that will best describe the real response of the sediment. The result is models that may work for a narrow band of frequencies but which will fail when an attempt is made to perform calculations over a more useful range.

The experimental techniques described in this paper were designed to maximize the amount of information that can be obtained pertaining to the sediment geoacoustic properties. They serve the dual purpose of checking the predictions of our theoretical models and providing a library of the geoacoustic response for a wide variety of different sediments to aid in the choice of simplified models to be used in acoustic calculations. Most of our experimental work has employed a linear array of receivers and a source, both deployed on the seafloor. Several different sources which apply a vertical pressure pulse to a small area of the seafloor have been used. These include explosives, an airgun, an 8-gauge seismic shell, and a 22-caliber blank cartridge of the type used in the construction industry to embed fasteners in structural members. In most cases these sources have been mounted on a helmet-shaped chamber with no bottom which focuses the energy into the seafloor and tends to suppress the bubble pulse in the water column. Earlier work with the first three of these sources is described in Stoll, Bryan, Mithal and Flood (1991) and Akal, Caiti and Stoll (1993). The 22-caliber source is described in this paper. These sources all tend to excite strong p- and s-wave refractions and strong, vertically polarized interface waves (Scholte waves) in the sediments immediately beneath the seafloor. The other type of source which we have recently developed applies a torsional pulse to a small, circular area of the seafloor. This excites horizontally polarized shear waves and Love waves in the near-bottom sediments.

2. Experimental Work

Several aspects of our basic experimental configuration are shown in Fig. 1. The receiving array consists of a number of gimbaled geophones, usually spaced 5 m apart, attached to the end of a long, reinforced cable. The geophone units contain standard velocity sensors suspended by a gimbal which may be chosen so that the unit responds to either vertical or transverse-horizontal motion. The geophones are attached to the main cable at molded takeouts by pigtails which help to isolate the sensors from spurious motion of the array.

When the array is deployed from a small vessel, the cable is usually paid out over the stern while the boat is moving against the prevailing current at about two or three knots. The best track for laying the cable depends on the prevailing wind, current and tidal conditions which are determined *apriori*. When the full array is extended, it is dragged a short distance to insure straight alignment and the cable is then slacked off to remove tension and prevent noise caused by motion of the ship. In some cases the source is attached to the same cable as the array while in others it may be lowered from the ship on a seperate cable. Both cases are shown in the schematic of Fig. 1 wherein a torsional source mounted on a self-righting sled is shown attached to the geophone cable while an impulsive vertical source is shown deployed directly from the stern of the ship. Each case has it's own advantages; when the source is attached to the cable, the exact range to each phone is fixed whereas when the source is deployed seperately, the cable may be retrieved in increments of about 100 m and several shots fired in order to obtain as many as 72 traveltime records, each at a different range. In some of the experiments performed by SACLANT Center the array is attached to a buoy which transmits data to the ship by radio link thus allowing free movement of the ship and source.

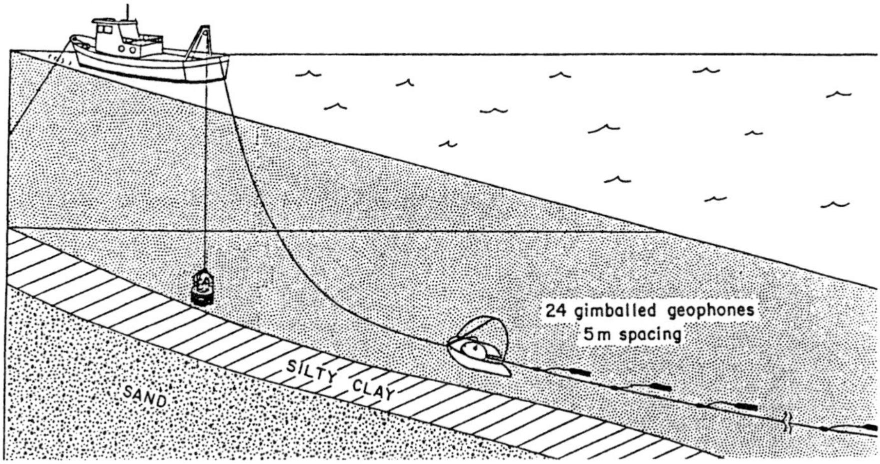

Fig. 1. Experimental configuration.

Our new torsional source applies a rotational pulse to a small circular area of the seafloor by utilizing energy stored in a rotating flywheel which is driven by a small, variable speed DC motor. The speed of the flywheel is slowly increased to the test level (typically 200 to 500 rpm) after which two diametrically opposed anvils are impacted by hammers located on the outer diameter of the wheel. The sudden torque caused by this impact is transmitted to the external base plate which rests on the seafloor. The torsional motion is coupled into the sediment by vanes or spikes attached to the bottom of the plate. The instant of impact is sensed by an inertial switch which sends a signal to one channel of the recording system. Schematics of the SH-wave generator and the 22-caliber source are shown in Fig. 2.

The shear wave generator may be deployed either as a stand-alone unit, as pictured in Fig. 2, or mounted on a self-righting sled as shown in Fig. 1. In the latter case, the sled can be moved from point-to-point along the bottom allowing a number of shots to be made along a line during a single deployment. SH-waves propagate away from the source in a radially symmetric pattern and the linear array of horizontal geophones measures horizontal particle velocity in a direction perpendicular to the azimuth of the array. Theoretically, a torsional source applied at a point on the bottom should produce a radially symmetric wavefield composed only of horizontally polarized shear waves; however, in reality, a part of the energy is converted to p-wave and sv-wave motion because of the vibrations from the impact of the hammers and the eccentricity of loading caused by the finite geometry of the unit.

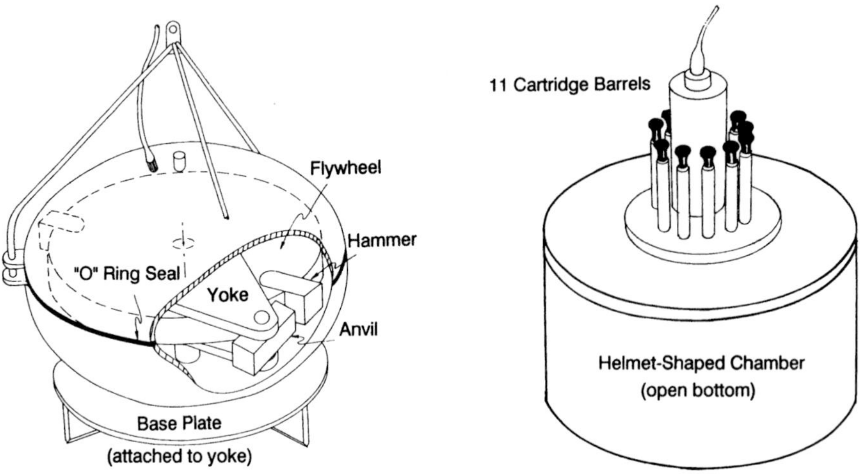

Fig. 2. Schematic of torsional and 22-caliber sources.

A set of seismograms generated by the torsional source in a sand bottom in the Gulf of Mexico is shown in Fig. 3. Both group and phase velocity dispersion curves may be derived from the traveltime curves of Fig. 3 and these dispersion curves may then be used as a basic for inversion to obtain geoacoustic models of the sediments near the seafloor.

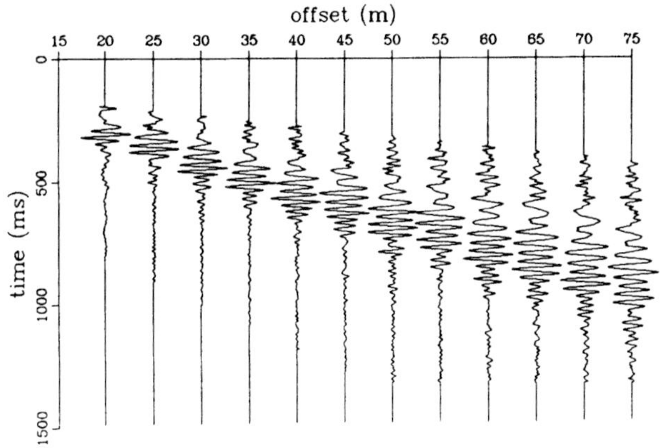

Fig. 3. Seismograms from Torsional Source in Sand Bottom

The 22-caliber source shown in Fig. 2 utilizes eleven blank cartridges, each confined in a seperate barrel by a frangible plastic disk sealed by an "O" ring. When a shell is fired, gas pressure builds until the plastic disk ruptures and the gas is vented downward into the helmet-shaped chamber that focuses the energy into the seafloor. In most of the experiments to date, this source has been mounted on the same self-righting sled used to deploy the torsional source, however the focusing chamber may be placed directly on the seafloor as an independent unit if desired. When this source is used, the receiver array consists of gimballed geophones that respond to vertical motion. Seismograms produced by this source are similar to those produced by the torsional source (see Fig. 3).

3. Data Interpretation

Data such as that shown in Fig. 3 may be inverted to derive a geoacoustic model by analyzing either the group or the phase velocity dispersion of the dominant interface mode. Group velocity dispersion corresponding to propagation from the source to any one of the receivers may be determined by the multiple filter technique described by Dziewonski, Bloch and Landisman (1969). Examples of the use of this technique are given in Stoll et al (1991) and Caiti, Akal and Stoll (1994). Phase velocity dispersion curves may be constructed using a variety of different techniques such as cross-multiplication and cross-correlation. The next step is to develop a forward modeling algorithm which allows the computation of interface-wave dispersion as a function of model parameters such as velocity, density and attenuation. One of the simplest schemes is to assume that the sediment inhomogeneity can be modeled on the basis of a series of horizontal, homogeneous layers with parameters that approximate the variation in the actual sediment. A characteristic equation can then be derived using a modified version of the Thomson-Haskell method to integrate the equations of motion subject to the appropriate boundary conditions at the water-sediment interface (Takeuchi and Saito, 1972).

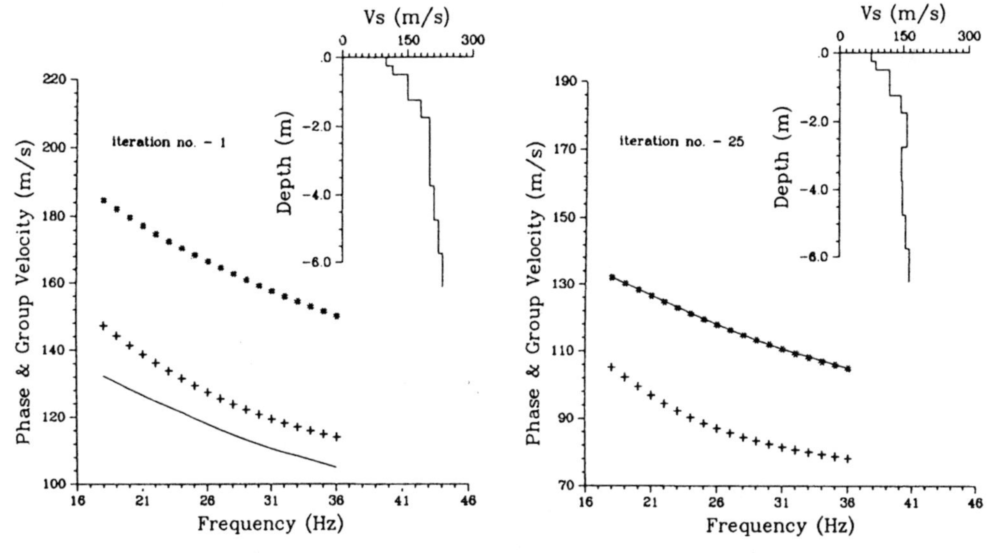

Fig. 4. Example of Inversion Based on Phase Velocity Dispersion

Once the forward problem is formulated, an inverse scheme can be devised that will iteratively solve for the model parameters, given the experimentally determined dispersion curves. The iterative procedure is based on the first order perturbation of the nonlinear implicit functional $\{c(\beta)\}$ around the layer parameters β_0 (shear-wave velocity and any other parameters not assigned constant values)

$$\{c\} = \{c(\beta_0)\} + [\partial c/\partial \beta] (\{\beta\} - \{\beta_0\}). \tag{2}$$

At any stage of the iteration this equation is used to reduce the difference between the experimentally determined phase or group velocity dispersion curve and the curve calculated for the partially developed model. The result is an overdetermined system of equations in which there are more equations (observations) than unknowns (model parameters) so that a least squares solution is appropriate. Since the generalized inverse of this system is ill conditioned owing to the fact that small changes in the data cause large changes in the model parameters, we use a constrained least squares inversion with singular value decomposition to solve for the model parameters. Details of the complete inversion process are given in Bautista (1994).

An example of the results of a typical inversion is shown in Fig. 4. In this figure, the solid line is the phase velocity dispersion curve determined by cross-correlation and then averaging the results for five different channel-pairs from the seismograms of Fig. 3. The curve formed by the symbols ($*$) is the phase velocity for the layered model shown in each panel and the curve formed by the symbols (+) is the group velocity calculated for the model.. The left panel shows the calculated dispersion curves for the trial model which was input to the inversion program and the right panel shows the calculated dispersion curves and the model achieved after 25 iterations.

4. Summary and Conclusions

Two new seismic sources , one a torsional source which generates sh-waves and the other a vertical, impulsive source which generates p- and sv-waves, have been described. Both of the new sources are particularly useful for obtaining high resolution information on the shear-wave velocity near the seafloor. With either source, both phase and group velocity dispersion are easily determined when a linear array of receivers is employed. Using these dispersion curves, sediment geoacoustic models may be derived using an interative, least-squares inversion program.

5. Acknowledgement

The work described in this paper was supported partly by the Office of Naval Research, Code 321OA and partly by the Coastal Benthic Boundary Layer Special Research Program, Naval Research Laboratory, Stennis Space Center.

6. References

Akal, T., Caiti, A. and Stoll, R. D. (1993) Remote Sensing in Shallow-Water Marine Sediments, *Proc. Inst. Acoustics*, 15, Pt. 2, 221-228

Bautista, E. O. (1994) Remote Determination of Insitu Sediment Properties using Love Waves, *PhD Dissertation, Columbia University.*

Bryan, G. M. and Stoll, R. D. (1988) The Dynamic Shear Modulus of Marine Sediments, *J. Acoust. Soc. Am.*, 83, 2159-2164.

Caiti, A., Akal, T., and Stoll, R. D. (1994) Estimation of Shear Wave Velocity in Shallow Marine Sediments, *IEEE Jour. Oceanic Engrg.*

Dziewonski, A., Block, S., and Landisman, M. (1969) A Technique for Analysis of Transient Seismic Signals, *Bull. Seismol. Soc. Am.*, 59, 427-444.

Stoll, R. D. (1990) *Sediment Acoustics*, Springer-Verlag, Berlin.

Stoll, R. D., Bryan, G. M., Mithal, R., and Flood, R. (1991) Field Experiments to Study Seafloor Seismo-acoustic Response, *J. Acoust. Soc. Am.*, 89, 2232-2240.

Takeuchi, H. and Saito, M. (1972) Seismic Surface Waves, in *Methods in Computational Physics*, 11, Seismology:Surface Waves and Earth Oscillations, Academic Press, New York., 217-295.

GLOBAL INVERSION OF ACOUSTIC FIELD DATA IN SHALLOW WATER USING GENETIC ALGORITHMS

DONALD F. GINGRAS and PETER GERSTOFT
SACLANT Undersea Research Centre,
Viale San Bartolomeo 400, 19138 La Spezia, Italy

ABSTRACT Precise knowledge about environmental parameters is highly desirable for precise acoustic modelling of the underwater channel. Simulations have shown that global inversion methods such as simulated annealing and genetic algorithms are effective for the estimation of both geometric and geoacoustic parameters. In this paper genetic algorithms are applied to acoustic field observations from a shallow water site in the Mediterranean Sea. A 20 minute sample of observations is considered, the results indicate that genetic algorithms are highly effective for the estimation of both geometric and geoacoustic parameters and that the estimates are stable over time.

1. Introduction

In order to model acoustic propagation accurately in a particular environment, precise knowledge of the properties of the ocean bottom is required. This is particularly true for shallow water environments where acoustic interaction with the bottom is an important factor. The use of global optimization methods for inversion of acoustic field data has been shown to be an effective means for estimating both water column and bottom properties [1-3]. In order to accurately estimate the bottom properties using full field inversion methods one must precisely know the geometric parameters (source location, receiver location and bathymetry) or include them as part of the inversion process. In this paper we apply genetic algorithms (GA) to the estimation of both geometric and geoacoustic parameters.

The nonlinear inverse problem is stated as an optimization problem; that is, find the model parameter vector $\mathbf{m} = \{m_i\}\ i = 1, 2, \ldots, p$ that minimizes an object function. The model vector $\mathbf{m}$ contains both geometric and geoacoustic parameters. The object function is a function of the vector of observations $\mathbf{d}_{obs}$ and the vector of forward model predictions $\mathbf{d}_{cal}(\mathbf{m})$. The object function used is the Bartlett processor,

$$\phi(\mathbf{m}) = 1 - \frac{|\mathbf{d}^*_{cal}(\mathbf{m})\mathbf{d}_{obs}|^2}{||\mathbf{d}_{cal}||^2\ ||\mathbf{d}_{obs}||^2}. \tag{1}$$

Genetic algorithms are based on an analogy with biological evolution. The basic principle of GA is simple: From all possible model parameter vectors, an initial population of q individuals is selected randomly. The fitness of each member (one minus the object function) is computed. Through a set of evolutionary steps the initial population evolves in order to become more fit. An evolutionary step consists of selecting a parental distribution from the initial population based on the individual's fitness. The parents are then combined in pairs; operators are applied to these pairs to form a set of children. Traditionally the crossover and mutation operators have been used. Finally the children replace a fraction of

O. Diaschok et al. (eds.), Full Field Inversion Methods in Ocean and Seismo-Acoustics, 317–322.

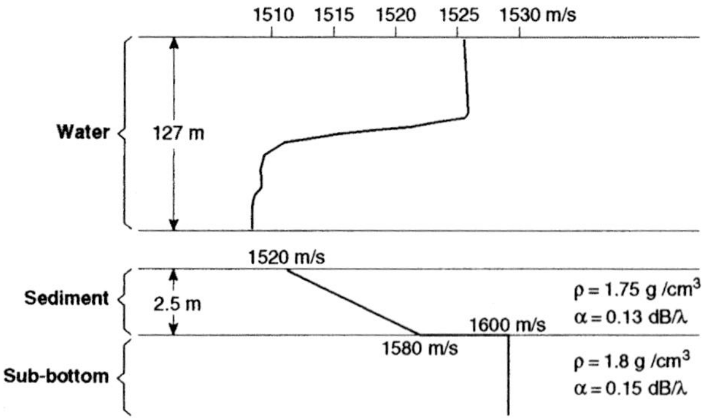

FIGURE 1. Historical bottom parameters with measured sound speed profile.

the initial population to get a more fit population. For a description of GA applied to the seismoacoustic parameter estimation problem, see Ref. [3].

In this paper we demonstrate that genetic algorithms (GA) are an effective tool for estimating both the geometric and geoacoustic parameters using shallow water data collected in the Mediterranean Sea.

2. Experimental Setup

The experimental data were collected at a site North of Elba Island, off the Italian west coast, where environmental conditions are known from earlier experiments [4, 5]. A vertical array was deployed in a bottom moored configuration. The array contained 48 hydrophones spaced at 2 m which spanned the depth interval of 18.7 m to 112.7 m. At the array site the water depth was measured to be approximately 127 m. The source was also bottom moored and deployed at a depth of 79 m in water that was approximately 130 m deep. The transmitted signal was a continuous time random noise sequence with center frequency of 170 Hz and bandwidth of 12 Hz.

The deployment position of the array and source with respect to global coordinates is only known to an accuracy of about 100 m. Thus, the range between the array and the source is expected to be 5600 ± 200 m. An acoustic navigation system was deployed with the vertical array to monitor array shape throughout the experiment. For the time period analyzed herein the array was essentially vertical.

The sound speed profile measured near the vertical array was a summer type profile, almost isovelocity down to 60 m and then a strong thermocline extending to 80 m and almost isovelocity to the bottom, see Fig. 1. Figure 1 also contains the historical description of the bottom parameters for the North Elba site, from Reference [5], which is partially based on a sediment core taken at a location about 7 km due East of the array site [4]. Figure 1 illustrates the basic parameterization that was used throughout the analysis. The channel was modeled using three layers: a water layer of constant depth and sound speed that only varies with depth; a sediment layer of constant thickness, density and attenuation, with linear compressional speed profile parameterized by an upper and lower value; and a semi-infinite bottom with constant density, attenuation and compressional speed. At 170 Hz this shallow water channel supports eight modes, the contributing modes are modes 2,4 and 7, at ranges greater than 3 km the contribution of mode 7 is very small.

For the acoustic field inversion a 20 minute observation was used throughout. This sample was collected from 15:44 to 16:04 on 26 October 1993. Prior to applying the GA, the data for each of the 48 hydrophones was transformed to the frequency domain using a time sample of 4.096 seconds, a sample was computed every minute over the 20 minute interval. While the source signal was broadband only a single 'bin' centered at 170 Hz was used in this investigation.

3. GA Parameter Estimation

There are a few GA parameters that must be defined for each GA application. Based on previous experience the following values were used in the analysis of the North Elba data, see [3, 6]. The population size was set to 64, the reproduction size was 0.5, the crossover probability was 0.8, a mutation probability of 0.05 was used, and the number of iterations (or forward model computations) was 2000. The forward model used was the SACLANTCEN normal mode propagation model, SNAP [7]. One 'inversion' consisted of 2000 iterations for each of ten independent parallel populations. Thus a total of 20,000 forward model computations were used for each inversion. At the completion of an inversion there were 320 model parameter vectors or population members resulting (population size times the number of parallel populations times the reproduction size). The CPU time for one inversion, that is 20,000 forward model computations, was 13 minutes on a DEC-3000/600 Workstation.

The optimization was carried out over a parameter space of 12 parameters, see Table 1, where the parameters are grouped into three sub-groups: geometric, sediment and bottom. There was no optimization of the water sound speed profile, a single sound speed profile based on a CTD measurement at the array site was used. The interpretation of the parameters of Table 1 is quite straightforward except for 'Receiver Depth'. The receiver depth is the depth of the deepest hydrophone, but the GA use this parameter to control the vertical position of the entire vertical array in the water column. The baseline value for receiver depth is based on the water depth and the physical characteristics of the array and mooring.

Inversion for the geometric and geoacoustic parameters was carried out for each of the 20 samples spaced at one minute intervals. For each of the 20 time samples the GA produced 320 model parameter estimates. A post process was applied to combine the 320 parameter estimates for each of the time samples into one collection of samples ordered according to their fitness. *A posteriori* probability distributions for each of the twelve parameters were formed from the ordered samples. Based on the *a posteriori* probability distributions the GA provide three estimates for the model parameters: those associated with the largest fitness (best), those based on the peak of the distribution (most likely) and those based on the mean of the distribution (mean). Table 1 contains the baseline parameters [4, 5], and the three GA estimates based on the *a posteriori* probability distributions. We note that the three model parameter estimates computed by the GA are quite similar which usually indicates that a stable solution has been found.

Comparing the baseline model parameters with the GA estimated parameters it is obvious that there are significant differences between the baseline values and the GA estimated values. In particular note that the water depth differs by about one and one-half meters, the location of the deepest hydrophone differs by about one meter, the GA estimated sediment compressional speeds are lower, and the GA estimated sediment attenuation is

TABLE 1. Baseline model parameters and GA parameter estimates based on 20 observations.

Model Parameter	Baseline	GA Best	GA Most Likely	GA Mean
Geometric				
Source Range (m)		5405	5405	5396
Source Depth (m)		74.5	74.5	74.6
Receiver Depth (m)	112.7	111.5	111.4	111.5
Water Depth (m)	127.0	128.5	128.5	128.1
Sediment				
Comp. Speed, Upper (m/s)	1520	1496	1487	1492
Comp. Speed, Lower (m/s)	1580	1576	1576	1562
Density (g/cm^3)	1.7	2.0	1.9	2.0
Attenuation (dB/λ)	0.13	0.02	0.04	0.07
Thickness (m)	2.5	2.6	2.3	3.0
Bottom				
Comp. Speed (m/s)	1600	1574	1574	1573
Density (g/cm^3)	1.8	2.1	2.1	1.8
Attenuation (dB/λ)	0.15	0.34	0.20	0.18

much lower. In the next section we compare the above model parameters using the Bartlett power as a measure of model match.

4. Performance Analysis

Since both the pressure vector of observations and forward model predictions are normalized to be unit norm vectors the Bartlett power is one when there is a perfect match. Thus, a useful measure of how well the estimated geometric and geoacoustic parameters match the actual parameters is the degree to which the Bartlett power is close to one. In this section the Bartlett processor, Eq. 1, is used to evaluate the closeness of the estimated parameters to the actual environment.

Figure 2 illustrates the Bartlett power as a function of time for a number of models. The Bartlett power was evaluated at one minute intervals for every point in a range interval of 4500 to 6500 m (128 samples) and a depth interval of 50 to 100 m (128 samples). The Bartlett power selected was the largest maximum on the range/depth space for each time sample.

Line (1) of Fig. 2 is the result of evaluating the Bartlett power using the baseline model parameters. For this case the sample mean of the Bartlett power over the 20 minute observation is -2.2 dB. Line (2) is the result when the 'geometric' parameter estimates from the GA Mean values of Table 1 are used in place of the baseline values. That is, the deepest receiver depth is 111.5 m (vice 112.7 m) and the water depth is 128.1 m (vice 127 m). For this case the sample mean of the Bartlett power is -1.3 dB. A significant improvement in the model fitness has resulted by the use of the GA Mean estimates for water depth and receiver depth. Line (3) is the result obtained when all of the GA Mean parameters are used except for compressional speeds in the sediment and bottom which are baseline values. In

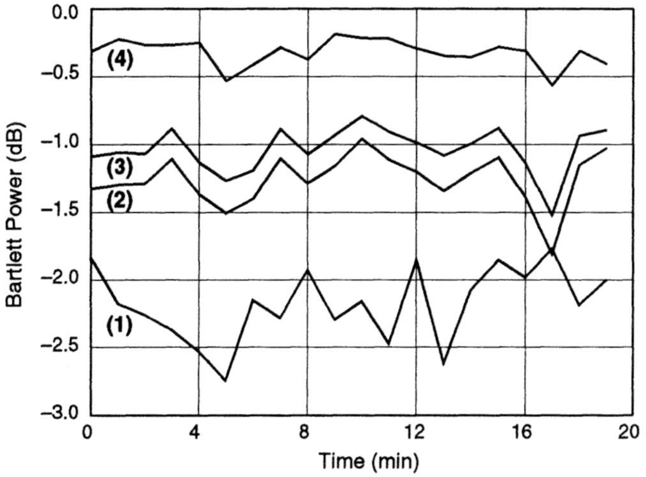

FIGURE 2. Bartlett power versus time; (1) baseline model parameters, (2) baseline parameters except receiver depth is 111.5 m and water depth is 128.1 m, (3) GA Mean parameters except compressional speeds for sediment and bottom are baseline values, and (4) all GA Mean model parameters.

this case the sample mean of the Bartlett power is −1.0 dB. Line (4) is the result when all of the GA Mean parameter estimates of Table 1 are used. In this case the sample mean of the Bartlett power is −0.3 dB, indicating that these parameters match the actual environment quite well. The increase in the mean Bartlett power from (3) to (4) indicates that the baseline values for compressional speed in the sediment and the bottom were not well matched to the actual environment at this site. The Bartlett power was also evaluated using the GA Best and GA Most Likely parameter estimates of Table 1, these results are not shown since they are essentially indistinguishable from that obtained using the GA Mean parameters. Comparing lines (1) and (4) note that the variability of the Bartlett power with time is significantly less for case (4). This implies that the variability over time observed with the baseline parameters, line (1), is not in fact due to temporal fluctuations in the data but due to the mismatch between the model parameters and the actual environment.

It is known that the inversion process is quite sensitive to uncertainties about the source location. In this analysis the source location parameters, range and depth, are included as part of the optimization. We now examine the use of source location estimates obtained using the Bartlett processor with fixed model parameters as another measure of performance. The source location estimates are obtained as the range and depth coordinates associated with the largest maximum Bartlett power on the range/depth space for each time sample. Figure 3 illustrates the source range and depth estimates over the 20 minute observation for the Bartlett processor using the baseline parameters and the GA Mean parameters. We see that the source range and depth estimates based on the GA Mean parameters are stable. The sample mean of the range estimates using the GA Mean parameters is 5400 m, which is the lower limit of expected source range. The sample mean of the depth estimates based on the GA Mean parameters is 75 m which is close to the expected source depth of 79 m. The range and depth estimates based on the baseline parameters are less stable, oscillating between two peaks on the range/depth surface, and are less accurate.

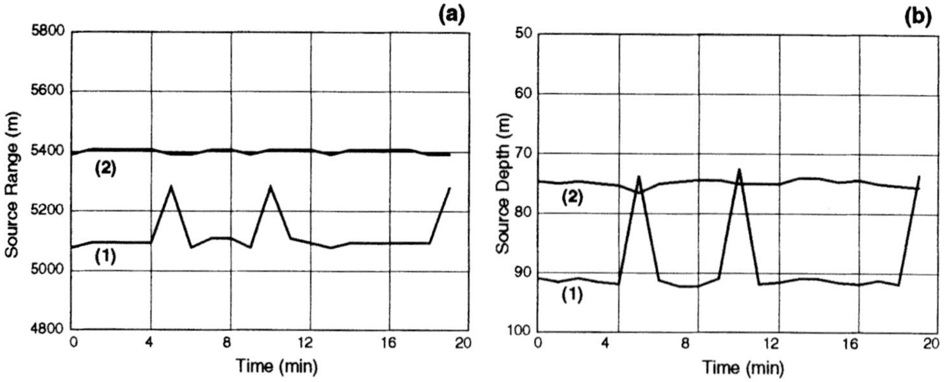

FIGURE 3. Source location estimates using the largest maximum of the Bartlett processor (a) Source Range, and (b) Source Depth. Line (1) baseline parameters and (2) GA Mean parameters.

When the GA Mean parameters were used the Bartlett power was very close to 0 dB (theoretical maximum) for all time samples, furthermore the range and depth estimates were quite close to the actual source locations. Thus, it appears that the GA model parameters provided estimates for the geometric and geoacoustic parameters that were quite close to the actual parameters for this site.

5. Conclusions

The GA have been applied to the problem of estimating geometric and geoacoustic parameters using shallow water vertical array data. Using the Bartlett power versus time as a measure of performance it has been shown that the GA can be used to significantly improve available historical knowledge about the geoacoustic parameters and simultaneously estimate the geometric parameters including source range and depth.

References

1. Collins, M.D., Kuperman, W.A., and Schmidt, H. (1992) Nonlinear inversion for ocean-bottom properties, *J. Acoust. Soc. Amer.*, **92**, 2770–2783.
2. Dosso, S.E., Yeremy, M.L., Ozard, J.M., and Chapman, N.R. (1993) Estimation of ocean-bottom properties by matched-field inversion of acoustic field data, *IEEE J. Oceanic. Eng.*, **OE- 18**, 232–239.
3. Gerstoft, P. (1994) Inversion of seismoacoustic data using genetic algorithms and *a posteriori* probability distributions, *J. Acoust. Soc. Amer.*, **95**, 770–782.
4. Akal, T., Gehin, C., Matteucci, B., and Tonarelli, B. (1972) Measured and computed physical properties of sediment cores, SACLANT Undersea Research Centre, SM-82, La Spezia, Italy.
5. Jensen, F.B. (1974) Comparison of transmission loss data for different shallow water areas with theoretical results provided by a three-fluid normal-mode propagation model, in *Sound Propagation in Shallow Water*, SACLANT Undersea Research Centre, CP-14, La Spezia, Italy.
6. Gerstoft, P. (1994) Global inversion by genetic algorithms for both source position and environmental parameters, *J. of Comp. Acoustics*, to appear.
7. Jensen, F.B. and Ferla, M.C. (1979) SNAP-The SACLANTCEN normal mode propagation model, SACLANT Undersea Research Centre, SM-121, La Spezia, Italy.

IMPROVING THE SENSITIVITY OF FULL-FIELD GEOACOUSTIC INVERSION FOR ESTIMATING BOTTOM SOUND SPEED PROFILES

T. C. Yang
Naval Research Laboratory
Washington, DC 20375

T. Yates
Vector Research Co.
Rockville, MD 20852

ABSTRACT. A critical issue in geoacoustic inversion of bottom sound speed profiles is the sensitivity of the inversion methods to the bottom parameters to be estimated. In many problems, the acoustic data received by hydrophones in the water column are dominated by waterborne (direct and refracted) signals. Consequently, the correlation coefficient of a replica field with the data varies little with the bottom sound speed variations. The bottom sound speed profile estimated by the matched field inversion methods can be different from the true bottom profile due to the sensitivity mentioned. For a more reliable estimate of the bottom profile one needs to improve the sensitivity of the inversion methods to the bottom parameters. For this purpose, we modify the matched field/mode inversion methods by including only the higher order modes or the deep angle paths in the replica field. Specifically, we use a mode filter and/or a vertical arrival angle filter; the latter is the matched beam processing which has been used to improve source tracking in real data. We show that this modified matched field/mode correlator is more sensitive to bottom sound speed mismatches. A SUS signal in Arctic shallow water is used in the analysis.

1. Introduction

Matched-field inversion of the geoacoustic properties of the ocean bottom has been demonstrated using simulated data and applied to field data with some success [1]. While the inversion scheme almost always yields a solution for the bottom parameters and the bottom sound speed profile, one is not sure the inverted result is correct until it has been verified by ground truth. Unfortunately, in most cases the ground truth is not available. In the worst case, one finds in some data that the source can not be correctly localized despite allowing a wide range of environmental acoustic parameters. As previously pointed out, a source can always be localized (focalized), but it may take an incorrect sound speed profile to do so [2]. If one were to apply matched field inversion to this data set, the inversion result is very much in doubt.

Matched field inversion searches for the environmental acoustic parameters by computing the correlation of the replica field with the data field at the known source location. In the simulation study for bottom inversion, the highest correlation is obtained when the correct environmental acoustic parameters are used. But real data has noise, fluctuations and other unknowns which may cause the inversion to produce the wrong result. For instance, varying certain bottom parameters does not seem to affect significantly the magnitude of the matched field correlation coefficient. This small sensitivity could easily be masked by the presence of noise and/or fluctuation in the data. In general a "valid" inversion requires the correlation coefficient to vary by, say, more than 10% over the range of the parameters to be searched for.

We address in this paper how to improve the matched field sensitivity, and hence the

O. Diaschok et al. (eds.), Full Field Inversion Methods in Ocean and Seismo-Acoustics, 323–328.

accuracy of inversion, by imposing a filter in the processor which enhances the correlation of the acoustic fields that are sensitive to the parameters to be inverted for. An example of such a filter is the mode filter as previously used in matched mode processing when low order modes are used to minimize environmental mismatches [3]. To increase the sensitivity of the acoustic field to bottom sound speed profile mismatches, one should filter out the waterborne (low order) modes and use only the bottom interacting higher order modes. Alternatively, one could use a filter in the vertical arrival angle domain in what is referred to as matched beam processing by correlating only the deep arrival, bottom interacting returns.

2. Geoacoustic Inversion With A Broadband Shot Signal

The data to be analyzed concern a broadband shot signal received on a vertical array of 32 phones in Arctic shallow water of 565 m depth. The hydrophones were spaced at 16 m with the top phone ~ 6 m below the surface. The shot (1.8 lb SUS) was dropped approximately 1.7 km away from the array at a nominal depth of 244 m. Figure 1a plots the signal spectral intensity as a function of time and depth as received on the vertical array and Fig. 1b plots the vertical beam intensity vs. time. From the (relative) arrival time of individual signals (rays), one can identify the direct path and bottom/surface reflected path returns. One can also identify two rays before the direct path arrival as the head wave traveling through the bottom. The bottom sound speed profile can be estimated using the (relative) arrival time of the head waves as done by Wolf [4]. The inverted bottom sound speed profile differs from the (nominal) bottom sound speed profile (see Fig. 5 below) estimated by Geddes [5] based on pervious acoustic and geoacoustic data by 20 to 150 m/s in the top 100 m of the bottom. Note that the head wave speed is determined primarily by sound speed in the top layers of the bottom.

To apply matched field inversion, one needs to know the source location as accurately as possible. To estimate the source location, we show the matched field ambiguity volume in Fig. 2a using all modes and Fig. 2b using only the waterborne modes for frequencies between 10 and 80 Hz. We see that the source range and depth are well estimated using all modes; the result is verified by the consistence with the waterborne mode result (which has less range and depth resolution). The estimated source range and depth are 1.7 km and 250 m respectively. The reason for the good localization result is that localization is determined primarily by the waterborne acoustic field except at very low (< 30 Hz) frequencies. In fact, the matched field signal gain is not very sensitive to the bottom sound speed variation. As shown in Ref. 4, from 10 to 110 Hz the signal gain differs by < 1 dB for the two bottom profiles mentioned above which have significantly different sound speed for the upper layers of the bottom. We noted that the exact location of the ambiguity surface fluctuates with respect to frequency. The inverted bottom profile could be very different depending on the signal frequency due to the sensitivity to the assumed source location and insensitivity to bottom sound speed. The question is which inversion result is more "correct" than the others?.

3. Correlation Using High Order Modes

Figure 1a showed many bottom reflected/refracted arrivals. They are identified in Fig. 1b by arrivals with grazing angle > 15 deg. These bottom interacting arrivals are of sufficient energy and will be sensitive to bottom sound speed profile variations. To improve the sensitivity of geoacoustic inversion, one needs a filter to extract bottom reflected/refracted arrivals from data and correlate with the model predictions. To demonstrate this, we use a mode filter [3] in this section and an angle filter [6] in the next section.

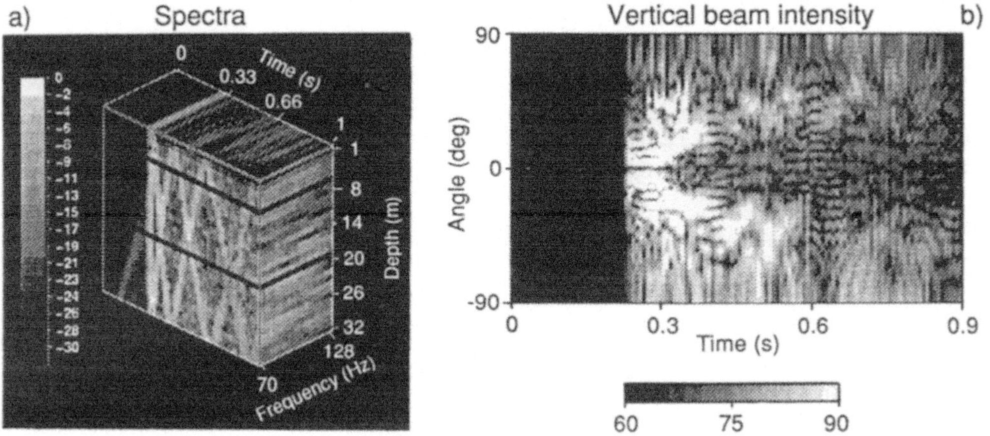

Fig. 1 (a) Signal spectral intensity vs. time and depth, and (b) vertical arrival pattern vs. time.

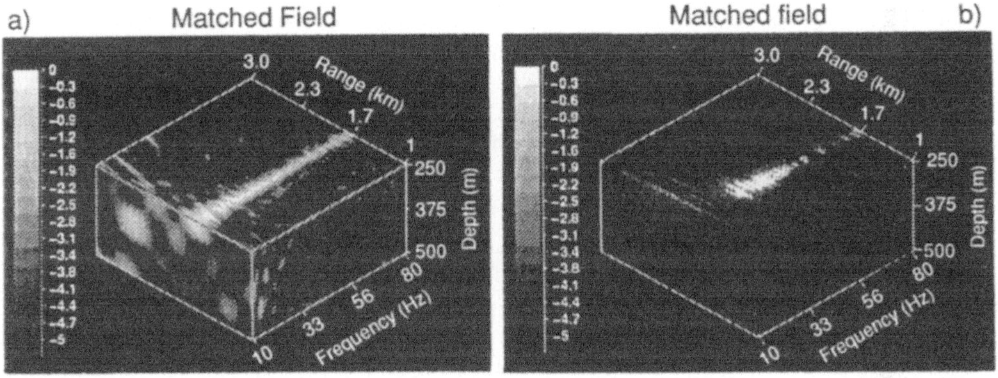

Fig. 2 Matched field ambiguity volume using (a) all modes and (b) waterborne modes only.

The mode filter can be implemented in matched field processing by restricting the modes used in the replica field. An alternative will be to use matched mode processing [7]. We show how in Fig. 3 the difference of the matched field correlation coefficients with and without bottom sound speed profile mismatch. The nominal bottom profile (Fig. 5) is incremented by 100 m/s uniformly over the bottom layers to create the bottom mismatch. The solid curve in Fig. 3 shows the differences in the matched field correlation coefficients with and without bottom mismatch as a function of frequency when all modes are included. The difference is on the average 5%. When only bottom interacting modes are used in the replica field, the difference in the matched field correlation coefficients has increased significantly (doubled over 30-50 Hz band).

4. Correlation Using Steep Angle Arrivals

As steep acoustic rays are bottom interacting rays, bottom inversion will be more effective if only steep vertical arrivals are used. This will be equivalent to correlating beams formed from the data and replica field corresponding to bottom and bottom/surface returns. To correlate the data and replica field in the angle domain, we define an ambiguity function B in terms of source range r and depth z,

$$B(r,z)=\frac{|\int_{|\theta|>|\Theta_0|}A^{rplc*}(\theta,r,z)A^{data}(\theta)d\theta|^2}{[\int_{|\theta|>|\Theta_0|}|A^{rplc}(\theta,r,z)|^2d\theta][\int_{|\theta|>|\Theta_0|}|A^{data}(\theta)|^2d\theta]} \tag{1}$$

where θ denotes the grazing angle, Θ_0 limits the correlation to steep angles, and $A^{data}(\theta)$, $A^{rplc}(\theta)$ are the spatial Fourier transform of the data and replica field respectively as defined by

$$A^{data}(\theta)=\sum_j e^{-ikz_j\sin\theta}p^{data}(z_j), \quad A^{rplc}(\theta,r,z)=\sum_j e^{-ikz_j\sin\theta}p^{rplc}(z_j,r,z), \tag{2}$$

where z_j is the receiver depth for the jth phone. Equation (1) is referred to as matched beam processing. We note that Eq. (1) without the angle cutoff reduces to the standard matched field processing for a large aperture array. Equation (1) has been used to improve source localization and tracking in real data by limiting the correlation to low grazing angles to avoid bottom mismatches.

Using Eq. (1), we calculate the matched beam correlation coefficients for the example used above. We plot in Fig. 4, the difference of the matched beam correlation coefficients with and without the bottom sound speed mismatch. The solid curves correlates returns from all angles, i.e., - 90 to + 90 degree whereas the dashed curve limits the correlation to grazing angle between ± 20 and ± 60 degrees. We see that the correlation coefficient is more sensitive to bottom sound speed profile mismatch when only bottom and bottom/surface returns are used.

5. Simulated Annealing

Geoacoustic inversion for the bottom sound speed profile is done using simulated annealing which searches effectively for a bottom sound speed profile that yields the highest correlation coefficient at the source location. For our applications, we use the matched beam ambiguity function, Eq. (1), summed over a frequency band as the cost function with a "temperature" parameter which is gradually lowered as the sound speed is perturbed. At high temperatures, many perturbations of the sound speed profile will be accepted to allow the search process to reach many maxima of the cost function. As temperature cools down, the searching process will narrow down to small perturbations around the true sound speed profile which yields the maximum of the cost function.

The bottom sound speed variations are parameterized using three empirical-orthogonal-functions allowing a maximum sound speed variation of ± 200 m/s. The KRAKEN normal mode program [8] is used to calculate the acoustic field for each sound speed profile. In actuality, we use a first order perturbation calculation of the mode wavenumber to sound speed variation to speed up the calculation; the perturbation results have been numerically checked with KRAKEN code. The simulated annealing method has been successfully tested with simulated data corresponding to the experimental environment described in Sec. 2. The inverted bottom sound speed profile is within 10 m/s of the true bottom sound speed profile.

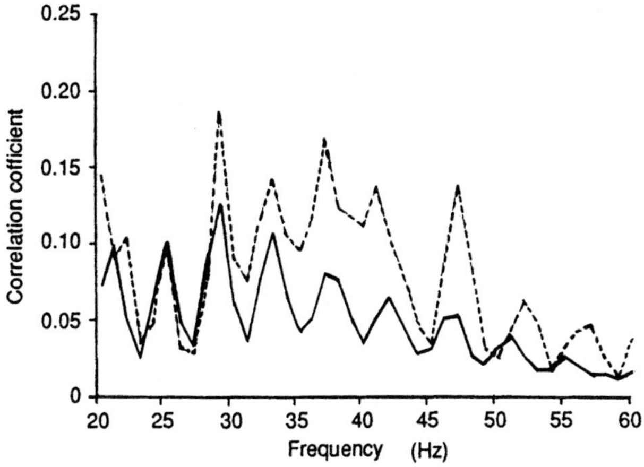

Fig. 3 Difference in the matched field correlation coefficients with and without bottom sound speed mismatch: using all modes (solid curve) and bottom interacting modes (dashed curve) with phase speed >1700 m/s.

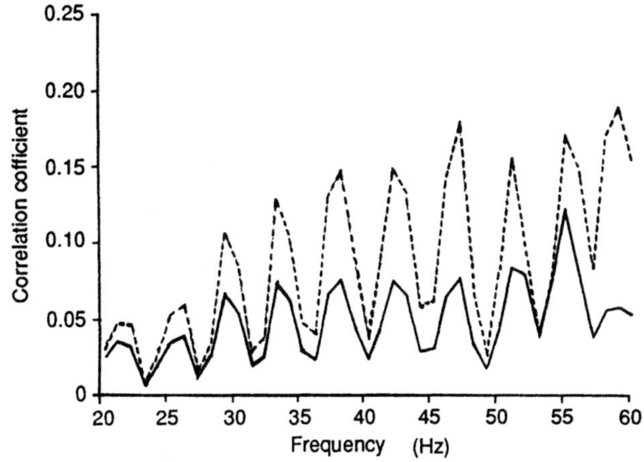

Fig. 4 Difference in the matched beam correlation coefficients with and without bottom sound speed mismatch: using all beams (solid curve) and high angle (±20 to ±60 deg.) beams (dashed curve).

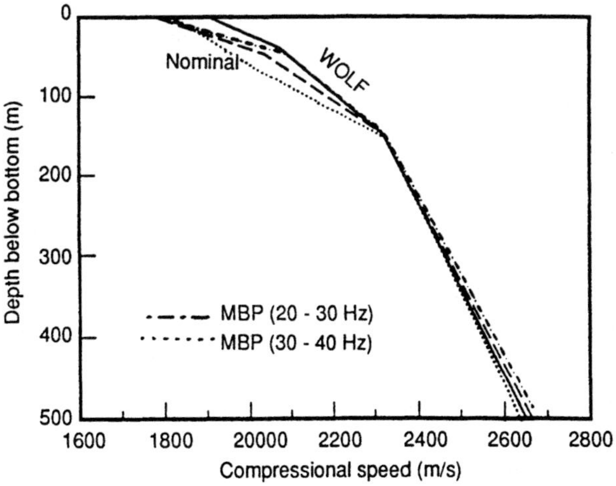

Fig. 5 Inverted bottom sound speed profile (SSP) using matched beam processing (MBP) compared with earlier results. Nominal SSP: dashed line; Wolf SSP: solid line, MBP: dotted line and dot-dashed line.

Next we apply the simulated annealing method to the shot data to invert the bottom sound speed profile. As previously pointed out [4], the normal mode acoustic field, integrated over the shot spectrum, successfully modeled the signal intensity as a function of time compared with the shot data despite the fact that the normal mode program ignores the continuum contribution. The continuum contribution will be neglected here. Applying the simulated annealing method to the shot data of 20-30 Hz band, we obtain a bottom sound speed profile which is plotted in Fig. 5. We note that the bottom profile for the 20-30 Hz band agrees with the (nominal) Geddes sound speed profile for the top 100 m and with the Wolf sound speed profile (estimated from the head waves) for the next 100 m layer. We also show the estimated bottom sound speed profile using 30-40 Hz data in Fig. 5. As remarked above, higher frequency (> 30 Hz) signals are less sensitive to the bottom sound speed variations in this environment.

6. Summary

The matched field inversion method which uses full field correlation between data and replica field may not be sensitive enough for inversion of the bottom sound profile due to the presence of predominant waterborne energy. By imposing a filter which focuses the correlation to bottom interacting signals, we show that the correlation coefficient becomes more sensitive to bottom sound speed mismatch. We demonstrated this by using a mode filter and a angle (beam) filter. Using the matched beam processing with beam angles limited to bottom returns, we estimated the bottom sound speed profile in a shallow Arctic water using a short range shot data. The inverted sound speed profile for the 20-30 Hz band is in close agreement with the bottom sound speed profiles estimated from in-situ sediment data [5] and from arrival times of the head waves [4].

1. A. Tolstoy, "Review of matched field processing for environmental inverse problems," Int. J. Mod. Phys. C 3, 691-708 (1992) and references cited therein.

2. M. D. Collins and W. A. Kuperman, "Focolization: Environmental focusing and source localization," J. Acoust. Soc. Am. 90, 1410-1422 (1991).

3. T. C. Yang, "Effectiveness of mode filtering: A comparison of matched-field and matched-mode processing," J. Acoust. Soc. Am. 87, 2072-2084 (1990).

4. J. Wolf, "Geoacoustic inversion in shallow Arctic water using broadband shot data," J. Ocean Eng. (to be published).

5. W. H. Geddes, "Geoacoustic model - Lincoln Sea," Geddes Geophysical Associates Tech. Rep. 3-90, 1990.

6. T. C. Yang and W. Cullop, "Source localization and tracking in a sloping littoral sea," unpublished (1994).

7. T. C. Yang, "A method of range and depth estimation by modal decomposition," J. Acoust. Soc. Am. 82, 1736-1745 (1987).

8. M. Porter, "The KRAKEN normal mode program," SACLANTCEN Mem. SM-245 (1991).

MODAL ANALYSIS OF DATA FROM SYNTHETIC APERTURE ARRAYS IN A SHALLOW WATER WAVEGUIDE USING A NON LINEAR MULTIDIMENSIONAL PARAMETRIC METHOD

E. Parastates
Technical Acoustics
Department of Vehicle Engineering
Royal Institute of Technology
S - 100 44 Stockholm, Sweden

ABSTRACT. Experimental data from a shallow water waveguide excited by a slowly moving CW source are analyzed by a nonlinear multidimensional estimator using modal subspace fitting techniques. Modal wave numbers are extracted from data sets with different excitation frequencies. The data quality is monitored by Hankel transformation. One dimensional MUSIC provides initial wavenumber values for the multidimensional search. It appears that a high number of modes can be resolved from short aperture lengths.

1. Introduction

A hydroacoustical experiment was performed by FOA (the National Defence Research Establishment) September 92. The purpose of the experiment was to determine seabed parameters using a technique based on estimation of modal parameters from synthetic horizontal apertures and mode matching with model predictions[1-4]. The experimental site was in Kalmarsund, the strait between the Öland Island and the mainland of Sweden. The propagation model assumes the medium to be laterally homogeneous, treating the seabed as a layered elastic solid[5]. The use of an estimator with the ability to resolve a high number of modes from short apertures reduces the risk of mismatch caused by range dependence. We have analyzed data sets with different excitation frequencies and source depths by first computing the complex amplitude of the observed pressure field and then using multidimensional MUSIC[9] (MD - MUSIC) for the wave number estimation.

2. Set - up of the experiment

A sequence of recordings was carried out where a CW source transmitting at a fixed frequency, specific for each run, was towed at constant speed along a straight line intercepting the array position. The data from a vertical array with eight hydrophones placed at depths (5, 10, 15, 20, 25, 30, 35, 40) m was transmitted to the ship by telemetry and was recorded there in parallel with a reference hydrophone signal. The position of the source was tracked continuously by radar ranging and the echo-sounder showed 39 - 40 m throughout the track.

3. The complex amplitude of the acoustic pressure field

A smooth approximation of the source receiver distance r(t) as function of continuous time t, needed for the synthetic aperture processing of the signals is obtained by least squares fitted B -

O. Diaschok et al. (eds.), Full Field Inversion Methods in Ocean and Seismo-Acoustics, 329–334.

splines[6]. The smooth function is then inverted by Hermite interpolation. With a monofrequent source and no interfering noise the emitted and the received signals will be

$$s_e(t) = A_e \cos(\omega t) - B_e \sin(\omega t) \tag{1}$$
$$s_j(t) = A_j(t)\cos(\omega t) - B_j(t)\sin(\omega t) \quad j = 1,\ldots,8 \tag{2}$$

where the carrier frequency ω equals the excitation frequency and the in - phase components $A_j(t)$ and quadrature components $B_j(t)$ are polynomial functions. After local maximum likelihood fitting (1) and (2) to data from a time interval centered around $t(r_l)$ the observed complex amplitude $x(r_l, z_j)$ at source receiver distance r_l and sensor depth z_j is obtained from

$$x(r_l, z_j) = \frac{A_j(t(r_l)) + iB_j(t(r_l))}{A_e + iB_e} \quad . \tag{3}$$

In a laterally homogeneous fluid - solid medium excited by a monofrequent source the complex amplitude is given by[7]

$$p(r,z) = \int_0^\infty \tilde{p}(\alpha,z) J_0(\alpha r)\alpha d\alpha = \frac{1}{2}\int_{-\infty}^{\infty} \tilde{p}(\alpha,z) H_0^{(1)}(\alpha r)\alpha d\alpha \tag{4}$$

where $\tilde{p}(\alpha,z)$ is an analytic function of α with poles at the wavenumbers of the normal modes of the wave guide. Wavenumbers of propagating modes are located close to the real axis in the complex α plane and thus peaks in $\tilde{p}(\alpha,z)$ as function of real α can be identified with the real part of wave numbers of propagating modes.

4. Spectral estimation by Hankel transformation

A non parametric method for the estimation of $\tilde{p}(\alpha,z)$ is provided by numerical evaluation of the Hankel transform

$$\tilde{p}(a,z) = \int_0^\infty p(r,z) J_0(\alpha r) r dr \quad . \tag{5}$$

J_0 is replaced by its asymptotic form for large arguments αr and a windowed (due to the finite aperture length) version of (5) is then evaluated by use of FFT. This method is computationally efficient but requires long apertures in order to resolve closely spaced wavenumbers.

5. Spectral estimation by parametric methods

If we neglect contributions from branch cuts and consider a fixed depth, omitting z, the residual theorem applied on the rightmost term of (4) gives the modal part of the field

$$p(r) = \sum_{k=1}^{K} \Gamma_k \alpha_k H_0^{(1)}(\alpha_k r) \approx \sum_{k=1}^{K} \Gamma_k \sqrt{\frac{\alpha_k}{r}} e^{i\alpha_k r} \; . \tag{6}$$

Introducing $\mathbf{d}(\alpha) = \left[e^{i\alpha r_0} \; \cdots \; e^{i\alpha r_N}\right]^{\mathrm{T}}, \mathbf{S} = \left[\mathbf{d}(\alpha_1) \; \cdots \; \mathbf{d}(\alpha_K)\right], \mathbf{g}_l = \left[\sqrt{\alpha_1}\Gamma_{1,l} \; \cdots \; \sqrt{\alpha_K}\Gamma_{K,l}\right]^{\mathrm{T}}$, and fitting $\sqrt{r}\,p(r)$ to $\mathbf{x}_l = \left[\sqrt{r_{l-1}}\,x(r_{l-1}) \; \cdots \; \sqrt{r_{N+l-1}}\,x(r_{N+l-1})\right]^{\mathrm{T}}$, $l = 1,\ldots,L$, leads to the well known maximum likelihood problem

$$\max_{a_1,\ldots,a_K} trace\left\{\mathbf{P}\mathbf{X}\mathbf{X}^H\right\} \tag{7}$$

where $\mathbf{X} = \left[\mathbf{x}_1 \; \cdots \; \mathbf{x}_L\right]$ is a data matrix, $\mathbf{S}^+$ is the Moore - Penrose pseudo inverse to $\mathbf{S}$ and $\mathbf{P} = \mathbf{S}\mathbf{S}^+$ is a projector to the modal subspace. After SV - decomposition of $\mathbf{X}$ a basis for the modal subspace is obtained from the set of eigenvectors $\mathbf{T}_M$, associated with the largest singular values. The object function for MUSIC[8] is obtained from (7) by replacing $\mathbf{X}\mathbf{X}^{\mathrm{H}}$ with $\mathbf{I} - \mathbf{T}_M\mathbf{T}_M^{\mathrm{H}}$, using a rank one projector $\mathbf{P} = \mathbf{d}(\alpha)\mathbf{d}^{\mathrm{H}}(\alpha)$ and inverting

$$Y(\alpha)_{MUSIC} = \left[\mathbf{d}^{\mathrm{H}}(\alpha)\left[\mathbf{I} - \mathbf{T}_M\mathbf{T}_M^{\mathrm{H}}\right]\mathbf{d}(\alpha)\right]^{-1}. \tag{8}$$

The object function for MD-MUSIC[8-9] is obtained from (7) by replacing $\mathbf{X}\mathbf{X}^{\mathrm{H}}$ with $\mathbf{T}_M\mathbf{T}_M^{\mathrm{H}}$.

$$Y(\alpha_1,\ldots,\alpha_K)_{MD} = trace\left\{\mathbf{P}\mathbf{T}_M\mathbf{T}_M{}^{\mathrm{H}}\right\} \; . \tag{9}$$

In practice, the computation of $\mathbf{P}$ is based on the QR decomposition of $\mathbf{S}$ by the modified Gram-Schmidt orthogonalization, - i.e., $\mathbf{S} = \mathbf{Q}\mathbf{R}$ and $\mathbf{P} = \mathbf{Q}\mathbf{Q}^{\mathrm{H}}$. The multidimensional search involves methods of nonlinear optimization and requires analytical expressions for the first and second derivative[10] of (9). Analysing real data we found that the rank of $\mathbf{T}_M$ was required to be substantially higher than the number of expected modes K, in order to obtain a reliable resolution in the MUSIC pseudo spectrum, Figure 6.2b. One explanation could be that the wavenumbers behave like narrow band spectral distributions due to range dependence effects. In that case the modal part of the covariance matrix estimate $\mathbf{X}\mathbf{X}^{\mathrm{H}}$ will be characterized by K sets of eigenvalues of decreasing magnitudes.

6. Experimental results

Three data sets have been analyzed. The (excitation frequencies, source depths) where (34 Hz, 37 m), (80 Hz, 20 m) and (160 Hz, 20 m). In Figure 6.1 we see the power of the Hankel transform $\left|\tilde{p}(\alpha,z_j)\right|^2$ for real wave numbers. The aperture lengths required to resolve closely spaced wavenumbers makes this method sensitive for range variations and broadens the modal peaks. Figure 6.2a, shows a MUSIC spectrum and we notice the sharp peaks indicating the wave numbers, frame b demonstrates the non sensitivity of the MUSIC spectrum to the choice of the

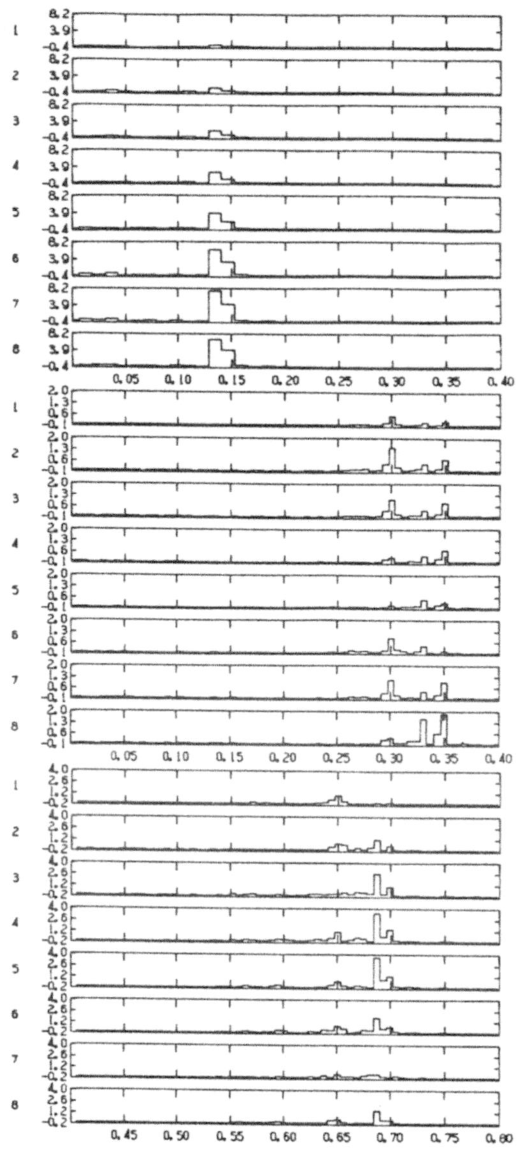

Figure 6.1.

The power spectrum of the Hankel transform $|\tilde{p}(\alpha,z_j)|^2$ of the complex amplitude as function of real wave numbers at depths $z_j, j = 1,\ldots,8$.

a (Top) : (34 Hz, 1700 m - 2100 m).
b (Middle): (80 Hz, 1600 m - 2400 m).
c (Bottom): (160 Hz, 1600 m - 2400 m).

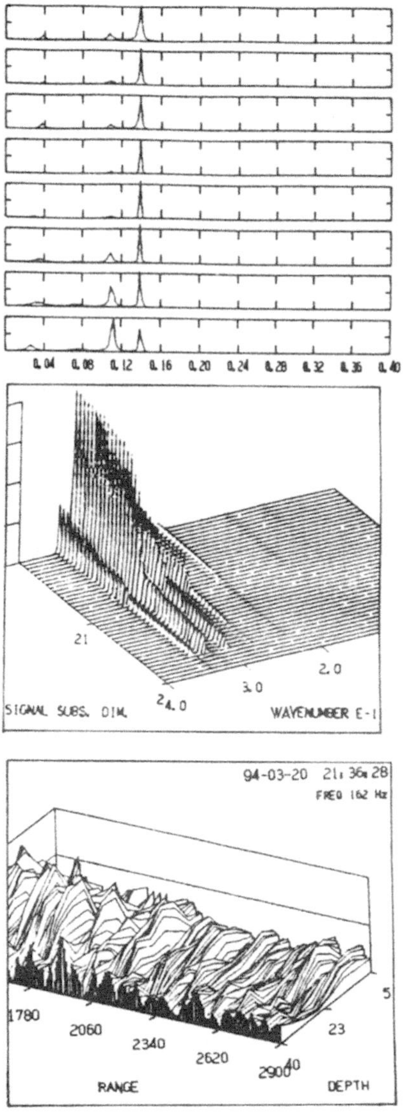

Figure 6.2.

a (Top): MUSIC spectrum at each receiver depth for (34 Hz, 1700 m - 2000 m).
b (Middle): MUSIC spectrum for one sensor as function of the modal subspace dim. for (80 Hz, 2300 m - 2900 m).
c (Bottom): Complex amplitude as function of range and depth for (160 Hz, 1500 m - 2900 m).

Table 6.1. Wavenumber estimates from MD-MUSIC

Freq.:	34 Hz			80 Hz			160 Hz		
Depth:	37 m			20 m			20 m		
Range	Mean	Dev.	Sensors	Mean	Dev.	Sensors	Mean	Dev.	Sensors
1400m	0.137	0.002	8	0.349	0.003	8	0.692	0.005	8
	0.114	0.003	8	0.330	0.006	8	0.664	0.009	8
to	0.069	0.006	8	0.289	0.016	7	0.634	0.010	6
	-	-	-	0.258	0.025	8	0.586	0.008	5
1700m	-	-	-	0.214	0.016	8	0.541	0.019	3
Range	Mean	Dev.	Sensors	Mean	Dev.	Sensors	Mean	Dev.	Sensors
1700m	0.138	0.001	8	0.347	0.002	7	0.696	0.004	7
	0.106	0.004	8	0.324	0.006	8	0.666	0.013	7
to	0.086	0.006	3	0.302	0.004	8	0.636	0.015	8
	-	-	-	0.275	0.003	8	0.589	0.013	8
2000m	-	-	-	0.218	0.003	5	0.556	0.008	8
Range	Mean	Dev.	Sensors	Mean	Dev.	Sensors	Mean	Dev.	Sensors
2000m	0.143	0.001	8	0.347	0.002	8	0.692	0.004	8
	0.108	0.004	8	0.329	0.004	8	0.663	0.011	8
to	0.068	0.003	7	0.296	0.002	8	0.632	0.015	8
	-	-	-	0.261	0.009	8	0.586	0.011	8
2300m	-	-	-	0.211	0.021	8	0.546	0.016	7
Range	Mean	Dev.	Sensors	Mean	Dev.	Sensors	Mean	Dev.	Sensors
2300m	0.142	0.001	8	0.353	0.002	8	0.692	0.003	8
	0.125	0.004	8	0.332	0.001	8	0.663	0.011	8
to	0.091	0.002	8	0.304	0.003	8	0.633	0.007	7
	-	-	-	0.271	0.004	8	0.586	0.007	8
2600m	-	-	-	0.218	0.004	8	0.544	0.007	8
Range	Mean	Dev.	Sensors	Mean	Dev.	Sensors	Mean	Dev.	Sensors
2600m	0.142	0.001	8	0.348	0.002	8	0.697	0.004	8
	0.108	0.005	8	0.332	0.004	8	0.672	0.008	8
to	0.081	0.002	4	0.310	0.008	5	0.639	0.008	8
	-	-	-	0.278	0.002	8	0.590	0.014	8
2900m	-	-	-	0.225	0.005	8	0.549	0.020	8

Table 6.2. Model predictions for the (80 Hz, 1700 m - 2000 m) aperture

Real part of wavenumbers 1/m:	Mode 1	Mode 2	Mode 3	Mode 4	Mode 5
	0.348	0.328	0.295	0.262	0.228

Table 6.3. Media parameters used for modeling of the (80 Hz, 1700 m - 2000 m) aperture

	Thickness m	P-vel. m/s	S-vel. m/s	Density kg/dm^3:	P-damping dB/λ	S-damping dB/λ
Water:	40	1415	-	1.0	0.0	-
Sediment:	13	1650	420	1.3	0.5	1.0
Rock:	∞	5000	3000	3.0	0.5	0.5

dimension number for the modal subspace and frame c shows the observed complex amplitude as function of range / depth. Comparisons of MUSIC and MD - MUSIC indicate a slightly better performance offered by the latter[10]. We have used MD - MUSIC for analyzing data from the range interval 1400 m - 2900m, Table 6.1. The aperture length was 300 m. Initial wavenumber values for the first search in each data set were obtained by MUSIC. The number of hydrophone signals for which the multidimensional method succeeded to estimate a certain wave number is given under the column 'sensors' and the average of the estimates is given under the column 'mean'. With known wavenumbers the modal amplitudes $\Gamma_{k,l}, k = 1,\ldots,K$ can be obtained from $\mathbf{g}_l = \mathbf{S}^{+}\mathbf{x}_l$. As an introduction to the inversion part of the experiment, a number of simulations[5] were manually carried out with the objective to suggest seabed parameters by mode matching. Wavenumber predictions from a first modeling effort of the 80 Hz case are given in Table 6.2 with the media parameters of the model as shown in Table 6.3. Our next step in the analysis of the experiment will be an investigation of optimization methods for acoustic field inversion.

Acknowledgments

I express my gratitude to Professor Ilkka Karasalo for his advice. The work presented in this paper has been carried out within the framework of UVC - KTH and is financed by FOA.

References

1. Frisk G. V.; Lynch J. F.; Shallow water wave guide characterization using the Hankel transform. J. Acoust. Soc. Am. 76: 205 - 216; 1984.
2. Frisk G. V.; Lynch J. F.; Subramaniam D. R. Determination of compressional wave speed profiles using modal inverse techniques in a range dependent environment in Nantucket Sound. J. Acoust. Soc. Am. 86: 1928 - 1939; 1989.
3. Cristini P.; Garnier L. J.; Gazanhes C. Analyse spectrale de la propagation sonore en eau peu profonde: application a la detection de modes presents dans une cuve experimentale. J. de Phys. C1 1053 - 1056; 1992.
4. Subramaniam D. R.; Saurav D. B. Evaluation of high-resolution frequency estimation methods for determining frequencies of eigenmodes in shallow water acoustic field. J. Acoust. Soc. Am. 93: 378-389; 1993.
5. Karasalo I. Exact finite elements for wave propagation in range independent fluid - solid media. J. Sound Vibr. To appear.
6. De Boor C. Package for calculating with B - splines. SIAM J. Numer. Anal. 6: 441 - 472; 1977).
7. Brekhovskikh L.; Lysanov Yu. Fundamentals of ocean acoustics. Springer Verlag. 122 - 129; 1982.
8. Schmidt S. D. A signal subspace approach to multiple emitter location and spectral estimation. Ph. D. thesis. Stanford Univ. 1981.
9. Cadzow J. A. A High resolution direction of arrival algorithm for narrow band coherent and incoherent sources. IEEE Trans. ASSP - 34: 965 - 979; 1988.
10 Parastates E. Modal analysis of data from synthetic aperture arrays in a shallow water waveguide. Report in preparation. Technical Acoustics, Dept. of Vehicle Engineering, Royal Institute of Technology, Stockholm.

MATCHED FIELD 3-D TOMOGRAPHIC INVERSION

A. TOLSTOY
U.S. Naval Research Laboratory
Acoustics Division
Washington, D.C. 20375-5000

ABSTRACT Matched Field Processing (MFP) is a full field signal processing technique that has enjoyed much recent success in applications to Acoustical Oceanography, mainly as a consequence of the high accuracy currently achievable in the propagation modeling of the phases and amplitudes of ocean acoustic fields. MFP is a generalization of plane wave beamforming and accounts for the non-planar, multipath character of the acoustic fields. The MFP technique compares array measurements of *relative* signal *phases and amplitudes* along the array to model predictions. It has been used primarily to solve the inverse source problem, i.e., to determine the unknown range, depth, and bearing of acoustic sources in a *known* ocean environment. However, the MFP approach has also been more recently applied to the environmental inverse problem, i.e., to determine the characteristics of an *unknown* ocean environment. The MFP tomographic technique is designed to use simple shot sources (either air or ship deployed) and either multiple vertical arrays, a single vertical array re-deployed several times (under slowly changing conditions), or horizontal arrays of sonobuoys. This paper will discuss such work with: (1) simulated data to determine (a) deep ocean sound-speed profiles for the Chiu ocean and (b) shallow water bottom parameters, e.g., sediment sound-speeds, densities, and layer thicknesses; (2) recent experimental data (collected on a a single vertical array deployed once) to determine average shallow water bottom properties in a region of the Mediterranean Sea.

1. Introduction

Matched field tomographic inversion is a relatively new method developed specifically for the estimation of environmental parameters. Work on this approach has its roots in the early environmental parameter sensitivity analyses done for source localization using matched field processing (MFP) which then progressed to the explicit application of MFP to environmental inverse problems (for a general review see [1]). The tomographic MFP approach began generating computational results in 1989 for the determination of deep ocean sound-speed profiles (see [2]) and has entailed multiple arrays of hydrophones widely dispersed over a region of interest plus multiple low frequency sources, e.g., shots, deployed around and through the region. The multiplicity of subsequent acoustic paths from sources to receivers samples the region in range, cross-range, and depth, and provides a tremendous amount of information about the region through which the energy has passed (see [3]). By means of carefully designed algorithms (see [4]) this plethora of information can be examined subject to constraints on possible parameter values to finally arrive at high resolution estimates of the unknown environmental parameters.

For multiple arrays or a single array multiply deployed, the parameters can be rapidly and efficiently determined in full 3-D, i.e., with variability in range, cross-range, and depth. For only a single array deployed once and in the absence of significant mode coupling, the

O. Diaschok et al. (eds.), Full Field Inversion Methods in Ocean and Seismo-Acoustics, 335–340.

parameters can be determined with depth variability *only* for each source to array path. That is, the range and azimuthal properties will be averaged over the path. Azimuthal variability may be subsequently examined by comparing estimated single path parameters for sources circling a lone array, and range variability for sources at different ranges along a track from the array.

2. Approach

Consider an ocean region of interest. The first step required by the MFP tomographic method is to grid the region into cells such that the unknown environmental parameters for each cell can be assumed to be constant throughout that cell. In the examples to follow these cells are assumed to be simple, uniform size squares but could be tailored to any given scenario in order to more finely sample areas of suspected high environmental variability.

Next, sources and receivers are distributed throughout the region in a configuration carefully designed by preliminary simulations to optimize the acoustic sampling of all the region cells. The size of the ocean region, the number of cells, the nature of the parameters being investigated, and the desired inversion accuracy all influence the selection of the source and receiver distributions.

Finally, one must select an appropriate frequency, i.e., a frequency high enough to excite many modes and yield sufficient resolution but low enough that attenuation and sensitivity to random and unknown perturbations is minimized. Then, in combination with a propagation model of choice, e.g., adiabatic normal modes, the inversion algorithm operates by comparing modeled fields with the measured fields. In particular, the algorithm seeks the family of parameters representing the unknown environment which *simultaneously* maximizes the correlation between measured and modeled relative acoustic amplitudes and phases across the arrays for *all* of the source-receiver paths.

Two important aspects of MFP tomographic inversion are that: (1) preliminary sensitivity analyses are crucial for the efficient design of the inversion algorithm, and (2) examination of ambiguity surfaces and their associated sidelobes as functions of the environmental parameters can rapidly convey a sense of the uniqueness (or lack of uniqueness!) of the inversion solution.

3. Deep Water Sound-speeds

Perhaps the most interesting simulated data set available to examine the deep ocean performance of tomographic inversion methods is that developed by Chiu (see [5]) representing a highly realistic, 1000km by 1000km by 5000m ocean region.

The MFP approach has been applied to this data set for region subsets 500km by 500km by 5000m, and results have been generated for a variety of simulated resources, e.g., for four 1000 m long vertical arrays of 15 phones and up to 174 shot sources (for full details see [6] and references therein). In general, the MFP inversion calculations at 20 Hz showed excellent sound-speed estimation accuracies, e.g., rms errors well less than 1.0 m/sec. Moreover, the calculations also showed that these errors were relatively insensitive to expected errors in source and array location (within 100 m), and in source depth (within 50 m), and that resolution degraded, i.e., errors increased, as the number of sources decreased, and as source frequency decreased.

Thus, the method appears to be quite stable and to offer potentially very high 3-D resolution when applied to the determination of deep ocean sound-speed profiles. The advent

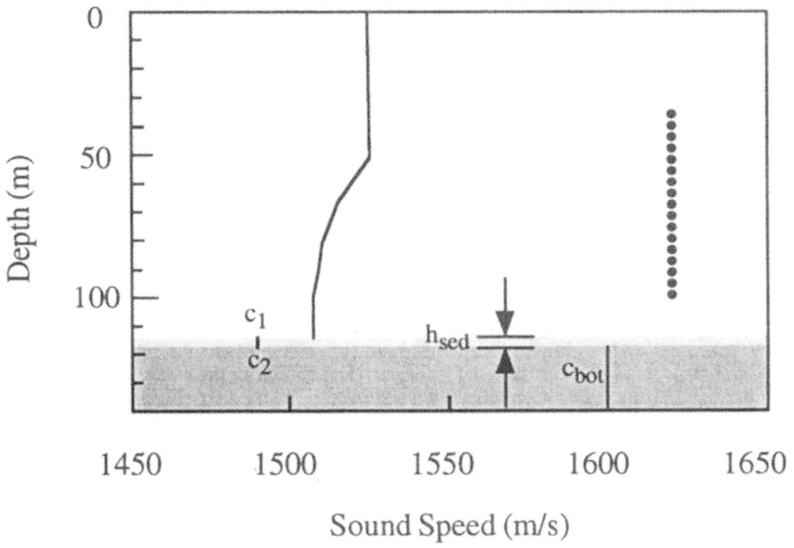

FIGURE 1. Nominal geoacoustic parameters for the Med Test 10/94.

of inexpensive, air-deployable vertical arrays of 1000 m in length would provide an important breakthrough in the attractiveness and efficiency of this method.

4. Shallow Water Bottom Properties

The most recent application of the MFP tomographic approach has been for the determination of shallow water bottom properties. Consider a shallow water environment with a bottom consisting of a sediment layer of thickness h_{sed} with a linear sound-speed profile varying from c_1 at the ocean bottom to c_2 at the sediment layer bottom overlaying a non-elastic half space with a constant sound-speed c_{bot}. Nominal bottom parameter values for the Mediterranean Sea experimental test (see [7] for details of the test) conducted in 10/94 are: $h_{sed} = 3.5$ m, $c_1 = 1490$ m/sec, $c_2 = 1490$ m/sec, $c_{bot} = 1600$ m/sec, based partly on values suggested in [8]. The water column sound-speed profile values were determined by CTD measurements and resulted in downward refracting energy. For this experiment a vertical array was deployed once in 114 m of water and consisted of 64 phones at non-uniform intervals. When considering the test data we shall use the full array while to reduce computation time for our 3-D simulation results we shall model only a subset of those phones, i.e., an array of 17 phones spaced every 4 m with the first phone at 35.72 m depth, the last at 99.72 m (see Fig. 1).

For the 3-D simulation results, consider a 20 km by 20 km region divided into 16 cells (each 5 km by 5 km in size), and let each cell have a different half-space sound-speed c_{bot}. Preliminary sensitivity analyses (not shown here) indicate that the MFP inversion will be highly sensitive to c_{bot} at 100 Hz. Let us assume the variability shown in Fig. 2a. Consider the 17 phone array "deployed" at four locations throughout the region (as shown by the black circles in Fig. 2b), 36 shot sources "deployed" around the perimeter (as shown by the Xs) at 30 m depth, and the fields modeled at 100 Hz using adiabatic normal modes (5 modes) to generate the "data" on those arrays. Then, after inversion we arrive at the estimated values $\hat{c}_{bot}$ shown in Fig. 2c with a maximum error of 1.6 m/sec. We note that

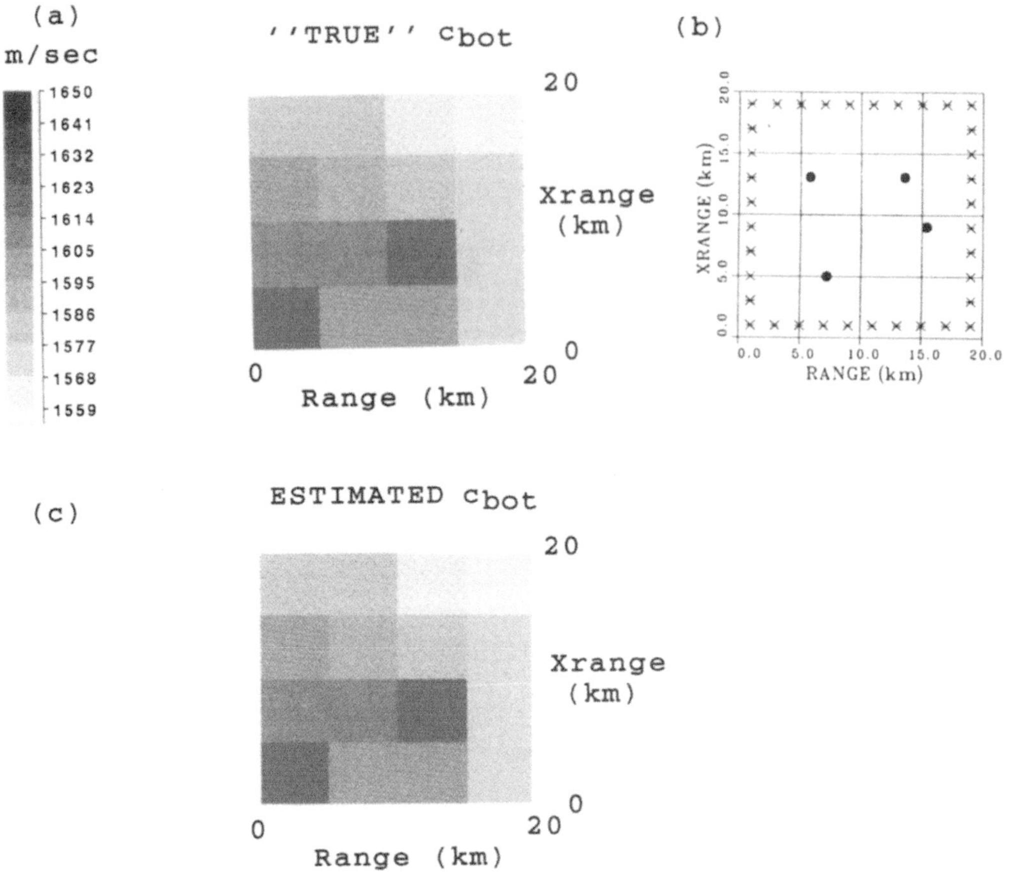

FIGURE 2. Simulated MFP tomographic inversion results. (a) "True" values of c_{bot} which vary with range over 20km by 20km region with values indicated by scale in m/sec; (b) configuration of 4 vertical arrays o and 36 shot sources X; (c) estimated values of c_{bot}.

we have assumed that we knew the source and array locations *exactly*. More realistically, we need to know the locations to within a wavelength, i.e., to within 15 m at 100 Hz. This point becomes critical for the analysis of the test data.

The experimental test data involved only one vertical array deployed once, and a variety of shots deployed at various ranges and azimuths. For each shot we can determine the *average* bottom properties over the range between the shot and the array. Consider shot 5 at a nominal range of 12.1 ± 0.4 km, depth of 18 ± 20 m. First, we processed the array data for that shot (using ALL the phones) and selected our frequencies of interest to which we applied standard FFTs. Next, assuming the nominal geoacoustic parameters of Fig. 1, we attempted to improve the source location estimates by generating the ambiguity surface (AMS) shown in Fig. 3a for the 100 Hz signal component indicating a peak Bartlett processor power of 0.70 (relative to a maximum possible value of 1.00) at 11.86km, 25m (the processor power was 0.25 at 12.1km, 18m) where "R" indicates candidate source range (km), and "Z" indicates candidate source depth (m). Then, using this "refined" source location value we applied the MFP tomographic inversion algorithm to calculate new estimates of

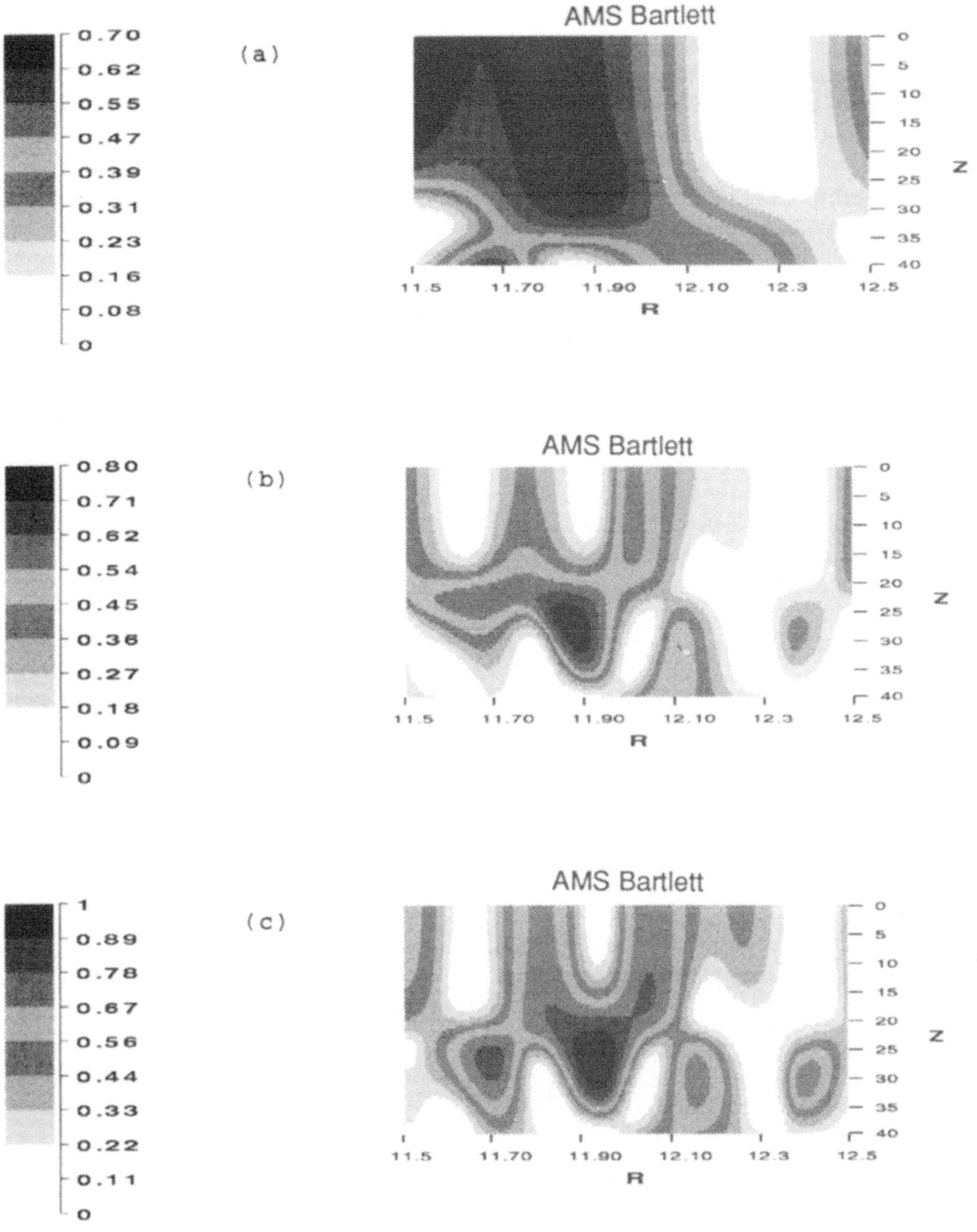

FIGURE 3. AMSs for: (a) test data vs. model data for nominal geoacoustic values. Peak = 0.70 at 11.86km, 25m; (b) test data vs. model data for estimated geoacoustic values. Peak = 0.80 at 11.87km, 27m; (c) simulated data for estimated geoacoustic values and source at 11.87km 27m.

the bottom parameters: $\hat{h}_{sed}$ = 3.5 m, $\hat{c}_1$ = 1550 m, $\hat{c}_2$ = 1470 m, $\hat{c}_{bot}$ = 1685 m. Using these values, we generated the AMS seen in Fig. 3b (comparing the test data with our new model values) with a higher peak value of 0.80 at 11.87km, 27m. Finally, using those same geoacoustic values, we generated simulated data ambiguity surfaces for a source at 11.87km, 27m as shown in Fig. 3c. We note that there is a strong resemblance between Figs. 3b and 3c suggesting that we may be on the right track.

However, we are finding that other frequencies, e.g., 200 Hz, suggest different source locations and result in different geoacoustic parameter estimates. At this stage we conclude that the source coordinates may need to be included in the unknown parameter search space *simultaneously* with the unknown geoacoustic parameters. This is a consequence of the need for higher resolution at frequencies like 100 Hz, e.g., we need to know the source locations to accuracies of a wavelength.

5. Conclusions

We conclude that MFP tomographic inversion has the potential to determine to within 1 to 2 m/sec both deep ocean sound-speed profiles and shallow water bottom sound-speeds in full 3-D. Most results to date have concerned simulated data. However, recent efforts are addressing a shallow water scenario (which used a single vertical array) to improve estimates of such bottom parameters as sediment thickness, sediment sound-speed, and half-space sound-speed. Results are promising but preliminary.

6. Acknowledgements

The author would like to express appreciation to D. Gingras and company for their generous and kind support in the generation and collection of the abovementioned test data, to J. Wolf and G. Vermillion for their invaluable assistance in acquiring, processing, and interpreting the data. Support by ONR.

References

1. Tolstoy, A. (1992) Review of matched field processing for environmental inverse problems, *J. Computat. Acoust.*, **3**, 691-708.
2. Tolstoy, A. (1989) Low frequency acoustic tomography using matched field processing, I and II, *J. Acoust. Soc. Am.*, **86**, S7.
3. Munk, W.H., and C. Wunsch (1979) Ocean acoustic tomography: a scheme for large scale monitoring, *Deep Sea Res.*, **26A**, 123-161.
4. Tolstoy, A. and L.N. Frazer (1993) Maximizing matched field processing output used in a new approach to ocean acoustic tomography, in *Computational Acoustics* Vol 2, Lee, Robinson, and Vichnevetsky, ed., North-Holland, pg. 79-84.
5. Chiu, C.S. (1992) Acoustical oceanography: Comparison of ocean tomography methods, *J. Acoust. Soc. Am.*, **92**, 2324.
6. Tolstoy, A. (1994) Performance of acoustic tomography via matched field processing, accepted by *J. Computat. Acoust.*.
7. Tolstoy, A. (1995) Matched field tomographic inversion to determine environmental properties, in *Current Topics in Acoustical Research*, Research Trends.
8. Jensen, F.B. (1974) Comparison of transmission loss data for different shallow-water areas with theoretical results provided by a three-fluid normal-mode propagation model, in SACLANTCEN Conf. Proc. No. 14.

EXPERIMENTAL HORIZONTAL WAVENUMBER SPECTRA AND IMPLICATIONS FOR FULL FIELD PROCESSING

J.A. DOUTT[1], R.J. CEDERBERG[2], R.B. EVANS[3], S.D. RAJAN[1], W.M. CAREY[4] and W.L. SIEGMANN[2]

(1) Woods Hole Oceanographic Instn., Woods Hole, MA 02543, USA

(2) Rensselaer Polytechnic Institute, Troy, NY, 12180-3590

(3) SAIC, 2 Shaw's Cove, Suite 203, New London, CT 06320

(4) ARPA (MSTO), 3701 N Fairfax Dr., Arlington, VA 22203-1714

ABSTRACT Acoustic pressure levels were recorded from a multifrequency source towed at mid-depth on three legs, both towards and away from a 24-element vertical hydrophone array. Horizontal wavenumber spectra were estimated using a synthetic aperture Hankel transform technique. Plots of modal amplitude as a function of depth and wavenumber reveal details of the propagation which are not readily apparent in plots of transmission loss. Two compressional speed profiles beneath the ocean bottom were derived for different legs from the modal wavenumbers at 50 Hz. Comparisons between the experimental data and SAFARI modelling using these profiles give excellent agreement both in amplitude and phase. KRAKEN and IFD/PE modelling using profiles based on nearby borehole data also give excellent agreement. Some aspects of the modal plots can be explained by lateral variability along with a double-ducted sound channel in the sediment below a depth of 75 meters. The multifrequency experimental acoustic data, measured sound speed profiles, and derived geoacoustic bottom profiles were used with variable success in matched field processing to obtain source localization. The high quality of this experimental data set makes it particularly useful for the testing of full field inversion techniques.

1. Introduction

The objective of this work was to supplement data from previous shallow water experiments by performing one under known and controlled conditions: a calibrated set of transmission data with minimal but known uncertainties in the acoustic and environmental measurements. Environmental parameters such as sound speed structure, sea state, and bottom depth were measured in an area where previous surveys had characterized the range-dependent bottom and sub-bottom layering. Issues addressed were the determination of the primary causes of sound transmission loss as a function of frequency (attenuation, shear waves, or scattering), and the feasibility of synthetic aperture processing as a shallow water survey method.

2. Description of the Experiment

The 1988 Hudson Canyon experiment was conducted off the New Jersey coast about 40 nautical miles SSW of the Hudson Canyon. Details of the experiment have been reported by Carey (1992). The analysis reported here deals with experiment TL2-2. Acoustic pres-

O. Diaschok et al. (eds.), Full Field Inversion Methods in Ocean and Seismo-Acoustics, 341–346.

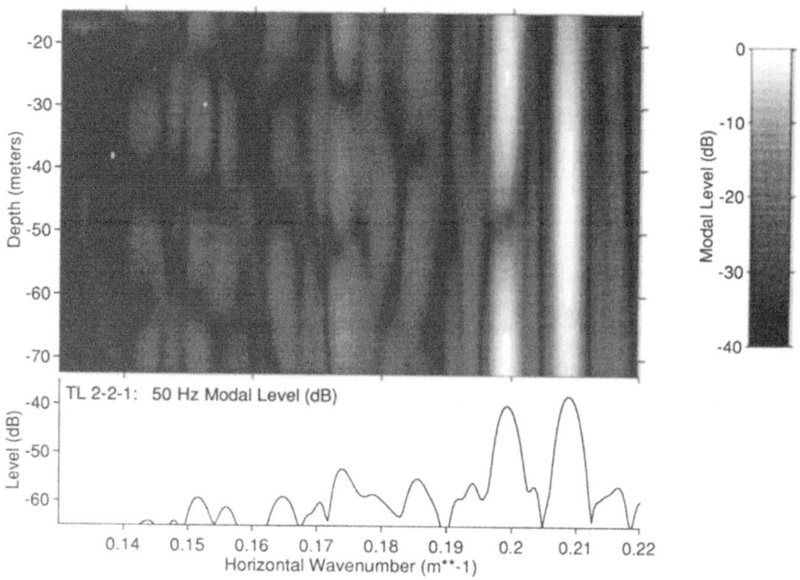

FIGURE 1. 50 Hz modal wavenumbers from experiment leg 221.

sure levels were recorded from a multifrequency source towed at mid-depth towards and away from a 24 element vertical array moored in 73 meters of water. The 24 hydrophones formed a real vertical array; as the source was towed behind the moving ship a horizontal synthetic aperture array was formed at the depth of each hydrophone. Thus a 2-dimensional slice of the acoustic field was sampled. The source emitted four frequencies simultaneously. These were extracted using an FFT, the precision range data was added, the signal was demodulated in range, and finally interpolated onto a 1 meter grid. The asymptotic Hankel transform was then used to obtain the horizontal wavenumbers. The wavenumber spectra for each hydrophone were plotted at the depth of the hydrophone and figures of modal level as a function of depth and horizontal wavenumber were generated using MATLAB to interpolate and plot the data. The line plot below each figure gives the sum in dB of the modal levels over all phones. (Figs. 1, 2).

3. Discussion

Figures 1 and 2 show the 50 Hz wavenumber spectra for legs 1 and 2 respectively. The single peak of the fundamental mode at a horizontal wavenumber (kr) of .2090 /m is clearly visible, as is the double peak of the second mode at .1994 /m. Peaks for higher-order modes are also apparent. Differences are apparent between the two legs: on leg 2 the modal peaks for the first two modes are better resolved than on leg 1. The fourth mode, at a kr of about .175, shows a splitting or echoing on leg 2 not seen on leg 1. Features like this are difficult to recognize in transmission loss plots, and are important for correctly deducing the fine-scale structure in the sediment.

The horizontal wavenumber information at 50 Hz was used with a perturbative inverse technique to derive the compressional wave speeds beneath the ocean bottom for legs 1

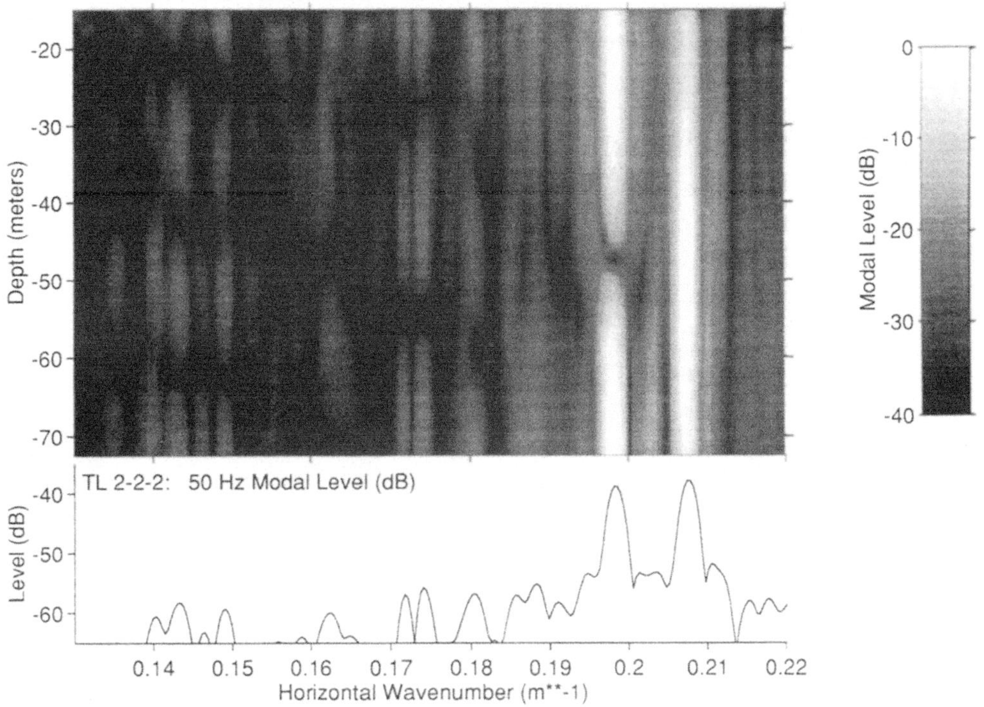

FIGURE 2. 50 Hz modal wavenumbers from experiment leg 222.

and 2. The two models are somewhat different (Fig. 3), which can be expected since the wavenumber spectra show differences. The pressure field generated by SAFARI for each model gives excellent agreement when compared to the corresponding experimental results. Figure 4 for leg 1 shows the agreement in both amplitude and phase at a hydrophone depth of 67.5 meters.

Additional layered bottom models were inferred from the density logs from AMCOR borehole 6010. Porosity values were computed, including factors for compaction due to overburden pressure. Biot theory was then used to convert the porosities into the bottom sound speed profile. The IFD implementation of wide angle PE of Claerbout was used to compute the acoustic pressure field as a function of both depth and range. The asymptotic approximation to the Hankel transform was then applied at each depth (in a similar manner to what was used on the experimental data) and the resulting wavenumber spectra were plotted using MATLAB to produce modal levels as a function of depth and horizontal wavenumber. The first two modes calculated for this profile agree well with those from the experimental data, but agree less well with higher-order modes. Two mechanisms were found which produced results like those seen in the leg 2 data. The first is the inclusion of a higher speed layer, ranging from 133 to 143 meters beneath the sea surface, which creates a duct in the sediment layer (Fig. 5).

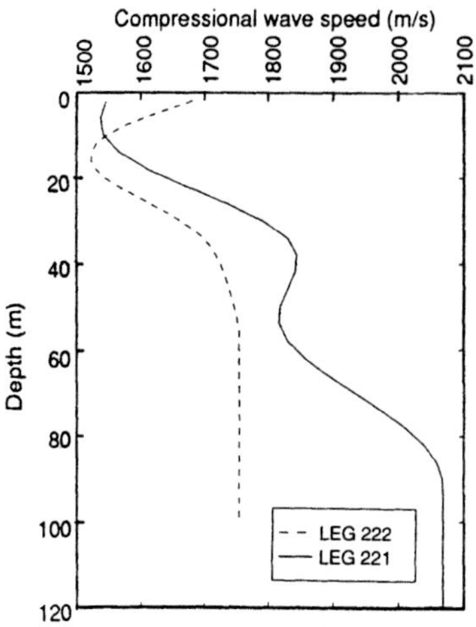

FIGURE. 3. Left: Sound speed profiles derived by perturbative inverse method.

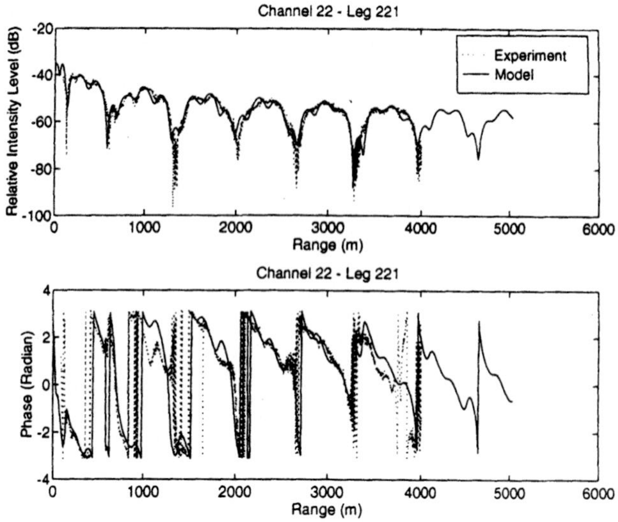

FIGURE. 4. Right: 50 Hz leg 221 fit between experimental data and SAFARI model – upper: relative intensity, lower: phase.

This was found to produce two wavenumbers very close together which had virtually the same modal structure within the water column (Fig. 6). This double peak appears similar to that seen in the experimental data. Although similar in the water column, the structure of these two modes does differ in the sediment. The introduction of range dependence to the bottom sound speed profile was found to produce sidelobes on the wavenumber peaks, causing multiple modal patterns to appear on the modal plots, much like those seen in the experimental data. Although many of the features of the data are well matched by this model, significant differences in the wavenumber positions and mode shapes exist for the higher-order modes.

Parameter sensitivity studies were performed using the IFD PE model in order to determine the extent to which environmental uncertainties affect predictions of the low frequency experimental results. A large uncertainty in the bottom sound speed structure was found to have the most significant affect on relative intensity predictions. It was found that only about the top 30 meters of sediment affect the intensity in the water. However, the higher order wavenumbers can be greatly affected by the deeper sediment sound speed values. The extreme sensitivity to water depth makes uncertainties in this parameter nearly as significant as the much larger uncertainties in the bottom sound speed. Variations in other parameters such as bottom density and attenuation, water sound speed, and source and receiver depth were less important. Given the parameter uncertainties, predictions of the horizontal wavenumbers corresponding to the propagating modes can be expected to be accurate to within about 0.005 /m.

Source localization matched field processing was attempted as a check of the validity of the bottom sediment models. The source localization scheme used a time average of the measured complex pressure vector on the vertical array and compared it to a hypothetical complex pressure vector due to a source at a known depth and range. The replica complex pressure vector was computed with the normal mode model KRAKEN (Porter, 1991). The comparison was made using the conventional linear beamformer, which is the correlation between the measured and replica complex pressure vectors. A range and depth window was swept and the true source location is expected to be where the comparison is optimum.

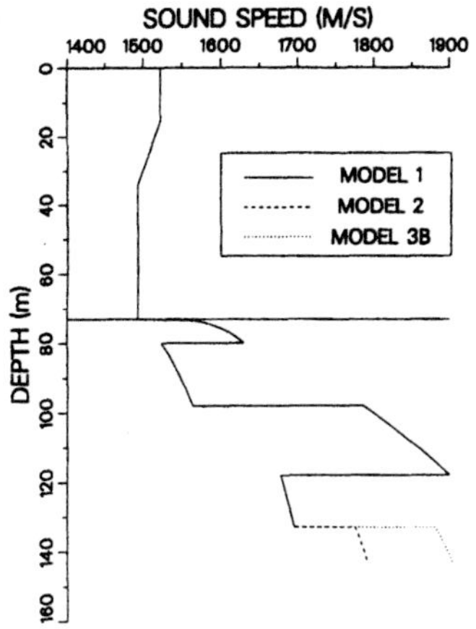

FIGURE. 5. Left: Double-duct sound-speed profile derived from borehole data with a decrease in porosity at 133 m.

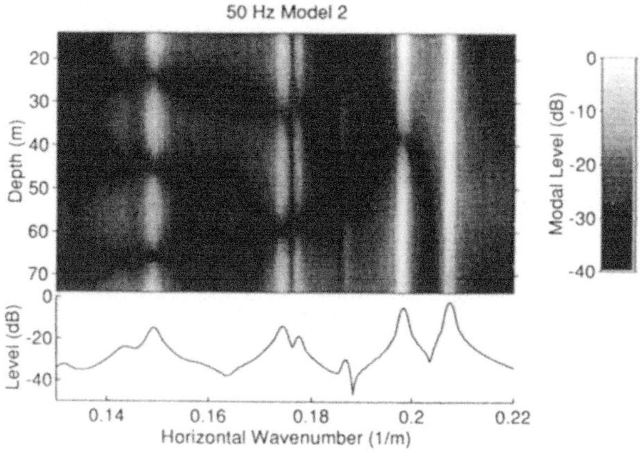

FIGURE. 6. Right: 50 Hz modal wavenumbers from profile in Fig.5.

The complex pressure data processed consisted of ten time snapshots at 50 and 175 Hz for the source at an range of 4.5 km and a depth of 36 m on the track TL 222 and another ten time snapshots at 75 and 275 Hz for the source at an range of 3.155 km and a depth of 36 m on the track TL 223. In both cases two sediment models, obtained by the perturbative (acoustic) inversion method at 50 Hz on the tracks TL 221 and TL 222, were used.

Only one of the single frequency calculations, based on the 175 Hz TL 222 data and the TL 221 sediment model, was successful in predicting the location of the source. In most of the other cases the source was located in one of the sidelobes of the beamformer, but the optimum location did not persist in frequency. To handle this situation, the ambiguity functions (at two frequencies) were added in dB (Porter, 1994) and this resulted in correct source localization for both ranges, using the TL 221 sediment model. The TL 222 sediment model was successful at one of the ranges.

4. Summary and Conclusions

For calm sea states, under the carefully controlled and measured conditions of this experiment, phase-stable results were obtained at multiple frequencies on repeated tows. This demonstrates that this synthetic aperture technique can be used at survey speeds and implies that sequential towed array measurements should yield comparable results. The measured transmission results were predictable using a variety of computer codes and bottom models derived from both borehole data and from the wavenumber spectra. A parametric analysis showed that the water depth, near-bottom sound speed gradient, and sub-bottom features were very important for accurately predicting the transmission loss. Modal plots are found to be more discriminating for matching models than transmission loss plots. Characteristics of the mode splitting and ghosting seen in the measured plots were seen to be consistent with the modal results from a double-ducted, range dependent model. The success of conventional linear matched field processing for source localization produced variable results. An ambiguity surface based on the product of the surfaces for specific frequencies gave the best results.

References

1. Carey, W., Dillman, L., Doutt, J., (1992) "Shallow-Water Transmission Measurements Taken on the New Jersey Continental Shelf", NUWC-NL Technical Document 10,025, Naval Undersea Warfare Center Detachment New London, CT.
2. Cederberg, R.J., Siegmann, W.L., Jacobson, M.J., Carey, W. (1993) "Predictability of Acoustic Intensity and Horizontal Wave Numbers in Shallow Water at Low Frequencies Using Parabolic Approximations", *J. Acoust. Soc. Amer.*, **94**, 1034–1043.
3. Cederberg, R.J., Siegmann, W.L., Jacobson, M.J., Carey, W. (1994) "Parameter Sensitivities in Two-Layer Isospeed Models of Shallow Water Channels", *J. Comp. Acoust.*, **1**, 469–486.
4. Frisk, G.V., Lynch, J.F. (1984) "Shallow Water Waveguide Characterization Using the Hankel Transform", *J. Acoust. Soc. Am.* **76**, 205–216.
5. Rajan, S., Lynch, J.F., Frisk, G.V. (1987) "Perturbative Inversion Methods for Obtaining Bottom Geoacoustic Parameters in Shallow Water," *J. Acoust. Soc. Am.* **82**, 998–1017.
6. Porter, M., (1991) "The KRAKEN Normal Mode Program," SACLANTCEN SM-254. La Spezia, Italy, SACLANT Undersea Research Centre.
7. Porter, M.B., (1994), "Source tracking in the Hudson Canyon," *J. Acoust. Soc. Am.*, **95**, (5, Pt. 2), 2981 (1994).

INVERTING AMBIENT NOISE IN SHALLOW WATER FOR THE BOTTOM GEO-ACOUSTIC PARAMETERS

Michael J. Buckingham†, Grant B. Deane and Nicholas M. Carbone
Marine Physical Laboratory, Scripps Institution of Oceanography
La Jolla, CA 92093-0213, USA

†also: Institute of Sound and Vibration Research
The University, Southampton, SO9 5NH, England

ABSTRACT For many years, oceanic ambient noise was treated exclusively as the background acoustic field against which signals of interest were to be detected. Only recently has it been recognized that, in fact, the noise also contains information about the ocean environment, which can be reliably extracted. To achieve a successful inversion, it is necessary to identify some robust feature of the noise that is consistently predictable. Intensity does not satisfy this requirement, since it depends on too many variable factors. Directionality or spatial coherence are better choices, particularly in shallow (continental shelf) water, where the vertical directionality of the noise is governed by the bottom boundary condition, which in turn is set by the geo-acoustic parameters of the bottom (*e.g.* the compressional and shear wave speeds). By inverting the directional density function of the noise in the water column, it should therefore be possible to determine the geo-acoustic parameters of the basement. As a simple example, consider a fast fluid bottom, which exhibits a critical grazing angle, say α_c, for sound incident from above. As a result the noise is distributed over an angular region in the vertical extending out to α_c on either side of the horizontal. (This is an over simplification, but serves to illustrate the point.) It follows that a direct measure of α_c is obtained by observing the noise power in the beam of a vertical line array as it is swept from the upward vertical, down through the horizontal, to the downward vertical. Assuming that the sound speed in the water column is known from an XBT deployment, the value of α_c, as determined from the noise, provides the value of the sound speed in the fluid sediment. In practice, it is more convenient to work with the (broad band) spatial coherence than the (narrow band) directionality of the noise, since the former requires only two sensors rather than a full array. A broad band noise inversion technique, based on the spatial coherence in the vertical, has been developed which yields the compressional and shear parameters of an elastic sea bed. The results of several such inversions, using real data from several sites around the UK coast, are consistent with the known values of the bottom geo-acoustic parameters, as obtained by more conventional methods.

1. Introduction

Acoustic reflection and refraction techniques are commonly used to obtain the geo-acoustic structure of the sea bed, including the sub-bottom stratification to a depth of two or three kilometres. Such techniques provide a high resolution view of the bottom, but at high cost: a sophisticated towed array of hydrophones is required, as well as an energetic source, which may be deployed from a second vessel. Bottom coverage tends to be slow and, typically, an exploratory cruise may last several weeks.

An alternative approach to the problem of surveying the sea bed and sub-bottom relies on the information contained in the ambient noise field in the water column[1]. In shallow water, over the continental shelf, where the water depth is nominally less than 200 metres, the noise undergoes multiple interactions with the bottom as it propagates through the channel. At each bottom reflection the noise field is modified by the boundary, suggesting

O. Diaschok et al. (eds.), Full Field Inversion Methods in Ocean and Seismo-Acoustics, 347–352.

that certain features of the noise must exist which embody the information imparted by the reflection process.

The level of the noise is one such feature, although the noise level is also influenced strongly by sea surface conditions and other factors that are highly variable, making it unsatisfactory as a tool for remotely sensing the bottom. A better choice is the structure (directionality or spatial coherence) of the noise field in the vertical, which, because it is almost exclusively governed by the bottom, is stable and predictable. Since the spatial coherence of the noise in the vertical is governed by the reflection coefficient of the sea bed, an inversion of measured coherence data should yield an estimate of the geo-acoustic parameters of the bottom.

A method for determining, from the noise in the water column, the compressional and shear wave speeds of an elastic bottom is described in this paper. At present the new technique provides information on the first few metres of the basement, but appears to have the potential for yielding an estimate of the sub-bottom structure to a depth of several hundred metres.

2. Fluid and elastic basements

Figure 1 illustrates the magnitude of the reflection coefficient as a function of grazing angle for two different bottom materials, one fluid, the other elastic. In the case of the (fast) fluid bottom, total reflection occurs at grazing angles less than the critical value, α_c, beyond which the curve falls sharply. For the elastic bottom, the magnitude of the reflection coefficient is less than unity for incident grazing angles below α_c, passing through a minimum when the coupling into shear waves is most efficient. Above α_c the decay of the curve is much like that for the fluid bottom. The pronounced difference between the two curves implies that the spatial structure of the noise in the two cases should show corresponding differences.

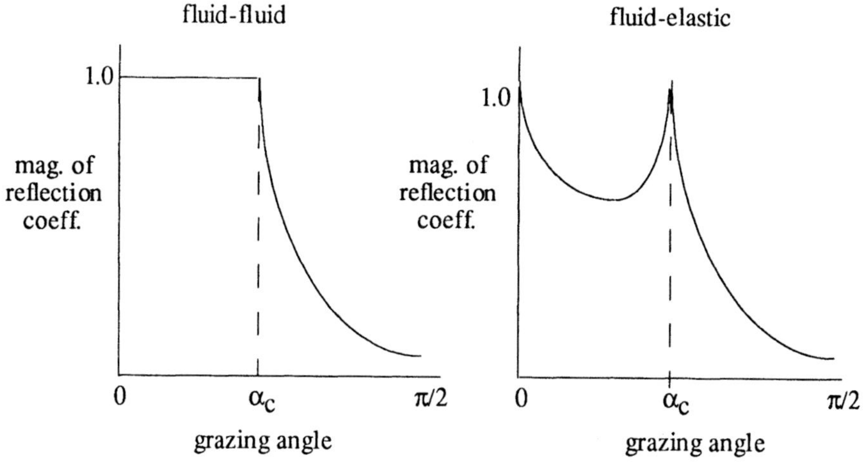

Figure 1. Schematic of reflection coefficient as a function of grazing angle for a fluid-fluid and fluid-elastic interface

As originally derived, the noise inversion technique[1] applied only to a fluid bottom. The (fluid basement) inversion procedure is based on a simple empirical model of the vertical directionality of the ambient noise: the directional density function is represented as uniform at angles within α_c of the horizontal and essentially zero at steeper angles. By converting this directional density function into a coherence function, through a Fourier transformation, and making a comparison with noise data gathered from two vertically aligned hydrophones in the water column, a fit is obtained from which the compressional sound speed in the bottom is estimated. This procedure was applied to data obtained from six sites around the UK coast whose bottom types were known. Good agreement between the estimated and expected (from prior information) values of compressional speed were found in five cases, where the bottom was known to be fluid. In the sixth case, the noise technique yielded a poor estimate of the sound speed, the reason being that the basement consisted of chalk, which is an elastic material supporting shear.

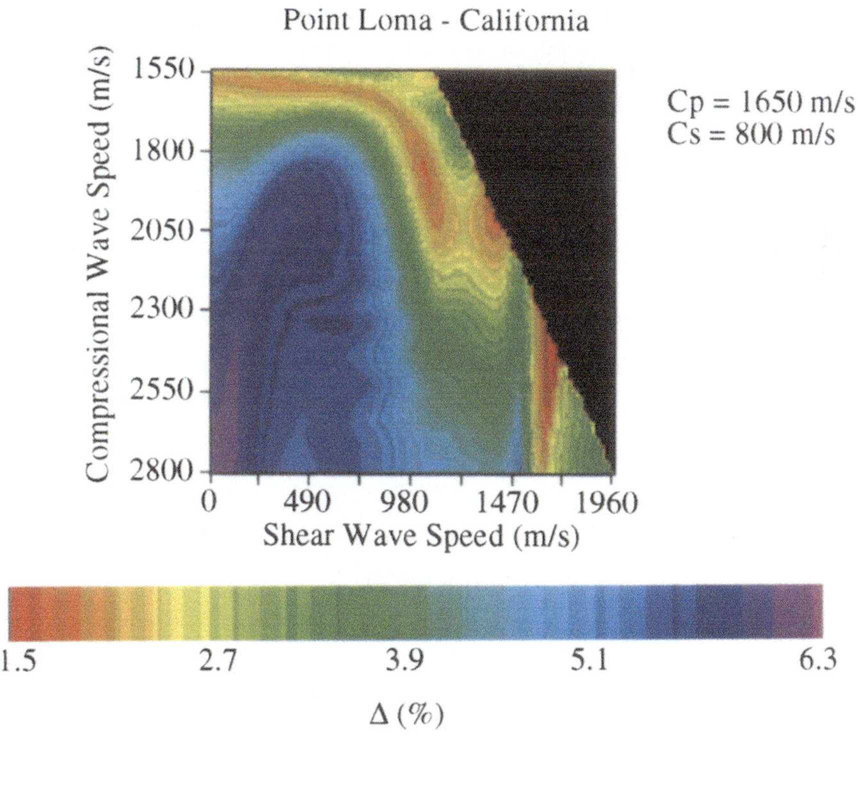

Figure 2. Ambiguity surface from noise data taken at Point Loma, California. The sea bed was modeled as a semi-infinite half-space.

3. Inversions for elastic geo-acoustic parameters

Recently, a new inversion procedure has been developed[2], also based on measurements of the spatial coherence of the noise between two vertically aligned hydrophones in the water column, but which includes the effects of shear in the bottom. The noise modelling procedure underpinning the inversion technique is based on a normal mode propagation code, combined with Carson's theorem to handle the stochastic nature of the sources, followed by an integration of the cross-spectral density from all noise sources on the surface of the ocean.

The inversion is not unlike matched field processing[3,4]. The model is used to generate the coherence function of the noise for a variety of bottom parameters and the source depth. These computed coherence functions are then compared with coherence data taken over the frequency band from 100 Hz to 2 kHz. A difference function is constructed and the set of parameter values corresponding to the minimum difference is taken as representative of the conditions in the channel and the underlying basement. The difference function is displayed as a contour map on an ambiguity surface.

When this inversion technique is applied to the data from the UK site with the chalk bottom, the shear and compressional speeds of the basement are estimated to be 908 and 2129 metres per second, respectively. These results are in good agreement with the values expected for the chalky bottom. For the other five UK sites, the new inversion technique yields essentially the same estimates for the compressional speed in the bottom as the older procedure. These estimated values are known to be representative of their respective basements.

Figure 2 shows and example of an ambiguity surface, constructed using ambient noise

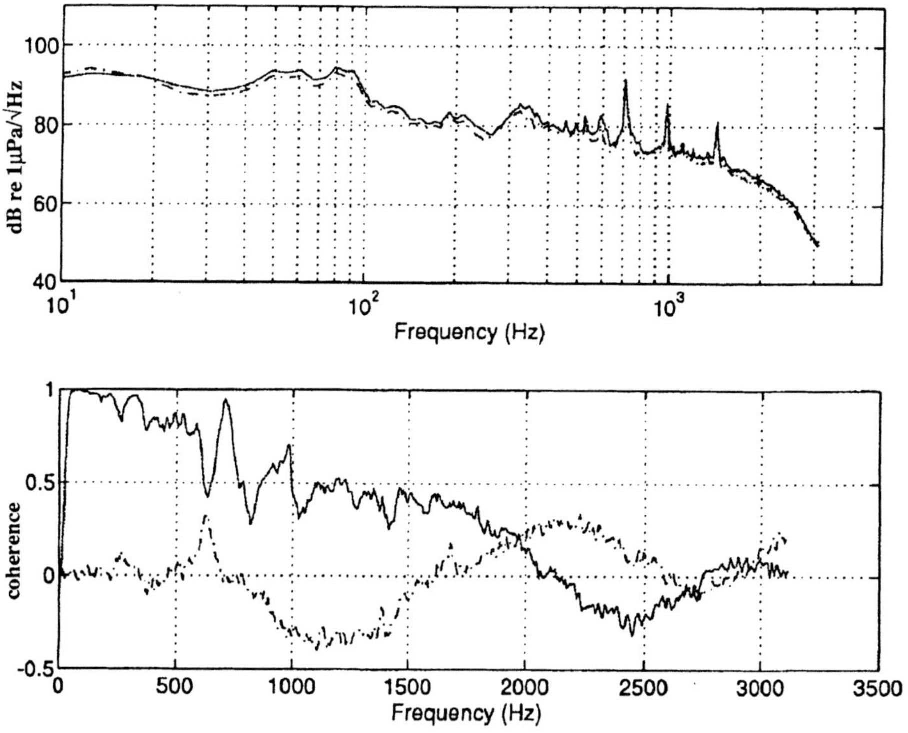

Figure 3. The power spectral density (top), and the real (solid line) and imaginary (broken line) coherence function (bottom), of noise measured off Point Loma, California.

data acquired in the Pacific Ocean, off San Diego, Southern California. The horizontal and vertical axis show the shear and compressional wave speeds, respectively. The minimum difference (red) corresponds to a compressional speed of 1650 m/s and shear speed of 800 m/s, values which are not unreasonable for the known bottom type at the measurement site. Nevertheless, for two reasons these numerical values should be treated with caution and the ambiguity surface in Figure 2 only regarded as an illustration of the output from the inversion: the data for Figure 2 were taken on the first trial deployment of a new vertical acoustic array; and surface shipping was abundant at the time of the experiment, heavily contaminating the noise field. Evidence of the shipping noise can be seen in Figure 3, showing the power and cross-spectral densities of the noise field, both of which exhibit well-defined shipping lines.

4. Extension to lower frequencies for penetration into the sub-bottom

At present the noise inversions yield estimates of the geo-acoustic parameters for the first few metres of the basement. To penetrate deeper into the bottom, allowing information on the sub-bottom stratification to be acquired, it is necessary to work at frequencies below 100 Hz, which is the lower limit of the frequency band currently employed. By shielding the sensors, to alleviate flow-noise problems[5], and taking other precautions, it should be possible to acquire uncorrupted noise data down to 5 Hz. Since noise in this frequency regime penetrates several hundred metres into the sediment, it should contain information on the sub-bottom structure down to a similar depth.

To extract the information requires a more subtle approach than that currently used. An iterative scheme seems feasible in which the first stage is an inversion using high frequency noise (> 100 Hz) to determine the bottom interface parameters, as at present. Somewhat lower frequencies would then be inverted to give the bottom structure deeper into the basement; and finally the lowest frequencies available would be used to yield information on the stratification as deep in the basement as possible. Since successive iterations in this procedure rely on previous estimates of the geo-acoustic parameters at shallower depths in the basement, the errors are likely to be cumulative; the deeper estimates would be less reliable than those of the interface values. Nevertheless, it should at least be possible to construct a qualitative picture of the bottom structure and perhaps after refinement of the technique to achieve improved depth resolution.

5. Concluding remarks

The noise inversion technique described in this paper provides an effective means of determining the geo-acoustic parameters of elastic basements in shallow water. Since the noise arrives at the sensors from all directions, and detection is over a radius of nominally 10 km, it is evident that the output of an inversion is some average of the bottom parameters taken over the footprint of the system. This suggests that an important application of the technique is in bottom surveying, that is, in identifying a bottom type as say sand rather than granite. Such a classification would be useful as input to ocean-acoustic propagation models; it also has application in connection with submarine cable laying and offshore construction. The rapid, large-scale, cost-effective coverage offered by the technique has appeal as a preliminary means of mapping the geology of vast areas of the ocean floor. Subsequent, higher resolution surveys could be conducted using more conventional methods.

References

1. M.J. Buckingham and S.A.S. Jones, "A new shallow-ocean technique for determining the critical angle of the seabed from the vertical directionality of the ambient noise in the water column", J. Acoust. Soc. Am. **81**, 938-946 (1987).

2. G.B. Deane, N.M. Carbone, and M.J. Buckingham, "Determination of the geo-acoustic parameters of an elastic seabed from the ambient noise field in the water column", J. Acoust. Soc. Am. **in preparation**, (1994).

3. H.P. Bucker, "Use of calculated sound fields and matched-field detection to locate sound sources in shallow water", J. Acoust. Soc. Am. **59**, 368-373 (1976).

4. A. Tolstoy, "Sensitivity of matched field processing to sound-field profile mismatch for vertical arrays in a deep water Pacific environment", J. Acoust. Soc. Am. **85**, 2394-2404 (1989).

5. M. Strasberg, "Nonacoustic noise interference in measurements of infrasonic ambient noise", J. Acoust. Soc. Am. **66**, 1487-1493 (1979).

INVERSION AND SANDY OCEAN SEDIMENTS

Nicholas P. Chotiros
Applied Research Laboratories (ARL:UT)
The University of Texas at Austin
Austin, Texas 78713-8029
U. S. A.

Inverting the acoustic wave speed of sandy sediments from its acoustic reflection coefficient, using viscoelastic wave theory, can lead to gross errors. The reflection coefficient of sandy sediments significantly deviates from viscoelastic wave theory predictions, with important ramifications for applications involving acoustic sediment classification, reflection and propagation loss predictions.

INTRODUCTION

In the viscoelastic theory of wave propagation, reflection coefficient is dependent on the ratio of acoustic impedances of the two media on either side of an interface. The impedance of each medium is determined by its density and sound speed. The impedance Z of two media, denoted by subscripts "w" and "s", are given in terms of the sound speed c and density ρ by,

$$Z_s = \rho_s c_s \quad \text{and} \quad Z_w = \rho_w c_w \tag{1}$$

At normal incidence, the reflection coefficient is given by,

$$R = (Z_s - Z_w)/(Z_s + Z_w) \tag{2}$$

The above equation has been used to compute the reflection coefficient of the interface between water and ocean sediments, from measured values of sound speed and density. It is found, however, that, for sandy sediments, the result is at odds with the directly measured value of reflection coefficient.

EXPERIMENTAL EVIDENCE

Experimental evidence can be found by comparing echo sounder measurements of reflection coefficient with the reflection coefficient computed from sediment sound speed and density. In spite of wide-spread use of the echo sounder, there are in fact very few published measurements of the reflection coefficient of sandy sediments. One reason is the difficulty of calibrating an echo sounder system for absolute measurements. Another is the large, random fluctuations in echo level from one ping to the next that is often observed. The calibration issue is a straightforward one. The fluctuation in echo level is due to scattering by the sediment surface roughness; it increases with carrier frequency and it is most noticeable above about 10 kHz. Nevertheless, with suitable incoherent averaging, as noted by Thorsos with regard to sea surface echo fluctuations[1], it is

O. Diaschok et al. (eds.), Full Field Inversion Methods in Ocean and Seismo-Acoustics, 353–358.

possible to obtain an accurate estimate of the reflection loss. Dodds[2] and Jones, Leslie and Barton[3] have published in-situ measurements of reflection loss of sandy ocean sediments using echo sounders. The laboratory measurements by Nolle, Hoyer, Mifsud, Runyan, and Ward[4] were made on cleaned and graded sand samples, and included both reflection measurements as well as sound speed and density.

In August 1993, in-situ reflection coefficient measurements were made and core samples collected and measured for sound speed and density on a sandy sediment off Panama City, Florida, jointly under the MCM Tactical Environmental Data System (MTEDS) Project and the Coastal Benthic Boundary Layer (CBBL) Special Research Project, under the support of the Office of Naval Research. This set of measurements is one of the most comprehensive, in which several acoustic measurements and core samples were taken within a uniform test area, and with sufficient sensitivity and accuracy to clearly show the discrepancy between the measured reflection loss and the value predicted by viscoelastic theory from sound speed and density. The maximum and minimum values of sediment sound speed and density were provided by Briggs and Richardson[5], based on measurements from numerous core samples; the sound speed was measured at 400 kHz using a pulse method. The predicted upper and lower bounds of reflection loss, using Eqs.(1) and (2) were computed. An echo sounder provided direct sediment reflection coefficient measurement in a continuous band of frequencies from 8 to 20 kHz in the same areas. The results of the echo sounder measurements are compared with the viscoelastic equation predictions in Fig. 1. The average of the measured reflection loss is in the region of 11 dB. The mid-point of the viscoelastic predictions is around 8 dB. In making this comparison, it is assumed that the sound speeds measured at 400 kHz and those in the 8 to 20 kHz band are not significantly different. This is not an unreasonable assumption. Hamilton[6,7] has measured sound speeds in sand cores at 200 kHz using a pulse method, at between 23 and 41 kHz using a resonance method in the laboratory, and in situ at 3.5, 7, 14 and 100 kHz using a pulse method, without reporting any frequency dependence. The data shown in Fig. 1 also shows no significant frequency dependence. The coherent ripples in the measured curves are due to random calibration errors and should be ignored. Although there is some overlap of the lower extreme of the viscoelastic predictions with the upper extreme of the experimental measurements, the average difference is significant. The mid-point of the two sets of values differ by approximately 3 dB. Therefore, one must conclude that there is an inconsistency between viscoelastic wave theory and the measured values of reflection loss, sound speed and density.

The range of measured reflection loss values and frequency band reported in previous publications are shown in Fig. 2. They span a broad band of frequencies from 100 Hz to 1 MHz. The viscoelastic wave theory prediction, based on typical sound speed and density values for water saturated sand, 1700 m/s and 2000 kg/m^3, respectively, is also shown for comparison. The August 1993 measurements from the Panama City site appear to be consistent with previous in situ measurements at lower frequencies and laboratory measurements at higher frequencies. The viscoelastic theory prediction is clearly inconsistent with the experimental data.

THEORETICAL MODEL

The correct theoretical model must be able to reconcile the reflection and sound speed measurements, and Biot's theory of acoustic propagation in porous solids is one that is

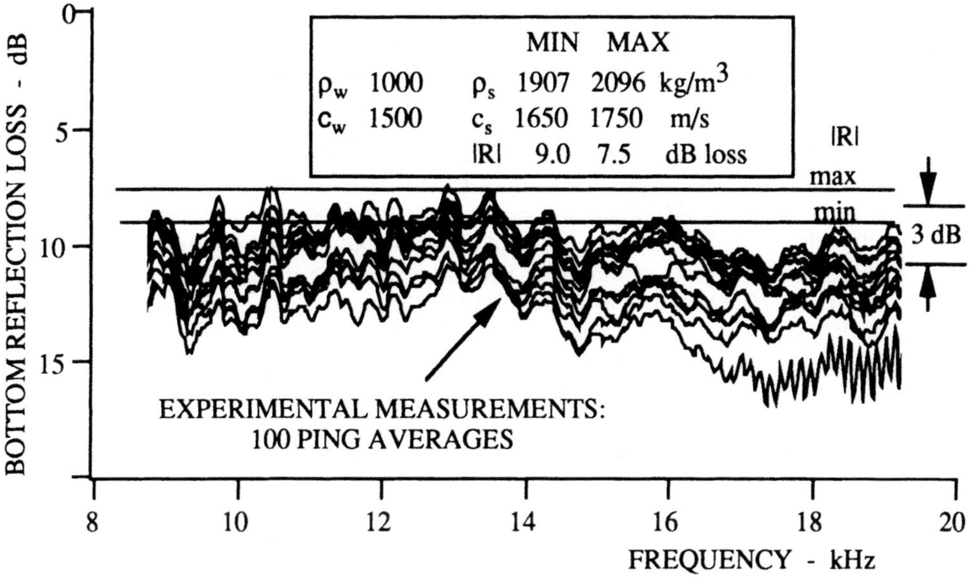

FIGURE 1. MEASUREMENT AND VISCOELASTIC MODEL PREDICTION OF REFLECTION LOSS OVER A SANDY SEDIMENT AT NORMAL INCIDENCE

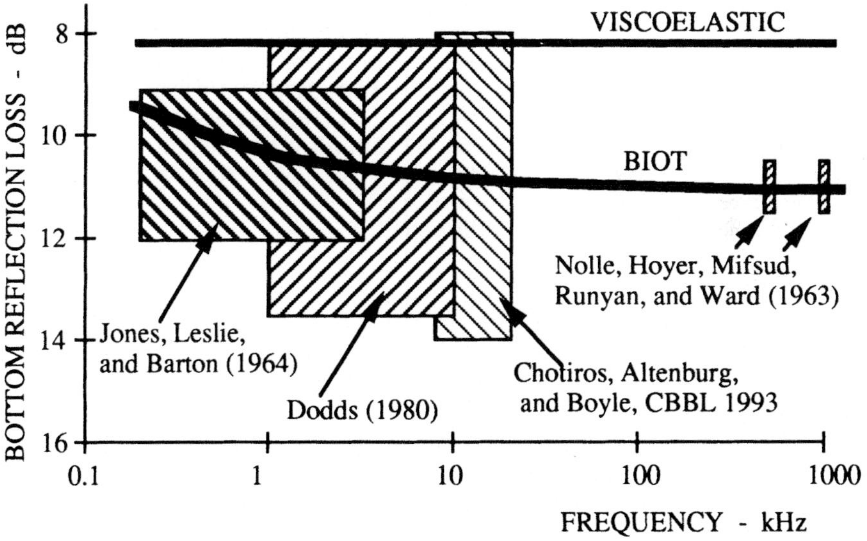

FIGURE 2. COMPARISON OF VISCOELASTIC AND BIOT MODELS OF REFLECTION LOSS WITH PUBLISHED EXPERIMENTAL DATA FROM WATER SATURATED SAND AT NORMAL INCIDENCE

able to do so. Estimates of the typical input parameters for a Biot model for sand are listed in Table 1. A justification of the values chosen is contained in a paper that has been submitted for publication[8]. A reflection loss curve computed using these values shows good agreement with the published data, as shown in Fig. 2. Also shown in Table 1 are two alternative sets of parameters for a viscoelastic model for the same sediment density. One set is based on the measured value of sediment sound speed, the other on that of the reflection coefficient at normal incidence. In the former, the sound speed is used to compute the reflection coefficient, and in the latter the process is reversed. The assumed shear speed and wave attenuations, in dB per wave length, are based on typical measured values for sandy sediments reported by Hamilton[9]. As an illustration, these values are used to compute reflection loss as a function of grazing angle at 3 kHz, shown in Fig. 3. The viscoelastic model computations were based on a plane wave reflection formulation from Brekhovskikh [10] which includes the effects of shear waves and attenuation. The Biot model computation was based on equations from Stern[11]. Clearly, there are considerable differences. The sound speed matched viscoelastic model shows lower reflection loss than the Biot model, while the reflection coefficient matched curve shows greater reflection loss with a sharp dip at a grazing angle of approximately 15°.

BIOT MODEL			VISCOELASTIC MODELS		
BULK PROPERTIES					
POROSITY	0.36	-	1. MATCHED WAVE SPEED		
GRAIN DENSITY	2650	kg/m^3	DENSITY	2056	kg/m^3
LIQUID DENSITY	1000	kg/m^3	ACOUSTIC WAVE SPEED	1700	m/s
GRAIN BULK MODULUS	7.00E+09	Pa	ATTENUATION	1.173	dB/λ
LIQUID BULK MODULUS	2.20E+09	Pa	SHEAR WAVE SPEED	214	m/s
LIQUID MOTION			ATTENUATION	2.06	dB/λ
VISCOSITY	1.00E-03	kg/m-s			
PERMEABILITY	2.90E-10	m^2	2. MATCHED REFLECTION LOSS		
PORE SIZE	1.27E-04	m	DENSITY	2056	kg/m^3
STRUCTURE CONSTANT	1.89	-	ACOUSTIC WAVE SPEED	1361	m/s
FRAME RESPONSE			ATTENUATION	0.752	dB/λ
SHEAR MODULUS	8.60E+07	Pa	SHEAR WAVE SPEED	214	m/s
LOG DECREMENT	3.47	dB/λ	ATTENUATION	2.06	dB/λ
BULK MODULUS	5.30E+09	Pa			
LOG DECREMENT	1.3	dB/λ			
RESULTING WAVE PROPERTIES					
FAST WAVE SPEED	1700	m/s			
ATTENUATION	1.173	dB/λ			
SLOW WAVE SPEED	1079	m/s			
ATTENUATION	4.36	dB/λ			
SHEAR WAVE SPEED	214	m/s			
ATTENUATION	2.06	dB/λ			

Table 1. Typical input parameter values for a Biot model of water saturated sand, and two corresponding viscoelastic models, one matching the acoustic wave speed and the other, the reflection coefficient.

SHALLOW WATER TRANSMISSION LOSS

These results will have an impact on shallow water transmission loss modeling over sandy sediments. Using the parameter values in Table 1, the transmission loss at 3 kHz of

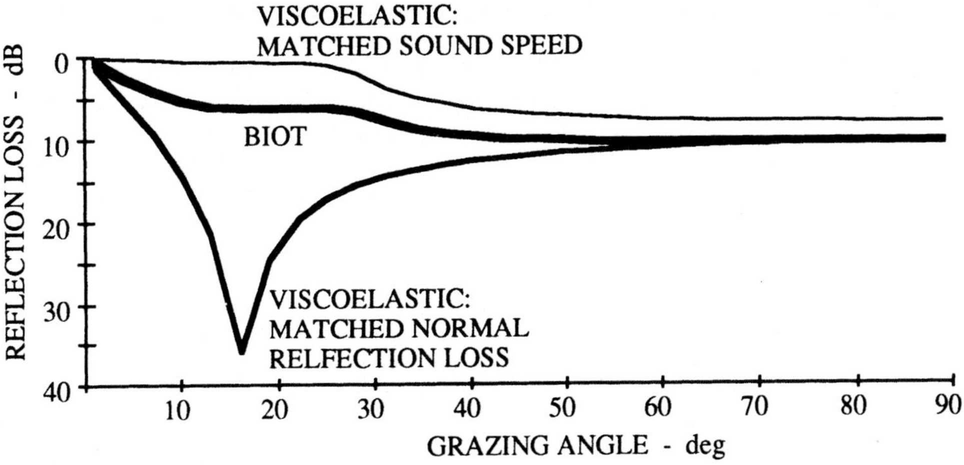

FIGURE 3. REFLECTION LOSS AS A FUNCTION OF GRAZING ANGLE PREDICTED BY BIOT AND VISCOELASTIC MODELS

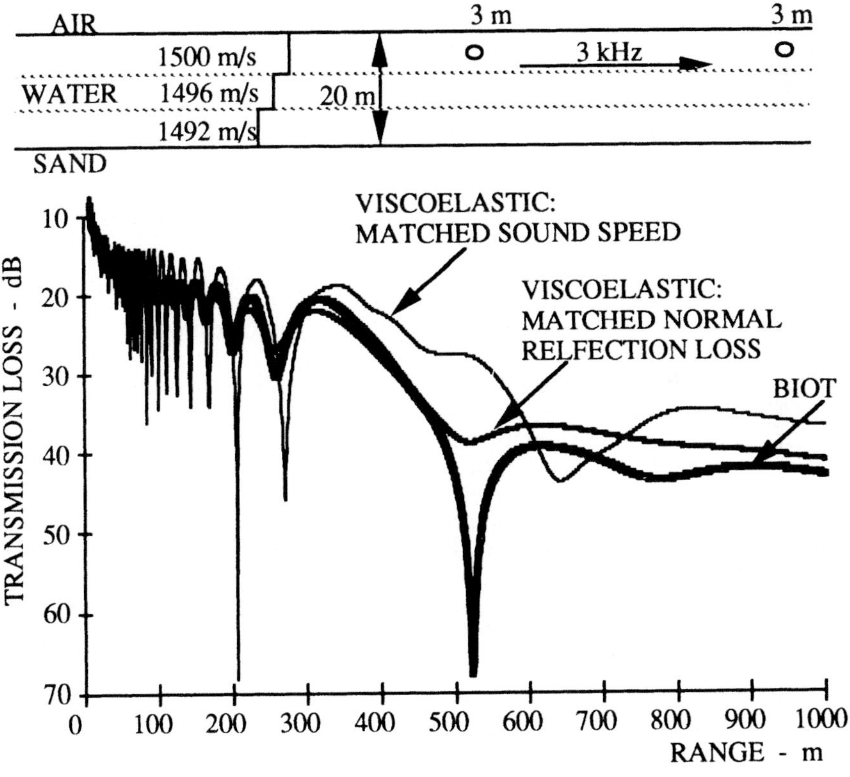

FIGURE 4. EXAMPLE OF SHALLOW WATER TRANSMISSION LOSS PREDICTED BY BIOT AND VISCOELASTIC MODELS

a downward refracting shallow sound channel, 20 meters deep, over a water saturated sand sediment, was modeled as a simple illustration; the source and receiver are at a depth of 3 m. A version of the fast field program OASES, which is a later version of SAFARI[12], was modified to solve the porous medium problem in a layered medium, using Biot's equations by Stern[11]. The results are shown in Fig. 4. There are significant differences between all three sets of predictions. The Biot model is the only one that is consistent with measurements of both reflection coefficient and sound speed of the sediment.

ACKNOWLEDGMENTS

The data collection and reflection coefficient analysis was supported by the MCM Tactical Environmental Data System Project, chief scientist, Sam Tooma, and the Coastal Benthic Boundary Layer Special Research Project, chief scientist, Mike Richardson, both at Naval Research Laboratories. Collection and processing of the echo sounder data was by Robert Altenburg and Frank Boyle, ARL:UT. The transmission loss modeling and the adaptation of the fast field program OASES for porous layered media by Prof. Morris Stern, was supported under a grant from the Office of Naval Research.

REFERENCES

1 Thorsos, Eric I. (1983) "Surface forward scattering and reflection", APL-UW 7-83, Applied Physics Laboratory, University of Washington, Seattle, WA.
2 Dodds, D. J. (1980) "Attenuation estimates from high resolution subbottom profiler echoes". Bottom-Interacting Ocean Acoustics, Kuperman, W. A. and Jensen, F. B. (ed.), 525-540, NATO Conference Series IV; Marine Sciences, Plenum Press, New York.
3 Jones, J. L., Leslie, C. B. and Barton, L. E., (1964) "Acoustic characteristics of underwater bottoms". J. Acoust. Soc. Am. 36(1), 154-7.
4 Nolle, A. W., Hoyer, W. A., Mifsud, J. F., Runyan, W. R. and Ward, M. B. (1963) "Acoustical properties of water- filled sands". J. Acoust. Soc. Am., 35(9), 1394-1408.
5 Briggs, Kevin and Richardson, Michael (1993) private communication
6 Hamilton, E. L. (1956) George Shumway, H. W. Menard, C. J. Shipek, Acoustic and Other Physical Properties of Shallow-Water Sediments off San Diego. Journal of the Acoustical Society of America, Vol. 28, No. 1, 1-15, January, 1956
7 Hamilton, E. L. (1970) Sound velocity and related properties of marine sediments, North Pacific. J. Geophys. Res. 75(23), 4423-4446.
8 Chotiros, Nicholas P. (1994) Biot model of sound propagation in water saturated sand, submitted to the Journal of the Acoustical Society of America
9 Hamilton, E. L. (1980) Geoacoustic Modeling of the Sea Floor. J. Acoust. Soc. Am. 68, 1313-1340,.
10 Brekhovskikh, L. M. (1980) Waves in Layered Media, pp 44, Academic Press
11 Stern, M., Bedford, A. and Millwater, H. R. (1985) Wave reflection from a sediment layer with depth dependent properties. J. Acoust. Soc. Am. 77, 1781-1788.
12 Schmidt, H. and Jensen, F. B. (1985) A full wave solution for propagation in multilayered viscoelastic media with application to Gaussian beam reflection at fluid-solid interfaces. J. Acoust. Soc. Am. 77, 813-825.

BOTTOM PARAMETER ESTIMATIONS IN SHALLOW WATER BY MEANS OF FIELD INVERSION METHODS

V. G. PETNIKOV
Inst. of General Phys., Russian Acad. of Sci.
Ocean Acoustic & Information Ltd.,
38 Vavilov St., Moscow, 117942, Russia

and

B. G. KATSNEL'SON
Voronezh State University,
Universitetskaya Sq.1, Voronezh, 394693, Russia

ABSTRACT Two methods are proposed for the determination of the sound absorption coefficient in the bottom. The methods are based on the best fit of theoretical predictions to measurements of 1) the loss of signal level with range, and 2) vertical interference structure. Experimental results obtained with these methods are comparable.

1. Introduction

In shallow water the sea bottom often consists of an extremely inhomogeneous medium with variable parameters (density, sound velocity, sound absorption coefficient). For calculation of long range sound propagation in the water column it is possible to apply the model of a fluid bottom with average compressional speed and absorption coefficient as a first approximation. The latter is defined as the imaginary part of the square of the refractive index in the bottom $n^2 = n_0^2(1 + i\alpha)$. We know [1] that α can vary by one to two orders of magnitude, depending on the nature of the bottom. In this work, two acoustic field inversion methods are proposed for determination of α.

2. Determination of absorption coefficient from horizontal interference structure

The first method is based on the measurement of the attenuation of sound in a shallow ocean waveguide as a function of range. The measured sound intensity is compared with the predictions given by the theory for several values of α. A quantity k is defined in such a way that, when k is maximized, the norm of the difference between measured and theoretically computed sound intensity is minimized. Let $\delta_e(r, z_1, z_0)$ be the experimental dependence of the relative average sound intensity on the distance r between the sound source and receiver:

$$\delta_e(r, z, z_0) = 10 \lg \left[\frac{\overline{I^e}(r, z, z_0)}{\overline{I^e}(r_0, z, z_0)} \right]. \tag{1}$$

Here $\overline{I^e}$ is the sound intensity averaged over a certain spatial interval Δr in the directions of the sound propagation; z_0 and z_1 are the depth of the source and receiver, respectively,

O. Diaschok et al. (eds.), Full Field Inversion Methods in Ocean and Seismo-Acoustics, 359–364.

and r_0 is a reference distance corresponding to 0 db ($r_0 \gg H$), where H is the depth of the waveguide. Let $\delta_t(r, z_2, z_0)$ be the theoretically predicted dependence for a fixed α. These functions are oscillating functions, since averaging smooths only the part of the spatial beats with period smaller than the averaging scale. If the averaging is carried out over an interval greater than the largest ray cycle in the waveguide, a smoothed curve representing the average decay law would be obtained. Accordingly we decompose the function $\delta(r)$ into two terms, one corresponding to the average decay law, the other describing the interference structure of the field. In this light, the coefficient k also consists of two parts, which independently describe the degree of closeness of the smooth and oscillating components of the field:

$$k = \overline{k} + \widetilde{k}, \qquad \delta = \overline{\delta} + \widetilde{\delta},$$
$$\overline{k} = \frac{\int_{r_0}^{r} \overline{\delta}_e \overline{\delta}_t dr}{\int_{r_0}^{r} \overline{\delta}_e^2 dr + \int_{r_0}^{r} \overline{\delta}_t^2 dr}, \qquad \widetilde{k} = \frac{\int_{r_0}^{r} \widetilde{\delta}_e \widetilde{\delta}_t dr}{\int_{r_0}^{r} \widetilde{\delta}_e^2 dr + \int_{r_0}^{r} \widetilde{\delta}_t^2 dr}. \tag{2}$$

Our problem is to select a coefficient α in such a way to maximize k.

This approach has been experimentally tested in one region of the Barents Sea. The relative sound intensity has been measured from two ships. The first ship towed two sound sources at the depth of 35–40 m. Pure-tone sound signals were radiated at two fixed frequencies of 100 and 230 Hz. As receiver, a vertical chain of several hydrophones and depth sensors was lowered from the second ship. Bottom relief along the path of the radiator-towing ship is shown in Fig. 1. Vertical profiles of sound speed at the beginning (1) and at the end (2) of the path are plotted in Fig. 2.

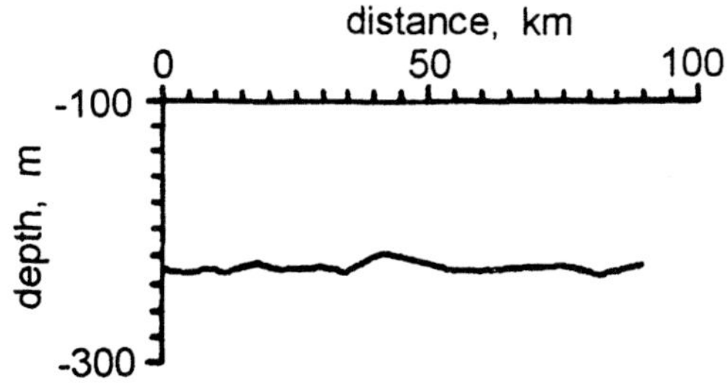

Fig. 1. Depth vs distance. (measured)

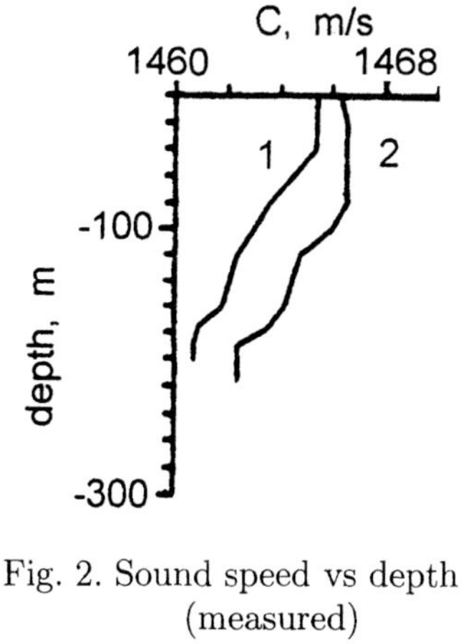

Fig. 2. Sound speed vs depth. (measured)

The relative intensity of the sound field was measured for two radiation frequencies and four reception depths. The computer signal-processing procedure corresponded to optimal separation of the useful signal in 300 s time interval and thus ensured a signal-to-noise ratio of at least 6 dB. This time interval corresponded to averaging of the intensity over a range of 750 m.

Adiabatic normal modes were used to calculate the sound field of the source. The sound pressure in the channel is sought in the form

$$\Psi(r,z) = \sum_n \left[\frac{\Psi_n(r,z)\Psi_n(r_0,z_0)}{\sqrt{\xi_n r}} e^{i\int_{r_0}^{r} \xi_n(r')dr'} \right], \tag{3}$$

where ξ_n and Ψ_n are the eigenvalues and eigenfunctions of the Sturm-Liouvile problem

$$\frac{d^2\Psi_n}{dz^2} + [\frac{\omega^2}{c^2(r,z)} - \xi_n^2]\Psi_n = 0, \tag{4}$$

$$\Psi_n(r,0) = 0, \qquad \frac{1}{\rho\Psi_n}\frac{d\Psi_n}{dz}|_{z=H} = 0. \tag{5}$$

Here $c(r,z)$ is the sound velocity, ρ is the density. The functions $\Psi_n$depend on r as on a parameter. The Cauchy initial-value method was used to solve the problem (4)–(5) numerically. This method has the advantage of being suitable for nonanalytical experimental profiles $c(z)$. It is convenient to treat the quantity $\Phi(\xi) = \frac{\Psi'}{\rho\Psi}|_{z=H}$ as a residual in the calculations. For an homogeneous fluid bottom,

$$\Phi(\xi_n) = \frac{1}{\rho_0\Psi_n}\frac{d\Psi_n}{dz}|_{z=H-0} + \frac{1}{\rho_1}\sqrt{[\xi_n^2 - \frac{\omega^2}{c_1^2}(1+i\alpha)]}, \tag{6}$$

where c_1 is the sound velocity in the bottom, $\Psi_n(z)$ is the numerical solution of Eq. (4) on $[0-H]$, ρ_0 and ρ_1 are the densities of the water and the bottom, respectively.

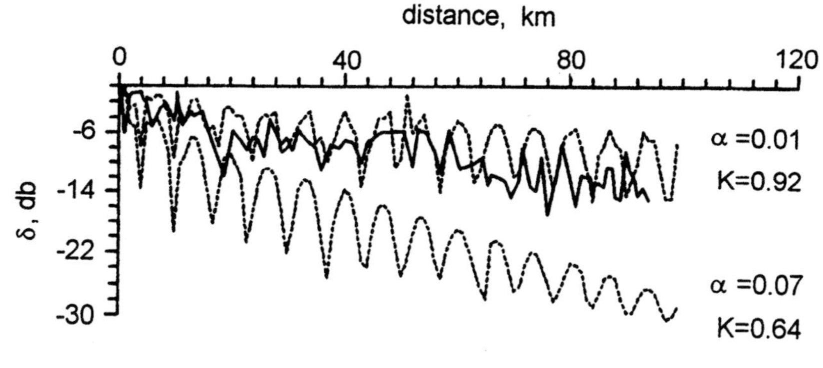

Fig. 3. Measured (solid line) and predicted relative signal levels vs range.

The comparison between the experiment and the theoretical calculations for several values of the coefficient α are shown in Fig. 3. Also shown are the values of the figure-of-merit. The solid curve in the figure represents the experimental results, and the dashed curves correspond to the numerical calculations. Two curves are shown for the numerical calculations: one corresponds to the optimum coefficient α(when k is maximum), and the other to a different value of α. It is clearly visible that the experimental and numerical curves diverge widely for nonoptimal values of α. The functions $k(\alpha)$ for various receiver depths are shown in Figs. 4 and 5 for frequencies of 100 and 230 Hz respectively. We see that many of the $k(\alpha)$ curves have a fairly distinct maximum, which is located in the same interval of α for different depths and even at different frequencies. This interval is $\alpha \approx 0.015 - 0.03$.The agreement of the results for different frequencies suggests that in the frequency interval considered, there is a linear frequency dependence of the absorption.

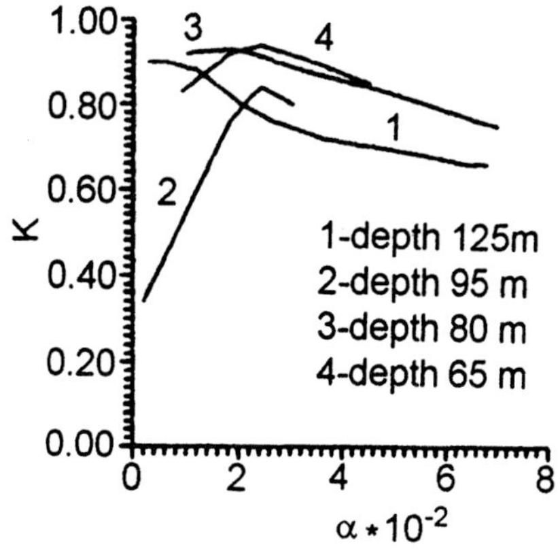

Fig. 4. K vs α at 100 Hz.

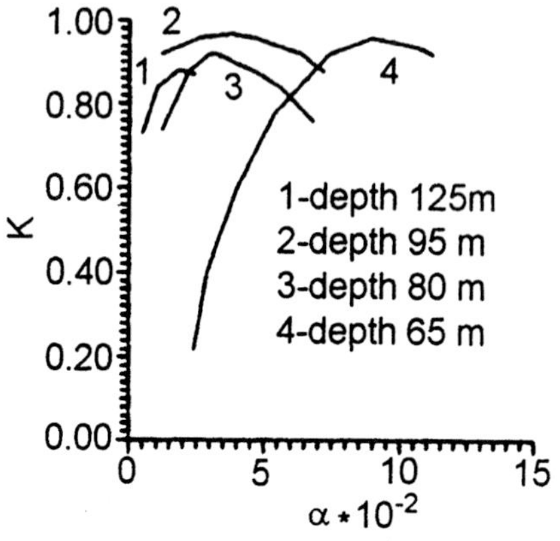

Fig. 5. K vs α at 230 Hz.

3. Determination of absorption coefficient from vertical interference structure

The second method for the determination of α is based on the vertical interference structure of the sound field in the water column. For measuring the vertical interference structure it is necessary to use an acoustic source and a vertical chain of receivers, that span the main part of the waveguide. It is important to note that in this procedure the positions of source and receivers have to be fixed.

This second method was used for the estimation of α on a stationary acoustic track in the Barents Sea [2]. The bottom profile along the track and the the sound speed profile are shown in Figs. 6 and 7, respectively.

The acoustic source and a vertical chain of ten equidistant hydrophones were lowered to the bottom and tied to two anchored mooring buoys at the beginning and end of the track. A pure-tone sound signal was radiated at a fixed frequency of 100 Hz.

The same (4)–(5) adiabatic mode theory was used for the calculation of the vertical interference structure in an inhomogeneous near-bottom sound channel with fluid absorbing bottom. However, to compare the calculated and experimental vertical interference structure we used the simpler criterion

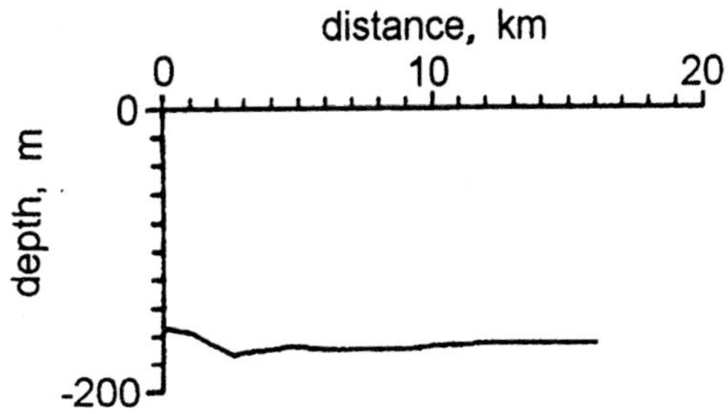

Fig. 6. Depth vs range.
(measured)

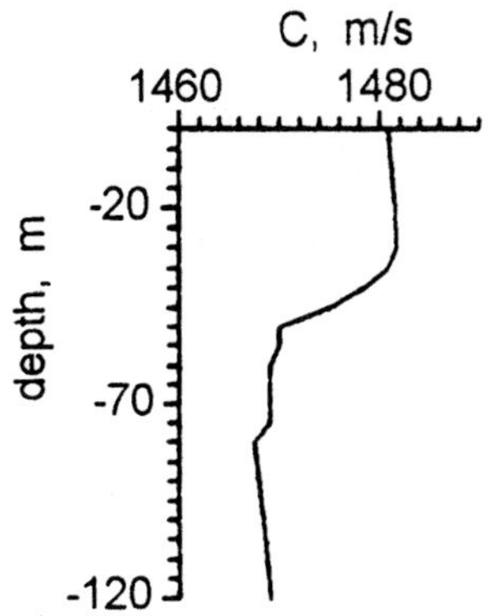

Fig. 7. Sound speed vs depth.
(measured)

$$\sigma = \frac{1}{L}\sqrt{\sum_k (I_k^e - I_k^t)^2}, \tag{7}$$

where L is the number of hydrophones, I_k^e and I_k^t are the experimental and theoretical intensities of the field at each (k-th) hydrophone.

The intensity is normalized in the form: $maxI_k^e = maxI_k^t = 1$. The minimum value of σ (ideally zero) ensures the best agreement between the calculations and the experiment. This agreement was found at $\alpha = 0.01 (\sigma \approx 0.01)$. Vertical interference structure for this value of σ is shown in Fig. 8. One can see reasonably good agreement between the calculated and experimental curves for an α which spans the same range of values [0.015–0.03] as for the previous experiment.

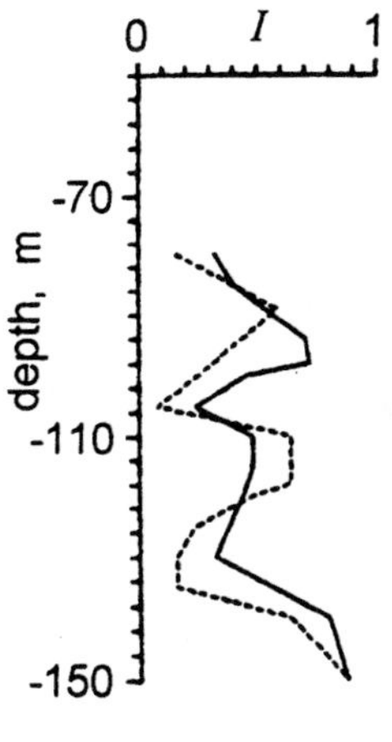

Fig. 8. Measured (solid line) and computed vertical interference patterns.

It is necessary to note that there are other waveguide parameters, that affect the vertical interference structure (for example waveguide depth) which were not known with reasonable certainty in our experiment. The effective values of these parameters were determined the same way as described above.

4. Conclusion

We conclude that the two methods for determination of the absorption coefficient described in this paper yield comparable results. The sensitivity of the method based on loss of signal level with range appears to be significantly greater than the sensitivity of the vertical interference structure method for the particular geometries and procedures of these two experiments. The resulting estimates of $\alpha = 0.015$–0.03 are considered reasonable for sediments consisting of sand and gravel.

5. Acknowledgement

We would like to thank L.G. Kulapin for his assistance with sound field calculations.

References

1. Kuperman, W.A. and Jensen, F.A.,(1981). *Eds.*, Bottom-Interacting Ocean Acoustics. New York, Plenum Press.
2. Katsnel'son, B.G., Kulapin, L.G., Migulin, A.A. and Petnikov, V.G, (1992). Influence of hydrodynamic fluctuations on the vertical interference structure of a sound field in a waveguide, *Sov. Phys.-Acoust.*, **38**(2), 164–168.

INFLUENCES OF SEDIMENT VARIABILITY ON BROADBAND ACOUSTIC WAVE PROPAGATION IN SHALLOW WATER

M. BADIEY
Office of Naval Research
Ocean Acoustics Program, ONR 321OA
Arlington, VA 22217–5660

ABSTRACT

Shallow water sediments often have highly variable range and depth dependent physical properties which can substantially affect the acoustic wave propagation. To assess the effects of the seafloor on propagation of broadband signals, acoustic experiments were conducted in a shallow water region with pre-existing sediment core data. Geoacoustic properties were inferred from the core data and transformed to compressional and shear wave speed, attenuation, and density profiles. The acoustic data show a strong variability of sound intensity with range and azimuth. This unique 3D geoacoustic data set can be used to model the range and azimuthal dependent effects of the geological features on the acoustic wave propagation.

1. Introduction

When broadband acoustic signals propagate in shallow water waveguides the result is a superposition of water-borne and bottom interacting signals. The propagation is geometrically and intrinsically dispersive and a complex function of range. The angular distribution, vertical and horizontal coherence properties, and higher order statistical properties of the acoustic field in shallow water environments are poorly understood [1].

In this paper we present results of recently conducted experiments aimed at better understanding the shallow water forward problem. The experiments were conducted in a location with pre-existing seafloor geological data [1]. The site, known as the Atlantic Generating Station (AGS), is located outside of Little Egg inlet, south of Atlantic City, New Jersey. This site was surveyed in the 1970's, to provide information for the foundation of a proposed nuclear power plant. The survey resulted in a detailed data set reporting the geological and engineering properties of the seafloor structure [1, 2]. Figure (1a) shows a map of the AGS drilling site with core locations [1, 3]. During the 1980's, this region was a site for several geoacoustic experiments aimed at remote measurements of the physical properties of the sediment layers pertinent to the acoustic and seismic wave interaction with the seabed [3, 4, 5, 6]. The work reported by Ewing et al. [3] was a comprehensive effort in obtaining the seafloor shear properties in this region. In addition, a more recent geological survey was conducted in 1992 to study the surficial sediments [7].

Calibrated acoustic experiments were conducted using a 20 in^3 airgun in a variety of source and receiver configurations [1]. The results of these tests along with modeling are used to assess the sensitivity of the acoustic field to the seafloor [8, 9].

O. Diaschok et al. (eds.), Full Field Inversion Methods in Ocean and Seismo-Acoustics, 365–370.

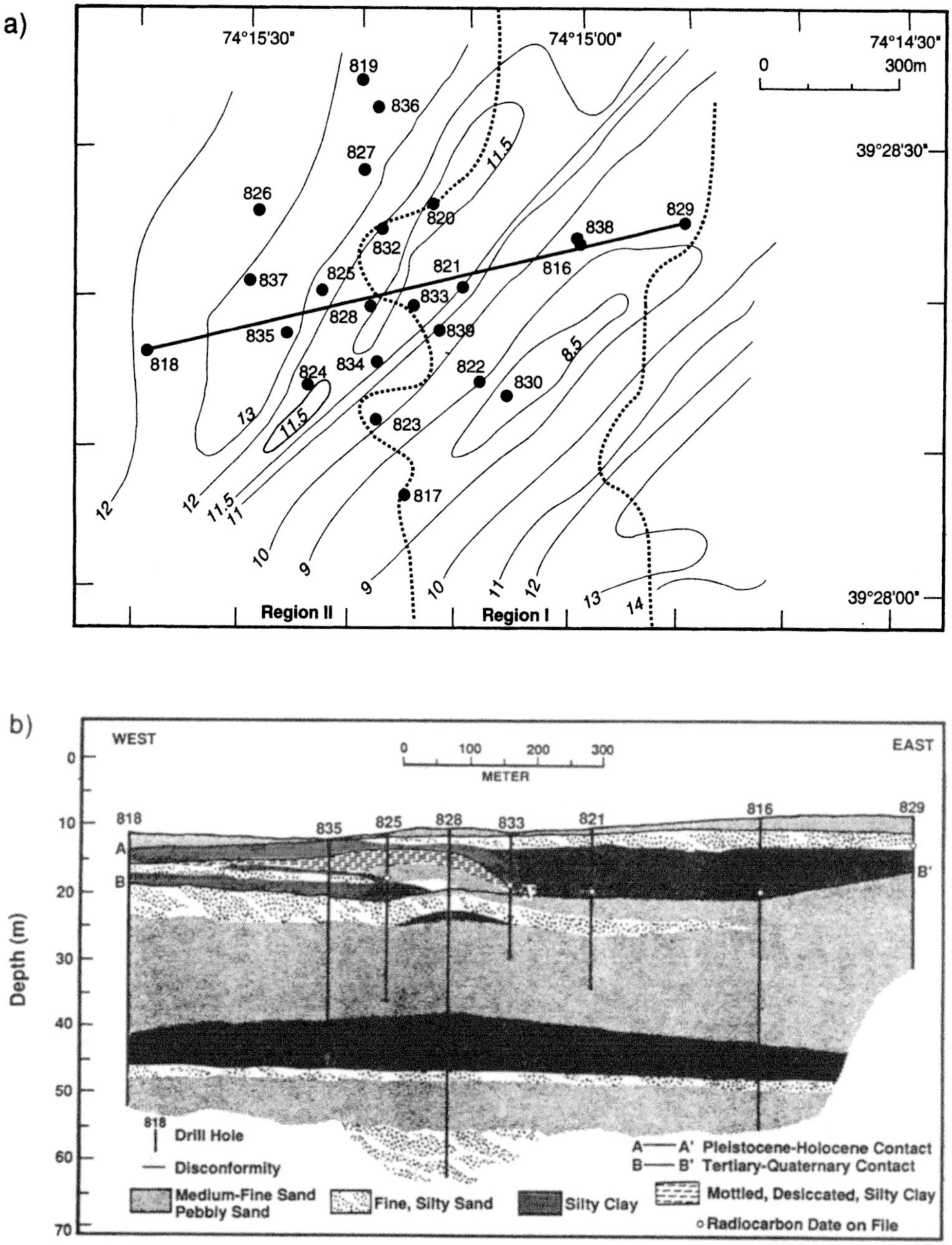

FIGURE 1. (a) Map of the Atlantic Generating Station (AGS) drilling site. The available core data, numbered 816 through 839, are shown with black circles. (b) A slected cross section of the AGS site representing sediment variations inside (Region I) and outside (Region II) of the buried channel.

2. Geology/geoacoustic data

The AGS site is located in water depths of 8.5 to 14 meters (see Fig.1a). As indicated in recent reports [3, 1], there exists a large buried channel underlying this area whose approximate boundary is depicted by dashed sinuous lines in Fig.1a. In the geological and seismic studies it was concluded that the sediments in the buried channel (which we call Region I) are different from the outside (which we call Region II) due to erosional transgression [1, 2, 3, 7]. Hence, the geology and geoacoustic properties of the inside and outside of the channel are different.

There are geological data for 23 cores numbered 816-839 shown by black circles in Fig. 1a. The solid line connecting boreholes 818 to 829 in this figure is a geological section presented by Stahl et al. [2] in which three major stratigraphic units were identified (see Fig. 1b). These layers can affect the bottom interacting acoustic field substantially. More detailed information about the geoacoustics properties of this region is given in [1].

3. Acoustic experiments

The acoustic experiments were conducted from June 8-12, 1992 [1]. The *R/V Cape Henlopen* was used as the receiving ship stationed at a fixed point with a vertical array of hydrophones, while the *R/V Skimmer*, equipped with an airgun source, moved to different locations. The hydrophones were spaced uniformly at one meter intervals and held in a vertical position. The source system consisted of a 20 in^3 airgun loaded with highly compressed air (2,000 psi) activated from the receiver ship using radio telemetry. The airgun was calibrated prior to the experiments and found to be quite repeatable. The distance between the source and the receiver was measured using a microwave ranging system with 1 meter accuracy. During the experiments, CTD records were obtained from both the source and receiver ships simultaneously to detect the spatial and temporal variabilities in the sound speed profile. The data indicated no significant variability in sound speed profiles during the experiment (ap proximately 2 hours) [1]. The sound speed profiles in the water were downward refracting.

Figure 2 shows an example of the experimental results. The hydrophone array, centered near core 833, received the airgun shots which were fired sequentially at the positions numbered (1) through (8) around a circle with radius of about 0.5 km. The received signals were processed using the wavelet transform and depicted for shots S-47 through S-76 in Fig. 2. In this figure, a strong azimuthal dependency of broadband signal is observed. This variablity of the shot signatures can be attributed to the underlying seafloor geoacoustic properties. For example, shots S-47,S-52 and S-56 which show similar time-frequency characteristics were fired in Region II and shots S-65, S-69 and S-74 were originated in Region I (see Figs. 1a and 1b). The shots with less similar features (such as S-60 and S-76) were either fired near the boundary between the two regions, or in a region with more complicated bottom structure (e.g. S-65). A more detailed explanation of this topic including acoustic and geoacoustic data is given in reference [1].

4. Effect of seabed variability on acoustic signature

As an example of the effect of bottom properties on the broadband siganls, two representative shots are considered as shown in Fig. 3. The corresponding geoacoustic profiles in the vicinity of these shots, i.e. the compressional C_p and shear wave speed C_s profiles for core

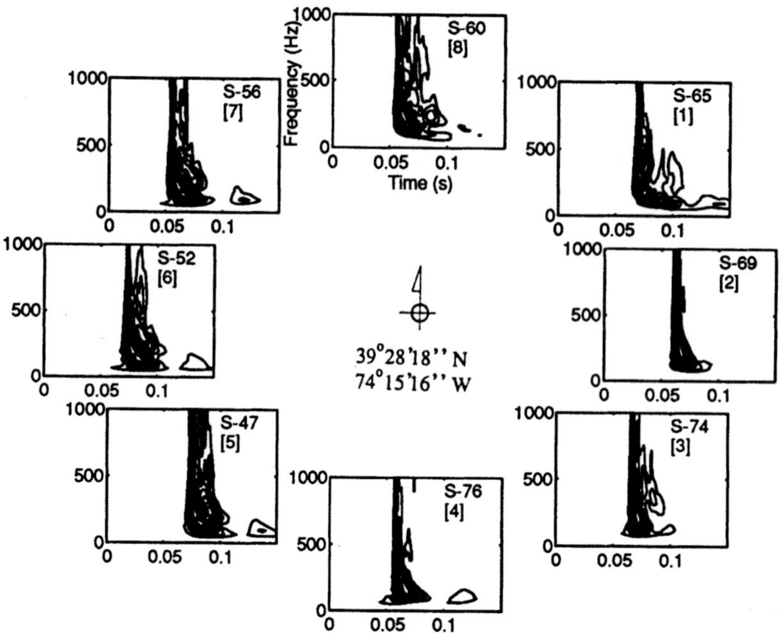

FIGURE 2. Time-Frequency transform of the received broadband acoustic signal for hydrophone number 5, located mid-water column, for different source positions. The coordinates of the receiver array, shown in the center, are near core 833, which is on the boundary between regions I and II (see Fig. 1a). The source has moved from postion [1] to postion [8] in a clockwise direction. The numbers S-47 through S-76 indicate shots.

829 (Region I) and core 818 (Region II) are shown in Figs. 4(a) and 4(b) respectively. The shots were chosen to show the difference between the two sides of the buried channel (see Figs. 1a, 1b, and 2). The time-frequency transform of shot S-65 and S-47 and their corresponding group velocity of mode one are shown in Fig. 3. The difference in the structure of the dispersion curves and the group velocities is an indication that the bottom interaction can cause the signature variation. It is noted that the sound speed profiles in the water column did not show variabilities during the experiment, and that the experiment was conducted during calm seas.

The general variability of the parameters in the region can be depicted by a composite diagram of all the available cores. For example, the compressional wave speed C_p versus depth in Fig. 4(c) shows that in the first 20 meters of sediments in this region, the compressional sound speed can vary from 1600 to more than 2200 m/s for different cores. This variation can profoundly affect the signal dispersion.

5. Conclusions

Results of recent acoustic measurements at the AGS site have shown considerable azimuthal variation of a broadband signal dispersion. The broadband acoustic signals in a layered environment for frequencies of up to 1 k Hz are dispersed, the degree of dispersion is believed to depend on the layer structure, the depth gradient of the sound speed profile, the bottom topography and subbottom inhomogeneties. The details of the dispersion characteristics related to the bottom and subbottom variabilities need to be understood prior to performing inversion in shallow water regions.

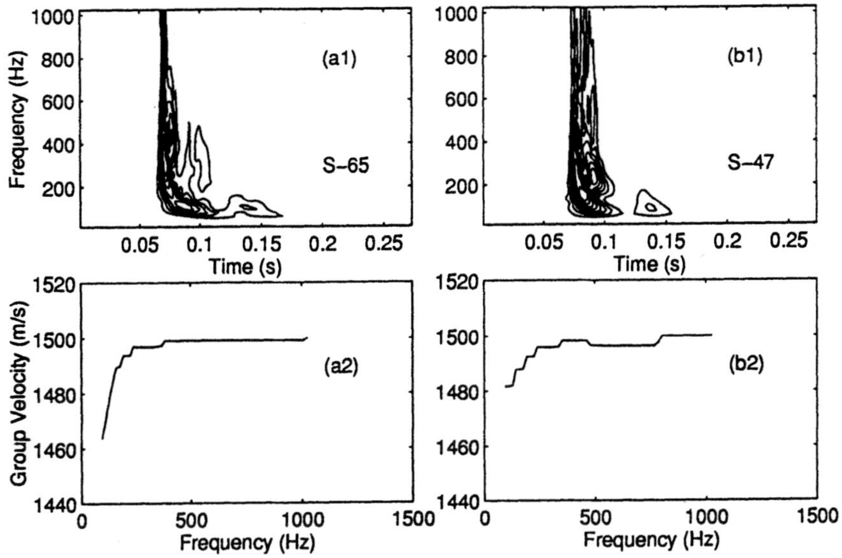

FIGURE 3. Time-Frequency transform (a1 and b1), and the corresponding mode 1 group velocities (a2 and b2) for shots S-65 and S-47 received by a hydrophone at the water depth 5 meters.

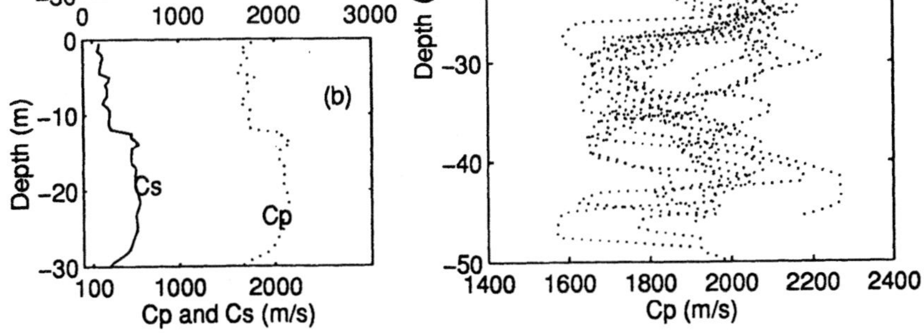

FIGURE 4. (a) Shear (C_s) and compressional (C_p) speed profiles of core 829 (region I, inside the channel), (b) core 818 (region II, outside), and (c) Variability of compressional sound speed versus depth for 23 cores.

6. Acknowledgements

The author would like to thank I. Jaya and M. Pulaski for their help in experiments and data processesing. This work was performed under SORP Program at ONR.

References

1. Badiey, M., Jaya, I., and Cheng, A. H-D. (1994). "Shallow water acoustic/geoacoustic experiments at the New Jersey Atlantic Generating Station Site," J. Acoust. Soc. Am. **96**(6), 3593–3604.
2. Stahl, L., Koczan, J. and Swift, D. (1974). "Anatomy of a shoreface-connected sand ridge on the New Jersey shelf: implications for the genesis of the shelf surficial sand sheet," Geology **2**, 117–120.
3. Ewing, J., Carter, J. A, Sutton, G. H. and Barstow, N. (1992). "Shallow water sediment properties derived from high frequency shear and interface waves," J. Geophys. Res., **92**, 4739–4762.
4. Stoll, R.D., Bryan, G.M., Flood, R., Chayes, D., and Manley, P. (1988). "Shallow seismic experiments using shear waves," J. Acoust. Soc. Am., **83**, 93–102.
5. Badiey, M., Yamamoto, T., and Turgut, A. (1987). "Laboratory measurements of geo-acoustic profiles of the New Jersey shelf sediments," J. Acoust. Soc. Am., Suppl. 1(81), 49.
6. Trevorrow, M.V., Yamamoto, T., Badiey, M., Turgut, A., and Conner, C. (1988). "Experimental verification of seabed shear modulus profile inversions using surface gravity (water) wave-induced seabed motion, " Geophysics. J. Int., **93**, 419–436.
7. Twichell, D. C. and Able, K. W., (1993). "Bathymetry, sidescan sonar image, and surficial geological interpretation of the inner shelf of Little Egg inlet, New Jersey," US Geological Survey Miscellaneous Field Studies Map MF–2221, 2 sheets, scale 1:15,000.
8. Jaya, I., Badiey, M., Cheng, A. H-D., and Pulaski, M. (1994). "Time-frequency energy distribution analysis of a shallow water acoustics experiment at the AGS Site," J. Acoust. Soc. Am. **95**, No. 5, Pt. 2, 2928.
9. Bongiovanni, K. P., Badiey, M., and Siegmann, W. L. (1994). "Shallow water sediment layer structure and composition effects on range-dependent transmission loss at the AGS site," (submitted to J. Acoust. Soc. Am.).

SHEAR WAVE PROPERTIES FROM INVERSION OF SCHOLTE WAVE DATA

HASSAN B. ALI and MICHAEL K. BROADHEAD
Naval Research Laboratory
Stennis Space Center, MS 39529-5004
USA

ABSTRACT. Scholte interface wave data from a deep-water site are inverted to estimate the ocean bottom sediment shear wave parameters. Techniques used include dispersion analysis, the spectral ratio method, and elastic modelling to produce shear velocity and attenuation estimates for the first tens of meters of a sediment bottom comprised largely of hemipelagic clays. The methodologies presented help provide insight into the first order shear properties of the sediments, and can be valuable in constructing initial guess models and bounds for automated inversion schemes.

1. Introduction

The propagation of sound in marine environments is frequently accompanied by interaction with the ocean bottom, particularly at low frequencies. In such cases, it is well known that the bottom compressional and shear wave properties significantly affect the propagation. We will limit our attention in this paper to the problem of estimating shear wave velocity and attenuation parameters, direct measurements of which are problematic in deep-sea areas. An alternative to *in situ* measurements uses interface (Scholte) waves to remotely sense the desired parameters; the feasibility of this stems from the strong dependence of these waves on the bottom sediment shear parameters.

To estimate shear velocities, we can exploit the dispersive nature of the Scholte waves, which is a manifestation of the frequency-dependent penetration of the energy into the variable velocity layered sediments of the sea bottom. For estimating shear attenuation, we can exploit spectral information obtained from the measured time series. In addition, elastic wave forward modelling can be used to constrain and refine these results. We present examples demonstrating these methodologies for a deep-water marine data set.

2. Dispersion Results From Recent Experiments

2.1 TEST SITE CHARACTERIZATION

The results to be discussed derive from Scholte wave measurements acquired in the marginal basin off the coast of California, using ocean bottom seismometers (OBS) and bottom explosive shots [1–3]. The test site is located in 3800 m water depth, where the seafloor sediments are characterized by low bathymetric relief (<10 m). Nearby Deep Sea Drilling Project Site 469

O. Diaschok et al. (eds.), Full Field Inversion Methods in Ocean and Seismo-Acoustics, 371–376.

reveals that the upper sediments are composed of hemipelagic clays, the first 42 m of which are characterized by a compressional speed of approximately 1510 m/s. Basement is located at a depth of 390 m below the seafloor, beyond the penetration depth of the interface waves used in this study.

2.2 DISPERSION ANALYSIS OF EXPERIMENT RESULTS

A typical OBS vertical component time series (seismogram) is shown in Fig. 1 for an 800 m range shot, showing the direct waterborne arrival, water column multiples, and Scholte wave arrivals. A dispersion analysis of the Scholte waves was performed by the well known multiple filter analysis (MFA) method [4]. Figure 2 shows the result of an MFA (Gabor matrix) obtained from a 1 km range time series; the response is dominated by the fundamental mode, with group velocities ranging from approximately 30–75 m/s. This is in general agreement with measurements by other researchers at a nearby test site [5]. Strong normal dispersion is evident in the MFA, which suggests a fairly steep positive shear wave velocity gradient. A weakly dispersed higher order mode, centered about a group velocity of approximately 55 m/s, is also evident.

2.3 MODELLING OF DISPERSION RESULTS

The solid curves superimposed on the MFA in Fig. 2 are the predicted modal dispersion curves, computed using an elastic normal mode program [6]. The shear velocity profile, Fig. 3, used to compute the dispersion curves was determined by iteration of various input models until an adequate match was achieved between the predicted dispersion curves and the peak amplitudes of the MFA. The associated predicted time series (synthetic seismogram), however, did not agree well with the measured time series, as shown for the 1190 m range case in Fig. 4. This mismatch has various contributing factors, but we made the reasonable presumption that it was largely due to incorrect input shear attenuations. This, in fact, gave us a strategy for estimating the attenuation profiles.

3. Estimates of Shear Attenuation

3.1 SPECTRAL RATIO METHOD

In order to have reasonable initial estimates and bounds on the attenuation profiles, we employed the spectral ratio method of attenuation analysis [3, 5], which makes use of the log spectral ratio of two time series, separated

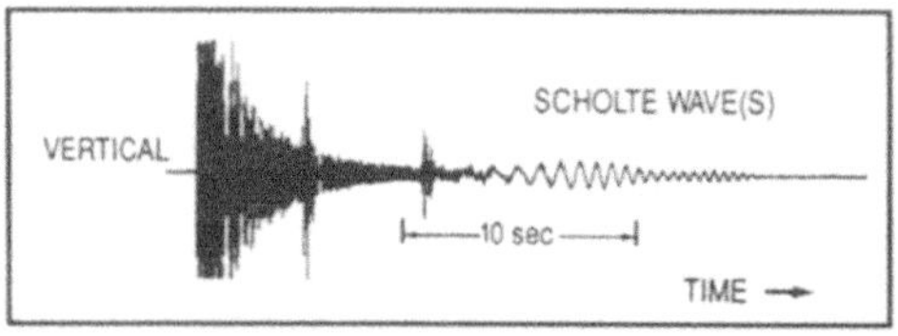

Figure 1. Vertical component OBS response to 800 m range shot.

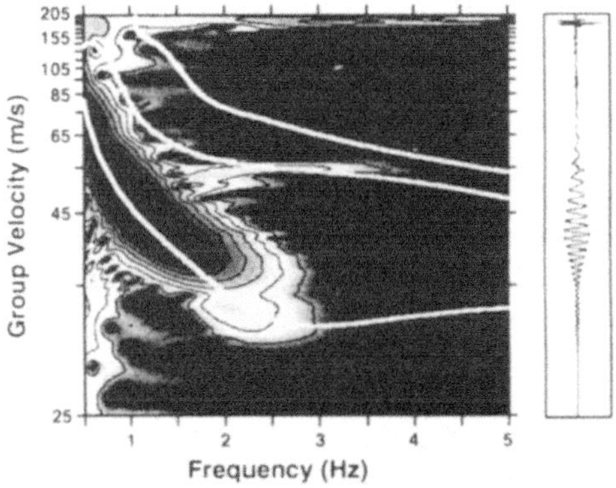

Figure 2. Dispersion analysis (Gabor matrix) of 1 km range time series (shown adjacent). Superimposed are computed dispersion curves from shear velocity model in Fig. 3 and shear Q model 1 (Fig. 6).

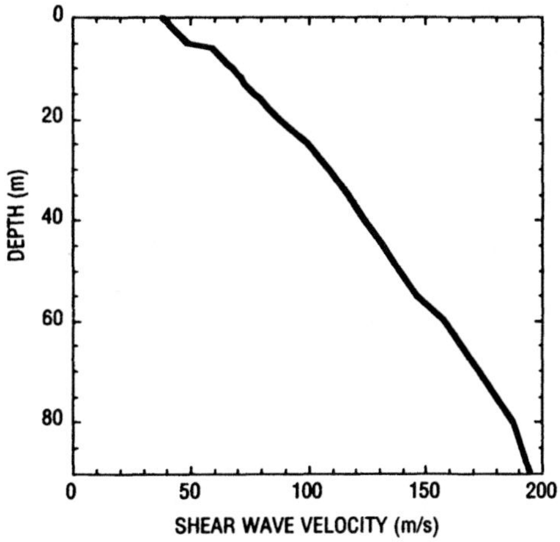

Figure 3. Shear velocity model.

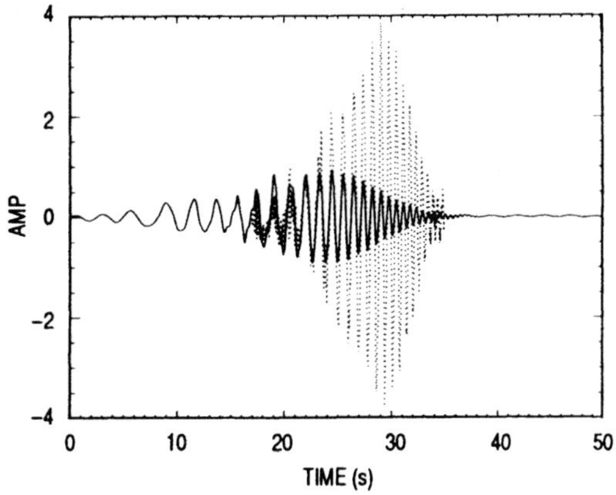

Figure 4. Simulated time series from model 1 (solid) and measured (dots) at 1190 m.

by an adequate distance, to provide estimates of the quality factor Q. Inversely proportional to attenuation, Q is defined here by the well known temporal form of the modal spectral attenuation term

$$G(\omega) = e^{-\omega t/2Q_t}. \tag{1}$$

The quality factor can alternatively be expressed in a spatial form Q_s, which is related to Q_t by the ratio of the phase velocity to the group velocity. For non-dispersive propagation, $Q_t = Q_s$. The spectrally derived Q's represent a weighted depth average of the layer (or material) Q's, the weighting being determined by the modal structure of the interface waves and the layer Q's in the medium. This type of analysis can provide gross information on the attenuation properties of the medium.

The spectral ratio of two range separated Scholte wave packets i and j can be written [5]

$$P_i(\omega)/P_j(\omega) = (R_i/R_j)^2\, e^{\omega\Delta t/Q_t} \tag{2}$$

where $\Delta t = t_j - t_i$ and R is a geometrical spreading loss factor. Here, we have made the assumption of range independent earth properties, which is reasonable at our test site due to the nature of the depositional processes. We also assume that the spectra from different shots are sufficiently similar. The above procedure assumes isolation of a single mode; in fact, our data is largely comprised of the fundamental mode.

For the cases analyzed, the log spectral ratio was approximately linear over the principal frequency range, and the slope was determined from a least squares fit. Note, Q is a function only of the slope and Δt, an estimate of which was obtained from differencing the arrival times of the envelope peaks of the Scholte wave packets [3]. Figure 5 shows the log spectral ratio and a linear fit for a range separation of 1390 m, giving an estimate of Q_t as 31±5. Estimates at two other range separations were also made, yielding: $Q_t = 29\pm7$ ($\Delta r = 820$ m) and $Q_t = 39\pm10$ ($\Delta r = 570$ m). These estimates are generally consistent with other published results, one of which was for a nearby test site [5, 7].

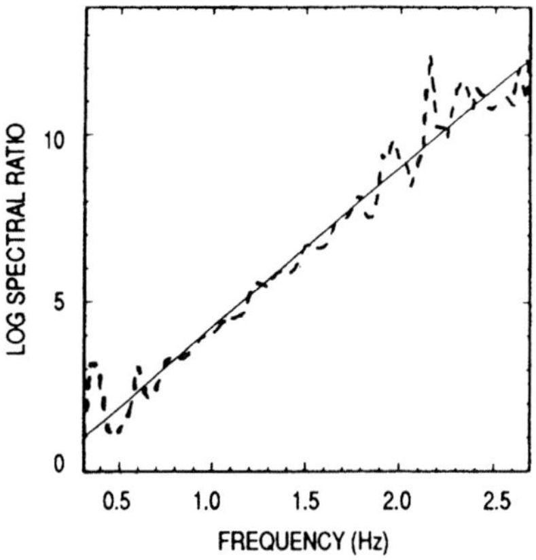

Figure 5. Measured log spectral ratio and linear fit.

3.2 INVERSION APPROACH FOR SHEAR ATTENUATION

Iterative inversion of the attenuation profile was approached by using the shear velocities shown in Fig. 3 and a series of layer shear Q profiles (Fig. 6). For each model a synthetic time series was computed with the elastic normal mode method and passed through a filter that simulated the effects of the recording instruments. In addition, log spectral ratios were determined from the simulations to compare with the data (Fig. 7). Model perturbations in Fig. 6 were based on physical intuition derived from comparing the measured and computed seismograms. For example, an excessive coda (tail) for the synthetic time series suggested that the surficial sediment model shear Q's were too high. A geoacoustic model was considered good if it produced acceptable matches to spectral ratios and measured seismograms and at more than one range. This minimizes the possibility that differences between the simulated and actual sound sources (delta function vs. explosion) lead to erroneous conclusions on the layer Q's. This trial and error approach gives insight into the effects of the geoacoustic parameters as a function of depth.

4. Selected Attenuation Results

Figure 7 compares the log spectral ratios resulting from three geoacoustic models of Fig. 6 with the data. Model 1, the initial Q profile used in Figs. 2 and 4, produces an overestimate of the spectral ratio, compared with the estimate from measured data (Fig. 5); also, the corresponding simulated time series (Fig. 4) does not match the data well. Several models (e.g. 8, 9, 12, 13,

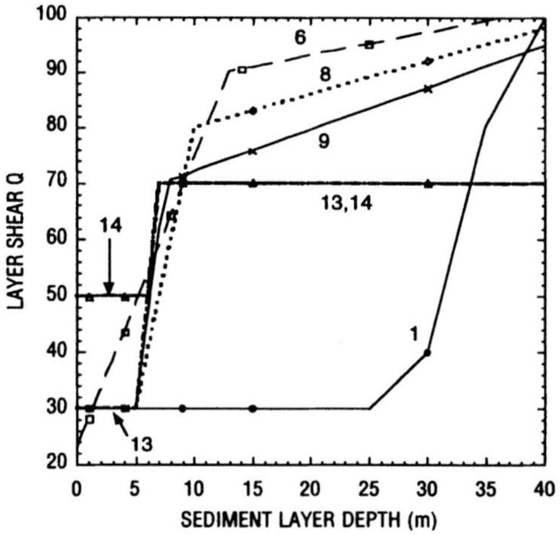

Figure 6. Selected sediment shear Q models.

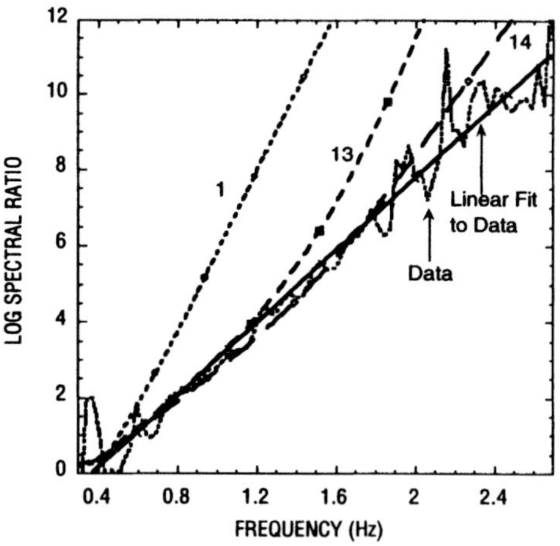

Figure 7. Computed spectral ratios corresponding to selected Fig. 6 models.

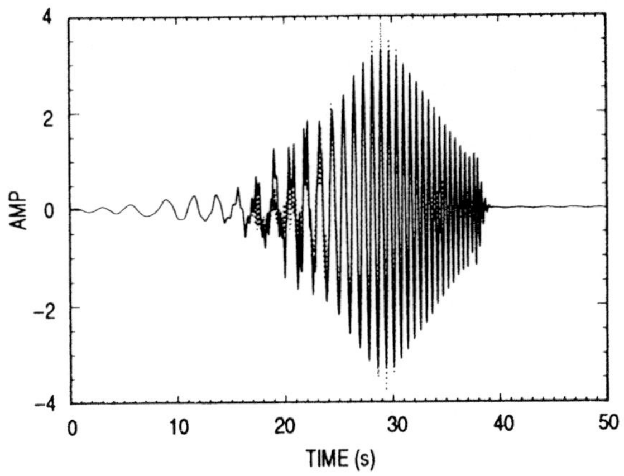

Figure 8. Simulated time series for model 14 (solid) and measured (dots) at 1190 m.

and 14) produced very similar spectral ratios and time series at one range, although the layer Q's below ~8m show significant departures. Model 14 produces a reasonable slope match between simulated and measured spectral ratios. The disagreement at the higher frequencies indicates a need for further refinement of model 14. Perhaps the surficial layer Q's in model 14 are still somewhat too low, but layer thickness and Q gradient may also need adjustment. Figure 8 shows a comparison of simulated and measured time series for model 14. This figure, along with Fig. 7, shows that the first order features of the data seem to be explained adequately by model 14.

The above results suggest that the deeper layers are not playing as significant a role as the surficial layers, at least in explaining the first order properties of the time series. The synthetic time series included the fundamental mode and the first two higher modes; however, comparison of synthetic seismograms using only the fundamental mode with those using the first three modes confirms our presumption that the higher order modes produce only a small contribution.

5. Discussion and Conclusions

The significant correlation between sediment stratigraphy and Scholte wave dispersion allows one to produce reasonable estimates of the bottom shear speeds. Iterative inversion can further refine these results and produce good fits to a dispersion analysis of the measured data. It is not surprising, however, that matching the measured seismograms requires additional constraints on the environmental parameters, particularly the shear Q profiles. Conclusions for our data set regarding attenuation include the following:

- The surficial sediments (< 10 m) play a large role in the first order propagation behavior; the low-speed interface waves are largely insensitive to the deeper layers.
- Significantly different Q models can produce very similar time series and spectral ratios, suggesting a degree of non-uniqueness.
- Matching the measured time series at one range without matching the spectral ratio will not produce a valid geoacoustic model; the spectral ratio provides an important constraint.
- A two-layer model seems to be adequate to explain the first order features of the attenuation properties of the data (cf model 14). In particular, both the thickness of the top layer, and the abrupt rapid Q increase at the termination of the layer were critical. The details of the profile thereafter seemed less important for the frequency range considered.

- Our results are consistent with the conventional wisdom [5, 7] that these sediment types possess the property of Q generally increasing with depth (i.e., attenuation decreases with depth).

The above remarks suggest that many basic features of the propagation can be explained by the type of analyses presented in this paper, but that inversion of Scholte wave data to obtain the geoacoustic parameters of bottom sediments probably requires information in addition to dispersion curves and spectral ratios. Full waveform inversion offers promise in this regard. At the least, such results as we have presented should be useful in providing initial guess models, and perhaps bounds, required by iterative type automated inversion schemes. Also, the role of higher order modes may become important in more uniquely specifying the model, especially as they tend to sample the deeper layers better.

6. Acknowledgments

This work was supported by the ONR Ocean Acoustics Program (Mohsen Badiey, Project Manager, Program Element No. 0601153N). We wish to thank Dr. L. Dale Bibee for making available an elastic normal mode code, and for valuable discussions on the subject of this paper. NRL contribution number NRL/PP/7173--93-0059.

References

1. Ali, H. B. and Bibee, L. D. (1993) The Influence of Sediment Layering and Geoacoustics on the Propagation of Scholte Interface Waves, *Proc. Oceans '93*, IEEE, **1**, 105–112.
2. Ali, H. B. and Bibee, L. D. (1993) The Influence of a Thin Sediment Layer on the Dispersion of Scholte Interface Waves, *Acoustics Letters*, **17**, 13–18.
3. Broadhead, M. K., Ali, H. B., and Bibee, L. D. (1994) Shear Q Estimates from Interface Waves in Marine Sediments, *J. Acoust. Soc. Am.*, **95**, 1650–1653.
4. Landisman, M., Dziewonski, A., and Sato, Y. (1969) Recent Improvements in the Analysis of Surface Wave Observations, *Geophys. J. Royal Astr. Soc.*, **17**, pp. 369–403.
5. Sauter, A. W., Dorman, L. M., and Schreiner, A. E. (1986) A Study of Sea Floor Structure Using Ocean Bottom Shots and Receivers, in *Ocean Seismo-Acoustics*, T. Akal and J. M. Berkson (eds.), (Plenum Press, New York), 673–681.
6. Gomberg, J. S. and Masters, T. G. (1988) Wave Form Modelling using Locked-Mode Synthetics and Differential Seismograms: Applications to Determination of the Structure of Mexico, *Geophysics. J.*, **94**, 193–218.
7. Jensen, F. B. and Schmidt, H. (1986) Shear Properties of Ocean Sediments Determined from Numerical Modelling of Scholte Wave Data," in *Ocean Seismo-Acoustics*, T. Akal and J. M. Berkson (eds.), (Plenum Press, New York), 683–692.

FULL WAVEFORM INVERSION OF SEISMIC INTERFACE WAVE DATA

L. DALE BIBEE
Naval Research Laboratory
Stennis Space Center, MS 39529-5004
USA

LEROY M. DORMAN
Scripps Institution of Oceanography
La Jolla, CA 92093
USA

ABSTRACT. A technique for inverting marine seismic interface wave data for shear wave attenuation parameters of ocean sediments is presented. Because of the difficulty in estimating spectral ratios of seismograms in the case when multiple modes are present, a linear perturbation technique that uses the recorded time series seismogram or its envelope as input data is proposed. The technique is stable for tests using synthetic data. Application of the technique to data collected in the Gulf of Mexico yields a model producing a good fit to the data. However, a simple source model is insufficient to correctly predict modal excitations, and the source must be parameterized as part of the inversion process.

1. Introduction

The shear wave velocity and shear wave attenuation of marine sediments remain the most poorly known of their geoacoustic properties. One remote sensing technique that has been used to estimate these parameters is the measurement and analysis of seismoacoustic interface waves. These are guided waves constrained near the sediment-water interface and are of several types. Scholte waves are easily generated by an explosive point source on the seafloor. Thus they are the easiest to produce during an experiment and are the focus of the analysis described in this paper.

To predict the sediment shear wave parameters from Scholte wave measurements requires a data inversion process. Since the problem is non-linear, a straightforward solution is not possible, but a number of techniques are applicable. These techniques may be interactive, where different sets of parameters are chosen intuitively and the corresponding field calculated and compared to data, or they may be automated, where data are input to a computer code which searches the parameter space for a "best fit" to the data, usually in a least squares sense. In this paper we explore automated techniques for performing the inversion, and then apply these techniques to a recently acquired data set.

Only a few investigators have reported application of automated inversion techniques to Scholte wave data. Caiti *et al.* [1] use an automated technique to generate shear velocity, but an interactive technique to find the shear attenuation profile. In addition to automated velocity inversion, Stoll *et al.* [2] report the inversion of spectral amplitude versus range data for shear attenuation. Fortunately, the analogous problem of Rayleigh wave inversion for land seismology has been extensively studied and reported (e.g. [3]). The standard approach has been to apply linear perturbation theory to the problem [4]. Measured data can be ordered into a data vector $\mathbf{d_{obs}}$. These data are the result of a set of model parameters that can be ordered into a model vector $\mathbf{m}$. The relation between $\mathbf{m}$ and $\mathbf{d_{obs}}$ is non-linear, but we assume that for small perturbation of the model parameters the resulting small perturbations in the data vector can be represented by a linear relationship. We then pick a starting model vector $\mathbf{m_{est}}$ and create a predicted data vector $\mathbf{d_{pred}}$. The linearization then reduces to:

O. Diaschok et al. (eds.), Full Field Inversion Methods in Ocean and Seismo-Acoustics, 377–382.

$$\mathbf{G}\,(\mathbf{m}\text{-}\mathbf{m}_{est}) = (\mathbf{d}_{obs}\text{-}\mathbf{d}_{pred}) \quad \text{or} \quad \mathbf{G}\ \delta\mathbf{m} = \delta\mathbf{d}$$

where **G** is a matrix of coefficients that represent the sensitivity of each data value to variation in each model parameter. The technique consists of solving the linear equation for $\delta\mathbf{m}$, calculating a new model estimate, and then iterating this process until $\delta\mathbf{d}$ stops getting smaller. In practice, the linear equation is under constrained because of the limited bandwidth and errors in the observed data. To stabilize the solution we use the singular value decomposition method , and eliminate small singular values. A potential problem with the technique is that one may find a set of parameters that locally minimizes the size of $\delta\mathbf{d}$, but other parameter combinations far from this model vector generate smaller $\delta\mathbf{d}$. To mitigate this problem, we use a variety of starting models and look for the minimum error solution.

While the focus of this paper is inversion for shear wave attenation parameters, a good estimate of the shear wave velocity profile is required before we can proceed. The linearized inversion of dispersion curves for shear wave velocity is a relatively well known technique having been done decades ago for Rayleigh waves (e.g. [4]) and more recently for Scholte waves [1, 2]. We have implemented these techniques to generate shear velocity-depth models from group velocity dispersion curves. To generate the dispersion curves, we calculate a Gabor matrix using a sliding time window technique [5].

A standard technique for determining attenuation parameters is through the use of ratios of signal spectra from seismograms at various ranges [2]. For a given mode, these ratios are a measure of the frequency dependent mode Q which is linearly related to the shear wave attenuation of the individual sediment layers. However, for data containing more than one mode, isolation of the part of the spectrum due to each mode is difficult. Therefore we present an alternative approach that inverts the waveform data rather than its spectra.

2. Data Set

The data we use consist of seismograms recorded on seafloor seismometers and generated by seafloor impulsive sources. The surface wave (Scholte) modes in the frequency range between 1 and 4 Hz are the significant signals for the inversion. Several experimental configurations have been used recently to generate such data [2, 6, 7, 8]. The data set that we will consider in section 5 in the paper was generated in intermediate water depths (250 m) using an airgun as a source and independent ocean bottom seismometers (OBS) as receivers. A 4.9 liter airgun was mounted on a steel plate and lowered to the seafloor with a deep-sea winch. An electrical firing cable and an air supply hose were payed out simultaneously. As the ship moved slowly through the field of OBS's with the airgun suspended just above the seafloor, the plate was periodically lowered to the bottom and the airgun was fired. Although the bubble oscillation frequency is very high at these depths (~80 Hz), source levels at frequencies of 1-4 Hz are sufficient to propagate Scholte waves over a distance of several hundred meters.

The difficulty in calculating spectral ratios for multi-modal data provides incentive for generating an inversion scheme that uses the actual time series directly. However, because of the very short wavelengths associated with the Scholte modes (generally a few meters), the phase of the signal is very sensitive to source-receiver range. Even under the best scenarios, errors of a few meters in source-receiver range are not unusual. For estimation of the attenuation, the phase is not as important as the shape of the signal envelope, which varies more slowly with range. Therefore we consider a data transformation that is the envelope of the time series (Fig. 1) as calculated by using the Hilbert transform [9]. Because the envelope function is in the time domain, it is easy to remove

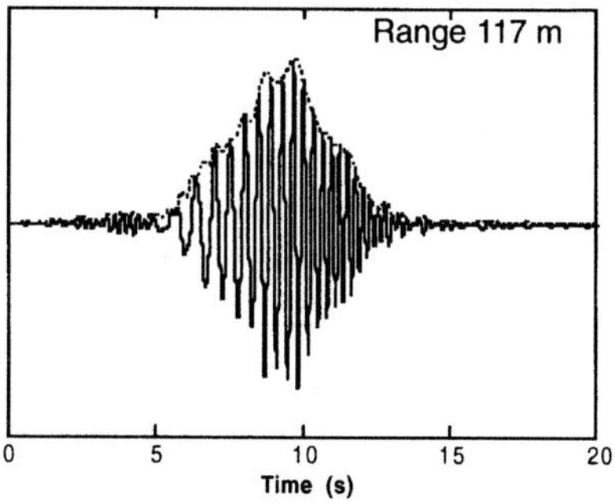

Figure 1. A Scholte wave time series and its envelope (dotted line) from a seafloor airgun shot recorded on a vertical seismometer.

interfering signals (i.e. noise transients or body wave phases) by ignoring the data at those times.

3. Forward Model

The inversion schemes that we consider require the ability to compute the data from a known geoacoustic model. Fortunately, for the 1-D inversions considered here, the elastic wave propagation problem has been well studied. However, few of the standard propagation codes in use in underwater acoustics produce ground motion rather than pressure outputs and handle the elastic problem correctly. SAFARI [10] can solve the problem, but is not efficient in that it computes the response over a whole frequency-wavenumber matrix, while only the elements of that matrix near the dispersion curves for the modes have significant contributions.

For the inversions addressed here, we have used a code originally designed for Rayleigh wave propagation to land seismometers [11] and modifed to add a water layer to compute Scholte wave modes [7]. This code requires an earth model consisting of a stack of homogeneous elastic layers, and it implements the efficient matrix methods of Woodhouse [12] to compute a modal summation. We have validated the model by computing time series from an identical geoacoustic model using SAFARI and this mode code.

4. Full Waveform Inversion

Linearized full waveform inversions of Rayleigh waves from earthquake sources have been reported for shear velocity structure [e.g. 11]. Here, we address the somewhat easier problem of estimating the shear wave attenuation profile with depth, assuming that we know the velocity profile with sufficient accuracy from a linearized inversion of the dispersion curves.

We begin by producing a data vector which consists of the data values at specified time sampling points for each seismogram to be used. A starting model for shear wave attenuation in each layer of the geoacoustic model is then assumed. We compute the seismograms associated with that model at the ranges of the observed seismograms. The partial derivatives of the time series points with respect to the layer Q's are then estimated by sequentially perturbing each layer Q by a small amount, calculating perturbed time series and differencing them from the unperturbed time series. We can then use linear perturbation theory to calculate model perturbations associated with these data differences. We iterate this process until the data fit does not significantly improve.

To test this algorithm, we have generated synthetic seismograms at three ranges with a geoacoustic model similar to that of our actual data set. A band width of 1-4 Hz was used in the inversion. Then, we used a different starting model of the attenuation-depth profile (but the known velocity profile) and calculated a new attenuation model using the inversion method. Fig. 2 compares the actual model used to generate the data, the starting model, and the recovered model after only 6 iterations. The essential features of the model were recovered with some errors associated with an inability of the band-limited data to resolve each layer parameter independently. In particular, the data have insufficient resolution to reproduce the sharp discontinuity at 1.5 m, and the inversion produces a smoothed profile with some overshoot at depths from 4-7 m.

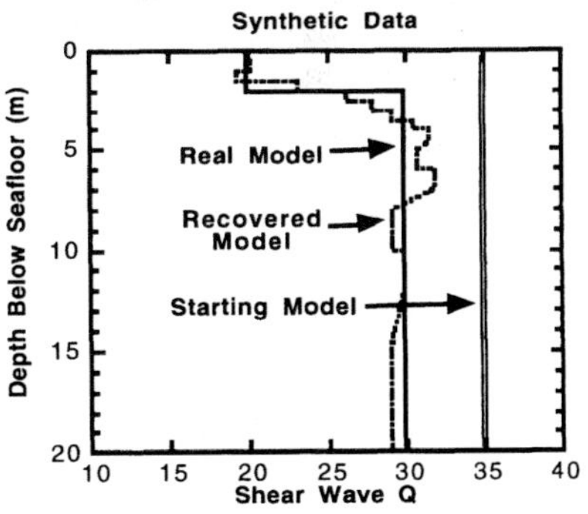

Figure 2. Using synthetic data and 6 iterations of the full waveform inversion, the recovered model shows the main features of the real model.

While this technique appears to be stable for synthetic data, problems can arise when the phase of the signal is imperfectly known either because the shear velocity inversion is inaccurate or the source-receiver ranges are slightly in error. To test this, we have generated data with a model that has a small difference in the shear velocity of the uppermost layer from that of the starting model. The phase of the signal is therefore slightly shifted. The stability of the solution is significantly affected and only one or two singular values can be used on each iteration. To deal with this problem, we have computed an envelope function from the data as described in section 2. Small shifts in phase will not significantly affect the envelope of the data. Inversion of the data envelope is stable for our test case and the salient features of the model were recovered.

Unlike for spectral ratio data, inversion of time series data presupposes that we know the source and receiver transfer functions. We are relatively comfortable with our knowledge of the receiver response. However, the source is more complicated and less understood. The airgun produces a bubble that oscillates near 80 Hz. That bubble interacts with the steel plate to which the airgun is mounted. The high strains near the source may push the sediments past their elastic limit which is very small for shear propagation. The finite size of the plate may introduce a filtering effect. Because the source pulse is generated in a very short time compared to the period of our waves, we would expect the source spectrum to be smooth over our frequency band, but the level and slope of that spectrum are difficult to predict *a priori*. Therefore we have chosen to parameterize the source spectrum and include those parameters in the inversion. We fit the source spectrum with a two parameter function in which the source amplitude is the product of a scaling parameter a_{src} and frequency to the power b_{src}; a_{src} and b_{src} are parameters to be determined in the inversion. Because the envelope function makes the inversion less sensitive to the phase of the signal, we assume zero phase for the source. Tests on synthetic data show that the inversion described above can recover both the source function and the attenuation profile.

5. Applications to Field Data

Fig. 1 shows an example of data collected on clay sediments from the Mississippi Fan (29.1°N, 88.5°W) during September 1993. Five OBS recorded signals from 27 airgun bottom shots during that experiment. For the discussion here we will be inverting data from a single shot as recorded on the 5 OBS. We must first estimate the shear velocity profile prior to inversion for attenuation. Group velocities estimated independently from each of the 5 receivers are virtually identical (< 5% variation). Fig. 3a shows the earth model resulting from the application of the linearized shear velocity inversion technique to the fundamental mode group velocity dispersion curve as described in section 1. The profile is characterized by very low seafloor shear velocities (11 m/s) and a very steep gradient (~7 s^{-1}) of velocity with depth. Velocities below 20 m depth were poorly resolved because of poor signal-to-noise ratio below 1 Hz.

The full waveform technique was then applied to the data to estimate shear attenuation. Initially, a white

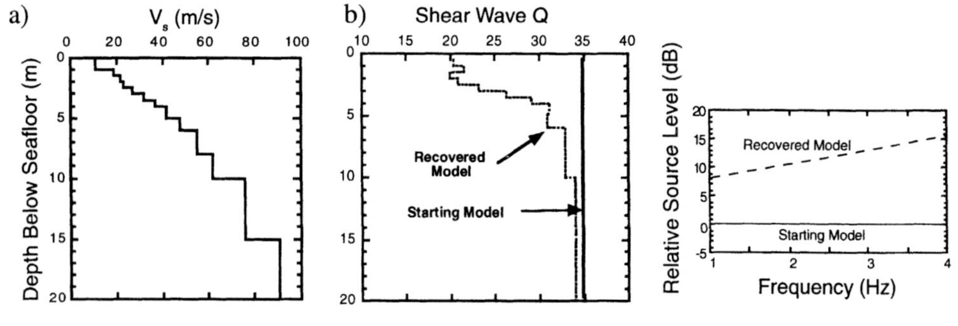

Figure 3. Results of inverting real data from the Mississippi fan. The shear velocity, (a), was generated by a linearized inversion of the group velocity dispersion curve. The attenuation inversion, (b), shows increased attenuation in surface sediments and a source spectrum which is approximately linear with frequency.

source spectrum was assumed and synthetic seismograms were generated with only the fundamental mode. Fig. 3b shows the results of the inversion process. The source spectrum recovered is essentially linear with frequency and increased attenuation is observed in the near surface sediments. Both the absolute amplitude and the envelope shapes of the data are well replicated by the computed synthetics (Fig. 4). We now allow the first overtone to be included in the synthetic seismogram. The recovered source spectrum is still similar, but the attenuation profile is significantly different with lower attenuations at depths near 3 m and higher attenuations at depths below 5 m (Fig. 5). We would expect the data fit to improve as more modes are added to the forward model calculation, creating a more accurate prediction of the data. However, the data residuals are larger for this inversion than for the one with only the fundamental mode. The first overtone predicted from the earth model that was generated with only the fundamental mode is significantly larger than that which is observed (Fig. 6). To fit the data, the attenuation profile is modified by the inversion process to try to suppress the overtone. This change degrades the overall fit to the seismograms. While such attenuation profiles are possible, we believe that it is more likely that the weak overtone energy is the result of complications at the source that affect the relative excitations of the modes. While it is clear from the horizontal seismometer data (not shown) that higher mode data is being generated, the relative amplitudes of the computed overtones may be in error. We have assumed an explosive point source geometry (i.e the moment tensor of the source function is diagonal). Because of the interaction of the air bubble with the plate, or the plate with the sediment, horizontal motions at the source may be significantly less than the vertical motions, resulting in higher fundamental mode excitation. To effectively use the higher mode energy in the inversion process, we will need to solve for the relative modal excitation in the inversion process and include the horizontal radial component seismograms in the data vector to constrain this additional complexity.

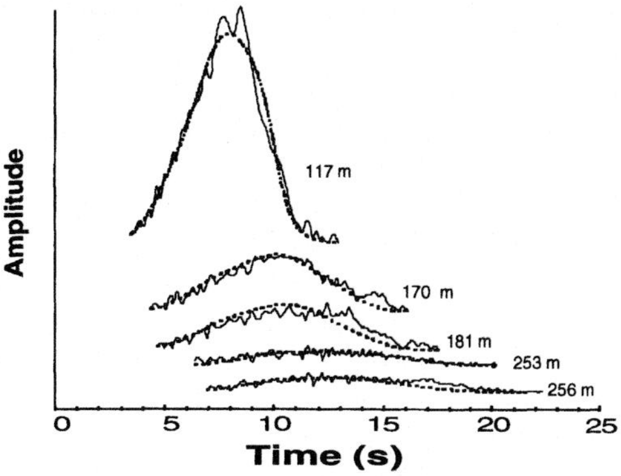

Figure 4. The envelope functions of the inverted data for the five shots at various ranges are indicated in the solid lines. The dotted lines are the predicted envelopes for the recovered model of Figure 3, assuming that only a fundamental mode is propagating. Both the shape and absolute amplitudes of the envelopes are well matched.

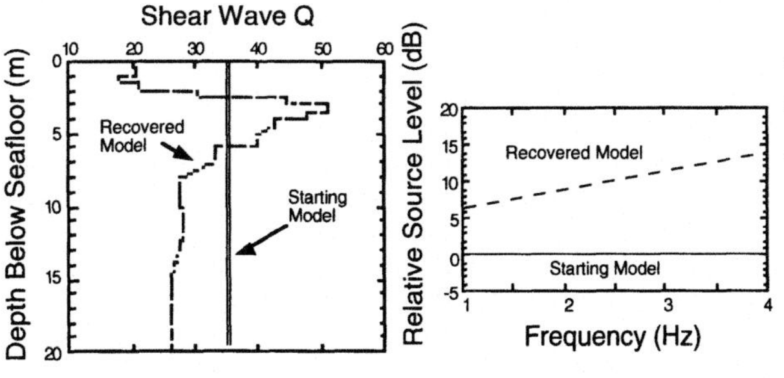

Figure 5. Results of inverting the Mississippi fan data for the case that overtone data is allowed in the forward model. The best fit model has significantly higher residuals than if only fundamental mode propagation is allowed.

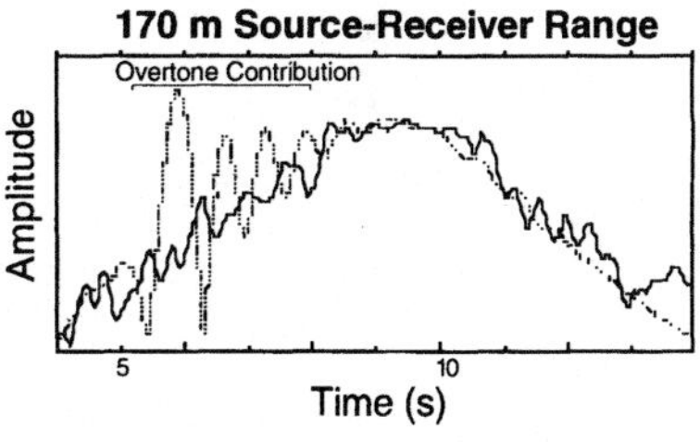

Figure 6. The observed (solid) and predicted (dotted) envelope for one of the seismograms in the Mississippi fan data is shown for the case that overtone propagation is allowed. The earth model is from Figure 3. The overtone propagation is clearly not seen in the data and is responsible for the degradation of the fit.

6. Conclusions

A technique has been implemented for inverting Scholte wave data for sediment shear wave attenuation using first order perturbation theory, requiring only minimal manipulation (an envelope) of the measured full waveform data. In theory, the technique will work on data that has significant contributions from more than one mode. Such data is desirable in that analysis of higher modes gives better constraints on deeper structure than analysis of the fundamental mode. Although the vertical component data that we have used in initial tests of the model are dominated only by the fundamental mode, radial component horizontal seismograms and seismograms from buried receivers [8] often contain strongly multi-modal data.

We have applied these techniques to synthetic data and also to a data set collected in the Gulf of Mexico on the Mississippi sediment fan. The inversion requires the shear attenuation to increase from Q_S near 20 (1.4 dB/λ) at the seafloor to near 30 (0.9 dB/λ) below 3 m depth. The data considered in this test can be fit well without higher mode contributions. However, when higher mode contributions are added to the forward model, the fit degrades and the model predicts more first overtone energy than is observed in the data. This suggests that description of the source geometry is inadequate and modal excitations will need to be parameterized as part of the inversion to adequately deal with multi-modal data.

7. Acknowledgements

This work was supported by the Interface Waves Project of the NRL 6.1 Core Program, Program Element 601153N (L. D. Bibee), and by ONR contract N00014-90-J-1275 (L. M. Dorman). We thank Joe Gettrust and Hassan Ali for helpful reviews. NRL contribution number NRL/PP/7432—93-0026.

References

1. Caiti, A., T. Akal and R.D. Stoll (1991) 'Determination of shear velocity profiles by inversion of interface wave data', in *Shear Waves in Marine Sediments*, Hovem, Richardson and Stoll (eds.), Kluwer, Dordrecht.
2. Stoll, R.D., G.M. Bryan, R. Mithal, and R. Flood (1991) 'Field experiments to study seafloor seismoacoustic response', *JASA*, **89**(5), 2232-2240.
3. Bolt, Bruce A. (ed.) (1972) *Methods in Computational Physics V.11 Seismology: Surface Waves and Earth Oscillations*, Academic Press, New York.
4. Dorman, J. and M. Ewing (1962) 'Numerical inversion of surface wave dispersion data and crust-mantle structure in the New York-Pennsylvania area', *J. Geophys. Res.*, **67**, 5227-5241.
5. Landisman, M., A. Dziewonski, and Y. Sato (1969) 'Recent improvements in the analysis of surface wave observations', *Geophys. J. R. astr. Soc.*, **17**, 369-403.
6. Ewing, J. J.A. Carter, G.H. Sutton, and N. Barstow (1992) 'Shallow water sediment properties derived from high-frequency shear and interface waves', *J. Geophys. Res.*, **97**, 4739-4762.
7. Sauter, A. W., L. M. Dorman and, A. E. Schreiner (1986) 'A study of sea floor structure using ocean bottom shots and receivers', in *Ocean Seismo-Acoustics*, Tuncay Akal and Jonathan M. Berkson (eds.), Plenum Press, New York, 673-681.
8. Bibee, L.D. (1993) '*In situ* measurements of seafloor shear-wave velocity and attenuation using seismic interface waves', in *Proceeding of the Institute of Acoutics Vol 15 Pt 2 (Acoustic Classification and Mapping of the Seabed)*, N.G. Pace and D.N. Langhorne (eds.), University of Bath, UK.
9. Taner, M.T., F. Koehler, and R.E. Sheriff, 'Complex seismic trace analysis', *Geophysics*, **44**, 1041-1063, 1979.
10. Schmidt, H. (1988) 'SAFARI, Seismo-acoustic fast field algorithm for range independent environments: Users guide', SACLANT ASW Research Centre Report SR-113.
11. Gomberg. J.S., and T.G. Masters (1988) 'Waveform modelling using locked-mode synthetic and differential seismograms: application to determination of the structure of Mexico', *Geophys. J. R. astr. Soc.*, **94**, 193-218.
12. Woodhouse, J.H. (1980) 'Efficient and stable methods for performing seismic calculations in stratified media', *Proceedings of the International School of Physics Enrico Fermi Course 78*, 127-151, North Holland, Amsterdam.

DETECTION OF HIGH SHEAR WAVE VELOCITIES IN MARINE SEDIMENT BY INVERSION WITH SIMULATED ANNEALING

DENNIS A. LINDWALL
Marine Sciences Division
Naval Research Lab, Stennis Space Center
MS 39529
U. S. A.

MRINAL K. SEN
Institute for Geophysics
8701 North MoPac Expressway
Austin, TX 78759
U. S. A.

JOSEPH F. GETTRUST
Marine Sciences Division
Naval Research Lab, Stennis Space Center
MS 39529
U. S. A.

ABSTRACT. Multichannel seismic data collected by NRL-SSC near the Blake Outer Ridge show regions of anomalous SVO (spectrum vs. offset) behavior which is modeled by thin sediment layers of high shear velocity. We apply a nonlinear inversion algorithm called simulated annealing (SA) to this problem with the aim of developing a tool for routine analysis of similar data with minimal user interaction. The use of a computationally intensive nonlinear scheme is necessary as the energy function is highly complicated, with many local minima and with large regions of very low gradient. This makes it impossible for iterative gradient-based inversion methods to work. To test the algorithm rapidly we generated synthetic (τ,p) gathers and used a classical Metropolis SA and a variant of it called Very Fast Simulated Annealing (VFSA). Both of the algorithms and four different types of error functions work extremely well. We are also able to invert the data in frequency-offset (ω,x) domain. Although forward modeling in the (ω,x) domain is slower than modeling in (τ,p) domain, transformation of most data into the (τ,p) domain without aliasing is very difficult due to limited spatial sampling in most observations. An inversion of thin layers in the (τ,p) domain requires high frequency data and many closely spaced receivers (or sources); inversion in the (ω,x) domain can resolve thin layer properties with a few more widely spaced receivers. Most field data have limited offset range and spatial sampling thus justifying the slower (ω,x) technique.

1. INTRODUCTION

Classical Simulated Annealing (SA) is an optimization method that randomly searches model space but controls the search by using a statistical selection criterion (Metropolis et al, 1953; Kirkpatrick et al., 1983). The search over model space is done by perturbing the current model, creating a new model, and then calculating synthetic data with a forward model routine. The synthetic data are compared to the real data using an objective function. There are a variety of objective functions that can be used such as cross correlation or residual (Sen and Stoffa, 1991). If the new model is a better "fit", resulting in a lower objective function for the new model, then this new model is chosen as the current model and the process is repeated. If the new model produces a higher objective function, it is occasionally chosen as the current model with the

O. Diaschok et al. (eds.), Full Field Inversion Methods in Ocean and Seismo-Acoustics, 383–388.

probability $P = e^{(E-E')/T}$ where E and E' are the objective function values and T is the "temperature". At a high T, the chance of choosing the solution with a higher objective function is much higher than at a low T. This process is begun at a high T and repeated hundreds or thousands of times as T is slowly reduced in a process that is mathematically analogous to annealing. If T is reduced slowly enough over the proper range, then the solution, or object, will "anneal" into a single crystal which is the lowest energy state possible and the synthetic data calculated will be the best possible fit to the data (van Laarhoven and Aarts, 1987). Here we apply this SA technique to automation of the SVO method (Gettrust and Rowe, 1989) for estimating shear wave velocities within marine sediments.

2. OUR VARIANT OF SIMULATED ANNEALING

Our SA algorithm is based on the classical Metropolis algorithm (Metropolis et al., 1953) but differs significantly in most steps, particularly in how the temperature parameter is used. Not only is the temperature definition different for each model variable, but the model search window is also a function of the temperature (Ingber, 1989). The layer dependent temperature fits the top layers first and is more efficient for a model with many layers as it results in a linear increase in the run time relative to the number of layers rather than a geometric or exponential increase in time. Using variable dependent temperatures, we can also fit the variables that have the strongest influence on the data first and then refine the model with the more weakly influencing variables. A temperature dependent search distribution gives a much finer search pattern at low temperatures when the solution is near the minimum.

Inversion of seismic or geoacoustic data using SA can be done in the τ-p domain (intercept time and ray parameter, a plane wave decomposition of the wave field), the t-x domain (time and offset, the usual domain of data collection), or the ω-x domain (frequency and offset) with any suitably defined objective function. Geoacoustic data are normally collected in the t-x domain and then transformed to τ-p or ω-x when desirable for analysis or inversion. However, a reliable τ-p transformation requires receiver (or source) spacing of less than 3 m for a 500 Hz signal in order to avoid spatial aliasing. This minimum spacing must be maintained over a range roughly equal to the depth to the layers under analysis. Data collected for inversion in the t-x or ω-x domain require only a few offsets, perhaps 10, and can be unevenly distributed over the same range. Data in the ω-x domain are also known as SVO (spectrum versus offset) and is very sensitive to shear wave velocity (Gettrust and Rowe, 1989).

We have tested the SA algorithm by inverting synthetic data. The synthetic data are simplified and noise free examples of what is expected from real geological sites collected by the sonobuoys and SUS charges as well as by multichannel seismic systems such as NRL's Deep Towed Acoustics-Geophysics System (Gettrust and Ross, 1990). Figure 1a is an example of data that might be collected over a thick sedimentary basin such as a continental shelf. These synthetic data are shown in the τ-p domain which make for a fast inversion calculation. Figure 1b shows the residuals or the differences between the "data" and synthetics calculated from the solution or fit found by the inversion calculation using the CC2 objective function (Sen and Stoffa, 1991). Figure 1c shows the cooling curve or the progress of the fit as the calculation proceeds from a high temperature on the right side to a low temperature on the left side. Figure 1d compares the model used to generate the "data" with the solution found by inversion. We have tested this method with noisy synthetic data and found that it is very robust even when the signal to noise ratio is less than one.

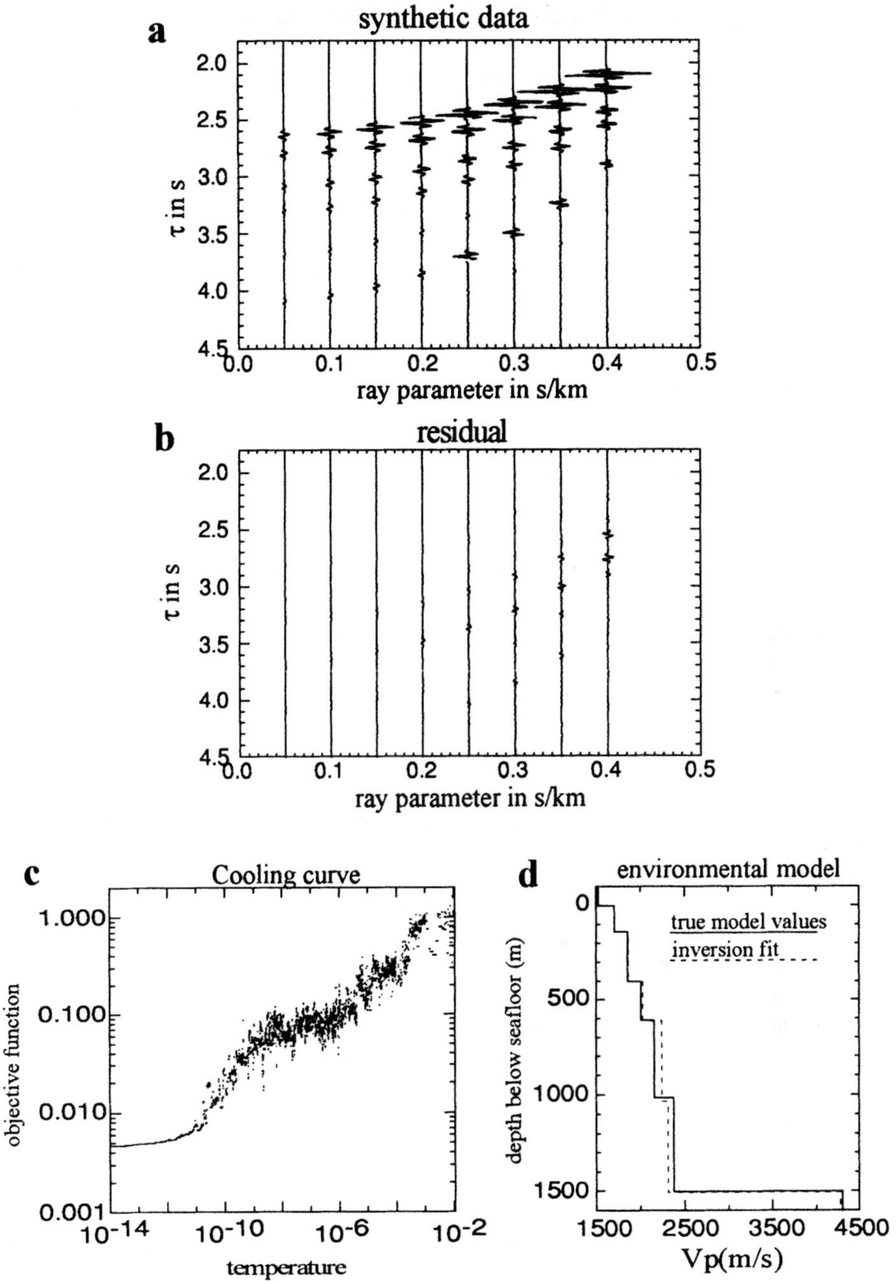

Figure 1. a. Synthetic data from thick continental shelf sediments in the τ-p domain. **b.** Residual (difference between data and solution) for the inversion fit. **c.** Cooling curve progress from the initial temperature of 0.01 and corresponding large objective function at the upper right to the final fit at the lower left. **d.** Comparison of the environmental model used to calculate the synthetic data with the final fit from the inversion.

3. AN INTERESTING NONLINEAR PROBLEM IN MARINE SEDIMENTS.

Figure 2 shows an interesting geophysical situation found with a multichannel seismic marine survey; where one or two thin high Vs layers lie over otherwise normal sediment. These data were initially analyzed and modeled using the SVO technique and several small regions were found where Vs at the sea floor was as high as 500 m/s (Lindwall and Gettrust, 1991). Figure 2a shows the synthetic "data" in t-x space. The offsets here are regularly spaced but need not be. Irregular spacing as from sonobuoys and SUS charges will work well in this domain. Figure 2c shows the cooling curve which shows a decreasing objective function (using CC2 again) at the lowest temperature suggesting that an even better fit could easily be achieved. However, the fit is very close and the residual is very small (Figure 2b). Figure 2d compares the model used to generate the "data" with the solution found by inversion. The resolution and frequency are much higher than in Figure 1 so that the thin layers can be resolved. The only variable searched over in the inversion was the shear velocity (Vs) giving only three variables. This inversion was done without layer stripping but with a temperature dependent search window. The inversion done in the t-x domain is very slow even though the search was only over three variables. Additional inversion runs in the τ-p domain searching over Vp, Vs, and layer thickness show that these data are sensitive to Vs and layer thickness but not Vp.

The advantage that SA (or other nonlinear inversion methods) has over linear methods is demonstrated in Figure 3 which is a gray tone contour plot of the objective function for Vs of the first and second layers for the model shown in Figure 2. The objective function used here is the negative of the cross correlation so that a perfect fit has a value of -1 while random models give an average value of 0. The black area is where the objective function has a value near -1 and the true Vs values are 500 m/s for layer 1 and 650 m/s for layer 2. The gray areas below it are local minima. The highest contour level on this plot is at -0.9 which is a fairly good fit. A linear inversion method will easily fall into one of the local minima and the user may be satisfied with an objective function of -0.92 to -0.94 but end up with the wrong solution. Our SA method easily found the global minimum every time, even when the search was done very rapidly. Other objective functions calculated for the same model give similar plots with multiple local minima. Any model with thin layers will be nonlinear and have local minima in their objective function meaning that a nonlinear search method such as our SA algorithm is needed to automatically find the optimum environmental model.

In our SVO analysis of this region, we found 3 patches about 50 m across in a line less than 3 km long that have very high Vs values like those shown in Figure 2. These patches behave as strong forward scatters but can not be identified from Vp measurements. They also indicate that there are significant small scale geological alterations to the geophysical properties of a seemingly very uniform sea bed.

4. SUMMARY

We have developed and tested a SA algorithm for inverting synthetic geoacoustic data. By making some modifications to the Metropolis SA algorithm which redefine the temperature as a function of each variable and make the search window a function of temperature we have improved the speed and accuracy of the method. This method works with three data domains that are either optimal for data collection or forward model calculation, but not both. We have demonstrated the effectiveness of the SA method to identify very high shear wave velocities in

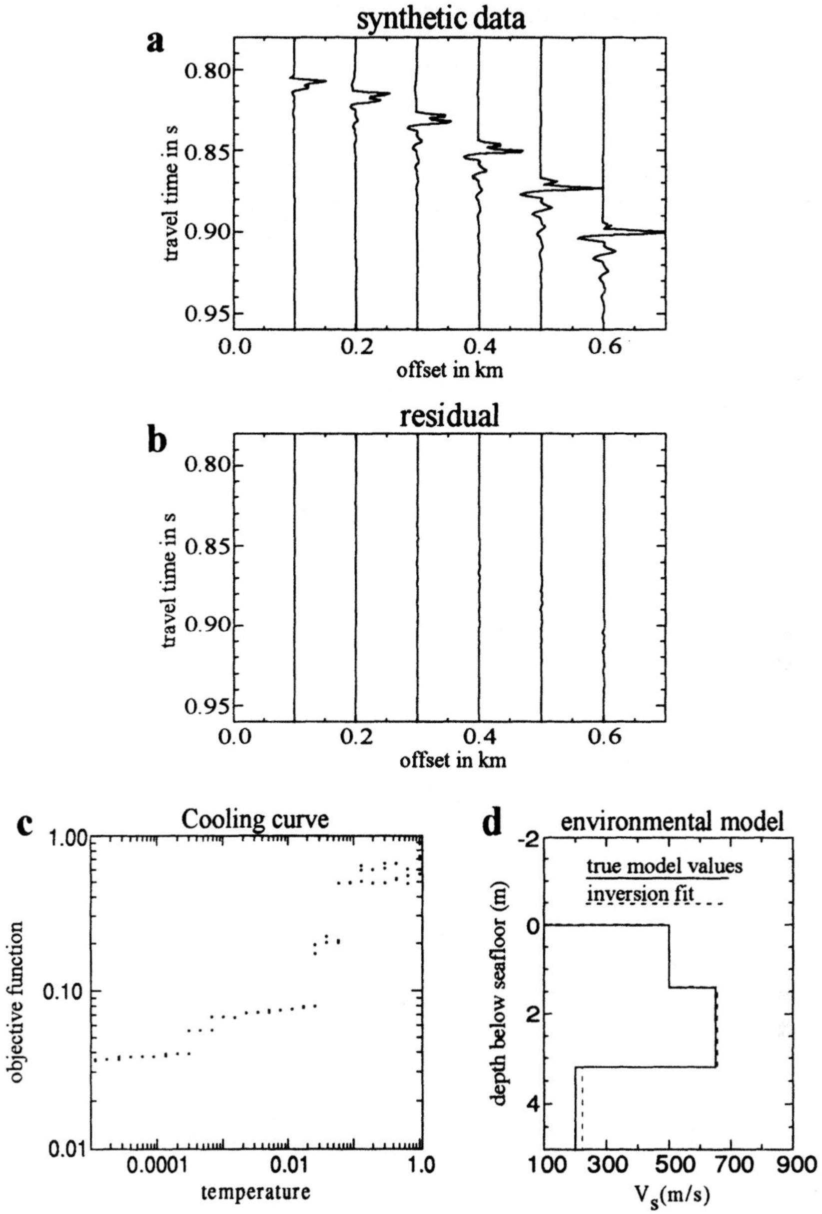

Figure 2. **a.** Synthetic data for a region of high Vs at the seafloor in the t-x domain. **b.** Residual (the difference between data and solution) for the inversion fit. **c.** Cooling curve progress from the initial temperature of 1.0 and corresponding large objective function at the upper right to the final fit at the lower left. **d.** Comparison of the environmental model used to calculate the synthetic data with the final fit from the inversion.

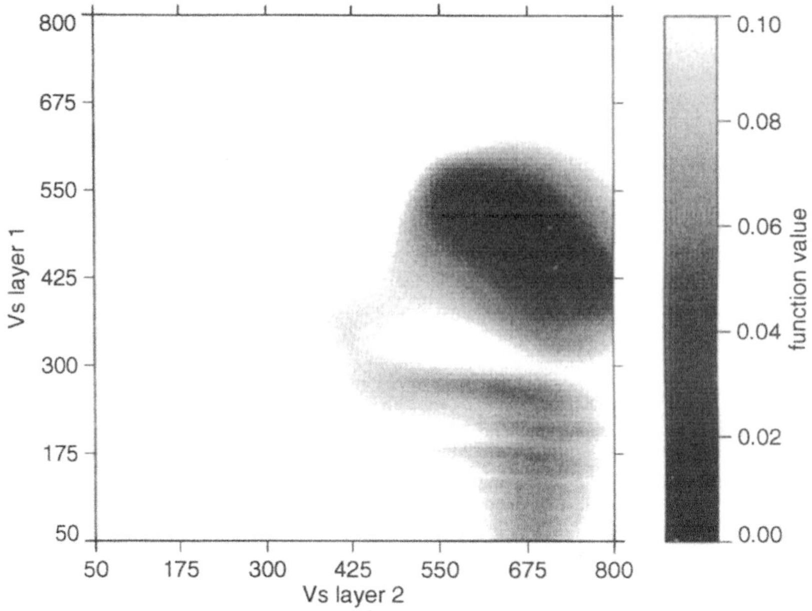

Figure 3. Contour plot of the objective function

marine sediments We also showed that this geophysical case is a nonlinear problem with several traps that foil other inversion methods.

5. REFERENCES

1. Gettrust, J. F. and J. H. Ross, 1990, Development of a low-frequency, deep-towed geoacoustics system , *Proceedings of Oceans '90*, Washington, D. C.
2. Gettrust, J. F. and M. M. Rowe, 1989, Constraints on shear velocities in deep-ocean sediments as determined from deep-tow multichannel seismic data, in *Shear Waves in Marine Sediment*, ed., Hovem, Richardson, and Stoll, Kluwer, Dordrecht
3. Ingber, L., 1989, Very fast simulated re-annealing, *Mathl. Comput. Modelling*, 12, 967-973.
4. Kirkpatrick, S., C. D. Gelatt Jr., and M. P. Vecchi, 1983, Optimization by simulated annealing, *Science*, 220, 671-680.
5. Lindwall, D. A. and J. F. Gettrust, 1991, Spectrum versus offset analysis of DTAGS data and small scale lateral variability of oceanic sediments, abstract, *EOS*, 72, 303.
6. Metropolis, N., A. Rosenbluth, M. Rosenbluth, A. Teller, and E. Teller, 1953, Equation of state calculations by fast computing machines, *J. of Chem. Physics*, 21, 1087-1092.
7. Sen, M. K. and P. L. Stoffa, 1991, Nonlinear one-dimensional seismic waveform inversion using simulated annealing, *Geophysics*, 56, 1624-1638.
8. van Laarhoven, P. J. M. and E. H. L. Aarts, 1987, *Simulated Annealing: Theory and Applications*, Kluwer Academic Publishers, 1409-1412.

MODELING AND INVERSION OF THE ELASTO-DYNAMIC EQUATIONS

BJØRN URSIN and HUASHENG ZHAO
The Norwegian Institute of Technology,
Division of Petroleum Engineering and Applied Geophysics,
N-7034 Trondheim, Norway.

ABSTRACT The equations of motion and Hooke's law for a stratified elastic medium can be transformed into a matrix-vector equation by a Fourier-Hankel transform using the symmetry properties of cylindrical coordinates. The solution for each frequency-wavenumber component is found by layer-recursively building up the generalized reflectivity function. The inverse problem is solved by a damped Gauss-Newton method in the transformed domain, where the unknown parameters are the density, the P-wave and the S-wave velocities in each layer and the layer thickness. The gradient of the objective function is computed from analytical expressions of the partial derivatives of the plane-wave reflectivity function. The truncation and discretization effects of the data in the time-space domain are considered in the plane-wave decomposition. Numerical example of synthetic data shows that the proposed inversion method satisfactorily reconstructs the elastic parameters of a stratified medium.

1. Introduction

Seismic wave propagation in the solid earth is governed by the elasto-dynamic equations. Numerical solutions of these equations are generally based on the approximations that have been applied to the medium. Finite-difference and finite-element techniques require only a few approximations, but the computational requirements are extensive. The dynamic ray tracing technique is very efficient, but only a limited number of rays can be computed and it is inefficient for media with thin layers. When the medium is approximated by a stack of homogeneous layers, the solution of the elasto-dynamic equations can be computed effectively in the frequency-wavenumber domain by the reflectivity method (Kennett, 1983).

The inversion of seismic data for a horizontally layered media has been studied by many authors during the past decade (Kolb, Collino and Lailly,1986; McAulay, 1986; Amundsen and Ursin, 1991; Kormendi and Dietrich, 1991). Zhao, Ursin and Amundsen (1993) have presented a method of elastic inversion in the frequency-wavenumber domain, where the gradient of the objective function is computed using the Jacobian matrix. The unknown parameters are the P- and S-wave velocities, the density in each layer and the layer thickness. The truncation and discretization effects of the data in time-space domain on the plane wave decomposition were also demonstrated. Ursin and Zhao (1994) applied the inversion scheme to a shallow seismic anomaly from the southern Barents Sea, where the anomalous reflector was interpreted as the reflection from the base of gas-hydrate sediment trapping underlying gas (Andressan, Hogstad and Berteussen, 1990)

The method of elastic modeling and inversion of a stratified medium in the frequency-wavenumber domain is briefly described and some related practical problems are discussed. A numerical example is given which demonstrates the feasibility of the method. The method has also been applied to real data by Ursin and Zhao (1994).

O. Diaschok et al. (eds.), Full Field Inversion Methods in Ocean and Seismo-Acoustics, 389–394.

2. Layer-recursive modeling

We consider wave propagation in a horizontally layered medium consisting of an acoustic layer at the top, and a stack of homogeneous isotropic elastic layers below. The entire medium is bounded above by a free surface and below by a half space. In a cylindrical coordinate system with the depth axis passing vertically downward through a point source, the equation of motion and Hooke's law can, after Fourier-Hankel transformation, be expressed by a matrix-vector equation in the frequency-wavenumber domain. The matrix-vector equation can further be transformed into equations of upgoing and downgoing waves, where the solution for each frequency-wavenumber component can be solved layer-recursively by building up the generalized reflection matrices (Kennett, 1983), with the boundary condition for the free surface being zero pressure and for the lower half space the presence of only downgoing waves.

The reflected part of the pressure response $P(\omega, k_r, z_r)$ at the receiver is given by (Amundsen and Ursin, 1991; Zhao et. al., 1993):

$$P(\omega, k_r, z_r) = aP_s(\omega, k_r)\frac{R_D^{PP}(\omega, k_r)}{1 + r_0 R_D^{PP}(\omega, k_r)}A(\omega, k_r, z_s)A(\omega, k_r, z_r), \tag{1}$$

where ω being the angular frequency and k_r the radial wavenumber. The factors $A(\omega, k_r, z_s)$ and $A(\omega, k_r, z_r)$ account for the source and receiver directivities, with z_s and z_r being the source and receiver depths, and

$$A(\omega, k_r, z) = \exp[-ik_{z,1}(z - z_0)] - r_0 \exp[ik_{z,1}(z - z_0)]. \tag{2}$$

$P_s(\omega, k_r)$ is the source wavelet, a is the scaling of the source function, $-r_0$ is the surface reflection coefficient and $R_D^{PP}(\omega, k_r)$ is the generalized downward PP reflection response of the stack of layers from the top of the water layer to the bottom of the stack, computed using a layer-recursive formula. The contribution of all shear waves, converted waves and intra-bed multiples are included in the computation of reflection response $R_D^{PP}(\omega, k_r)$.

A useful feature of the generalized reflection matrix method is the ability to separate different components of the wave field. In particular, the contributions of the compressional, shear, and converted waves can be computed as partial solutions to the complete response, and the orders of the multiple reflections can be chosen arbitrarily. To model the marine seismic environment, special considerations has to be taken into account for the reflection and transmission at the acoustic/elastic interface where the vertical components of the displacement and the stress are continuous (the shear stress being zero), while the radial component of the displacement is discontinuous.

3. Least-squares inversion

The seismic waveform inversion consists of retrieving model parameters from observed data by minimizing the least-squares misfit between the reference and predicted data. The data in the frequency-wavenumber domain are represented by a vector of dimension equal to the number of frequencies times the number of wavenumbers. The parameters of the stratified medium are defined by the parameter vector $\mathbf{m}$ which consists of four sub-vectors containing the P-wave velocity vector α, the S-wave velocity vector β, the density vector ρ, and the layer thickness vector $\Delta\mathbf{z}$, i.e., $\mathbf{m} = [\alpha^T, \beta^T, \rho^T, \Delta\mathbf{z}^T]^T$, where ^{T} denotes the transpose.

A damped least-squares inversion method is used to compute the unknown parameters by a Levenberg-Marquardt iterative search in the transformed domain. The gradient of the objective function is computed from analytical expressions of the Jacobian matrices as derived directly from the modeling equation. The partial derivatives of the reflection response with respect to the model parameters are computed simultaneously with the reflection response by solving a set of expanded recursive formulas (Zhao et al, 1993).

One of the advantages of the Levenberg-Marquardt search scheme is that it is a compromise between two classical search procedures: the steepest descent and the Gauss-Newton method, depending on the value of the damping factor. Good initial convergence properties of the steepest descent in the region far away form the solution combined with fast convergence of the Gauss-Newton method in the region close to the solution, make the algorithm very attractive in inversion problems when the size of Hessian matrix is relatively small.

Data given in time and space need to be transformed to the frequency-wavenumber domain. The transformation will introduce edge and aliasing artifacts in the transformed data due to the limited aperture and coarse sampling in space. These artifacts should be accounted for in order to have a successful full waveform inversion (Zhao et al., 1993). This is done by convolving the modeled data and the Jacobain matrices with a wavenumber filter. Figure 1 shows the wavenumber filter for the sampling interval and the parameters used in the seismogram shown in Figure 2.

4. Numerical example

The seismogram shown in Figure 2 was generated by the reflectivity method, including all primaries, multiples, and converted shear events. A Ricker wavelet with center frequency of 20 Hz was used, and the surface reflection coefficient was chosen to be 0.9. The data consist of 51 traces with the near offset being zero and the maximum offset 1000 m, and a trace interval of 20 m. The reference data are transformed to the frequency-wavenumber domain by a discrete Fourier-Hankel transform (Zhao et al., 1993) within the bandwidth of the source, and for wavenumbers from zero to the maximum wavenumber.

The model parameters represent a 14-layer model, and are displayed in Figure 3a in solid lines with the P-wave velocity to the left, S-wave velocity in the middle and density to the right. The initial guess of P-wave velocity and layer thickness are given by perturbing the true model, while S-wave velocity and density are initially constant (500 m/s and 1.75 g/cm^3, respectively), as plotted in dashed lines in Figure 3a. The parameters in the water layer are assumed to be known, i.e., the velocity and density and the thickness of the water layer as well as the positions of the source and receivers. Estimates of these parameters are in general accessible from marine seismic acquisition. We then seek to determine the parameters of solid layers.

The results of the inversion after 1, 5, 10 and 20 iterations are shown in Figures 3b-e. We observe that all the parameters fit the true model with a reasonable accuracy after the final iteration. The low frequency trends for P-wave velocity, S-wave velocity and density are found after 5 iterations, but the density shows faster convergence than the S-wave velocity.

From the results of the final iteration we see that both the parameters of the lower half space have not been estimated properly. This behavior may be expected since there are no upgoing waves coming from the lower elastic half-space as defined by the radiation condition. These parameters are therefore only estimated from the last reflection coefficient. No travel time information about the velocity in the lower half-space is available. A relatively high

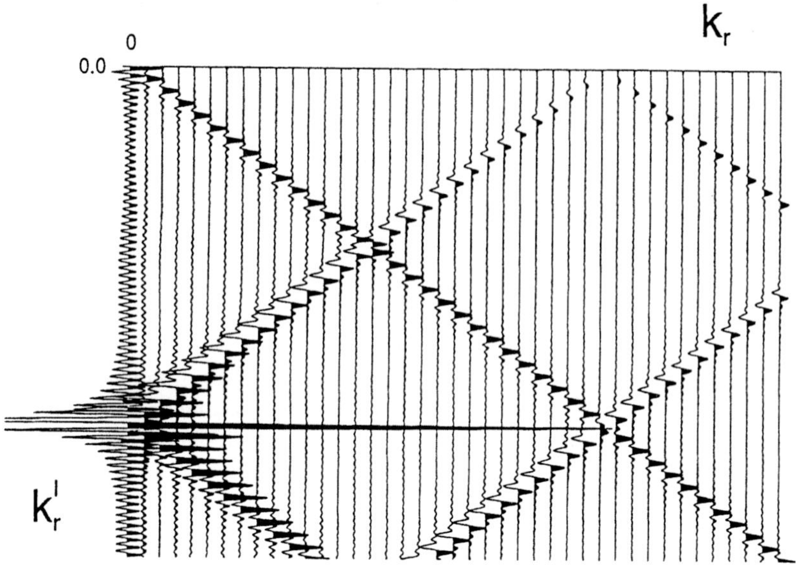

FIGURE 1. The wavenumber filter.

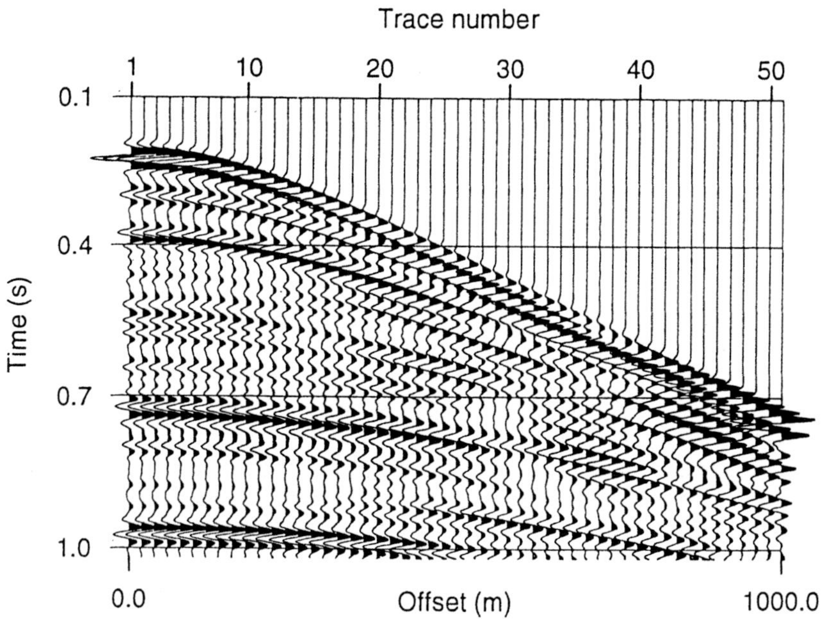

FIGURE 2. The reference data in the time and space domain.

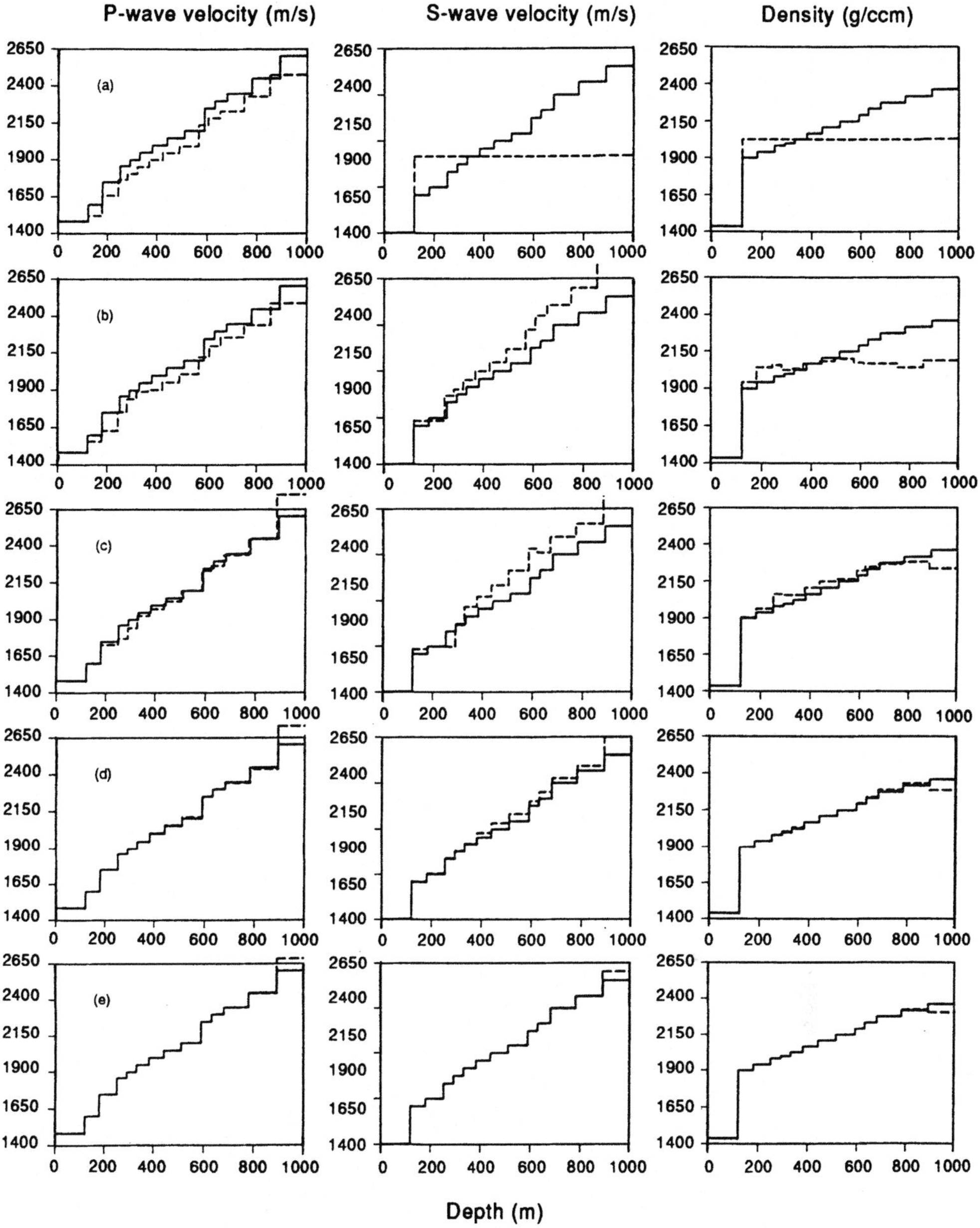

Figure 3. Full elastic inversion of the 14-layer model. The exact model is displayed as a solid line and the estimated model as a dotted line. The P-wave velocity is shown to the left, the S-wave velocity in the middle and the density to the right. (a) Initial model. (b) Result after 1 iteration. (c) Result after 5 iterations. (d) Result after 10 iterations. (e) Result after 20 iterations.

value of the P-wave velocity may be compensated for by a low value of the density as long as the P-wave impedance fits the reference model.

5. Conclusions

A method for solving the non-linear inversion problem for a horizontally layered elastic medium in the frequency-wavenumber domain has been presented. The truncation and discretization effects of the data in the time-space domain are compensated for by a wavenumber filter. The numerical example demonstrated the feasibility of the method. The unknown elastic parameters and the layer thicknesses were estimated from the bandlimited marine data with good resolution.

References

1. Amundsen, L., and Ursin, B. (1991) Frequency-wavenumber inversion of acoustic data: *Geophysics*, **56**, 1027-1039.
2. Andreassen, K., Hogstad, K., and Berteussen, K.A. (1990) Gas hydrate in the southern Barents Sea indicated by a shallow seismic anomaly: *First Break*, **8**, 235-245.
3. Kennett, B.L.N. (1983) *Seismic wave propagation in stratified media*: Cambridge Univ. Press.
4. Kolb, P., Collino, F., and Lailly, P. (1986) prestack inversion of 1-D medium: *Proc. IEEE*, **74**, 498-508.
5. Kormendi, F. and Dietrich, M. (1991) Nonlinear waveform inversion of plane-wave seismograms in stratified elastic media: *Geophysics*, **56**, 664-674.
6. McAulay, A.D. (1986) Plane-layer prestack inversion in the presence of surface reverberations: *Geophysics*, **51**, 1789-1800.
7. Ursin, B. and Zhao, H. (1994) Elastic inversion of a shallow seismic anomaly in the southern Barents sea: *Submitted to First Break*.
8. Zhao, H., Ursin, B., and Amundsen, L. (1993) Frequency-wavenumber elastic inversion of marine seismic data: *Submitted to Geophysics*.

LINEARIZED INVERSION IN VISCOELASTIC MEDIA

JOAKIM O. BLANCH

The Rice Inversion Project, Department of Geology and Geophysics
Rice University, P.O. BOX 1892, Houston TX 77251-1892
e-mail joakimb@rice.edu

and

WILLIAM W. SYMES

The Rice Inversion Project, Department of Computational and Applied Mathematics
Rice University, P.O. BOX 1892, Houston TX 77251-1892

ABSTRACT

Real earth media disperse and attenuate propagating waves. This anelastic behavior can be well described by a viscoelastic model. Here we introduce a viscoelastic linearized inversion routine developed through the adjoint state technique. It is based on a viscoelastic model of standard linear solids in parallel and built upon a recently developed finite difference simulator. The routine enables linearized inversion in viscoelastic media. We will show that the reproduction of reflectors can be very poor if anelastic effects are neglected.

1. Introduction

Viscoelasticity provides a good model for representing real earth media [2]. Attenuation of propagating waves is often measured in the quality factor Q. Sedimentary rocks have been shown to have a nearly constant Q over the seismic frequency range. This behavior is easily described with a simple viscoelastic model using standard linear solids in parallel, [1], [2], [4]. As a matter of fact, it is in many applications sufficient to use only one standard linear solid [1],[2]. We have developed a reliable forward simulator which solves the viscoelastic wave equation using the finite difference method. The forward simulator has already been used in scattering studies and been subject to several convergence tests [3], [10]. We will start by briefly discussing the forward simulator. We will then use this forward simulator to derive an inversion routine through the adjoint state technique. This algorithm is different from a similar algorithm developed by Martinez and McMechan, [8]. Last we will show from inversion results that it is necessary to take anelastic effects into account, to be able to find the correct reflectors.

2. Forward Modeling

The one dimensional viscoelastic wave equation for one standard linear solid can be written,

O. Diaschok et al. (eds.), Full Field Inversion Methods in Ocean and Seismo-Acoustics, 395–400.

$$\begin{cases} p_t = -K(1+\tau)v_x - r \\ r_t = -\dfrac{1}{\tau_\sigma}(r + K\tau v_x) \\ v_t = -\dfrac{1}{\rho}p_x \end{cases} , \tag{1}$$

which is an initial value problem, i.e., $p = v = r = 0, t \ll 0$. K is the bulk modulus, ρ is the density and τ is roughly proportional to the inverse of Q and a measure of attenuation. τ_σ is a relaxation time and determines the frequency range where there is most attenuation. p is the pressure (stress) and v is the particle velocity. r is called a memory variable, which reflects the 'memory' of the viscoelastic medium. Waves are initiated by sources, $f(x_s, t)$, which are added to the right hand side of equation (1). The fields (p, v, r) are sampled at receiver locations, x_r, to produce a seismogram, $p(x_r, t)$. We solved equation (1) with the finite difference method.

3. Inversion

The inversion routine is derived by first perturbing the parameters determining the media (K, ρ, τ) and the fields describing the viscoelastic waves (p, v, r). The linear terms in the perturbations are kept to produce a linearized equation. This linearized "forward map" is identical to a Born approximation, and is used to produce simulated seismograms in the inversion routine. To perform inversion for the parameters describing the medium it is necessary to find the gradient. The gradient for updating the parameters can be found through the adjoint state technique [6]. This involves solving a final value equation system. The new equation system is the adjoint of the linearized equation.

$$\begin{cases} u_t = -\left(\dfrac{1}{\rho}q\right)_x - u0 \\ s_t = \dfrac{1}{\tau_\sigma}(s + u) - s0 \\ q_t = -\left(K(1+\tau)u - K\tau s\right)_x - q0 \end{cases} \tag{2}$$

$u = q = s = 0, t \gg 0$. q, u, and s are the adjoint fields. The sources, $q0(x_r, t)$, $u0(x_r, t)$, and $s0(x_r, t)$, in equation (2) are the differences between the measured seismogram, $p^{obs}(x_r, t)$, and the forward simulated seismogram. Finally to find the gradient for updating the parameters the adjoint fields, equation (2), must be cross correlated with the background fields, equation (1).

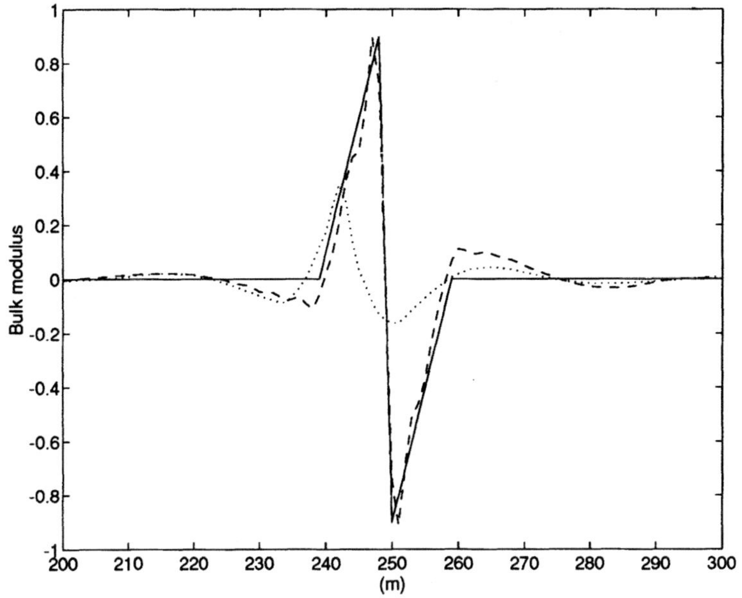

FIGURE 1. The true reflector with two inversion estimates. Solid - True reflector, Dashed - Viscoelastic estimate, Dotted - Elastic estimate.

$$\begin{cases} \delta K = \int_0^T -(1+\tau)v_x u - \tau v_x s \; dt \\ \delta \tau = \int_0^T -K v_x u - K v_x s \; dt \\ \delta \rho = \int_0^T \frac{1}{\rho^2} p_x q \; dt \end{cases} \tag{3}$$

This gradient can then be used in an optimization algorithm or the linearized forward model and the adjoint can be used to solve the normal equation for a linear(ized) inverse problem. It is necessary to derive the adjoint equation using the discretized system to find the correct discrete adjoint equation.

4. Results

We produced a synthetic 1-D seismogram by viscoelastic (realistic) modeling ($\tau = 0.05$ corresponds to $Q \approx 40$, and velocity $c = \sqrt{K/\rho} = 2.0\, km/s$). The synthetic model consisted of one distinct reflector in perturbed bulk modulus, (Figure 1). We used this seismogram as input to our inversion routine and performed viscoelastic ($\tau = 0.05$) and elastic inversion ($\tau = 0$). The result from the two inversions is shown in Figure 1, where the original reflector is compared to reflectors estimated by elastic and viscoelastic inversion. It is clear that the

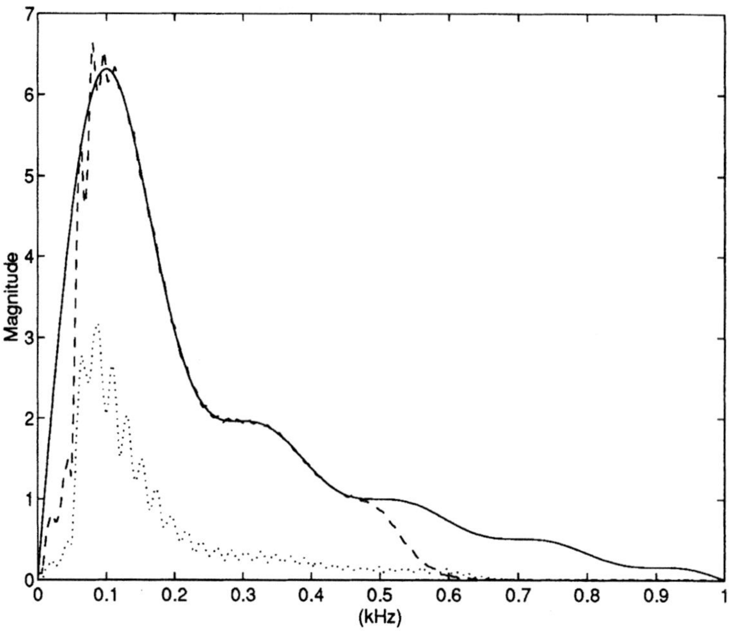

FIGURE 2. Frequency content of the true reflector and the two estimates. It is possible to switch between the frequency and wavenumber domain through the relation $\lambda f = c$, since the velocity, $c = 2.0\,km/s$, in the medium is constant, λ-wavelength and f-frequency. Solid - True reflector, Dashed - Viscoelastic estimate, Dotted - Elastic estimate.

result from the elastic inversion is quite poor. The wavenumber spectra of the reflectors yield more insight to the differences and why the viscoelastic estimate does not perfectly match the model. The wavenumber spectra of the three reflectors are shown in Figure 2. We can see an almost perfect match in the range where the reflector is illuminated by the source wavelet for the viscoelastic case but poor reconstruction in the elastic case.

5. Analysis of estimation results

Both modeling and inversion were done with a constant reference velocity in 1-D. This enables us to easily switch between wavenumber and frequency domain, which makes it possible to compare the wavenumber response of the differently estimated reflectors (Figure 2) with propagation effects, due to the viscoelastic medium, derived in the frequency domain. These propagation effects are the likely cause for the incorrect estimate found from the elastic inversion using the viscoelastic seismogram.

For 1-D constant velocity/attenuation medium it is possible to derive a simple expression for the attenuation of different frequency components in the seismogram. By transforming equation (1) to the frequency-wavenumber domain, it is possible to express the frequency or the velocity of waves as a function of wavenumber (dispersion relation). The velocity will be a complex quantity due to attenuation, [5], [2]. The exponential of the imaginary part of the velocity determines the amplitude decay of the different frequency components (the expression for the velocity is inserted in the expression for a plane wave). By approximating,

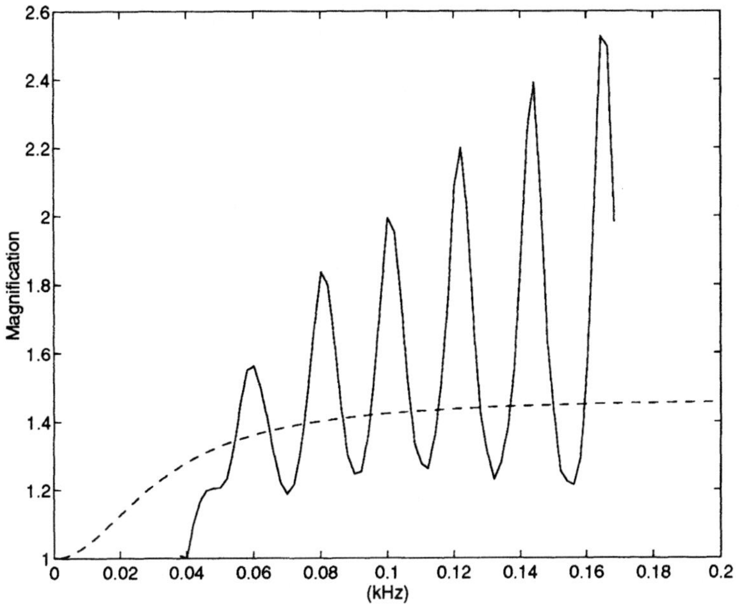

FIGURE 3. Ratio of the wavenumber spectra of two estimates and the predicted effect from attenuation in medium, equation (4). Solid - Ratio of the wavenumber spectra of two estimates, Dashed - The inverse of the predicted effect for decay.

$\sqrt{1+x} \approx 1 + x/2$, for the expression of the imaginary part of the velocity and correcting for two-way traveltime, the different components in the frequency spectra decays in time according to

$$e^{-\omega t \frac{\tau \omega \tau_\sigma}{1+\omega^2 \tau_\sigma^2}}, \tag{4}$$

for a viscoelastic medium with a constitutive relation corresponding to a standard linear solid.

The above effect should be possible to track in the inverted results if an incorrect value for the attenuation in the media was used, e.g., elastic inversion for viscoelastic (realistic) seismogram, and could in other words be the cause for erroneous estimation of reflectors. Dividing the wavenumber spectra for an elastically estimated reflector with the wavenumber spectra for a viscoelastically estimated reflector should yield a rough approximation to the reciprocal of the function in equation (4), and since there is only one reflector it is possible to approximate the time t in equation (4) by dividing the distance from the source/receiver to the reflector by the velocity in the medium. This is shown in Figure 3. The ratio between the two wave number spectra has only been calculated on the interval where the spectra have a good support. The ratio of the wavenumber spectra is oscillating around the predicted curve which suggests that the erroneous result is due to neglect of the propagation effects (attenuation and dispersion) for the elastic inversion.

6. Summary

We have described a viscoelastic linearized inversion routine. The error in estimates of reflectors can be severe if significant anelastic effects are neglected. Immediate areas of research are effect on AVO (Amplitude Versus Offset analysis, [7]) and inversion for smoothly varying Q in the medium.

7. Acknowledgements

This research was supported by ONR grant N00014-89-J1115. We would also like to thank our dear friend Johan Robertsson, who had an active part in the development of the forward simulator.

References

1. Blanch, J. O., Robertsson, J. O. A., Symes, W. W., (1994), Modeling of a constant Q: Methodology and algorithm for an efficient and optimally inexpensive viscoelastic technique, *Technical Report 94-14, Department of Computational and Applied Mathematics, Rice University,* also submitted to Geophysics.
2. Blanch, J. O., Robertsson, J. O. A., Symes, W. W., (1993a), Viscoelastic Finite Difference Modeling, *Technical Report 93-04, Department of Computational and Applied Mathematics, Rice University.*
3. Blanch, J. O., Robertsson, J. O. A., Symes, W. W., (1993b), *Viscoelastic Finite Difference Modeling*: SIAM Proceedings 1993 Mathematical and numerical aspects of wave propagation, eds. Kleinman, R., Angell, T., Colton, D., Santosa F., Stakgold, I.
4. Bourbie, T., Coussy, O., Zinzner, B., (1987), *Acoustics of porous media*: Gulf Publ. Co.
5. Carcione, J. M., Kosloff, D., and Kosloff, R., 1988a, Wave propagation simulation in a linear viscoacoustic medium: Geophys. J. Roy. Astr. Soc., **93**, 393-407.
6. Lions, J. L., (1971), *Optimal control of systems governed by partial differential equations,* Springer-Verlag.
7. Martinez, R., D., (1993), Wavepropagation effects on amplitude variation with offset measurements: A modeling study, *Geophysics*, **58**, 534-543.
8. Martinez, R., D., McMechan, G., A., (1991), τ-p Seismic data for viscoelastic media - part 2: linearized inversion, *Geoph. Prosp.*, **39**, 157-181.
9. Tarantola, A., (1988), Theoretical background for the inversion if seismic waveforms, including elasticity and attenuation, *in* Scattering and attenuation of seismic waves, *PAGEOPH.*, (Editor: A Aki, R. S. Wu) Birkhäuser Verlag, Basel, 365-399.
10. Robertsson, J. O. A., Blanch, J. O., Levander A., Webb, H. F., Goff, J. A., (1993), 2-D and 3-D viscoelastic seafloor reverberation modeling: *Abstract for the 1993 ARSRP meeting at Scripps Institute, La Jolla, CA.*

INVERTING SEA BED ACOUSTIC PARAMETERS FROM REVERBERATION DATA

J.L. BISHOP, M.T. SUNDVIK, and D.W. GRANDE
Naval Undersea Warfare Center Detachment
New London, CT 06320
U.S.A.

ABSTRACT. The extraction of sea bed acoustic parameters from measured reverberation is a non linear inverse problem. We present a robust solution to this problem utilizing monostatic sonar geometries to characterize unknown bottoms. Two methods of solution are reported in this paper. The first exploits a priori knowledge about local propagation conditions to obtain special case solutions using algorithms developed for their suitability in real time applications. The second method employs a general non-linear least squares solution with only a small compromise in computational efficiency.

The algorithms were tested during sea trials conducted on the New England continental shelf in October 1993. The bottom loss and bottom scattering strength functions obtained from reverberation data were benchmarked carefully in simulation and compared favorably with historical measurements, with geophysical maps of sea bed composition, and with results derived using more general inverse methods not suitable for real time applications. In order to quantify the model and system dependent features of bottom properties derived in this manner, we have explored the effect of mismatch between prior knowledge presumed by the inversion algorithms and the actual conditions which prevailed during the experiment.

1. Introduction

We would like to derive the underlying mid frequency (3 kHz) ocean bottom scattering and reflection functions from bottom reverberation measured in shallow water by means of an inverse procedure which is well matched to the task. Given the inherent variability of shallow water, what we expect to obtain are the average scattering and reflection properties of the water-sediment interface.

Our prescription for inversion is based upon the following:

1. Propagation conditions which ensure that bottom interaction plays a central role in determining the overall acoustic picture, where under downward refracting conditions the data permit limited angular resolution of the pertinent boundary properties.

2. Efficiency in obtaining an average solution will be considered more important than generality of the solution. Consequently, we have chosen to represent the acoustic field in terms of a multipath expansion [1]. The formulation is summarized in Figure 1.

O. Diaschok et al. (eds.), Full Field Inversion Methods in Ocean and Seismo-Acoustics, 401–406.

$$\frac{\text{BOTTOM REVERBERATION}}{\text{SOURCE LEVEL}} = \int_{\text{AREA}} \sum_{\text{TRN}} \sum_{\text{RCV}} B_T \left[V_{T_{\text{bot}}}^{N_{\text{bot}}} V_{T_{\text{srf}}}^{N_{\text{srf}}} A_T \right]^2 S_{T \times R} \left[V_{R_{\text{bot}}}^{N_{\text{bot}}} V_{R_{\text{srf}}}^{N_{\text{srf}}} A_R \right]^2 B_R \rho d\rho d\varphi$$

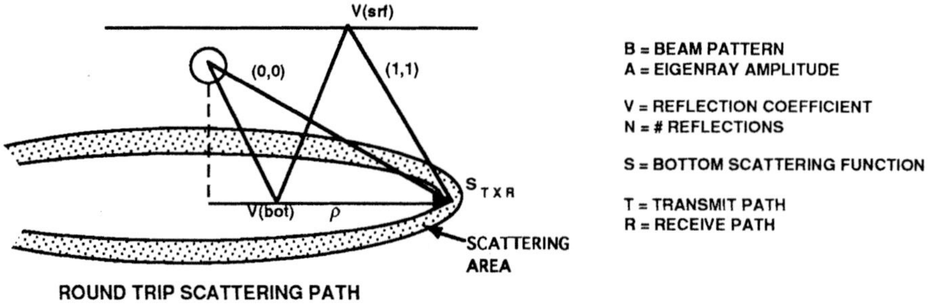

Figure 1. Formulation of Multipath Reverberation Problem

3. Stability and reproducibility of the solution will be considered more important than resolution of detail.

2. Nature of the Inverse Problem

Three approaches are available to us.

Direct extraction. We can exploit constraints imposed by downward refracting conditions to simplify equation (Figure 1) so the desired bottom properties can be extracted directly without resorting to an inverse procedure.

Least-squares inversion. We can assume a physically plausible representation for the boundary functions (e.g., a Lambert's Law Scattering function with an adjustable scattering coefficient, a Rayleigh reflection coefficient with adjustable sediment properties) and employ an iterative non-linear least squares algorithm to determine the unknown parameters.

Full inversion. We can permit the scattering function S and the reflection coefficient V to be unconstrained general vectors and tackle the degeneracy problem which is central to inverse theory [2]: how to devise a regularizing function which eliminates physically unreasonable solutions from the infinite array of achievable solutions now permitted by the unconstrained minimization problem.

These methods have been outlined in order of increasing execution time, increasing generality, and increasing complexity. This paper reports the results obtained to date in working through stages 1 and 2 of the problem.

3. Inversion Methods and Experimental Results

Direct extraction. Under downward refracting conditions, round trip scattering paths with simple multipath structure dominate early time reverberation. Time windows defined by pure multipath contributions simplify the double sum in equation (Fig. 1) so scattering and reflection functions can be extracted directly from experimental data. A 10ms cw signal transmitted over sonar depression angles ranging from 0-30 degrees is referred to as an environmental probe pulse.

Review of reverberation data collected on the New England continental shelf from sand and mud bottoms (Fig. 2) indicates 1) the apparent scattering strength of sand is higher than that of sand/silt/clay (hereafter referred to as "mud") and 2) reflection from mud involves more rapid signal attenuation than reflection from sand (difference in decay rate).

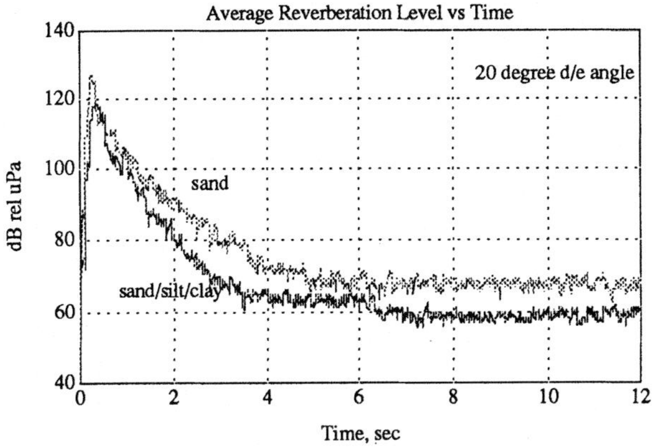

Figure 2. Reverberation from Mud and Sand for Beam Steering Angle 20°

In order to decipher the shallow grazing angle information available at late times, a simple measure of level and decay rate of reverberation is assessed from the slope and intercept of the reverberation tail. These numbers can be related to scattering strength and bottom loss, provided that the slope and intercept are evaluated over regions of the reverberation which are dominated by the limiting grazing angle characteristic of downward refraction.

The effectiveness of the extraction procedure is demonstrated on simulated data in Figure 3. Vertical steering direction, vertical beam width and channel refraction combine to determine grazing angles which dominate the reverberation. As the sonar is steered down, steeper angle multipaths are excited. As the sonar looks out towards long range, the span of grazing angles excited by an otherwise similar beam pattern is narrowed by refractive focusing. Slope/intercept results are consistent with probe pulse results. Specifically, the Lambert's Law scattering coefficient which fits the reverberation decay "intercept" is consistent with the probe pulse.

Results derived from measured reverberation in both sand and mud environments at nominal water depths of 60-80 meters are displayed in Figures 4 and 5. The inset table shows how the direct extraction algorithm was implemented. For the sand bottom example, starting with a previous solution allows direct extraction in a single iteration. Six iterations on bottom loss were required for mud, when a previous result was unavailable. Re-running the reverberation model with the extracted parameters replicates the observed reverberation decay, except in the earliest time frames (ranges less than 500 yards.)

Since extraction depends on input system and environmental parameters (e.g., system beam patterns, sound velocity profiles, water depth) errors in measuring these quantities prior to extraction can lead to mismatch effects in the extracted output. For 20° sonar depressions in sand, the effects of slightly shallower water depth and high scattering strength produce reverberation which overloads the receiver. This condition distorts and underestimates the scattering strength function. Calibration issues allow only relative results for sand and mud scattering strengths.

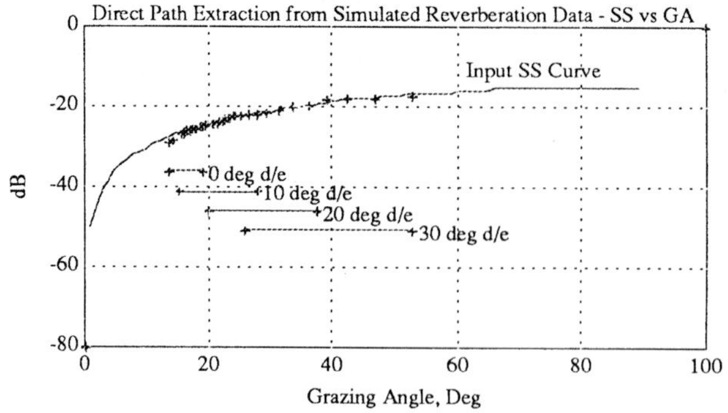

(a)

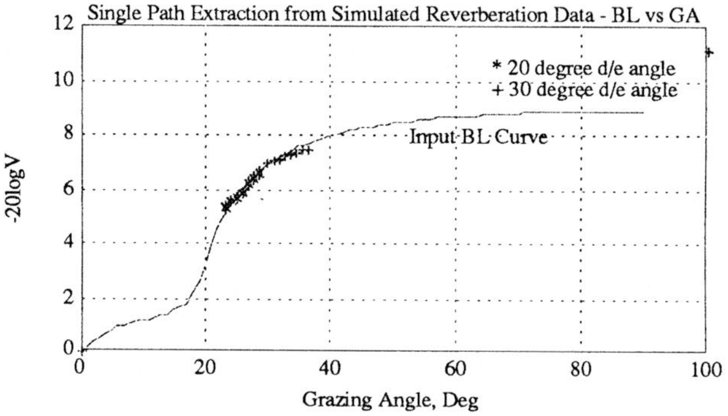

(b)

Figure 3. Results for Simulated Reverberation
a) scattering strength extraction from simulated data within the direct path time window, accumulated from multiple depression angles.
b) bottom loss extraction from simulated data within the first bottom bounce path time window, accumulated from multiple depression angles. Note that only the 20° and 30° d/e angles allow sufficient separation of multipath reverberation to isolate the first bottom bounce return.

Least squares method. The approach which has become standard for non-linear least squares is the Levenberg-Marquardt algorithm [3]. This algorithm is regulated by an internal adjustable parameter which steps towards a minimum in χ^2 by the method of steepest descents when the solution is far away from the minimum and switches continuously to an inverse Hessian method close to the minimum. By contrast to direct extraction, this method fits the parameters of specific models to determine bottom functions of interest. Thus, generality is sacrificed for stability. The detailed grazing angles which actually contribute to determining the unknown parameters must be constructed a posteriori. The parameters from sands on the New England Shelf obtained by Cole and Podeszwa are listed in Table 1 [4] [5]. We have employed modeled results to verify our ability to extract these parameters from idealized data, as indicated in the table.

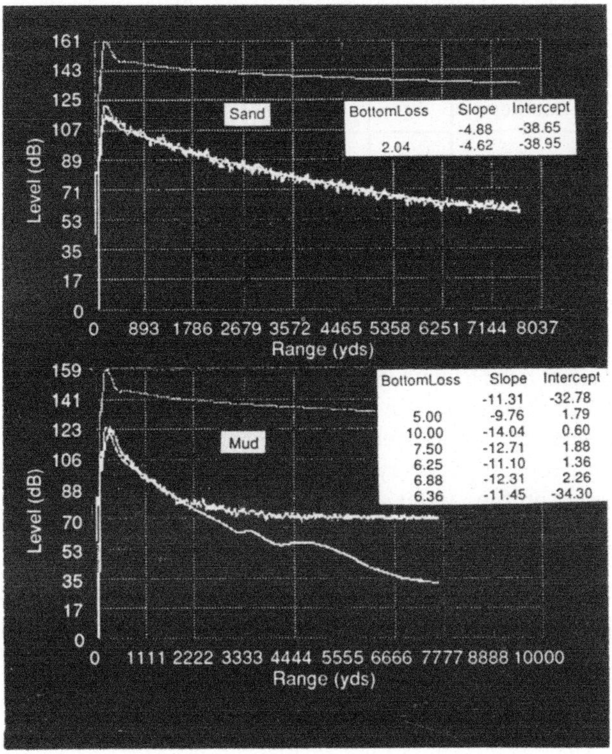

Figure 4.
Slope-intercept extraction from sand and mud regions, depression angles of 10°. The displays show modeled reference reverberation (top), reverberation data (scattered points), and reverberation modeled using extracted bottom loss and scattering strength. Inset tables show extracted parameters.

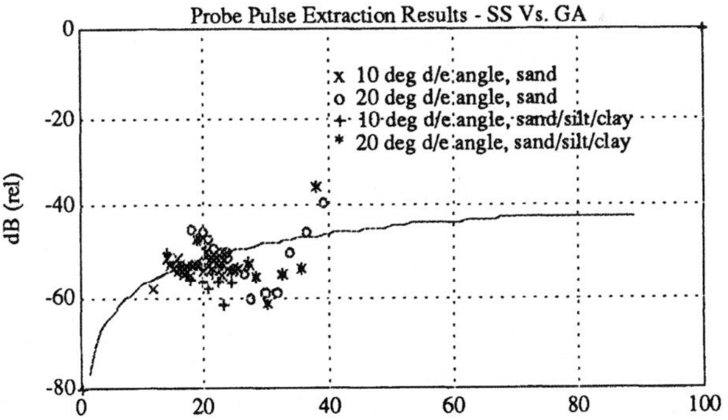

Figure 5. Results for Measured Reverberation
Scattering strength extraction for mud and sand from accumulated multiple depression angles (relative levels).

Table 1. Bottom Properties Measured on the New England Continental Shelf

	Scattering Strength Coefficient	B_L	Grazing Angle	Sediment/Water Density Ratio	Sediment/ Water SSP Ratio	Sediment Attenuation (db/λ)
Cole & Podeszwa	-23	1.3	13	2.06	1.059	.656
Value extracted in simulation	-23	N/A	N/A	2.06	1.059	.656

5. Summary

We have demonstrated a reliable method for extraction of bottom loss and scattering coefficient from two direct methods, suitable for use in shallow water under downward refracting conditions. Of these, the slope/intercept fit to reverberation decay is more useful. "Probe pulse" direct extraction, completely reliable in theory, is presently limited only by practical uncertainty in knowing sound speed profile, water depth, and beam patterns with sufficient accuracy. Reasonable scattering strength results were obtained in the shallow water mud area (~70m water depth), while a shallower (55m) sand site showed the effects of receiver overload. This technique may be better in deeper waters. The technique has practical application in detecting differences in bottom acoustic parameters for real time sonar applications. Future work on generalizing beyond direct methods to a least-squares algorithm will lead towards a more robust method which employs results from both to achieve a best solution.

6. Acknowledgments

The Authors gratefully acknowledge the support of D. Small of the AEAS program office. Manuscript preparation was assisted by J.E. Turner and S. Morgan. Programming support and system development work of S. Trudel, C. Tavares, and M. Gervais is gratefully acknowledged. This work was supported by the U.S. Navy Office of Naval Research.

7. References

[1] Weinberg, H. and Burridge, R. (1977) "Horizontal rays and vertical modes," in Wave Propagation and Underwater Acoustics, ed, by J. B. Keller, J.S. Paradakis, Lect. Notes Phys., Vol. 70 (Springer, Berlin, Heidelberg), pp. 86-152.

[2] Press, W.H., Teukolsky, S., Vetterling, W., and Flannery, B. (1988) Numerical Recipes in C, Cambridge University Press, Cambridge, pp. 804-818.

[3] Marquart, D.W. (1963) Journal of the Society for Industrial and Applied Mathematics, Vol. 11, pp. 431-441.

[4] Cole, B.F., and Podeszwa, E.M. (1967) "Shallow-water propagation under downward-refraction conditions," Journal of the Acoustical Society of America, Vol. 41, pp. 1479-1484.

[5] Cole, B.F., and Podeszwa, E.M. (1974) "Shallow-water bottom reverberation under downward refraction conditions," Journal of the Acoustical Society of America, Vol. 56, pp. 374-377.

LIST OF PARTICIPANTS

B. Alessandrini
Istituto Nazionale di Geofisica
Via di Vigna Murata, 605
00143 Roma
Italy
Tel : *+39 6 51860422*
Fax : *+39 6 5041181*
E-mail :

H.B. Ali
Naval Research Laboratory
Oceanography Division (Code 7173)
Stennis Space Center
MS 39529-5004
USA
Tel : *+1 601 688 5736*
Fax : *+1 601 688 5763*
E-mail

G.A. Athanassoulis
National Technical Univ. of Athens
Dept. Naval Archt. & Marine Eng.
P.O. Box 64070, Zographos 157.10
Athens
Greece
Tel : *+30 1 7705535*
Fax : *+30 1 7774478*
Telex : *262860 RCC GR*
E-mail : *mathan@esprit.naval.ntua.gr*

J. Arvelo
Alliant Techsystems, Inc.
Advanced Technology Center
1911 Ft. Myer Dr. Suite 400
Arlington, VA 22209
USA
Tel : *+1 703 558 9425*
Fax : *+1 703 558 9424*
E-mail : *Juan_Arvelo@ATK.com*

M. Badiey
Office of Naval Research
Ocean Acoustic Program
Code 321OA
800 North Quincy Street
Arlington, VA 22217-5660
USA
Tel : *+1 703 696 6994*
Fax : *+1 703 696 2710*
E-mail : *badiey@ultima.onr.navy.mil*

A.B. Baggeroer
Massachusetts Institute of Technology
Cambridge, MA 02139
USA
Tel : *+1 617 253 4336*
Fax : *+1 617 253 2350*
E-mail : *abb@boreas.mit.edu*

L.D. Bibee
Naval Research Laboratory
Stennis Space Center, MS 39529
USA
Tel : *+1 601 688 5459*
Fax : *+1 601 688 5752*
E-mail : *bibee@nrlssc.navy.mil*

G. Bienvenu
Thomson Sintra ASM Activites Sous-Marine
525, route des Dolines, B.P. 157
06903 Sophia-Antipolis, Cedex
France
Tel : *+33 92 96 43 12*
Fax : *+33 92 96 40 32*
E-mail :

J.L. Bishop
Naval Undersea Warfare
Center Detachment
New London, CT 06320
USA
Tel : *+1 203 440 5312/4671*
Fax : *+1 203 440 6696*
E-mail :

J.O. Blanch
Rice University
Department of Geology and Geophysics
MS-126, Box 1892
Houston, TX 77251-1892
USA
Tel : *+1 713 527 8101 ext. 2243*
Fax : *+1 713 285 5214*
E-mail : *joakimb@rice.edu*

G. Boehm
Osservatorio Geofisico sperimentale
P.O. Box 2011 (Opicina)
34016 Trieste
Italy
Tel : *+39 40 2149219*
Fax : *+ 39 40 327307*
Telex : *460329 OGS I*
E-mail : *walter@reds.ogs.trieste.it*

V.V. Borodin
N.N. Andreyev Acoustic Institute
4, Shvernik str.
Moscow 117036
Russia
Tel : *+7 095 126 9004*
Fax : *+7 095 126 8411*
E-mail : *bvp@asu.acoins.msk.su*

D. Bradley
SACLANT Undersea Research Centre
Viale San Bartolomeo, 400
19138 La Spezia
Italy
Tel : +39 187 540233
Fax : +39 187 524600
E-mail : bradley@saclantc.nato.int

L.M. Brekhovskikh
Acoustic Wave Propagation Lab.
P.P. Shirov Oceanology Institute
23 Krasikova St.
Moscow 117851
Russia
Tel : +7 095 124 5983
Fax : +7 095 135 1568
E-mail :awpl@sovam.com (attn. brekhovskikh)

M.J. Buckingham
Marine Physical Laboratory
SCRIPPS Institution of Oceanography
La Jolla, CA 92093-0213
USA
Tel : +1 619 534 7977
Fax : +1 619 534 7132
E-mail : mjb@mpl.ucsd.edu

A. Caiti
DIST, Dept. Comm., Comp., System Sciences
University of Genova
Via Opera Pia, 13
16145 Genova
Italy
Tel : +39 10 3532727
Fax : +39 10 3532948
E-mail : andy@dist.unige.it

J.V. Candy
University of California
Lawrence Livermore National Laboratory
P.O. Box 808, L-495
Livermore, CA 94550
USA
Tel : +1 510 422 8675
Fax : +1 510 422 3358
E-mail : candy@fudd.llnl.gov

W.M. Carey
Advanced Research Projects Agency
DARPA/MSTO
Department of Defense
3701 North Fairfax Drive
Arlington, VA 22203-1714
USA
Tel : +1 703 696 2339
Fax : +1 203 434 6394
E-mail : wcarey@arpa.mil

N.R. Chapman
Defence Research Establishment Pacific
FMO Victoria
VOS 1BO, B.C.
Canada
Tel : +1 604 363 2194
Fax : +1 604 363 2856
E-mail : chapman@orca.drep.dnd.ca

C.S. Chiu
Naval Postgraduate School
Department of Oceanography
Code OC/Ci
Monterey, CA 93943
USA
Tel : +1 408 656 3239
Fax : +1 408 656 2712
E-mail : chiu@oc.nps.navy.mil or chiu@usw.nps.navy.mil

N.P. Chotiros
Applied Research Laboratories,
The University of Texas at Austin
Austin, TX 78713-8029
USA
Tel : +1 512 835 3512
Fax : +1 512 835 3259
E-mail : chotiros@msmailgwl.arlut.utexas.edu

M.D. Collins
Code 7140
Naval Research Laboratory
Washington DC 20375
USA
Tel : + 1 202 767 9037
Fax : +1 202 404 7732
E-mail : collins@abyss.nrl.navy.mil

W. Denner
EOS Research Associates
200 Camino Aguajito
Suite 202
Monterey, CA 93940
USA
Tel : +1 408 373 1576
Fax : +1 408 373 1578
E-mail : OMNET: W.Denner

F. Desharnais
SACLANT Undersea Research Centre
Viale San Bartolomeo, 400
19138 La Spezia
Italy
Tel : +39 187 540352
Fax : +39 187 540331
E-mail : francine@saclantc.nato.int

O. Diachok
SACLANT Undersea Research Centre
Viale San Bartolomeo, 400
19138 La Spezia
Italy
Tel : *+39 187 540317*
Fax : *+39 187 540331*
E-mail :

J. Doutt
209 A Bigelow Bldg.
Woods Hole Oceanographic Institution
Woods Hole, MA 02543
USA
Tel : *+1 508 457 2000 2201*
Fax : *+1 508 457 2194*
E-mail : *jdoutt@whoi.edu*

L. Dozier
Alliant Techsystems, Inc.
Advanced Technology Center
1911 N. Ft. Myer Drive
Suite 601
Arlington, VA 22209
USA
Tel : *+1 703 558 9430 (work)*
+1 703 573 7326 (home)
Fax :
E-mail : *lewis_dozier@atk.com*

D.D. Ellis
Defence Research Establisment
Atlantic
P.O. Box 1012
Dartmouth, N.S. B2Y 3Z7
Canada
Tel : *+1 902 426 3100*
Fax : *+1 902 426 9654*
E-mail : *ellis@bonnie.drea.dnd.ca*

H. Essen
SACLANT Undersea Research Centre
Viale San Bartolomeo, 400
19138 La Spezia
Italy
Tel : *+39 187 540349*
Fax : *+39 187 540331*
E-mail : *essen@saclantc.nato.int*

L. Fishman
Applied Mathematics Sciences
136 Wilhelm
Ames Laboratory
IOWA State University
Ames, IA 50011
USA
Tel : *+1 515 294 9310*
Fax : *+1 515 294 9683*
E-mail : *fishman@ameslab.gov*

G.V. Frisk
Woods Hole Oceanographic Institution
Woods Hole, Ma 02543
USA
Tel : *+1 508 457 2000 ext 2283*
Fax : *+1 508 457 2194*
Telex : *951679*
E-mail : *gfrisk@whoi.edu*

P. Gerstoft
SACLANT Undersea Research Centre
Viale San Bartolomeo, 400
19138 La Spezia
Italy
Tel : *+39 187 540302*
Fax : *+39 187 540331*
E-mail :*gerstoft@saclantc.nato.int*

D. Gingras
SACLANT Undersea Research Centre
Viale San Bartolomeo, 400
19138 La Spezia
Italy
Tel : *+39 187 540 303*
Fax : *+39 187 540 600*
E-mail : *gingras@saclantc.nato.int*

J. Glattetre
Norwegian Defence Research
Establishment
Division for Underwater Defence
P.O. Box 115
N-3191 Horten
Norway
Tel : *+47 33 04 2081*
Fax : *+47 33 04 7834*
E-mail :

O.A. Godin
Acoustic Wave Propagation Lab.
P.P. Shirov Oceanology Institute
23, Krasikova St.
Moscow 117851
Russia
Tel :
Fax : *+7 09 5124 5983*
E-Mail : *ampl@sovam.com(att Godin)*

T.G. Goldsberry
Office of Naval Research
800 North Quincy Street
Arlington, VA 22217/5660
USA
Tel : *+1 703 696 0805*
Fax : *+1 703 696 8423*
E-mail :

W. Greene
Naval Undersea Warfare Center
Newport, R.I.
USA
Tel : *+1 401 841 4302*
Fax : *+1 401 841 2146*
E-mail :
greenewe@medeiros4.npt.nuwc.navy.mil

V.T. Grinchenko
Institute of Hydromechanics
of Ukrainian Academy of Science
Zhelyabov Street, 8/4
Kiev, 252057
Ukraine
Tel : *+7 044 446 4313*
Fax : *+7 044 446 4229*
E-mail : *vgr@inm.kiev.ua*

J. Guilbot
Laboratoire d'Acoustique,
Systèmes, Signaux et SOnar (LASSSO)
ICPI Lyon
25, rue du plat
69288 Lyon
France
Tel : *+33 72 32 5137*
Fax : *+33 78 37 8054*
E-mail :

M.V. Hall
DSTO Aeronautical and Maritime
Research Laboratory (AMRL)
Maritime Operations Division
17 Jones Bay Road
Pyrmont NSW 2009
Australia
Tel : *+1 61 2 692 1470*
Fax : *+1 61 2 660 0019*
E-mail : *hall@modsyd.mrl.dsto.gov.au*

J.P. Hermand
SACLANT Undersea Research Centre
Viale San Bartolomeo, 400
19138 La Spezia
Italy
Tel : *+39 187 540353*
Fax : *+39 187 540331*
E-mail : *hermand@saclantc.nato.int*

W.S. Hodgkiss
Marine Physical Laboratory
Scripps Institution of Oceanography
University of California at San Diego
San Diego, CA 92152-6400
USA
Tel : *+1 619 534 1798*
Fax : *+1 619 553 0764*
E-mail : *wsh@mpl.ucsd.edu*

J.M. Hovem
University of Trondheim
The Norwegian Institute of Technology
Division of Telecommunication/Acoustics
N-7034 Trondheim
Norway
Tel : *+47 73 59 1418*
Fax : *+47 73 59 1412*
E-mail : *hovem@tele.unit.no*

Q. Huynh
Naval Undersea Warfare Center
Code 8212
Newport, R.I. 02841
USA
Tel :*+1 203 440 5312/4671*
Fax :*+1 203 440 6696*
E-mail :

J. Impagliazzo
Naval Undersea Warfare Center
Newport, R.I. 02841
USA
Tel : *+1 401 841 4369*
Fax : *+1 401 841 2146*
E-mail :
impagliazzo@medeiros4.npt.nuwc.navy.mil

S. Ivansson
FOA, National Defence Research
Establishment
S-17290 Sundbyberg
Sweden
Tel : *+46 8 706 3629*
Fax : *+46 8 706 3543*
E-mail : *sveni@swe.sto.foa.se*

F. Jensen
SACLANT Undersea Research Centre
Viale San Barolomeo, 400
19138 La Spezia
Italy
Tel : *+39 187 540270*
Fax : *+39 187 540331*
E-mail : *jensen@saclantc.nato.int*

S.M. Jesus
Universidade Do Algarve
Unidade de Ciencias Exactas e Humanas
Campus de Gambelas,
8000 Faro
Portugal
Tel : *+351 89 800 951*
Fax : *+351 89 818560*
E-mail : *sjesus@mozart.si.ualg.pt*

N.A. Kampanis
Institute of Applied
and Computational Mathematics
FO.R.T.H. P.O. Box 1527
71110 Heraklion
Greece
Tel : *+30 81 210034*
Fax : *+30 81 238868*
E-mail : *modulef@idomen.cc.uch.gr*

I. Karasalo
Hydroacoustics and Seismology
National Defence Research Establishment
S-17290 Sundbyberg
Sweden
Tel : *+46 8 7063627*
Fax : *+46 8 7063543*
E-mail : *ilkka@sto.foa.se*

Å. Kristensen
STATOIL
Postut tack
N-7004 Trondheim
Norway
Tel : *+47 584011*
Fax : *+47 73584665*
E-Mail :

M. Lambert
Equipe Eletromagnétisme
Laboratoire des Signeaux et Systemes
Plateau de Moulon
91192 Gif-sur Yvette, Cedex
France
Tel : *+33 69 41 80 40*
Fax : *+33 69 41 30 60*
E-mail : *elissalt@esev01.supelec.fr*

R. Laterveer
TNO Defence Research
Physics and Electronics Laboratory
Oude Waalsdorperweg 63
2597 AK The Hague
The Netherlands
Tel : *+31 70 3264221*
Fax :
E-mail : *laterveer@fel.tno.nl*

M. Letiche
Thomson Sintra ASM
BP 53
F-06801, Cagnes-Sur-Mer
Cedex
France
Fax :
Tel :
E-mail :

D.K. Lewis
Lawrence Livermore National Laboratory
7000 East Ave. Mial stop L-495
Livermore, CA 94551-0808
USA
Tel : *+1 510 422 7959*
Fax : *+1 510 422 3358*
E-mail : *lewis@inept.llnl.gov*

D.A. Lindwall
Naval Research Laboratory
Marine Science Division
Stennis Space Center
MS 39529
USA
Tel : *+1 601 688 5306*
Fax : *+1 601 688 5752*
E-mail : *lindwall@nrlssc.navy.mil*

F. Magand
Laboratoire d'Acoustique
Systemes Signaux et Sonar
ICPI/CPE Lyon
25 Rue du Plat,
69288 Lyon
France
Tel : *+33 72 32 5137*
Fax : *+33 78 37 8034*
E-mail :

N.C. Makris
Code 7140
Naval Research Laboratory
Washington D.C. 20375
USA
Tel : *+1 202 767 9037*
Fax : *+1 202 404 2732*
E-mail : *makris@i7140w.nrl.navy.mil*

B.E. McDonald
Naval Research Laboratory
Code 7104
Washington D.C. 20375
USA
Tel : *+1 202 767 9037*
Fax : *+1 202 404 7732*
E-mail : *mcdonald@i5160n.nrl.navy.mil*

J.J. McCoy
School of Engineering
The Catholic University of America
Washington, D.C. 20064
USA
Tel : *+1 202 319 5160*
Fax : *+1 202 319 4499*
E-mail : *mccoy@cua.edu*

J. Meunier
IFREMER Centre de Brest
B.P. 70
29280 Plouzane
France
Tel : *+33 98 22 4146*
Fax : *+33 98 22 4135*
E-mail : *jmeunier@ifremer.fr*

P. Mignerey
Naval Research Laboratory
Acoustic Division
Washington D.C. 20375
USA
Tel : *+1 202 767 2230*
Fax : *Autovon 297 2230*
E-mail : *mignerey@wave2i.nrl.navy.mil*

M. Morrison
NCCOSC RDT&E Division
Code 705
San Diego, CA 92152-5000
USA
Tel : *+1 619 553 3084*
Fax : *+1 619 553 3079*
E-mail : *morrison@nosc.navy.mil*

L. Mozzone
SACLANT Undersea Research Centre
Viale San Bartolomeo, 400
19138 La Spezia
Italy
Tel : *+39 187 540285*
Fax : *+39 187 524600*
E-mail : *mozzone@saclantc.nato.int*

C. Noël
SEMANTIC TS
8 Av. du 11 Novembre
83150 Bandol
France
Tel : *+33 94 29 4125*
Fax : *+33 94 32 5631*
E-mail :

R.M. Oba
Naval Research Laboratory
Acoustic Simulation Section
Code 7181
Stennis Space Center,
MS 39529-5004
USA
Tel : *+1 601 688 4797*
Fax : *+1 601 688 5049*
E-mail : *roba@nvo.navy.mil*

G.J. Orris
Naval Research Laboratory
Washington D.C. 20375
USA
Tel : *+1 202 7679037*
Fax : *+1 202 4047732*
E-mail : *orris@chaos.nrl.navy.mil*

E. Parastates
Technical Acoustics
Department of Vehicle Engineering
Royal Institute of Technology
S-100 44 Stockholm
Sweden
Tel : *+46 8 790 8927*
Fax : *+46 9 790 6122*
Telex : *103 89 kthb s*
E-mail :

V. G. Petnikov
Inst. of General Physics
Russian Academy of Sciences
and Ocean Acoustics & Information Ltd.
38 Vavilov St.
Moscow 117942
Russia
Tel : *+7 095 1287833*
Fax : *+7 095 135 0277*
or *+7 095 135 0278*
Telex : *411074 LIMEN SU*
E-mail : *petniko@az.gpi.msk.su*

M.B. Porter
Center for Applied Mathematics
and Statistics
New Jersey Institute of Technology
Newark, NJ 07102
USA
Tel : *+1 201 596 2995*
Fax : *+1 201 596 6467*
E-mail : *miport@alba.njit.edu*

J.R. Preston
Penn State University.
P.O. Box 30
State College, PA 16804
USA
Tel :
Fax : *+1 414 865 3105*
E-mail :

S. D. Rajan
Woods Hole Oceanographic Institution
Woods Hole, MA 02543
USA
Tel : *+1 508 457 2000 (ext 2317)*
Fax : *+1 508 457 2194*
E-mail : *rajan@arcac.whoi.edu*

D. Rauch
USEA
Via delle Pianazze
19136 La Spezia
Italy
Tel : +39 187 982750
Fax : +39 187 980748
E-mail :

M.J.D. Rendas
Lab.13S CNRS URA 1376. Bat. 4
250 Ave. Albert Einstein
Sophia Antipolis
F-06560 Valbonne
France
Tel : +33 92 942898
Fax :
E-mail : rendas@mimosa.unice.fr

W. I. Roderick
SACLANT Undersea Research Centre
Viale San Bartolomeo, 400
19138 La Spezia
Italy
Tel : +39 187 540300
Fax : +39 187 540331
E-mail : roderick@saclantcn.nato.int

H. Schmidt
Massachusetts Institute of Technology
Cambridge, MA 02139
USA
Tel : +1 617 253 5727
Fax : +1 617 253 2350
E-mail : henrik@keel.mit.edu

H. Schneider
Forschungsanstalt der Bundeswehr
für Wasserschall und Geophysik
Klausdorfer Weg 2-24
24148 Kiel
Germany
Tel : 431 720 4103
Fax : 431 720 4150
E-mail :

P. Scrimger
JASCO Research Ltd.,
Marine Technology Center
9865 W. Saanich Rd.
Sidney, B.C., V8L 3S1
Canada
Tel : +1 604 656 0656
Fax : +1 604 656 5833
E-mail : jresearch@amtsgi.bc.ca

G.C. Searing
Sonar Systems Dept.
Defence Research Agency
Portland
Dorset, DT5 2JS
U.K.
Tel : +305 820311 6031
Fax : +305 820311 6838
E-mail :

D.G. Simons
TNO Defence Research
Physics and Electronics Laboratory
Oude Waalsdorperweg, 63
2597 AK The Hague
The Netherlands
Tel : +31 70 326 4221
Fax :
E-mail : simons@fel.tno.nl

C. Siedenburg
Alliant Techsystems, Inc.
Mukilteo
Washington 98275
USA
Tel : +1 206 356 3909
Fax : +1 206 356 3185
E-mail : sieden@atk.com

M.M. Slavinsky
Institute of Applied Physics
Russian Academy of Sciences
46, Uljanov Str.
603600, Nizhny Novgorod
Russia
Tel : +7 831 236 6617
Fax : +7 831 236 2061
or 236 5745
E-mail : slavin@appl.nnov.su

Y. Stéphan
Centre Militaire d'Océanographie
Epshom BP 426
F-29200
France
Tel : +33 98 22 0684
Fax : +33 98 22 1864
E-mail : stephan@shom.fr

R.D. Stoll
Lamont-DohertyEarth Observatory
Of Columbia University
Palisades, N.Y. 10964
USA
Tel : +1 914 365 8392
Fax : +1 914 365 8156
E-mail :

F. Sturm
MS/ETMA
Université de Toulon et du Var
83957 La Grade, Cedex
France
Tel : *+33 94 34 9682*
Fax : *+33 94 34 9923*
E-mail :

E.J. Sullivan
Naval Undersea Warfare Center
NUSC Code 82101
Newport, RI 02841
USA
Tel : *+1 401 841 2011*
Fax : *+1 401 841 5045*
E-mail :

M.T. Sundvik
Naval Undersea Warfare Center
Environmental and Underwater
Acoustics Division
New London, CT 06320
USA
Tel : *+1 203 440 4069*
Fax : *+1 203 440 6696*
E-mail : *sundvik@ai.vsdec.navy.mil*

M. Taroudakis
University of Crete
Dept. of Mathematics
714 09 Heraklion
Crete
Greece
Tel : *+30 81 210034*
+30 81 210013
Fax : *+30 81 238868*
E-mail : *taroud@knosos.cc.uch.gr*

D. Tollefsen
Norwegian Defence
Research Establishment
Division for Underwater Defence
P.O. Box 115
N-3191 Horten
Norway
Tel : *+47 33 04 2081*
Fax : *+47 33 04 7834*
E-mail :

A. Tolstoy
4224 Waialae Ave.
Suite 5-260
Honolulu, Hawayii 96816
USA
Tel : *+1 808 735 8760*
Fax : *+1*
E-mail : *tolstoy@bixi.com*

B. Ursin
University of Trondheim
Division of Petroleum Engineering
and Applied Geophysics
S.P.Andersens V. 15A
N. 7034 Trondheim
Norway
Tel : *+47 59 49 25*
Fax : *+47 94 44 72*
E-mail : *zhao@ipt.unit.no (Zhao)*

A. Warn-Varnas
Naval Research Laboratory
Stennis Space Center
Ms 39529
USA
Tel : *+1 601 688*
Fax : *+1 601 688*
E-mail :

R. Weatherburn
SACLANT Undersea Research Centre
Viale San Bartolomeo, 400
19138 La Spezia
Italy
Tel : *+39 187 540377*
Fax : *+39 187 524600*
E-mail : *weatherburn@saclantc.nato.int*

H. van Asselt
AESN
Via Hermada, 6/b
16154 Sestri Ponente (GE)
Italy
Tel : *+39 10 6546574*
Fax :
E-mail :

A. Vesnaver
Osservatorio Geofisico Sperimentale
P.O. Box 2011
34016 Trieste
Italy
Tel : *+39 40 2140219*
Fax : *+39 40 327307*
E-mail : *aldo@ reds.ags.trieste.it*

A. Voronovich
NOAA/ERL/ETL, R/E/WP1
Environmental Technology Laboratory
325 Broadway Boulder
Colorado, 80803
USA
Tel : *+1 303 497 6464*
Fax : *+1 303 497 6978*
E-mail : *agv@wave.wpl.erl.gov*

Z. Wéber
Department of Geophysics
Eötvös University
Ludovika tér 2
H-1083 Budapest
Hungary
Tel :
Fax : *+36 1 210 1089*
E-mail :

W.T. Wood
Naval Research Laboratory
Code 7432
Stennis Space Center
MS 39529
USA
Tel : *+1 601 688 5311*
Fax : *+1 601 688 5752*
E-mail : *wwood@nrlss.navy.mil*

P. Zakarauskas
Defence Research Establishment Pacific
Victoria. B.C.
FMO VOS-180
Canada
Tel : *+1 604 363 2879*
Fax :
E-mail : *zak@orca.drep.dnd.ca*

M. Zakharia
LASSO/ICPI Lyon
25 rue du Plat
69288 Lyon Cedex 02
France
Tel : *+33 72 32 50 74*
Fax : *+33 78 37 80 34*
E-mail : *manell.zakharia@sp1.y-net.fr*

INDEX

MODERN APPROACHES IN GEOPHYSICS

formerly *Seismology and Exploration Geophysics*

---

1. E.I. Galperin: *Vertical Seismic Profiling and Its Exploration Potential.* 1985 ISBN 90-277-1450-9
2. E.I. Galperin, I.L. Nersesov and R.M. Galperina: *Borehole Seismology.* 1986 ISBN 90-277-1967-5
3. Jean-Pierre Cordier: *Velocities in Reflection Seismology.* 1985 ISBN 90-277-2024-X
4. Gregg Parkes and Les Hatton: *The Marine Seismic Source.* 1986 ISBN 90-277-2228-5
5. Guust Nolet (ed.): *Seismic Tomography.* 1987 ISBN 90-277-2521-7
6. N.J. Vlaar, G. Nolet, M.J.R. Wortel and S.A.P.L. Cloetingh (eds.): *Mathematical Geophysics.* 1988 ISBN 90-277-2620-5
7. J. Bonnin, M. Cara, A. Cisternas and R. Fantechi (eds.): *Seismic Hazard in Mediterranean Regions.* 1988 ISBN 90-277-2779-1
8. Paul L. Stoffa (ed.): *Tau-p: A Plane Wave Approach to the Analysis of Seismic Data.* 1989 ISBN 0-7923-0038-6
9. V.I. Keilis-Borok (ed.): *Seismic Surface Waves in a Laterally Inhomogeneous Earth.* 1989 ISBN 0-7923-0044-0
10. V. Babuska and M. Cara: *Seismic Anisotropy in the Earth.* 1991 ISBN 0-7923-1321-6
11. A.I. Shemenda: *Subduction.* Insights from Physical Modeling. 1994 ISBN 0-7923-3042-0
12. O. Diachok, A. Caiti, P. Gerstoft, H. Schmidt (eds.): *Full Field Inversion Methods in Ocean and Seismo-Acoustics.* 1995 ISBN 0-7923-3459-0

KLUWER ACADEMIC PUBLISHERS – DORDRECHT / BOSTON / LONDON

GPSR Compliance

The European Union's (EU) General Product Safety Regulation (GPSR) is a set of rules that requires consumer products to be safe and our obligations to ensure this.

If you have any concerns about our products, you can contact us on ProductSafety@springernature.com

In case Publisher is established outside the EU, the EU authorized representative is:

Springer Nature Customer Service Center GmbH
Europaplatz 3
69115 Heidelberg, Germany

Batch number: 10888021

Printed by Printforce, the Netherlands